The 'Mystical' TCM Triple Energizer

The 'Mystical' TCM Triple Energizer

Its Elusive Location and Morphology Defined

Water transmutes into Fire in the Triple Energizer

Dr Louis Gordon
(Chinese Medicine)

Copyright © 2016 by Dr Louis Gordon.

Library of Congress Control Number: 2016914001
ISBN: Hardcover   978-1-5245-1693-2
          Softcover   978-1-5245-1692-5
          eBook   978-1-5245-1691-8

All rights reserved. No part of this book may be reproduced or transmitted in any form or by any means, electronic or mechanical, including photocopying, recording, or by any information storage and retrieval system, without permission in writing from the copyright owner.

Any people depicted in stock imagery provided by Thinkstock are models, and such images are being used for illustrative purposes only.
Certain stock imagery © Thinkstock.

Published by Xlibris.
Xlibris is a self-publishing and on-demand printing services provider founded in 1997 and based in Bloomington, Indiana, USA. In 2000, the *New York Times* stated it was the foremost on-demand publisher. Parent company: Author Solutions.

Cover graphic © Depositphotos.com/[Photo by Palinchak].
In the 31st Difficult Issue of Unschuld's (1) translation of the *Nan Ching*, Yeh Lin relates, '*When fire meets water*, a *transformation into influences* takes place' (emphasis is mine).

Print information available on the last page.

Rev. date: 09/21/2016

To order additional copies of this book, contact:
Xlibris
1-800-455-039
www.Xlibris.com.au
Orders@Xlibris.com.au
726310

CONTENTS

The Aim of this Book xi
Introduction.. xiii
WHO Standardization of Traditional Medicine Terminology xvii

Chapter 1: Introduction to the Traditional Chinese Medicine (TCM) Concept of the Triple Burner 1
Chapter 2: Several Aspects of the Triple Burner Are Controversial 13
Chapter 3: Controversy About the Triple Energizer Having No Form 20
Chapter 4: Controversy Over the Anatomical Location of the Triple Energizer... 42
Chapter 5: Principal Features of the Three Heaters..................... 52
Chapter 6: The Triple Burner Governs the Movement of Qi Throughout the Entire Body 55
Chapter 7: The Triple Burner as a Threefold Division of the Body 62
Chapter 8: Correlation Between the Triple Burner and the Pericardium (Palaces/Depots and Their Internal/External Relationship) 71
Chapter 9: Relationship of the Triple Burner With the Stomach 76
Chapter 10: The Manufacture of Blood by the Middle Heater 82
Chapter 11: Formation of Nutritive Qi in the Middle and Upper Heaters 100
Chapter 12: The Triple Burner is Associated with the Residence of the Minister Fire 112
Chapter 13: The Triple Burner Encompasses the Whole Body 117
Chapter 14: The Relationship Between the San Jiao, the Cou Li, and the Bladder 131
Chapter 15: The Triple Heater Has a Direct Relationship with the Source Points... 152

Chapter 16: The Triple Energizer in Relation to the Shao Yang.....158
Chapter 17: The Diaphragm (Ke) Membrane Screening off the Turbid Influences... 163
Chapter 18: The Triple Burner Is Energized by Fire and Light From the Sun 166
Chapter 19: Introduction to the Nature of Water and the Triple Energizer 177
Chapter 20: Do Other Scientific Authorities Agree with Pollack's EZ Water Theory?.. 211
Chapter 21: Further Information about EZ Water and Biology 228
Chapter 22: Further Information About EZ Water and Fascia 242
Chapter 23: Biophotons and Biological Functions........................249
Chapter 24: Biophotons and the Highly Ordered Nature of Cellular Communication..261
Chapter 25: Biophotons Emitted from DNA With Laser-Like Coherence 272
Chapter 26: Biophotons and the Power of Light to Heal 277
Chapter 27: The Effect of Biophotons on Foods...........................284
Chapter 28: Practical Application of Biophotons 291
Chapter 29: How Biophotons Energize the Triple Energizer.......297
Chapter 30: Morphogenetic Fields Bathe the Universe Independent of Time and Space.................................304
Chapter 31: The Triple Burner as a System of Waterways..........310
Chapter 32: Acupuncture Meridians Exist Within an Omnipresent Liquid Crystalline Membrane System of Collagen 320
Chapter 33: Relationship Between the Primo Vascular System and the Connective-Tissue Metasystem 340
Chapter 34: Western Medical Description of Body Cavity Membranes... 410
Chapter 35: The Connective Tissues (Membranes) of the Body are the Major Component of the Triple Energizer 415
Chapter 36: The Omnipresence of Connective-Tissue Membranes Throughout the Body 434
Chapter 37: Incredible Physical Properties of Connective-Tissue Membranes (Fascia) in the Body 441

Chapter 38: Ming Men, Yuan Qi and the Origin and Inauguration of the Triple Energizer 457
Chapter 39: Relationship Between the Triple-Energizer Metasystem, the Connective-Tissue Metasystem, the Primo Vascular System, the Acupuncture Meridian System, and the Kidneys 503
Chapter 40: Taste Receptors in the Stomach Determine Necessary Nutrient Absorption 538
Chapter 41: Taste Receptors in Numerous Organs Necessary for Biological Control Throughout the Body by the Zang and Fu 552
Chapter 42: How the Gut Microbiome Complements the Triple Energizer 562
Chapter 43: How the Enteric Nervous System Complements the Triple Energizer 605
Chapter 44: How the Lymphatic System Complements the Triple Energizer 608
Chapter 45: How Brown Adipose Tissue (BAT) and White Adipose Tissue Complement the Triple Energizer 627
Chapter 46: How Earthing or Grounding Complements the Triple Energizer 645
Chapter 47: The Many Forms of Qi Associated with the Triple Energizer 671
Chapter 48: Practical Application of Acupuncture Points to Treat Problems Associated with the Triple Energizer 685

Glossary of Chinese Medical Terms Used Throughout This Book 701
References 707
Index 727
About the Author 765

I would like to thank and dedicate this book to my wife, Linda Jean Gordon. Without her constant support and unconditional love, I could not have completed the manuscript, which was more than four years in the making. I also wish to dedicate this book to the memory of my mother, Margaret Jaeckel Gordon, who instilled positivity into me, and my father, William Francis McMillan Gordon, who encouraged my interest in science. I also wish to sincerely thank Dr Paul U. Unschuld for the selfless and tireless work he has committed to make many ancient Chinese medical classics available in English for study and research. My book is based predominantly around his scholarly work *Nan-Ching: The Classic of Difficult Issues*. I also wish to sincerely thank Professor Unschuld for permission to use citations of his translation in my book.

The Aim of this Book

Financial gain was not the reason that I wrote this book. I wish! This book has been written solely for educational, study, and research purposes for a select group of individuals who are interested in the philosophy, theory, and practical application of traditional Chinese medicine (TCM). I illuminate how many of the comments regarding physiology and biology that are presented in the ancient classic the *Nan Ching* are only now being substantiated by modern science.

Individuals wishing to know more about newly discovered biological systems—including the Primo Vascular System, the gut microbiome, taste and sensory receptors spread throughout the body, and the endocrine function of adipose tissue, etc.—will find the book very illuminating and practical.

Regarding the so-called mysterious San Jiao or Triple Burner (aka the Triple Energizer), there is an enormous amount of information and differing, prejudiced personal opinions and preconceptions that have been written over millennia and especially in the last 50 years by scholars who are inclined to apply their own interpretation to the original ancient TCM texts. While I do in part compare and contrast this plethora of views and renditions, my aim is to explain what I believe to be the actual location, the composition and morphology, and the traditionally described functions of the Triple Energizer as expressed by modern scientific discovery. I present scientific evidence to show how the Triple Energizer works, and why it has remained hidden all this time. Those who believe that it has *a name but no form* will be truly amazed as to its actual location and its actual form.

While I freely admit that I do not have the full understanding about the absolute composition of the Primo Vascular System, the Triple-Energizer Metasystem, the Connective-Tissue Metasystem, or the Acupuncture Meridian Metasystem, I believe that the research findings and comments I present in the book will elicit much discussion about these interconnected systems, and I am open to all suggestions about the clarification of their respective compositions and their intercommunication.

Introduction

The 31st Difficult Issue of *Nan Ching* states, 'When Fire meets Water in the body, a transformation into Qi takes place.' There is no doubt that in spite of the scientific enlightenment of today, numerous aspects of the human anatomy are still not understood. I remember being ridiculed when I graduated from acupuncture college more than three decades ago because I used acupuncture points on the Triple-Burner meridian, whereby cynics and critics pointed out that the Triple-Burner organ we used in our treatment was a fictitious fabrication that 'did not even exist' because its existence in the human body had not been elucidated. I have based much of my research on the 2,500-year-old medical classic *Nan Ching* or *Nan Jing*. George Soulié de Morant (2), in the book *Chinese Acupuncture*, notes that the *Nan Jing*, or *Classic of Difficulties*, is attributed to Bian Que and is still used to this day. It attempts to clarify the most obscure parts of the *Nei Jing*. Bian Que was born in Cheng (present-day Xin Cheng) in Henan near Kaifeng.

Unschuld (1), in page 51, cites *Ku Te-Tao on the Nan-Ching* (1979), stating, 'The authorship of the *Nan-ching* has been ascribed to Ch'in Yüeh-jen (Pien Ch'io); however, this has been doubted by many in the past.'

Today, in the West, acupuncture is generally accepted as a complementary form of medical treatment. It is ironic that acupuncture and Traditional Chinese Medicine (TCM) have a history of 3,000 to

5,000 years, while the infant Western medicine is only a little over a hundred years old, yet it is TCM that is relegated as complementary.

Like most TCM graduates, over the years, I used to ponder the *San Jiao* (or Three Heaters/Triple Burner) organ and its actual location. About four years ago (2012), it dawned on me that the San Jiao is an actual organ with a defined abode, be it polymorphous and multilocational in nature. It is not my aim to analyze every single step in the massively detailed derivation of all the Qi and fluid forms and flows as outlined in the classic Three-Heater manufacture of Qi, Blood, *Jing*, *Jin*, and *Ye*. That is not my agenda.

My aim is to demystify the supposedly mysterious San Jiao with all the apparent contradictions found between the classic ancient texts, including the *Nan Jing* and the *Nei Jing*. I intend to prove using recent scientific research findings, along with deduction, that the San Jiao does actually exist and that it does have an individually defined polymorphous form yet is like none of the established and known organs, which have a fixed shape and defined anatomical location.

I aim to establish the nature of the Western medical equivalent of this enigmatic TCM organ, along with many defined auxiliary anatomical components that cooperate in allowing the Triple Energizer to perform the TCM functions ascribed to this mysterious organ complex. I have collated numerous recent scientific references and results of about 60 diverse biological and biochemical functions of the complicated gut microbiome organ system that classical TCM attributes to the San Jiao.

While I do not believe that the gut microbiome constitutes the Triple-Energizer organ per se, it is highly likely that these diverse functions of the gut microbiome intimately complement the San Jiao and its TCM-attributed functionality. The logical scientific conclusion is that this polymorphous and multilocational gut microbiome organ has always been there, but only with the substantial investigative research of modern science has this semisolid, polymorphous, multilocational organ been shown to be a profoundly complex and sophisticated organ that has defied identification because of its very semisolid metamorphosing composition, which is replenished on an hourly and daily basis.

I also discuss at length the recent scientific findings of another newly discovered organ, the Primo Vascular System, and how it is intimately associated with the Connective-Tissue Metasystem. I suggest that my original view and interpretation of the Triple Burner presented here will substantially add to the understanding of this so-called mysterious organ system and that all TCM practitioners will find its practical application immensely beneficial in the treatment of numerous diseases. I aim to show that the San Jiao causes 'a transformation into Qi . . . when Fire meets Water in the body'.

WHO Standardization of Traditional Medicine Terminology

The 2007 *WHO International Standard Terminologies on Traditional Medicine in the Western Pacific Region* (3) nominated that the San Jiao be called *triple energizer* (三焦), and defined it as 'a collective term for the three portions of the body cavity through which the visceral qi is transformed, also widely known as triple burners'. The upper energizer (上焦) is defined as 'the chest cavity, i.e., the portion above the diaphragm housing the heart and lung, also known as upper burner'. The middle energizer (中焦) is defined as 'the upper abdominal cavity, i.e., the portion between the diaphragm and the umbilicus housing the spleen, stomach, liver and gallbladder, also known as middle burner'. The lower energizer (下焦) is defined as 'the lower abdominal cavity, i.e., the portion below the umbilicus housing the kidneys, bladder, small and large intestines, also known as lower burner'.

When discussing the WHO-defined *triple energizer* throughout the book, I will use the original terminology of the author. Many variations do indeed exist, including *San Jiao, sanjiao, Triple Heater, Triple Burner, Tri-Heater, Triple Energizer, Three Heaters, Three Burning Spaces*, etc. Personally, I prefer the term *Triple-Energizer Metasystem* as I believe this terminology truly shows the Triple Energizer in its true light as a ubiquitous, omnipresent physical system that literally pervades the entire body.

CHAPTER 1

Introduction to the Traditional Chinese Medicine (TCM) Concept of the Triple Burner

Introduction
In the 2002 book by Giovanni Maciocia (4) titled *The Foundations of Chinese Medicine: A Comprehensive Text for Acupuncturists and Herbalists*, on pages 117–120, the author writes that there are several interpretations of the nature of the San Jiao depending on the context of ancient texts and to the understanding and interpretation of those texts. Some authorities see the San Jiao as a Yang organ. Other authorities see the San Jiao more as a collection of functions rather than being an organ, and the San Jiao is thought to be where the *Yuan* (original) Qi flows. To others, the San Jiao is thought to be three distinct divisions of the body as follows:

A. The San Jiao as One of the Six Yang Organs
Maciocia cites *The Simple Questions* from the *Su Wen*: 'The San Jiao is the official in charge of irrigation and it controls the water passages.' In this context, the San Jiao does have a physical form similar to the other organs. It assists in receiving food, digesting food, transforming and transporting food, and then excreting the waste materials. The San Jiao is responsible for moving fluids in the Upper Jiao through the defensive Qi, through the Middle Jiao as Nutritive Qi,

and through the Lower Jiao as body fluids. The functional capacity of the Stomach, Lungs, Kidneys, and Bladder to disperse their fluids is reliant on control from the San Jiao. Disharmony of the San Jiao can manifest as sneezing in the Upper Jiao, abdominal distension in the Middle Jiao, or retention of urine in the Lower Jiao.

B. The San Jiao Is Where the Yuan Qi Flows
This theory supposes that the San Jiao has no form and is not considered to be an organ but, rather, an assortment of functions. The *Classic of Difficulties* states that the Yuan Qi is located between the Kidneys and diffuses to the *Zang Fu* via the San Jiao; it then enters the twelve meridians and emerges at the *Yuan* source points. The Yuan Qi can only make possible all body functions through its spreading by the San Jiao. Therefore, it greatly effects the warming for digestion and excretion. 'The triple heater is sometimes specifically linked to the kidneys forming one of the two fu corresponding to the zang of the kidneys (with the bladder)' (*Ling shu*, 2 and 47).

C. The San Jiao as Three Divisions of the Body
This theory comes from both the *Spiritual Axis* and the *Classic of Difficulties*. The Upper Jiao is from the diaphragm up (Heart, Lungs, Pericardium, throat, and head), the Middle Jiao is from the diaphragm to the umbilicus (Stomach, Spleen, and Gallbladder), and the area below the umbilicus is the Lower Jiao (Liver, Kidneys, Intestines, and Bladder).

Each of these three interpretations is discussed in detail throughout this book. I personally believe that facets of each of these concepts coexist and that the Triple Energizer is an omnipresent organ complex that is represented throughout the entire body—from the deepest of organs internally (Spleen and Kidneys) and permeating our body externally to the tips of our toes and fingers—such that one of the many fluid products biosynthesized by the Triple Energizer includes the oily film we leave on surfaces we touch, which is called our fingerprints. I suggest that such oily residue is released through some of the smallest cavities within the body that constitute the Triple-Energizer organ complex, the *couli*. Just where did the Sanjiao fit into ancient TCM theory?

1.1 Chinese Culture Has High Regard for the Legendary Irrigation Official Named Yu

In the article entitled 'Cultural Reference for Increased Understanding of the San Jiao' regarding the legendary irrigation official named Yu, the author, Glenn Grossman (5), states, 'This legend gives a cultural context for understanding the high regard that Chinese culture has placed on the legendary irrigation official named Yu.'

1.2 Functions of the Viscera and the Bowels According to TCM

In the article entitled 'Cultural Reference for Increased Understanding of the San Jiao' regarding the anatomical location of the San Jiao components, the author, Glenn Grossman (5), notes that TCM often compares the viscera and the bowels in the body to that of twelve upright officials in an empire, with each one having an important dedicated role. He summarizes by stating, 'These twelve officials should not fail to assist one another.'

It is very interesting to note that many of the medical and scientific concepts reported in ancient TCM theory exactly mirror concepts mentioned in the Holy Bible. It is a pivotal concept throughout the TCM scriptures that for good health, each of the organs must cooperate with every other organ and work harmoniously and that each organ and its ascribed function is critical for the health and well-being of the entire body and that no single organ should be minimalized. Note what the following biblical scripture from 1 Corinthians 12:12–27 states about the importance of cooperation of all the body parts.

1.3 Primary Scriptures Show that All the Individual Parts of the Body Coexist with the Rest

At 1 Corinthians 12:12–26, the *New World Translation of the Holy Scriptures* (6) says:

> 12 For just as the body is one but has many members, and all the members of that body, although being many, are one body, so also is the Christ. 13 For truly by one spirit we were all baptized into one body, whether Jews or Greeks,

whether slaves or free, and we were all made to drink one spirit. 14 For the body, indeed, is not one member, but many. 15 If the foot should say: 'Because I am not a hand, I am no part of the body,' it is not for this reason no part of the body. 16 And if the ear should say: 'Because I am not an eye, I am no part of the body,' it is not for this reason no part of the body. 17 If the whole body were an eye, where would the [sense of] hearing be? If it were all hearing, where would the smelling be? 18 But now God has set the members in the body, each one of them, just as he pleased. 19 If they were all one member, where would the body be? 20 But now they are many members, yet one body. 21 The eye cannot say to the hand: 'I have no need of you'; or, again, the head [cannot say] to the feet: 'I have no need of YOU.' 22 But much rather is it the case that the members of the body which seem to be weaker are necessary, 23 and the parts of the body which we think to be less honorable, these we surround with more abundant honor, and so our unseemly parts have the more abundant comeliness, 24 whereas our comely parts do not need anything. Nevertheless, God compounded the body, giving honor more abundant to the part which had a lack, 25 so that there should be no division in the body, but that its members should have the same care for one another. 26 And if one member suffers, all the other members suffer with it; or if a member is glorified, all the other members rejoice with it.

1.4 The Triple Burner: Where Does It Start, and Where Does It End?

So let's start from the very beginning. That's a very good place to start. I feel like I should have quotation marks around that! Regarding the nuts and bolts of the Triple Energizer, the 31st Difficult Issue on page 347 of Unschuld's (1) translation of the *Nan Ching* states, 'The Triple Burner: how is it supplied and what does it generate? Where does it start and where does it end? And where, in general, [are its disorders] regulated? Can that be known?'

In the commentaries on the 31st Difficult Issue on pages 352–353 of Unschuld's (1) translation of the *Nan Ching*, Hsü Ta-ch'un states:

> The [treatise] 'Ku k'ung lun' of the Su[-wen states]: 'The through-way vessel starts from the "street of influences."' The treatise 'Jung wei sheng hui' of the Ling[-shu] states: 'The upper [section of the Triple] Burner emerges from the upper opening of the stomach. It ascends together with the throat. It penetrates the diaphragm and spreads into the chest. It proceeds to the armpits, and follows the great-yin Section [of the conduits]. Then it returns to the yang-brilliance [conduit] and ascends to the tongue. Again, it descends and [meets with] the foot-yang-brilliance [conduit]. Normally, [its influences, i.e., the protective influences,] proceed together with the constructive influences . . . The central [section of the Triple] Burner, too, is associated with the center of the stomach. It emits [its influences] upward after the upper [section of the Triple] Burner [has done so]. The influences received by the [central section of the Triple] Burner are gushing dregs and steaming liquids. The essential and subtle [portions] of these [dregs and liquids] are transformed and flow upward into the vessel [associated with the] lung. *There they are transformed into blood. Nothing is more valuable concerning the maintenance of life in one's body than the [blood]. Hence, it alone may move through hidden conduits. It is named 'constructive influences'.* The lower [section of the Triple] Burner separates [the essential from the dregs and transmits them to] the coiled intestine, from which [the liquid portions] leak into the bladder. Hence, water and grains are normally present in the stomach simultaneously. They become dregs and move down together. When they reach the large intestine, they enter the [realm of the] lower [section of the Triple] Burner. [The liquid and the solid dregs] leak downward together. The liquid [portions] are then strained off; they follow the lower [section of the Triple] Burner and leak into the bladder.' It is also said that the constructive [influences] emerge from the central

> [section of the Triple] Burner, while the [influences] of the stomach emerge from the lower [section of the Triple] Burner. The Su[-wen treatise] 'Ling lan mi tien lun' states: 'The Triple Burner is the official responsible for the maintenance of the ditches. The waterways emerge from there.' If one takes all these textual passages into consideration, the meaning [of the Triple Burner] becomes even more obvious. (Emphasis is mine)

The contents of the above citation will be discussed in detail throughout the book. However, note that the 'essential and subtle portions' of the liquids and grains (food and drink) from the Stomach are sent upwards to the Lungs (to be oxygenated) and become the 'constructive influences'—that is, the life-giving and life-sustaining blood and other fluids. The text categorically states, 'Nothing is more valuable concerning the maintenance of life in one's body than the [blood].' Note how this sentiment agrees with the Holy Bible (6) again. At Leviticus 17:14, it says, 'For the soul of every sort of flesh is its blood by the soul in it. Consequently I said to the sons of Israel: "You must not eat the blood of any sort of flesh, because the soul of every sort of flesh is its blood."' Because there are currently 33 recognized different defined blood groups and over six hundred different blood group antigens, due to the subsequent combinations and permutations, no two people on earth have identical blood, and our blood is as individual as our fingerprints. As our *Shen* resides in our Heart and permeates the entire body through the blood in our vascular system, it is critical that our blood is kept vital and clean to ensure that optimal wellness pervades our very soul.

In the Holy Bible, the word *soul* is derived from the word *nephesh* in the Hebrew scriptures and *psyche* in the Christian Greek scriptures. Both of these words mean 'a breathing thing'. Nephesh, for example, occurs over seven hundred fifty times in the Old Testament. For example, in the Holy Bible (6), Genesis 2:7 says, 'And Jehovah God proceeded to form the man out of dust from the ground and to blow into his nostrils the breath of life, and the man came to be a living soul.' In the Holy Scriptures, death or the disappearance of the soul is described as the breath ceasing from an individual, as can be seen in the following two verses from Genesis. Genesis 35:17–18 says, '17 But so it was that while she had difficulty in making the delivery the midwife said to her: "Do not be afraid, for you will have this son also." 18 And the result was that as her soul was going out (because she died) she called his name Ben-o'ni; but his father called him Benjamin.' Thus, it can be shown that a living being is a breathing being.

Note further what Hsü Ta-ch'un states when he says, 'The lower [section of the Triple] Burner separates [the essential from the dregs and transmits them to] the coiled intestine, from which [the liquid portions] leak into the bladder.' He actually mentions that 'the liquid portions are then strained off . . . and leak into the bladder' twice. This is an interesting comment in light of the recent (2012) finding of Kwang-Sup Soh (7) in page 29 when he reported:

> In the midline of the abdominal wall of a rat, there is a band of adipose tissues which we named the conception vessel (CV) fat line. Along this CV fat line, we can see a large vein and artery running from the xiphoid through the navel to the bladder. According to the chart of human acupuncture meridians, there is a CV meridian, and the WHO nomenclature named the acupoints on this meridian as CV 14 at the xiphoid and CV 8 at the navel; other points between these two acupoints are located at equal distances. Primo nodes at CV 12, 10, 8 were observed, and basic histological study with H&E and Mason's trichrome revealed that they were different from lymph nodes. By injecting FNP [fluorescent nanoparticles] into the primo nodes, we traced the flow of nanoparticles along the CV line to the ligament wrapping the bladder in the primo vessels. Thus, we established the presence of extravascular primo vessels along the blood vessels just outside the connective tissues of the blood vessel. In this experiment, the PVS ran along the CV fat line to the bladder.

So we have modern scientific confirmation that fluids in the Middle Heater can be traced within the Primo Vascular System and that they follow the lower section of the Triple Burner to the Bladder, exactly as Hsü Ta-ch'un stated above when he said, 'The liquid portions are then strained off; they follow the lower section of the Triple Burner and leak into the bladder.' Do they possibly 'steam' from here and then ascend via the San Jiao pathway?

I will discuss later in the book why I believe that the Triple Energizer develops very early after conception, but should the unborn baby die in utero for whatever reason, obviously the Triple Energizer would also cease to function. Based on the above and for other reasons discussed later in the book, I propose that should the pregnancy continue to term, the Triple Energizer commences functioning independently only after the first breath is taken after parturition.

1.5 Another Ancient View of the Triple Burner Based on the 31st Difficult Issue

In the commentaries on the 31st Difficult Issue on pages 353–355 of Unschuld's (1) translation of the *Nan Ching*, another scholar, Yeh Lin, relates:

> The [treatise] 'Ling lan mi-tien lun' of the Su-wen states: *'The Triple Burner is the official responsible for the maintenance of the ditches. The water-ways originate from there.'* That is [what is] meant here. As to its location on the 'street of influences,' the ch'i-chieh, [holes] are located on both sides of the [center of the] hairline. They represent holes on the foot-yang-brilliance [conduits]. *That is the root and the origin of the Triple Burner; it is the location of the influences. It is a fatty membrane emerging from the tie between the kidneys.* The Triple Burner is associated with the residence of the minister-fire. The nature of fire is to ascend from below. Hence, the [treatise] 'Ching-mai pieh-lun' of the Su-wen states: 'Drinks enter [the organism] through the stomach where their essential influences float off, moving upward to the spleen.' That is a reference to the central [section of the Triple] Burner. 'The influences of the spleen distribute the essence which ascends [further] and turns to the lung.' That is a reference

to the upper [section of the Triple] Burner. 'From there they penetrate into and regulate the passageways of water, moving downward to the bladder.' That is a reference to the lower [section of the Triple] Burner. *But why are only drinks emphasized in this discussion of the influences of the upper, central, and lower [section of the Triple] Burner? [Anybody posing such a question] does not know that the influences are transformed from water.* Through the inhalation of the heavenly yang, the water of the bladder follows the fire of the heart downward to the lower [section of the Triple] Burner. There it evaporates like steam and is transformed into influences moving up again, where they become the chin [liquids], the yeh [liquids], and the sweat. All of that rests on the principle that when fire meets water, a transformation into influences takes place. *The meaning is that heavenly yang [i.e., the influences of the sun] enters earthly yin [i.e., the water in the soil].* The [latter], following the movement of the yang influences ascends and become clouds and rain. (Emphasis is mine)

Here, I propose that the 'heavenly yang [i.e., the influences of the sun]' are the biophotons generated due to the solar irradiation from the sun that enter the body and which then 'ascends and become clouds and rain' that are circulated upwards to permeate the entire body just as clouds and rain cover the entire terrain (i.e. to rain on the terrain). Hold on! What are biophotons? I will discuss biophotons in great detail later in the book. It is very interesting to note that while the word *sun* appears nine times in Unschuld's (1) translation of the *Nan Ching*, six of those times the sun is mentioned in connection to the Triple Energizer in the same text. I believe that is because the Sun imbues food and water that we consume with absorbed biophotons, which help to drive the Triple Energizer's energy production system.

1.6 The Upper Burner Warms the Skin and Flesh
In the commentaries on the 31st Difficult Issue on page 349 of Unschuld's (1) translation of the *Nan Ching*, Yang states, '[The region] from the diaphragm upward is called the upper [section of the Triple]

Burner. It masters the emission of yang influences, providing warmth to the space between the skin and the flesh. That resembles the gentle flow of fog.'

Yang advises here that the Lungs, which are located above the diaphragm, are responsible for circulating the yang protective influences, including oxygenated blood to the literal 'space between the skin and the flesh'. Just as 'the gentle flow of fog' moistens everything that it comes in contact with, so too does the blood and body fluids warm, nourish, and moisturize all the tissues that it comes in contact with. This 'space between the skin and the flesh' is shown here to be an outer extension of the Triple Energizer organ complex and actually constitutes the couli, which will be discussed in a dedicated chapter later in the book.

1.7 The Upper Burner Also Sends Nourishment to All the Organs
In the commentaries on the 31st Difficult Issue on pages 349–350 of Unschuld's (1) translation of the *Nan Ching*, Yü Shu states:

> Tan-chung (CV 17) is the name of a hole. It is a hole situated exactly in the center between the two breasts. The influences of the controller vessel are emitted from here. The Su-wen states: 'The tan-chung is the emissary among the officials.' It masters the distribution of influences into the yin and yang [sections of the organism]. *When the influences are balanced, and when one's mind reaches into the distance, happiness and joy originate. That is [what is] meant by 'distribution of influences.'* Hence, [disorders in the upper section of the Triple Burner] are regulated through [a hole located] in the center [between the breasts].
>
> The upper [section of the Triple] Burner is responsible for the entry of water and grains [into the organism]. It takes in but it does not discharge. The Ling-shu ching states: 'The upper [section of the Triple] Burner resembles fog.' That is to say, when it passes the influences, that resembles mist

> gently flowing into all the conduits. In other words, the influences of the stomach and the influences distributed by the tan-chung are poured downward by the lung into all the depots. The [Nei-]ching states: 'The lung passes the influences of heaven.' That is the meaning implied here. (Emphasis is mine)

This Upper Burner is associated with the properties of the Sea-of-Qi point, CV17, and is associated with the oxygenation of the blood in the Upper Burner in the Lungs, and it helps to move the qi produced in the lungs downwards. So the Upper Burner causes the beneficial influences of water and grains—that is, complex fluid containing all the necessary nutrients from the foods eaten—to permeate the entire body with nutrient-rich, resuscitating, life-giving blood that has been oxygenated in the Lungs, like a 'mist' that permeates the entire body. For example, when mist appears in a valley or on a mountain range, it is all encompassing, and it moistens and affects everything in its path. No area is spared from its effects.

1.8 The Meaning of the Term *Distribution of Influences*

The term *distribution of influences* occurs many times throughout the *Nan Ching*. What does the term mean? In the commentaries on the 31st Difficult Issue on page 349 of Unschuld's (1) translation of the *Nan Ching*, commenting on the meaning of the *distribution of influences*, Yü Shu states:

> Tan-chung (CV 17) is the name of a hole. It is a hole situated exactly in the center between the two breasts. The influences of the controller vessel are emitted from here. The Su-wen states: 'The tan-chung is the emissary among the officials.' It masters the distribution of influences into the yin and yang [sections of the organism]. *When the influences are balanced, and when one's mind reaches into the distance, happiness and joy originate. That is [what is] meant by 'distribution of influences.'* (Emphasis is mine)

The *distribution of influences* involves all the life-sustaining biological functions performed by each of the organs under the control of the Triple Burner. Likewise, the nutrients of food are passed from Stomach to the Spleen to the Lungs and to the Entire body. Note too that when the distribution of influences is successfully accomplished throughout the entire body, the mind is at peace, and happiness and joy are the outcome. Obviously, when people are healthy, active, and well, then there will be peace of mind and contentment. This shows that the Triple Energizer is responsible for elevating the mood of individuals along with maintaining physiological balance throughout the entire body. So now that I have introduced the nature and general location of the Triple Energizer, let's consider the controversial aspects related to this very important *Fu* organ.

CHAPTER 2

Several Aspects of the Triple Burner Are Controversial

Introduction
Regarding the nature and composition of the San Jiao, Yongping Jiang (8), the author of the *Journal of Chinese Medicine* article titled 'The San Jiao: Returning to the Nei Jing (A Modern Explanation of Original Theory)' states, 'The San Jiao is arguably the most disputed organ of traditional Chinese medical (TCM) theory. Despite decades of investigation and research, TCM practitioners and scholars have yet to agree on the true identity of the San Jiao.' His personal belief regarding the components of the San Jiao are very unique, as you will soon discover.

To make matters even worse, no other organ has so many synonyms as the Triple Energizer. Depending on the author, it is also known as the San Jiao, Sanjiao, Three Heaters, Triple Heater, Triple Warmer, Tri Heater, or the Triple Burner. None of these renderings do justice to the TCM San Jiao concept. It is thus no wonder that many Western researchers have ridiculed the Traditional Chinese Medicine (TCM) concept of the San Jiao organ because TCM practitioners can't even agree about its preferred name, let alone its location, nature, and form. The current WHO standard term is *Triple Energizer* (TE). I feel this terminology does not do justice to such an important and omnipresent organ complex, and subsequently, I refer to the good old *Triple Burner* throughout my book as the Triple-Energizer Metasystem. So where exactly did all this controversy about the Triple Burner begin?

2.1 Unschuld's Opinion Concerning the Various Concepts about the Triple Burner

In the commentaries on the 31st Difficult Issue on page 355 of Unschuld's (1) translation of the *Nan Ching*, with reference to the controversy surrounding the Triple Burner, Unschuld states:

> Throughout history, commentators have voiced all kinds of different opinions concerning the Triple Burner as one of the six palaces. Most important was the argument over whether the [Triple Burner] represents an entity with a name and no form, or with both name and form. In addition there were [authors] proposing [that the Triple Burner] occupies three locations in the body's cavity, and others who referred to the lower [section of the Triple] Burner as simply a waterway penetrating the six palaces.

2.2 Chapters 25 and 38 of *Nanjing* Initially Proposed that the Sanjiao Has No 'Form'

In the 2010 article 'Gross Conception of Anatomical Structure of the Triple Burner in *Huangdi Neijing*' (9) regarding the location of the three burners, the authors state:

> Ever since Chapter 25 and Chapter 38 of Nanjing brought forth the hypothesis that the Triple Burner is an amorphous structure, with only a name and no real structure, people have been disputing about the nature and significance of this important organ in Traditional Chinese Medicine (TCM). As something difficult to describe, the Triple Burner has made people who pursue medicine doubt the scientific significance of TCM. In order to define the existence of the Triple Burner, or the lack thereof, some have created the notion that the organs in TCM are not

comparable with those in Western Medicine, further separating TCM from modern science.

2.3 Varying Opinions about the Functional Aspect of the Triple Burner Even in the Classics

Regarding the processing of foods and fluids in the Triple Burner, in the 2011 article titled 'The Triple Burner (2)', Giovanni Maciocia (10) notes the differing viewpoints between two of the classic literature sources, the *Nei Jing* and the *Nan Jing*. He states that the *Nan Jing* emphasizes the functions of receiving, rotting and ripening, and excretion of foods and fluids, while the *Nei Jing* highlights the function of the Triple Burner in its letting out role, seeing the three Burners as three paths of excretion or letting out.

2.4 Mystery Associated with the Triple Energizer Continues to This Very Day

In the appendix of the book *Heart Master Triple Heater*, on page 122, while summarizing aspects of the Three Heaters, Elisabeth Rochat de la Vallée and M. Macé (11) state, 'San jiao, the triple heater, is one of the most difficult concepts to grasp in Chinese medicine, not only because as an entity it has no equivalent in Western medicine, but also because in China itself it has not been clearly and plainly defined.' They continue:

> In fact the triple heater can be presented equally as a concrete and localised organ, such as the pipes for evacuating urine or the cavities of the stomach, or as very general functions for the animation and irrigation of the whole body. The triple heater appears in the texts of the Nei jing as one of the six fu with the stomach, the two intestines, the bladder and the gallbladder. With the exception of the gallbladder, (Su wen 11) they form the set of fu for transmission and transformation ... in charge of digestion, assimilation and elimination.

2.5 The San Jiao Has Been Poorly Understood and Has Been a Topic of Disagreement for Centuries Even in China

In agreement with the former comments, in the article entitled 'Cultural Reference for Increased Understanding of the San Jiao' regarding the understanding of the San Jiao, the author, Glenn Grossman (5), states:

> The San Jiao has been poorly understood and a topic of disagreement for centuries in China. No wonder that it is hard to understand in English. Furthermore, it can have different degrees of importance depending on the way it is studied. It can be studied as a fu organ, a meridian system or as a major TCM pattern of differentiation.

2.6 Controversial Aspects Regarding the Numerous Functions of the San Jiao

Yongping Jiang (8), the author of the *Journal of Chinese Medicine* article titled 'The San Jiao: Returning to the Nei Jing (A Modern Explanation of Original Theory)' believes that 'the original theory of the San Jiao has been lost due to misinterpretation of the original texts' and proposes that the structure of the San Jiao is essentially made up of the esophagus, Stomach, and Small Intestine. Based on his interpretation of the San Jiao in the *Nei Jing*, he rejects 'the widely held notion that the San Jiao contains the zang organs and their functions'. He feels the modern description of the San Jiao 'is too large, involves too many organs, and has such numerous functions attributed to it that the result is a vague and confusing organ with no clear functions of its own'.

2.7 Diverse Descriptions of the Origin and Functionality of the Triple Burner

In the article titled 'The Kidney Network and Mingmen: Views from the Past', the author (12) has assembled six various references that cite interpretations of what the different authors believe regarding the origin and functions of the Triple Burner. The author states:

From Tang Zonghai, *A Refined Interpretation of the Medical Classics (Yijing Jingyi)*, Qing Dynasty: The root of the triple burner is in the kidney, more precisely right between the two anatomical kidneys. Right there is a greasy membrane that is connected with the spine. It is called mingmen, and constitutes the source of the three burners.

From Zhang Shanlei, *A Revised Edition of Master Zhang's Treatise on the Organ Networks (Zhang Shi Zangfu Yaoshi Buzheng)*, ca. 1918: The triple burner is really a name for the function of the body's ministerial fire. It is the process of disseminating original qi from mingmen, which is in charge of ascending and descending, and absorbing and excreting.

From Sun Yikui, *Mysterious Pearls of Wisdom (Chi Shui Xuan Zhu)*, 1584: The so-called triple burner is embedded in the greasy membrane of the diaphragm, that is the hollow space between the five zang/six fu organs and the connective pathway through which food and grain must pass.... The regions that it reaches are labeled according to their location, that is why we speak of the upper burner, the middle burner, and the lower burner. Although the triple burner does not have any structural reality to it, it has a distinct location that is determined by the structural entities surrounding it.

From Shen Jin'ao, *Illuminating Lantern on the Origins of Complex Diseases (Zabing Yuanliu Xizhu)*, 18th century: What we call the triple burner is actually the corridor above and below the stomach. The triple burner and its associated regions thus entirely belong to the stomach, and what it oversees is primarily the functioning of the stomach. The triple burner qi is utilized to ferment and cook the food.

From Li Dongyuan, *Illuminating the Science of Medicine (Yixue Faming)*, 13th century: The triple burner is an entity that has a name but no structural form. It is in charge of all bodily qi, and it is a functional manifestation of the three treasures [jing, qi, shen]. All of the body's physiological movements, its unobstructed ins and outs and ups and downs, therefore, rely on the triple burner-the process of breathing in and breathing out, the ascending and descending motion of qi, and the absorption and excretion of food and water.

From Chen Nianzu, *The Three Character Classic of Medicine (Yixue Sanzi Jing)*, Qing Dynasty: The term triple burner refers to the qi that circulates in the upper, middle, and lower burners. Burner means heat. Only when the entire body cavity is permeated with hot qi can the body's water ways be open and regulated. The triple burner is the fu organ that forms a zang/fu pair with the pericardium, and thus belongs to the phase element fire. . . . It is for this reason that the triple burner is called the official in charge of uninhibited water flow.

Tang Zonghai, cited above (12), believes that the 'root of the triple burner is in the kidney, more precisely right between the two anatomical kidneys. Right there is a greasy membrane that is connected with the spine'. Sun Yikui, on the other hand, believes that 'the so-called triple burner is embedded in the greasy membrane of the diaphragm'. At least, they have a greasy membrane in common. Shen Jin'ao proposes that 'the triple burner is actually the corridor above and below the stomach'. Li Dongyuan states, 'The triple burner is an entity that has a name but no structural form.' Chen Nianzu believes the Triple Burner affects the entire body and 'is called the official in charge of uninhibited water flow'. There is such diversity of opinion concerning the origin, nature, and function of the Triple Burner from these learned authorities. How is one to comprehend the true nature of the Triple Burner? I hope my book will set the matter straight.

2.8 There Is Even Doubt about the Interior/Exterior Relationship between the Pericardium and Triple-Burner Organs

In the article entitled 'Cultural Reference for Increased Understanding of the San Jiao' regarding the internal/external relationship between the San Jiao and the Pericardium, the author, Glenn Grossman (5), states:

> Also difficult to understand is the relationship between the San Jiao and, its Zang partner, the Pericardium. One of the standard texts for use in TCM schools worldwide, Chinese Acupuncture and Moxibustion (CAM), contains no mention of the Pericardium/San Jiao in its section on 'The Relationship Between the Zang and Fu Organs'. To complicate matters further 'some Chinese teachers and doctors go so far as saying that the Pericardium and Triple Burner organs are not interiorly exteriorly related as the other organs are'. Others state that the San Jiao and Pericardium's interior exterior relationship is important only for acupuncture.

2.9 Hooray! There Is General Agreement about a Major Role of the Triple Energizer

The Triple Burner performs a major role in all the various phases of digestion and assimilation and administers the body fluids and their distribution throughout the body. The association with the transformation of liquids and the formation of body fluids is constantly reaffirmed in the physiology and pathology of the San Jiao. This intimate relationship with water in its diverse forms is accentuated by the traditional titles attributed to each of the three heaters—*Wu*, mist, humid vapors for the upper heater; *Ou*, maceration for the middle heater; and *Du*, canal, conduit for the lower heater. So fortunately, in this regard, there is general agreement that the Triple Energizer regulates the various body fluid forms throughout the body.

Now what about the elusive form or morphology of the Triple Energizer? The *Nei Jing* said the Triple Energizer had a form. Then along came the *Nan Ching*, which said that the Triple Energizer 'had a name but no form'. It cannot be both ways. The organ complex either does have a form or does not have a form, so which is it? Chapter 3 discusses this matter.

CHAPTER 3

Controversy About the Triple Energizer Having No Form

Introduction
Yüeh-jen supposedly wrote the *Nan Ching* during the Chin dynasty (221–206 BC). About one-third of the medical treatise was devoted to acupuncture and moxibustion practice, which were already considered parts of a single medical science. While Yüeh-jen was regarded as the foremost master of acupuncture by later generations, many scholars have called him careless, believing he made many translations and interpretations from the classics that were not very clear and open to extremely opposite interpretation. For example, he introduced the cat amongst the pigeons, when he suggested that the Triple Burner 'has a name but no form'.

3.1 Historical Development and Conceptualization of the Nature of the Triple Energizer
Regarding the history of the belief that the Triple Burner has *no form*, in the Notes for the 25th Difficult Issue (page 316), Unschuld (1) expressed his view by stating:

> This difficult issue marks the beginning of a controversy that has not been settled even today. The heart-master, also called heart-enclosing network, may originally

have been a concept developed to meet the number six for the depots, if they were to correspond to the three yin and three yang subcategories. The Triple Burner may have been conceptualized in correspondence to environmental symbolism. In the last centuries B.C., the entire physiological organism was seen as a mirror image of the state and its economy. The terms 'depot,' 'palace,' 'conduits,' the bureaucratic hierarchy of the organism, and so on reflect this understanding. In this context the assumption of some heating device in the organism—corresponding, for instance, to the most important economic functions of the saline and iron works—may have been a stringent consequence.

While an interesting concept, I personally believe that in the light of modern scientific research and discovery, the anatomical descriptions and functions pertaining to the Triple Burner in the *Nan Ching* are perfectly accurate and not manufactured to fit some contrived paradigm.

In the notes for the 31st Difficult Issue (pages 355–356), Unschuld (1) again discusses the controversial aspects of the Triple Burner when he states:

> The character of the so-called Triple Burner has remained a controversial issue for as long as we can trace this concept in medical literature. Not unlike their counterparts in ancient Greek medicine, ancient Chinese thinkers assumed the existence of some kind of a heat source in the organism. Hence, they conceptualized the ruler-fire and the minister-fire, as well as the Triple Burner. Obviously, the Nei-ching documents the development of the Triple Burner from a designation of functions [see the various *Su-wen* passages quoted by the commentators] to the designation of a tangible entity [see the *Ling-shu* treatise quoted by Hsü Ta-ch'un in his commentary on sentences 1 through 6]. In the Nan-ching, in contrast to both the Su-wen and the Ling-shu, the Triple Burner—with a name

but no form itself [see difficult issue 38]—appears to be considered a functional description of the upper, central, and lower groups of organs of the body. . . . Current textbooks in the People's Republic of China offer differing opinions as to whether one should interpret the Triple Burner as an anatomical entity or simply as a functional description.

Regarding the historical development and conceptualization of the nature of the Triple Energizer, in the Notes for the 25th Difficult Issue (page 316), Unschuld (1) further states:

See also Medicine in China: A History of Ideas, chapter 3.3. Obviously, it was apparent even during the Han era that no anatomical entity corresponded to the concepts of 'heart-master'/'heart-enclosing network' and 'Triple Burner' in the same way that a real liver corresponds to the concept of the liver. Hence, the compromise approached here assigned a function to the heart-master/heart-enclosing envelope and to the Triple Burner, but no anatomical substratum.

I believe that due to misunderstanding, many erudite scholars read the *Nan Ching* and believe that it rejects the fact that the Triple Burner does have a distinct form and that it relegates it to being merely a collection of biological functions subdivided into three sections—the upper, middle, and lower burners. I believe that this was never the intention of Yüeh-jen, the author of the *Nan Ching*. I suggest that he understood that the Triple Burner existed within the body as a highly structured, omnipresent physical complex but without a succinct and defined, recognizable morphological form. I ask you, does the immune system have a form? What about the recently discovered highly complex gut microbiome? Do either of these organ systems have an easily recognized morphological form, like the kidney or gallbladder? No, of course not! That definitely does not mean that they do not exist and that they are merely 'functional entities'.

3.2 Ancient Defenders of the Belief that the Triple Burner Does Have Form

Regarding the same issue, in the commentaries for the 25th Difficult Issue on page 312 of Unschuld's (1) translation of the *Nan Ching*, Hsü Ta-ch'un says:

> [The text] states that the Triple Burner has no form. That cannot be. It states [further] that the hand-heart-master has no form, but such a doctrine definitely does not exist. The heart-master is the network enclosing the heart, it consists of a fatty membrane protecting the heart. How could it have no form? It is not called a depot because the heart-master acts on behalf of the heart. In itself, it does not store anything. Hence, it is not called a depot.

Based on scriptures throughout the *Nei Ching*, Hsü Ta-ch'un insists that for the Triple Burner to have *no form* simply cannot be. With respect to the tangible quality of the Triple Burner, in the Notes for the 45th Difficult Issue (page 439), Unschuld (1) made a personal comment when he stated:

> My rendering here corresponds to the interpretation of this sentence by a number of commentators who read it as . . . *san chiao, wai*. . . . Others, including Hsü Ta-ch'un . . . have interpreted *san chiao wai* as 'outside of the Triple Burner.' *Hsü Ta-ch'un appears to have preferred such an interpretation since, as a conservative commentator who gave priority to the sayings of the Nei-ching, he believed in a tangible quality of the Triple Burner* (see also his comments on difficult issue 31, where he quotes the respective passage from the *Ling-shu*). The *Nan-ching* itself, in contrast, did not consider the Triple Burner to be a tangible entity; in difficult issue 38, it states: 'It has a name but no form.' (Emphasis is mine)

This is one of the reasons that there is confusion regarding many aspects pertaining to the Triple Energizer. Many original words and statements are open to interpretation based on the knowledge base and preconceptions (and misconceptions) of the translator.

Yeh Lin also believes that there is no doubt that the Triple Burner has a physical form. However, he does feel that the fluid and energy metamorphoses attributed to the Fu organ are difficult to understand. In the commentaries on the 38th Difficult Issue on page 396 of Unschuld's (1) translation of the *Nan Ching*, Yeh Lin states, 'The Triple Burner has a form! That has already been discussed in sufficient detail in the commentaries on Difficult Issue 25. . . . *It can be proven that the Triple Burner has material form*. But the transformation of influences through the Triple Burner is difficult to perceive. Hence, [the text] says: "It has a name but no form".' (Emphasis is mine)

In the commentaries on the 25th Difficult Issue on pages 312 to 315 of Unschuld's (1) translation of the *Nan Ching*, Ting Chin's loquacity is amply but eloquently expressed. I have emphasized his main points. When you read each sentence and analyze its content with regard to what was stated in the original medical classics, it is obvious that Ting Chin masterfully disputes all criticisms regarding the triple heater having *no form* when he states:

> *This paragraph states that the heart-master and the Triple Burner constitute outside and inside, and that both have a name but no form. Because of the two words 'no form,' people in later times who did not check the meaning of the [Nei-]ching have engaged themselves in highly confused argumentations.* They not only criticized the [alleged] mistakes of Yüeh-jen but also criticized [what they considered to be] erroneous interpretations forced [on this passage] by [Wang] Shu-ho. *Over the past three thousand years, this has never been settled finally.* I always think that the *Nan-ching* was not yet distant from antiquity. Of all the authors who appeared [in later times to comment on the ancient scriptures, Yüeh-jen] was the very first. Also, one must base [one's understanding of the *Nan-ching*] word for word on the *Nei-ching*. *Why should misunderstandings*

and a deception of mankind be created just for the two key [concepts] of the [heart-]enclosing network and the Triple Burner? There is no other way to elucidate [their meaning] except by comparing the meaning in the *Nei-ching* with that in the *Nan-ching*. (Emphasis is mine)

So please deeply consider what I am about to say, and then read the continuing rebuttal from Ting Chin. I am sure you will see where the confusion lies, and you will understand the true nature and misunderstood morphology of the Triple Burner.

3.3 Defining the Elusive Morphology of the Triple Burner

I ask you to describe the form and shape of the average cloud. Just because clouds do not have a standardized describable shape, size, and form, that does *not* mean that clouds do not exist. Likewise, the form of the Triple Burner does not possess a standard shape and form as a heart, stomach, or liver does. But let's get back to human body associated analogies.

Figure 1. Philosophy concept. By kraphix. © Depositphotos.com.

1) Where is your circulatory system? Point to it! Actually, it is a *body-wide network* of blood and blood vessels powered by the heart that takes blood to every nook and cranny in your entire body. The circulatory system is *ubiquitous and omnipresent within the body.*
2) Where is your muscular system? Point to it! Actually, it allows your body to move every part in very precise ways. Have you ever observed the intricacy of the movements of a ballerina or an acrobat? Your muscular system is composed of about

seven hundred major named muscles which make up about half of your body weight. Muscle tissue is also present inside your heart, digestive organs, and blood vessels. The arrector pili muscles are the smallest muscles in your body. They attach to your hair follicles and cause goose bumps when the hairs stand on end. So the muscular system *encompasses your entire body* right down to the movement of each of your hairs. How many hairs do you have?

3) Where is your endocrine system? Point to it! Your endocrine system is composed of many glands that secrete powerful hormones directly into your circulatory system. These substances are carried to distant target organs. Your endocrine system is an information transfer system that affects every cell in your entire body. It is *ubiquitous and omnipresent* within the body.

4) Where is your digestive system? Point to it! It begins at your lips and terminates at your anus. Internally, your digestive system is made up from a group of organs cooperating to convert ingested food and fluids into essential energy and nutrients to *nourish and support the entire body*.

5) Where is your skeletal system? Point to it! The 206 bones of your skeletal *system protect and support the entire body* along with giving it shape and form.

6) Where is your nervous system? Point to it! Your nervous system transmits signals to *every single part of your entire body* and coordinates all the voluntary and involuntary actions within every single part of your entire body. Stick an acupuncture needle into any part of the body and jiggle it vigorously, and the patient will let you know they feel it, thanks to the *omnipresent nervous system*.

7) Where is your lymphatic system? Point to it! 'Lymph is the fluid that is formed when interstitial fluid enters the initial lymphatic vessels of the lymphatic system' (13). The *lymphatic system is omnipresent within your body*. Burn yourself anywhere, and a blister will form. The blister is filled with straw-colored fluid lymph.

8) Where is your immune system? Point to it! The immune system defends you against the numerous bacteria, fungi,

viruses, toxins, and parasites that would love to invade your body. Once again, *your immune system extends throughout the entirety of your body.* It is composed of special blood cells and chemical proteins called antibodies that fight infection and neutralize toxins.

9) Where is your microbiome? Point to it! Your microbiome is composed of billions of beneficial commensal microflora that are concentrated in your gut but, likewise, *covers every square millimeter of your skin,* mouth, lungs, vagina (females only), sweat glands, etc. If you are not aware how extremely important this newly discovered organ is, I cover it in great detail in another Chapter of this book.

10) Where is your connective-tissue system? Point to it! *Connective tissue is found everywhere throughout your entire body,* including the central nervous system. Connective tissue is located in between all other tissues. Connective tissue happens to be the most abundant, most widely distributed, and most varied type of tissue within the body. It is also *ubiquitous and omnipresent within the body.*

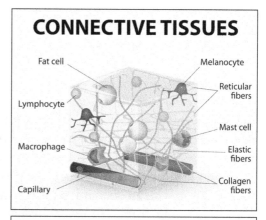

Figure 2. Connective tissue. By edesignua. © Depositphotos.com.

The ten body systems discussed above are all accepted by Western allopathic medicine and are integral for doctors to determine which pharmaceutical medications to prescribe to treat the associated problem symptomatically. Because of their global and convoluted intricacies within the body, none of these ten *systems* could be dissected out of your body and placed in a container, for example. The point I am making is that all ten of these defined, accepted, complex body systems involve and affect your entire body internally and externally. What about the Triple Energizer? From

TCM, we can see that the Triple Energizer involves water and food metabolism and energy production *throughout the entire body*. Many of their defined features actually overlap with several of the ten systems mentioned above (e.g. circulatory, endocrine, lymphatic, immune and connective tissue systems).

However, the major ten stand-alone body organs—including the heart, small intestine, spleen, stomach, lungs, large intestines, kidneys, bladder, liver, and gallbladder—can all be dissected out of the body as totally complete organs and placed into a container. However, the same *cannot* be said for the Triple Energizer. Does it mean that just because the Triple Energizer does not have a distinct dissectible form, it has only a function and subsequently has *a name but no form*? No! That is why this *unique Fu* organ is called the *palace of uniqueness*, or *fu of uniqueness*.

3.4 Seeing the Fu of Uniqueness as a Metasystem and *Not* as an Organ Like the Spleen

However, if we think of the Triple Energizer not as an *organ* per se but, rather, as an omnipresent metasystem in the same light as the former ten acknowledged and accepted body metasystems, it is very easy to accept the existence, the actual physical presence, and the reality of the Triple-Energizer Metasystem. Hereafter, in the book, I predominantly address the San Jiao as the *Triple-Energizer Metasystem*. While I have not yet described what I believe the Triple-Energizer Metasystem to be, with the forgoing in mind, now please continue to read the rebuttal of Ting Chin with an open mind to see where the rationale of his pure logic is coming from. However, instead of thinking of the Triple-Energizer Metasystem, or (as Ting Chin calls it) the Triple Burner, as a dissectible *organ*, please visualize the Triple Burner system as you would when reading about any one of the ten body metasystems described above. Bear in mind that 'depots' are the zang solid organs and the 'palaces' are the fu hollow organs. Ting Chin continues:

> Thus, the *Nei-ching* states that all the five depots have form and color, that the five palaces, too, can be measured in chang and feet, and that the water and the grains with

which they are filled can be recorded in amounts of pints and pecks. If the [heart-]enclosing network and the Triple Burner had a form, why would they be the only ones with colors, sizes, and capacities that are not clearly recorded? Well, one should look at what the *Nan-ching* says about the [heart-]enclosing network and pick its meaning from the term 'enclosing,' and [one should look] also at what [the *Nan-ching*] says about the Triple Burner and pick its meaning from the term 'triple.' Thus, in the *Ling*[*-shu*] and in the *Su*[*-wen*], the treatise 'Pen-shu' states: 'The Triple Burner is a palace [acting as] central ditch; the passageways of water emerge from it. It is associated with the bladder and it constitutes the *palace of uniqueness*.'

The treatise 'Pen-tsang' states: 'When the pores are sealed tightly and when the skin is thick, the Triple Burner and the bladder are thick [too].' The treatise 'Chüeh-ch'i' states: 'The upper burner [is responsible for] emissions; it disperses the taste [influences] of the five grains, [a process] resembling the gentle flow of mist. [What is distributed] is called 'the influences'. The central burner receives influences. It absorbs the juices, transforms them, and turns them red. That is called the blood.' In the treatise 'Ying wei sheng hui', it is stated: 'The constructive [influences] emerge from the central burner; the protective [influences] emerge from the lower burner.' It states further: 'The upper burner resembles fog; the central burner resembles foam the lower burner resembles a ditch.' The discussion in the [treatise] 'Wu lung chin-yeh pieh-lun' states: 'The Triple Burner emits the influences in order to warm the flesh and to fill the skin.' The treatise 'Hsieh-k'o' of the *Ling-shu* states: 'The heart is the great master of the five depots and six palaces. As a depot it is strong and firm. Evil [influences] cannot settle in it. If they do settle in it, the heart will be injured; the spirit will leave and [the respective person] will die. Hence, it is stated that if evil [influences] are present in the heart, they are always in the network enclosing the heart.'

All the lines quoted above from the *Ling*[*-shu*] and from the *Su*[*-wen*] describe the Triple Burner as completely enclosing the five depots and six palaces. The [heart-] enclosing network has the meaning of enclosing only the heart. The 'palace [acting as] central ditch' is the 'palace of uniqueness.' If it were not for the fact that the Triple Burner enclosed the [organism] externally, how could [this Palace] have this singularly honorable designation? It was said further that 'when the pores are sealed tightly, and when the skin is thick, the Triple Burner is thick [too].' Now, if the inside of the skin and the flesh of the entire body were not supported by the Triple Burner, how could their thicknesses correspond to each other? It was said further that 'the upper burner disperses the taste [-influences] of the grains; the central burner receives influences, absorbs the juices, transforms them, and turns them red.' If the Triple Burner did not enclose all the body's depots and palaces, how could all the body's depots and palaces share in the influences of the Triple Burner in order to [further] diffuse and transform them? It was said further that 'the constructive [influences] emerge from the central burner; the protective [influences] emerge from the lower burner.' The constructive [influences] become the blood because they are [generated from] the essence of the taste[-influences] of the grains. The protective [influences] are [volatile] influences [because they are] generated from the [volatile] influences of the grains. All these [transformations occur] because of the [activities of the] stomach. But how could the stomach be stimulated to perform these transformations if it were not for the fact that the Triple Burner externally completely encloses [the stomach] and manages the movement of the influences? It was stated further: '[The upper burner] resembles fog; [the central burner] resembles foam; [the lower burner] resembles a ditch.' Above, [the upper burner] gives orders concerning emissions; below, [the lower burner] manages the passageways of water. How could this be if it were not for the fact that the Triple Burner externally encloses all the body's depots and palaces, exerting complete control over them? It was stated further: '[The Triple Burner] emits the influences in order to warm the flesh and fill the skin.' That is a clear indication of the fact that the Triple Burner constitutes a layer supporting the skin and the flesh from inside. It was stated further: 'Whenever evil [influences] are present in the heart, they are always in the network enclosing the heart.' That is a clear indication of the fact that the enclosing network constitutes a layer holding the heart from outside.

Later readers of these texts were to say, if the Triple Burner has no form, how can passageways of water emerge from it? How can it be thick or thin? How can it be like mist or fog or foam or a ditch? How can it emit influences in order to supply warmth to the flesh? And if the enclosing network [of the heart] has no form, how can all the evil [influences] settle in this network enclosing the heart? *Why is it the only [entity] that cannot be seen?* Why does it lack color, width, and length? They obviously did not know that *the [heart-] enclosing network is a small bag providing a network internally and an enclosure externally.* Thus, the name already states that it is an *'enclosing network.'* Its form does not have to be described in terms of big or small, feet or inches. *The Triple Burner is a large bag supporting [the organism] from outside and holding it inside.* The uniquity of its holding [function] is described fully by nothing but the term 'triple.' *The term 'burner' fully describes the provision of the entire [body] with influences.* Hence, the name already states that it is a Triple Burner. Again, its form does not have to be described in terms of large or small, chang or feet. Anybody who hitherto has harbored some doubts can have them resolved now if he follows this [argumentation]. Also, if one matches this small bag resembling a depot and [therefore] constituting a separate depot with that large bag resembling a palace and [therefore] constituting a separate palace, that is the principle of heavenly creation and earthly organization.

> *Yüeh-jen stated the two words 'no form' here, and again in the thirty-fourth difficult issue. An examination reveals that they are highly appropriate; an analysis shows that they are quite correct.* How could the people of later times grasp but the hair on the skin of the *Nei-ching* and then criticize exemplary men of former times? Often [enough, *their statements] reveal only the dimensions of their ignorance.* (Emphasis is mine)

Wow! Ting Chin pulls no punches here. Note that I have separated Ting Chin's text above into four paragraphs for clarity. That is not how Professor Unschuld presented them. Note above how Ting Chin explains:

- that the stomach is completely enclosed by the Triple Burner, which orchestrates the influences and functions attributed to the stomach;
- that the Triple Burner orchestrates all the influences and functions attributed to all the zang and fu of the three different Burners;
- the fact that the Triple Burner 'emits the influences' that 'warm the flesh and fill the skin' throughout the entire body;
- that because the Triple Burner surrounds and encloses all the organs of the entire body, like a bag, it obviously does exist, but its form, its elusive morphology, is not able to be defined like any other organ. It truly is a *palace of uniqueness* because it is a *unique Fu*, like no other Fu or zang for that matter.

3.5 The Triple Burner Absolutely Does Have a Morphological Form

These arguments confirm that the Triple Burner must have a form, and he sums it up when he says, 'Anybody who hitherto has harbored some doubts [about the Triple Burner having a form] can have them resolved now if he follows this argumentation.'

Note that Ting Chin agrees with Yüeh-jen when Yüeh-jen stated the two words *no form* while describing the *physicality* of the Triple Burner system in the *Nan Ching* Difficult Issues 25 and 34. Regarding Yüeh-jen's choice of the *two words,* Ting Chin says they were 'highly appropriate; an analysis shows that they are quite correct'. Ting Chin shows that Yüeh-jen believed that the Triple-Energizer Metasystem was a real, existing organ system, but it was just not as tangible as the major ten stand-alone body organs—the heart, small intestine, spleen, stomach, lungs, large intestines, kidneys, bladder, liver, and gallbladder—and thus had *no form.* Indeed, what is the form of the endocrine system, the microbiome system, or the connective tissue system, etc.? While those systems definitely exist, they also have no form.

Note the closing words from the third paragraph above, which states, 'If one matches this small bag resembling a depot and therefore constituting a separate depot with that large bag resembling a palace and therefore constituting a separate palace that is the principle of heavenly creation and earthly organization.' The 'small bag resembling a depot' is, of course, the Pericardium, which is just like a wrapping bag that encloses and protects the Heart. The author said previously the 'heart-enclosing *network* is a small bag providing a *network internally* and an *enclosure externally*' (emphasis is mine).

In a like manner, Ting Chin states, 'The Triple Burner is a large bag supporting the organism from outside and holding it inside. The uniquity of its holding function is described fully by nothing but the term "triple." The term "burner" fully describes the provision of the entire body with influences", and it is obvious that the Triple Burner too is a tangible organ "network" resembling a large bag that functions by "providing a *network internally* and an *enclosure externally*"' (emphasis is mine). I will confirm in later Chapters just how the Triple Burner supports or contains the entire organism from outside (just under the skin) while holding and providing a functional network for the entire organism from the inside.

When Ting Chin states 'If one matches this *small bag resembling a depot* and [therefore] *constituting a separate depot* with that *large bag resembling a palace* and [therefore] *constituting a separate palace,* that is the principle of heavenly creation and earthly organization' (emphasis is mine), he is confirming that the 'small bag resembling a depot' (i.e. the zang Pericardium) is matched with or is internally/

externally related to the 'large bag resembling a palace' (i.e. the fu Triple Burner). Further to that, he is saying that they *constitute a separate depot* and *constitute a separate palace* respectively, where the *depot* and the *palace* represent a separate zang and fu respectively. Importantly, when he likens the Triple-Burner organ system to a separate palace, he is not comparing it to formless palaces but to other real palaces, or hollow fu organs in the body, all of which definitely have a form.

In the final paragraph, Ting Chin expresses his annoyance at the ignorance of some TCM scholars who misunderstood Yüeh-jen. Yüeh-jen never stated that the Triple-Energizer organ system did not *exist but rather was a combination of functions*. Ting Chin's comment that 'their statements reveal only the dimensions of their ignorance' pertains to scholars who have missed the point altogether and incorrectly believe that Yüeh-jen was saying that the Triple Burner was not an organ but only a group of functions. This misunderstanding is what has led to 'confused argumentations' in later times.

Interestingly, while the initial statement of having *no form* pertained to *both* the heart-enclosing network (Pericardium) and the Triple Burner, today the general belief by TCM practitioners is that the Pericardium does indeed have form. In the original writings of the *Nei ching*, there was never any indication that the Triple Burner had *no form*. Further, the *Nan Ching* never said that the Triple Burner did not exist at all; rather, unlike all the other tangible dissectible morphic organs, the Triple Burner system is essentially amorphous, encompassing the whole body, and its actual shape and indefinable morphology is highly dependent on the size, height, and shape of the individual.

Hsü Ta-ch'un sums it up very aptly in the commentaries on the 38th Difficult Issue on page 396 of Unschuld's (1) translation of the *Nan Ching*, where he states:

> The Ling[-shu] and the Su[-wen] discuss the Triple Burner
> more than once. These are individual statements the

grammatical styles of which are more or less transparent. [The Triple Burner] is called a palace because it emits and takes in, because it links and because it spreads. That makes it obvious that [the Triple Burner] has the function of storing and draining. How could one say that it has no form! It is spread out all around the upper and the lower [parts of the body], enclosing [all the other] depots and palaces. *Its form differs from that of the [remaining] five palaces, each of which has its distinct body. Hence, its appearance cannot be defined [in the same way].* But to say 'it has no form'—that is impossible. (Emphasis is mine)

Regarding 'a name but no form', interestingly, the early developing human embryo, while being a living entity, does not have a defined form or morphology. The embryo starts from the union of a single sperm and an egg and then subdivides into a continually changing strange-looking life form that only by about week 16 has a form that is recognizable as a small human being. We do not say that the embryo is not human just because it doesn't have a fixed and constantly defined structure in the early stages of development.

3.6 Practitioners Today Still Believe the Triple Heater Has No Form

In the 1998 book *Heart Master Triple Heater* (page 52), regarding the nature of the Triple Burner, Rochat de la Vallée (11) states:

> But this original unity of water and fire within the kidneys renders them imperceptible, and *the triple heater is also imperceptible*. Fire and water, at the level of the kidneys, are like the primitive and fundamental unity of the living being. It is quite impossible and intangible—*without form*, but through the triple heater fire and water can produce effects. *They do not have a form* but there is the possibility of transformation and diffusion which will make all manifestation possible, and possibly the evolution of all the organs, if you are looking at it from a genetic point of view. But that is speculation. (Emphasis is mine)

Interestingly, Larre does not disagree with Rochat de la Vallée in this matter so must also believe that the Triple Energizer has a name but no form.

In the 1998 book *Heart Master Triple Heater* (page 117), while discussing the three heaters having a name but no form in *Nan Jing* Difficulty 62, Elisabeth Rochat de la Vallée (11) says, 'This is an explanation of the source points as it is given in the *Nan Jing*, along with the very close relationship with everything *that we call the triple heater, which has a name but no form. It has no form because it is the intermediary between that which has no form and that which is formulated and becomes a form*' (emphasis is mine).

The 2008 website article entitled *Fire Element: San Jiao Theory, Meridians & Points* (14) states:

> The San Jiao [Triple Heater, Triple Burner, Triple Energiser, Three Heater] is one of the Yang Organs of the Fire Element. The other is the Small Intestine. Considerable ambiguity and debate is associated with the San Jiao Organ. Its exact nature is not clear from the classical texts. It is said to 'have a name but no shape' or 'is located separately from the Zang Fu Organs and inside the body'. It may be understood as the functional relationship between the various organs which regulate water metabolism—Lungs, Spleen, Stomach, Kidneys, Small Intestine, Large Intestine and Bladder. *It is possibly best thought of as the overall unity, or pathways, which make these Organs a complete system—in terms of fluid metabolism. It does not exist as an entity outside these other Organs.* (Emphasis is mine)

The above information is generally what many TCM practitioners believe about the mysterious San Jiao. Most believe that the San Jiao is a functional complex rather than an individual stand-alone organ. I suggest the final sentence also sums up what most TCM practitioners believe. However, notice what the author quoted initially: 'It [i.e. the San Jiao] is said to "have a name but no shape" or "is located separately from the Zang Fu Organs and inside the body".' Note that both of these quotes allow for an actual physical (yet formless) organ system—for example, an immune system, a gut microbiome organ system, an intricately interwoven connective-tissue system, etc. Excluding bias due to present-day peer pressure, the second quote absolutely accommodates the presence of a tangible organ system when it says the San Jiao 'is located separately from the Zang Fu Organs and inside the body'.

In 2010, Dharmananda (15) stated that as discussion about the triple burner progresses over time, the divergence in agreement about the form and function of the triple burner will become obvious. He acknowledged that the sanjiao has been recognized in the Chinese system as *'having less physicality than the other defined organs'*. Further in the article, he states:

> The lack of shape and 'physicality' of the triple burner was disturbing to many Chinese medical scholars, and so they interpreted some of the *Neijing* and *Lingshu* statements about the triple burner *as pointing to a membrane that encased the entire torso,* having three parts: one covering the chest cavity; one for the upper/middle abdomen; and one for the lower abdomen. *However, such projection of a 'form' seems out of touch with the primary descriptions of the triple burner, and represents part of the struggle to figure out its nature and qualities.*

I cannot understand how Dharmananda believes that many Chinese medical scholars were out of touch with the primary descriptions of the triple burner when they suggested that the triple burner was an actual organ system encased by physical membranes. Regarding the 31st Difficult Issue in Unschuld's (1) translation of the *Nan Ching*, in the commentaries (page 355), Huang Wei-san said: 'The Triple Burner encloses all the depots and palaces externally. *It is a fatty membrane covering the entire physical body from the inside.*' Another example is in the commentaries on the 38th Difficult Issue on page 396 of Unschuld's (1) translation of the *Nan Ching*, where Li Chiung states, '*The Triple Burner represents nothing but membranes* attached to the upper, central, and lower opening of the stomach' (emphasis is mine).

3.7 There Is No Acknowledged Western Anatomical Counterpart for the San Jiao

Regarding the location of the San Jiao, in the article entitled 'Cultural Reference for Increased Understanding of the San Jiao', the author, Glenn Grossman, L.Ac. (5), states:

> *There is no anatomical counterpart for this fu organ.* Of the twelve Zang-fu the other ten are associated directly with anatomical organs, i.e., heart, lungs, liver, spleen, kidneys, gall bladder, urinary bladder, large and small intestines. The eleventh, the pericardium, is a type of space like its externally related partner the San Jiao. The pericardium, however, is recognized in western anatomical and diagnostic studies whereas the San Jiao is not. (Emphasis is mine)

Simply because the San Jiao has not been recognized and acknowledged in Western medical literature does not mean that it does not exist. The majority of Western medical practitioners and surgeons do not yet recognize and acknowledge the immense importance of the newly discovered organ complex called the gut microbiome or the very first embryological organ complex called the Primo Vascular System. This is out of ignorance of the knowledge of their existence. In spite of such ignorance, the gut microbiome, the Primo Vascular System, and the San Jiao do actually exist, and each one possesses a unique physicality. They just do not possess physicality like the heart, liver, stomach, or small intestine. Why should they?

3.8 Possible Reasons Contemporary Practitioners Believe What They Believe

Dr Edward Neal has been practicing and teaching Chinese medicine for over 20 years. Originally trained as a Western allopathic physician, he first studied traditional acupuncture with Dr Anita Cignolini of Milan, Italy. He is currently the director and senior researcher for Nei Jing Studies at the Xinglin Institute, a multidisciplinary research-and-educational institute dedicated to the study of early Chinese medical texts in order to find solutions to global health problems.

Following the provocative, paradigm-shaking series of three articles written in the *Journal of Chinese Medicine* (issues 100, 102, and 104), the editor of the journal, Daniel Maxwell, interviewed Dr Neal in order to tease out some of the threads arising from his rendering of the classical scriptures. On page 46 of Issue 105 of *The Journal of Chinese Medicine*, Dr Edward Neal (16) clearly explains himself when he states:

> If we examine the global situation today, most practitioners have never read the medical classics—and many discount them as being outdated. Western practitioners who have read the classics typically read a translation. Although the situation has improved in recent years, most translations are inadequate and many core ideas found in the classics are simply too difficult to engage with without examining the original Chinese source text. A smaller number of people have read the texts in their original language and possess the requisite understanding of classical Chinese grammar. Of these, *the vast majority try to interpret the text through the contemporary lens of TCM. This desire to evaluate classical texts through one's contemporary experience has been a stumbling block for commentators throughout the centuries. This is a critical impairment, as it is simply impossible to investigate a vessel-based system with its different anatomical descriptions, different circulation pathways, different theoretical approaches and different clinical techniques from the viewpoint of another system such as TCM.* This is just a non-starter, yet it is by far the most common approach used by people studying the classical approach today. Others have the prerequisite language skills and a background in sinology but have little or no clinical experience. This is also a problem—in general sinologists should not be teaching clinical medicine. Of people who have read the texts in Chinese, understand classical Chinese grammar, have clinical experience and have studied the text on its own terms, few know how to put this information into direct clinical practice and support what they do from text passages. Of

these, a smaller percentage will be willing to share their knowledge with you in a language you can understand. So from a field of roughly a million practitioners, we have now narrowed things down to a relatively small number of individuals who understand this material. So to find a physician who knows all of this and is willing to share their knowledge can be quite difficult. (Emphasis is mine)

Neal (16) continues, 'When I work with these texts I first try to empty my mind of any preconceptions—my goal is to try and understand a perspective of people from a very different culture and time, not to make the pieces fit into something familiar.'

Regarding the aspect of the Triple Energizer supposedly having a name but no form, I suggest that many modern-day scholars have done exactly what Neal discusses above when he stated that these scholars 'interpret the text through the contemporary lens' and they 'evaluate classical texts through one's contemporary experience'. In this case, I believe that Elisabeth Rochat de la Vallée, Claude Larre, and Giovanni Maciocia—in spite of their truly immense knowledge of sinology—have perpetuated the concept of the Triple Energizer not having a form, because that is the presently held paradigm. Sorry to say that, guys. I really love your hard work and appreciate you sharing so much knowledge that would otherwise not be accessible. I know that I am going to be stepping on toes, proposing that the Triple Energizer actually exists and has a defined structure and form that is individual to every single person. I take comfort in what Albert Szent-Györgyi, Nobel laureate (1893–1986), wisely stated, when he said, 'Discovery consists of seeing what everybody has seen and thinking what nobody has thought.'

3.9 Recent Defenders of the Belief that the Triple Burner Has Precise Physicality

Regarding the origin of the concept that the San Jiao *has a name but no form*, Yongping Jiang (8), the author of the 2009 *Journal of Chinese Medicine* article titled 'The San Jiao: Returning to the Nei Jing (A Modern Explanation of Original Theory)' states, 'The *Nan*

Jing was the first text to state that the San Jiao has "a name but no form" ("You ming er wu xing").' He continues, 'This statement started centuries of disputes. Why would the *Nan Jing* state that the San Jiao has no form after the *Nei Jing* has already clearly described its precise physicality?'

3.10 The Definition of the Triple Burner Being a Gross Organ Is Obvious

In the 2010 article 'Gross Conception of Anatomical Structure of the Triple Burner in Huangdi Neijing' (9), regarding the Triple Burner as a gross organ in *Huangdi Neijing*, the authors state, 'Huangdi Neijing mentions the Triple Burner in a total of seventy-six places, each occurrence describing it with respect to a different concept, such as organs, meridians, disease etiology, or qi metaplasia, etc. Nonetheless, the definition of the Triple Burner being a gross organ is obvious.'

So, it can be seen that there is still controversy about whether the Triple Energizer exists as a physical system or if it simply defines a group of functions. In the next Chapter, I will discuss the controversy associated with the individual anatomical locations of each of the three burners, and what different authors believe the Triple Energizer actually is.

CHAPTER 4

Controversy Over the Anatomical Location of the Triple Energizer

Introduction
The authoritative, classic ancient Chinese medical text, the *Nan Jing*, outlines the functions of receiving, rotting and ripening, and excretion of food and fluids, regarding digestion as a process of *Qi transformation*, which is activated by the Yuan Qi and accomplished via the Three-Heater organ complex.

4.1 General Description of the Triple Burners
Accordingly, the Shang Jiao (Upper Heater) traditionally occupies the upper chest cavity, and controls the intake of the only three primary components necessary for sustaining human life—namely air, food, and fluids. The Upper Heater synchronizes the functions of the *Fei* (Lungs) and *Xin* (Heart), governs respiration, and regulates the distribution of protective energy (*Wei Qi*) to the body's external surfaces.

The *Zhong Jiao* (Middle Heater) occupies the upper abdominal cavity, and controls digestion by harmonizing the functions of *Wei* (Stomach), *Pi* (Spleen), and pancreas. Note that the pancreas was not a defined organ in TCM, with its function coming under the influence of Earth—Stomach and Spleen. The Middle Heater is responsible for transformation of food intake; extracting nourishing energy, aka Nutrient Qi (*Ying* Qi), from food and fluids; and distributing it via the meridian system to the Lungs and other parts of the body. The manufacture of Blood commences here.

The Lower Heater occupies the lower pelvic cavity, and is responsible for separating the pure from the impure products of digestion, absorbing nutrients, and eliminating solid and liquid wastes. The Lower Heater includes and harmonizes the functions of *Gan* (Liver), *Xiao Chang* (Small Intestine), *Da Chang* (Large Intestine), Shen (Kidneys), and *Pang Guang* (Bladder), and also regulates sexual and reproductive functions. Some authors place the *Dan* (Gall Bladder) here, while others exclude it, as the Gall Bladder is considered to be one of the Extraordinary Yang Organs. It's because, unlike the other Yang organs, it is not directly involved with the processing of food and drink, and similar to the storage property of the Yin Organs, it stores bile, which is a refined essence. The other Extraordinary Yang Organs are the Uterus, Brain, Marrow, Bones, and the Blood Vessels.

4.2 Variations in the Anatomical Placement of the Triple Energizer
Most TCM practitioners generally agree with the former description of the Triple Energizer. However, there is some variation in the placement of the organs within each of the Heaters. Note that some authors (17) believe the Spleen is preferably called the 'nutrient-uptake system', including (in western physiological terms) the Hepatic Portal System, various secretory pancreatic cells, perhaps the duodenum, portions of the lymphatic system carrying chyle, and the catabolic and transaminoic functions of liver cells, all of which are associated with digestion.

However, there is a lot of variation as to where each Heater commences and terminates. This can be seen from the following references.

One source (14) states, 'The triple burner refers to specific body areas: According to this view, the organs of the human body are distributed among three segments, referred to as the upper, middle, and lower burners. The upper burner includes the head and chest, the heart and the lung. The centre burner corresponds to the upper abdomen (the area above the umbilicus) and includes the stomach and the spleen. The lower burner corresponds to the lower abdomen (inferior to the umbilicus) but includes the liver and kidney.'

In the Notes on the 31st Difficult Issue on page 356, Unschuld (1) advises that the *Chung-i-hsüeh kai-lun* of 1978 (Shanghai, page 23) states, 'The upper, central, and lower sections of the Triple Burner are assumed to refer to the heart and lung, the spleen and stomach, and the liver, kidneys, and bladder, respectively.'

In the 35th Difficult Issue on page 375 of Unschuld's (1) translation of the *Nan Ching*, it is stated, 'The small intestine is called red intestine; the large intestine is called white intestine; the gall is called virid intestine; the stomach is called yellow intestine; the bladder is called black intestine. They belong to the governing district of the lower [section of the Triple] Burner.' I believe that this passage indicates that the organs are Fu organs associated with the five phases (hence, they're called intestines). It's not suggesting that all of them are located in the Lower Burner, but, rather they are all lower than 'the Heart and the Lung, which are located far away'.

In the Appendix of the book *Heart Master Triple Heater* (page 125), regarding the location of the triple heaters, Elisabeth Rochat de la Vallée and M. Macé (11) state, 'This unity is fully expressed in the texts of the Song dynasty (Zhongzang jing of the alleged Hua tuo) or the Yuan dynasty (Dongyuan shishu of Li Gao) where the three heaters incorporate all the functions and circulations of the body, and extend from the top of the head to the heart (upper heater), from the heart to the navel (middle heater), and from the navel to the feet (lower heater).'

4.3 Differing Opinions Regarding the Start and Finish of Each of the Triple Burners

On page 347 of Unschuld's (1) translation of the *Nan Ching* with commentaries, the 35th Difficult Issue is discussed. It asks the question 'The Triple Burner . . . Where does it start and where does it end? ... Can that be known?' Unschuld then states, 'The upper [section of the Triple] Burner extends from below the heart downward through the diaphragm [and ends] at the upper opening of the stomach. ... The central [section of the Triple] Burner is located in the central duct of the stomach; it does not extend further upward or downward. ... The lower [section of the Triple] Burner [begins] exactly at the upper opening of the bladder [and extends downward].' Unschuld's locations of each of the energizers, as discussed in the 31st Difficult Issue above, is in complete harmony with the descriptions of Maciocia (18) in his blog at Blogspot.

However, another version of the *Nan Ching* (19) states, 'The energies of these Three Burners (Upper, Middle and Lower) take as origin, respectively, the cardia, the fundus, and the pylorus.'

In Chapter 18 of the Lingshu, Huangdi asked the Baron of Qi about the anatomical locations of the Triple Burners. In the article 'Gross Conception of Anatomical Structure of the Triple Burner in Huangdi Neijing', the authors (9) state that that the Baron of Qi advised that the Upper Burner occupies space from the upper orifice of the stomach, and included above the pharynx, extended across the chest, included the underarms, and that the upper extremity was the tongue. The Baron of Qi responded further, advising that the Middle Burner is also in the stomach, and that the qi that it receives includes 'the excretion of chyme and the vaporization of saliva'. He stated that the fine essence from the Middle Burner flows into the Lung meridian and is transformed into Blood. The Baron of Qi concluded that 'the Lower Burner leaves the ileum and flows into the Bladder by osmosis'. He advised that grains and water present in the Stomach turn into chyme that then flows into the Large Intestine.

A fascinating side point regarding The Baron of Qi—or Ch'i Po, as he is otherwise known—is presented in the book *Gua Sha: A Traditional Technique for Modern Practice* by Arya Nielsen (20). In the book, she states:

> In *Celestial Lancets*, Lu Gwei-djen and Joseph Needham (1980) note that the Chinese medical classics, the *Nei*

Ching and early Han texts of the 2nd and 1st centuries BC, correspond in large measure with the Hippocratic Corpus. The *Nei Ching* (≈ 200 BC) is a record of a constructed conversation between the Yellow Emperor Huang ti and his physician, Ch'i Po. Nigel Wiseman (personal communication, 2000) has noted that in early Chinese, a mono- and disyllabic language, the name 'Hippocrates' (≈ 400 BC), would have been pronounced as 'chee-po', or the softer 'hee-po', which is remarkably akin to the name of the *Nei Ching* physician 'Chi Po'. This correspondence was noted in 1685 by Willem ten Rhijne (1647–1700). According to Baldry (1989), ten Rhijne is credited as the first person to give the modern Western world a detailed account of Chinese acupuncture and moxibustion.

Regarding how Chapter 18 of the *Ling Shu* describes the location of the Three Burners within the body, in his blog at Blogspot entitled 'The Triple Burner as a System of Cavities and a Three-Fold Division of the Body', Maciocia (18) states, 'The Upper Burner comes out from the mouth of the stomach, it runs along the gullet, passes the diaphragm and spreads in the chest. The Middle Burner comes out at the stomach. The Lower Burner comes out at the lower end of the small intestine and pours into the bladder.'

In their Glossary section under *San Jiao*, the Sheng Zhen organization (21) states that the 'upper Jiao is from the diaphragm up (Heart, Lungs, Pericardium, throat and head), mid Jiao is from the diaphragm to the umbilicus (Stomach, Spleen, and Gallbladder), and the area below the umbilicus is the lower Jiao (Liver, Kidneys, Intestines, and Bladder)'.

4.4 Some Interesting Concepts as to the Location of the Triple Heater

For some authors that do believe that the Triple Energizer is an actual organ, there are some very interesting views as to where they believe the Triple Energizer is situated in the body.

Some authors (22) believe that the Triple Heater is linked to the hypothalamus, the part of the brain responsible for regulating appetite, digestion, fluid balance, body temperature, heartbeat, blood pressure, and other basic autonomic functions. This connection is via the vagus nerves, which are the tenth pair of cranial nerves. This view is likely held because the vagus nerves connect to numerous parts of the body and send messages between the brain and various organs, including the throat (where these nerves connect to the muscles that control swallowing), the heart (as part of the autonomic nervous system that controls the way the heart works), and organs in the chest and abdomen (including the stomach and intestine, where they help to control the function of the bladder and bowels, and also the lungs). About 80% of the vagus nerve fibers are afferent, meaning that they relay messages from the body to the brain. In this way, the brain receives incoming messages from numerous organs and tissues of the body, and these messages are, in turn, transmitted to different areas within the brain.

Regarding several possible proposed locations of the three sections of the three burners, the Abstract of the article 'Gross Conception of Anatomical Structure of the Triple Burner in *Huangdi Neijing*' (9) suggests:

> 1. The Triple Burner is a gross anatomical structure. 2. The Triple Burner is a cavity of the gross anatomical structure. 3. The Upper Burner is the craniospinal cavity, the Middle Burner the thoracic cavity, perchance containing the stomach, while the Lower Burner the abdominopelvic cavity, maybe including the reproductive cavity. 4. It is speculated that the organ to which the Upper Burner corresponds is the brain, in spite of whatever this organ has been named in the past.

4.5 The Triple Burner Pertains to the Uterus
Due to the fact that *Huangdi Neijing* does not define the location of the Triple-Burner cavity organ explicitly, the authors (9) state, 'Many different views regarding the Triple Burner have arisen in the past.'

They continue, 'Some others have chosen to believe in a myriad of other things, the wildest one of which includes the assignment of the Triple Burner to the uterus; their goals are, understandably, simply to search for an organ in the human body that they can draw a parallel to Triple Burner as described in TCM.'

In my opinion, this is indeed a strange concept as that would obviously mean that males would not possess a Triple Burner.

4.6 Two Contradicting Conclusions from the Same Source Make the *Nan Ching* Untrustworthy

Regarding the inconsistency of descriptions about the three burners in *Nanjing*, where Chapter 25 and Chapter 38 brought forth the hypothesis that the Triple Burner is an amorphous structure (without form), in the article 'Gross Conception of Anatomical Structure of the Triple Burner in *Huangdi Neijing*', the authors (9) state, 'On the other hand, Chapter 31 of *Nanjing* describes the morphology of the Triple Burner. According to normal logic, if two contradicting conclusions originate from the same source, then the source is not credible and should not be used.'

4.7 Grossman Suggests that the San Jiao Pertains to the Body's Interstitial Fluids

In the article entitled 'Cultural Reference for Increased Understanding of the San Jiao', regarding the nature of the San Jiao, the author, Glenn Grossman (5), expressed dissatisfaction in thinking of the San Jiao as 'a name without substance' and developed a theory whereby the San Jiao was 'a way to account for the body's interstitial fluids'.

4.8 The San Jiao Is Essentially Made Up of the Esophagus, Stomach, and Small Intestine

Yongping Jiang (8), the author of the *Journal of Chinese Medicine* article titled 'The San Jiao: Returning to the Nei Jing (A Modern Explanation of Original Theory)', proposes that the structure of the San Jiao is essentially made up of the esophagus, Stomach, and Small

Intestine and also includes the water passage between the Small Intestine and Urinary Bladder. He suggests that due to confusion, and the inability to anatomically locate the San Jiao, generations of TCM practitioners consider the San Jiao to be a combination of other organs rather than being an independent physical organ.

4.9 Diverse Nature and Composition of the San Jiao

Regarding the form and composition of the San Jiao, Yongping Jiang (8), the author of the *Journal of Chinese Medicine* article titled 'The San Jiao: Returning to the *Nei Jing* (A Modern Explanation of Original Theory)' states:

> Some authors have asserted that the San Jiao has a name but no form (e.g. Hua Shou, Li Ting and Zhang Shanlei), some have stated that it corresponds with the lymphatic system (e.g. Lu Yuanlei, Zhang Taiyan and Zhu Weiju) or the "fat-membranes" (e.g. Tang Zonghai and Zhang Xichun), whilst other authorities have put forth the view that the San Jiao actually includes all of the zangfu organs (e.g. Sun Simiao), or even the entire body (e.g. Xu Jun).

4.10 Location of the San Jiao Components According to Linda Barbour

Regarding the anatomical location of the San Jiao, in the article entitled *San Jiao*, the author, Linda Barbour (23), notes that each Burner is in a separate body cavity surrounded by distinct membranes that constitute part of the San Jiao. She states that the upper jiao comprises the lungs, pericardium, and the heart; the middle jiao comprises the stomach, spleen, liver, and gallbladder; and the lower jiao comprises the kidneys, bladder, and intestines. Barbour notes that many sources locate the liver in the lower jiao. Regarding the state of the fluid components of the San Jiao, Barbour (23) states, 'The upper burner is like a "mist"; the middle burner is analogous to a "maceration chamber" or a "muddy pool", and the lower burner is like a swamp, or a "drainage ditch".'

4.11 Locations of the Three Components of the San Jiao According to Glenn Grossman

In the article entitled 'Cultural Reference for Increased Understanding of the San Jiao', regarding the anatomical location of the San Jiao components, the author, Glenn Grossman (5), states:

> Clinically, the terms upper, middle and lower jiao are often applied to generalize the function of the internal organs of the chest and abdominal cavity. Above the diaphragm is the upper jiao which include the heart and lung; between the diaphragm and the umbilicus is the middle jiao which include the spleen and stomach; and below the umbilicus is the lower jiao which includes the kidney, intestines and bladder.

4.12 Location of the Triple Burner Components Based on Chapter 31 of the *Nan Jing*

Regarding the location of the Triple Burner components, in the 2011 article titled 'The Triple Burner (2)', Giovanni Maciocia (10) cites Chapter 31 of the *Nan Jing* and describes the Triple Burner as the 'avenue of water and food.' He states that the Upper Burner ranges from below the heart and diaphragm and extends up to the 'mouth of the stomach'. He states that the Middle Burner is positioned at the central duct of the stomach (*Zhongwan*), and it extends no further up or down. Maciocia then advises that the Lower Burner commences above the upper opening of the bladder.

4.13 Western Medical Perception Regarding Location of the San Jiao

Regarding the Triple Energizer, as of 2014, this mysterious organ energy construct is still not recognized in developed Western physiology. In TCM theory, the San Jiao is called the Minister of Dykes and Dredges, and it is in charge of the movement and transformation of numerous solids and fluids throughout the entire body. The San Jiao also has the role of producing and circulating nourishing energy (Ying Qi) and protective energy (Wei Qi) throughout the body. While each of the other eleven recognized organs are located in defined specific locations, the San Jiao is portrayed as having more of a functional role in the body, controlling the energy and biological functions of several other organ systems. The San Jiao is composed of three different components known as heaters or burners. The anatomical sites under control of the San Jiao are made up of the three main cavities in the body: the upper thoracic cavity, the middle abdominal cavity, and the lower pelvic cavity.

Obviously, there is no agreement exactly where each component of the Triple Energizer even starts and finishes. There is not even agreement on which organs are included in the three different burners. Note that while these authoritative sources differ in their exactness as to the location and delineation of the boundaries of the Three Heaters and the organs therein, their descriptions really are not that much different. It is essentially relative to the perspective of the discussion at hand. In the Classics, often discussion about the San Jiao appears to be focused on the trunk of the body—for example, when the individual Jiao's are referred to. On other occasions, the body in its entirety is being referred to. I am sure that whichever way these ancient and modern authors perceived the pathologies in their patients, they would have generally had positive outcomes in their treatment programs. Forgetting about Three-Heater boundary and locational semantics, I consider that these sages displayed remarkable wisdom and foresight in acknowledging this amazing, complex organ system thousands of years before modern enlightened medical science realized that it actually existed and had been too blind to see it. I think the exact borders really don't matter one iota. The underlying Three Heaters' functions work extremely efficiently and independently of where we might think they should be to satisfy our personal belief system based on the university we attended, what resources we have read, or our own personal interpretation probably flavored by personal beliefs and prejudices.

In the next Chapter, I will discuss the principal features of the Triple-Energizer organ complex, and then in subsequent Chapters, I aim to prove how the Triple Energizer works, using modern scientific findings to support my theory.

CHAPTER 5

Principal Features of the Three Heaters

Introduction

What do we know about the Three Heaters from the ancient classic texts? Let's compare and contrast the major features of the Three Heaters from a classic TCM point of view compared to what is being established by global scientific research. The San Jiao is amongst the most confusing issues in TCM. As in the past, there are still many differing theories regarding its nature and location. Regarding the rendering of the San Jiao, world-renowned authority on TCM classic literature Giovanni Maciocia (24) notes in his blog at Blogspot that *burner* is translated from the word *jiao*, which denotes being 'burned' or 'scorched'. Others call it the Triple Warmer, while some authors refer to it as its untranslated terminology, the San Jiao. Maciocia advises that there are two differing views amongst Chinese medical practitioners, with some believing that the San Jiao has no form, a concept which originates primarily from the *Nan Jing*. Conversely, the *Nei Jing* teaches that it actually does possess a form. So I will now discuss many aspects of the mysterious Fu.

5.1 Seventeen Major Aspects of the San Jiao (Triple-Energizer Metasystem)

In the article, Maciocia (24) believes that there are six major aspects to consider when discussing the intricacies of the San Jiao. He is an expert in the area of Chinese medicine. I have analyzed those aspects in line with modern scientific research. While I have modified the order in which he discussed them, he believes those six aspects include:

1. the Triple Burner as a system of cavities and Membranes
2. the Triple Burner as the activator of the Yuan Qi
3. the Triple Burner as a system of waterways
4. the Triple Burner governing movement of Qi
5. the Triple Burner as a threefold division of the body
6. internal/external relationship between Triple Burner and Pericardium.

I believe that there are 11 other major aspects to consider, and they include:

1. the Triple Burner having a name but no form
2. the relationship of the Triple Burner with the Stomach
3. the Triple Burner, where Blood is manufactured
4. the formation of Nutritive Qi in the Middle and Upper Heaters
5. the Triple Burner associated with the residence of the minister fire
6. the Triple Burner encompasses the whole body
7. the relationship between the Triple Burner, the Couli, and the Bladder
8. the Triple Burner having a direct relationship with the Source points
9. the Triple Burner in relation to the *Shao Yang*
10. the Diaphragm (*Ke*) membrane screening the turbid influences
11. the Triple Burner being energized by the Sun.

5.2 Several Forms of Qi Circulate throughout the Body via the Triple-Energizer Metasystem

While these Qi forms have different functions and properties, their synergism ensures that all the specific requirements and demands of each organ are adequately supplied so that good health and wellness reign. These Qi forms circulate and disperse the various liquids and the vital nourishing substances, combining as required to perform their different specific functions in all the organs, as described by the *Ling shu* (36) and *Nan jing* (31).

Modern translations of Chapter 8 of the *Su Wen* translate the ancient classical text to include terms like *waterways, water passages, watergates, water channels, irrigation, ditches,* and *sluices* which are all obviously associated with the reticulation of water and fluids. So the Triple-Energizer Metasystem is obviously intimately associated with water and fluid distribution throughout the body.

Interestingly, while the San Jiao is a Fu of the Fire Element, along with the Small Intestine, it clearly embraces waterways and body fluids too, along with Fire. Undeniably, it is this intricate combination of water and fire which allows the Triple Heater to circulate the various forms of Qi, including Yuan Qi and liquids in the specific proportions required to invigorate and sustain all the other organs distributed through the three cavities and to fulfill all the diverse functions within the three cavities necessary to sustain life.

I believe that San Jiao fluids that flow through the ditches and sluices embody the inclusion of all extracellular fluids, including blood and lymph and the interstitial fluids. Other body fluids—such as cerebrospinal fluid (CSF), synovial fluid, peritoneal secretions and pleural secretions, digestive juices, mucosal secretions, amniotic fluid, etc.—are all under the control of the Triple-Energizer Metasystem.

CHAPTER 6

The Triple Burner Governs the Movement of Qi Throughout the Entire Body

Introduction
In his *Clinical Tip* (September 2011), Maciocia (25) discussed the role of the Triple Burner as governing the movement of Qi throughout the entire body. He cites Chapter 38 of the *Nan Jing*, which says, 'The Triple Burner stems from the Yuan Qi; it governs all Qi in the body, it has a "name but no form", it belongs to Hand Shao Yang, it is an external Fu.' And he points out that this small passage discusses five salient features of the Triple Burner. He notes that (1) the Triple Burner stems from the Yuan Qi, (2) it governs all the Qi throughout the body, (3) the Triple Burner has a name but no form, (i.e. it is not an actual organ but a function, (4) it pertains to the Shao Yang with the Gallbladder, and (5) it is a Fu.

Maciocia (25) points out that the movement of Qi by the Triple Burner is called the *Qi Mechanism* (*Qi Ji* 气 机). He states further that the Qi Mechanism involves the ascending and descending of Qi throughout the body and especially where Qi enters and exits different locations and different organs. He further notes that each organ has a specific direction of flow of Qi (e.g. Spleen Qi has an ascending direction, while Stomach Qi descends). Because of the defined flow of each acupuncture channel, Qi flows in an upward or downward direction. Maciocia then points out that acupuncturists are generally unfamiliar with the horizontal movement of Qi throughout the body.

6.1 Acupuncturists Are Generally Unfamiliar with the 'Horizontal' Movement of Qi

Maciocia (25) points out that acupuncturists use ascending/descending movement of Qi every time we insert a distal and a local point to treat a particular problem along a channel—for example, to treat a shoulder problem using LI-1 *Shangyang* and LI-15 *Jianyu*. However, he further notes that acupuncturists are less familiar with the horizontal movement of Qi—that is, the entering and exiting of Qi into various structures and organs, including the space between skin and muscles, the membranes, the joint capsules, and all the other cavities of the body. He states, 'Moreover, Qi "enters and exits" in every part of the body among the Tai Yang, Shao Yang and Yang Ming and among the Tai Yin, Jue Yin and Shao Yin.'

6.2 Methodology to Treat Perverse Horizontal Movement of Qi within Different Body Areas

Regarding the possible horizontal movement of Qi within different body areas, these horizontal Avenues of Qi (*Qi Jie* 气 街) are described in chapter 52 of the *Ling Shu*. Maciocia (25) notes that Qi has 'streets' in the head, in the chest, in the abdomen and that Qi even has streets in the lower legs. He then cites Chapter 52 of the *Ling Shu* and notes that if there is a medical condition in one of these locations, then stop the Qi at that location. This involves the use of Back-Shu points for Qi problems in the chest and the use of Back-Shu points and abdominal *Chong Mai* points for abdominal problems. Likewise, ST-30 is used to treat problems at the front of leg, and BL-57 is used to treat problems at the back of the leg. As can be seen, these conditions are treated, taking the horizontal element into the picture. I suggest that these horizontal streets are made up of conduits within the Primo Vascular System and/or the Connective-Tissue Metasystem and are yet to be elucidated.

6.3 The Production and Transformation of the Numerous Types of Qi via the Triple Burner

Regarding the production and transformation of the numerous types of Qi via the Triple Burner, Maciocia (25) advises that the word *tong* (通) is very frequently used in Chinese literature to aptly describe the function of the Qi Mechanism of the Triple Burner. The word *tong* means 'free passage', 'to pass through', or 'to penetrate'. This perfusion and penetration property of the Qi Mechanism ensures that blockages are overcome and that the Triple Burner disperses the numerous Qi forms, including Nutritive Qi (Ying Qi) and Defensive Qi (Wei Qi), throughout the body as required. Further, optimal performance of the Qi Mechanism ensures that Blood and Body Fluids moisturize and nourish the entire body.

To further confirm that the Triple Burner sends its influences throughout the entire body at all levels, Maciocia (25) cites the *Central Scripture Classic* (*Zhong Zang Jing*, Han dynasty), which says, 'Qi passes freely into interior, exterior, left, right, above and below. The Triple Burner irrigates the body, harmonizes interior and exterior, benefits the left and nourishes the right, it conducts upwards and descends downwards.'

6.4 The Triple Burner Controls the Movement/Passage of Qi Forms in All Three Burners

Maciocia (25) reiterates that Chapter 66 of the *Nan Jing* also endorses the fact that the Triple Burner controls the movement of Qi in general, and he quotes, 'The Triple Burner makes the Yuan Qi separate [into its different functions] and it controls the movement and passage of the 3 Qi [of the Upper, Middle and Lower Burner] through the 5 Zang and 6 Fu.' He then states:

> The '3 Qi' are the Qi of the Upper, Middle and Lower Burner. Apart from referring generally to all the types of Qi in each Burner, this passage also refers specifically to the Gathering Qi (Zong Qi) in the Upper, Nutritive-Qi (Ying Qi) in the Middle and Defensive-Qi (Wei Qi) in the Lower Burner. Although the Defensive-Qi exerts its

influence primarily in the Upper Burner and the superficial layers of the body (the space between skin and muscles), it originates in the Lower Burner from the Ming Men. Chapter 18 of the *Ling Shu* says: 'Ying Qi originates from the Middle Burner; Wei Qi originates from the Lower Burner.'

6.5 The Triple Burner Controls the Movement/Passage of Qi throughout the Entire Body

To lend even more support to the Triple Burner controlling the manufacture and distribution of Qi throughout the entire body, citing from Chapter 38 and Chapter 31 of the *Nan Jing*, Maciocia (25) states that the Triple Burner is derived from the Yuan Qi and that it supports all the Qi forms throughout the body. He stresses that the Triple Burner is responsible for the free movement of Qi in all the channels and structures, including the cavities of the body.

In the commentaries on the 38th Difficult Issue on page 397 of Unschuld's (1) translation of the *Nan Ching*, regarding the influence that the Triple Burner has in the body, Liao P'ing states, 'That is to say, its territory is *much broader than that of the remaining depots.*' In the same commentaries, Li Chiung says, '[Hence, the Triple Burner] controls the *influences of the entire body*' (emphasis is mine).

Maciocia (25) makes the interesting comparison of the Triple Burner's function in controlling and governing the movement of Qi with that of the Liver's property of maintaining the free flow of Qi to ensure harmony throughout the body. He notes further, however, that the Liver plays no part in the regulation of Qi entering and exiting the body cavities, whereas the Triple Burner does.

Rochat de la Vallée agrees with Maciocia that the Triple Energizer governs the transmission and transmutation of qi and fluids within the body. In the book *Heart Master Triple Heater*, Rochat de la Vallée (11), notes at page 50 that the Triple Heater unifies all the processes between entry and exit.

6.6 The Triple Energizer Governs the Movements of External Impacts to Become Myself

With reference to the properties and functions of the Triple Heater, on page 49 of the book *Heart Master Triple Heater*, Rochat de la Vallée (11) explains that when the *Neijing Jingyi* teaches that the Triple Heater is the passageway for the entrance and exit of liquids and cereals, it means that even during the process of digestion, the Triple Heater possesses the characteristics of reception, transformation, diffusion, and elimination. She eloquently notes that beyond the processes of swallowing and excreting, the omnipresent processing mechanism intrinsic to the Triple Heater pertains to transforming external things that are not 'myself' becoming internalized and actually being absorbed by the body and becoming a part of 'myself'. Similarly, the Triple Heater eliminates everything that cannot be assimilated into 'myself'.

This is a very important aspect of the Triple Heater. It means that the Triple Heater is the medium that causes external elements and influences (drinks, grains, air, odors, sounds, perceptions, and even timekeeping) to enter our body and mind and take up permanent residence within our very being. This is shown in explicit detail in the commentaries on the 66th Difficult Issue on pages 567–568 of Unschuld's (1) translation of the *Nan Ching*, where Yang states, '*The two kidneys are distinguished by their storing the essence of the sun and of the moon* . . . Hence the text states "origin" is an honorable designation for the Triple Burner" because the Triple Burner's influences are accumulated in the kidneys' (emphasis is mine).

I believe that the fact that 'the two kidneys are distinguished by their storing the essence of the sun and of the moon' is intimately involved with the 24-hour-long circadian rhythm associated with solar days and the 'Chinese Clock', whereby each of the twelve organs have their maxima for two hours and their minima 12 hours later, and that the 29-day Lunar cycle's influence directly affects life forms on Earth, which is obviously evidenced by the female menstrual cycle, with both the Moon and harmonized females having 13 cycles per solar year. The kidney is well known for its 7-year life cycles in females and 8-year life cycles in males. I suggest that it is the influence of the

Triple Burner on the two Kidneys that causes these cyclical events and that subsequently, '"'origin' is an honorable designation for the Triple Burner" because the Triple Burner's influences are accumulated in the kidneys'. This Triple Burner–initiated circadian effect on each individual certainly plays a major role in what makes me 'myself'.

6.7 Qi Courses Harmoniously through All 12 Meridians over a Defined 24-Hour Period

This controlling of Qi through the body allows for a harmonious circadian rhythm, which allows the Qi (energy) to flow through all the twelve meridians in a defined order over a 24-hour period. The website *index-china.com* (26) states:

> The distribution of qi through the pathways is said to be as follows (based on the demarcations in TCM's Chinese Clock): Lung channel of hand taiyin to Large Intestine channel of hand yangming to Stomach channel of foot yangming to Spleen channel of foot taiyin to Heart channel of hand shaoyin to Small Intestine channel of hand taiyang to Bladder channel of foot taiyang to Kidney channel of foot shaoyin to Pericardium channel of hand jueyin to San Jiao channel of hand shaoyang to Gallbladder channel of foot shaoyang to Liver channel of foot jueyin then back to the Lung channel of hand taiyin. Each channel occupies two hours, beginning with the Lung, 3AM–5AM, and coming full circle with the Liver 1AM–3AM.

This accounts for the fact that many people tend to wake around 3 a.m. because this is a time of the completion of one Qi circuit throughout the body and the commencement of the next cycle, starting at the Lung meridian at Lung 9. It is a highly energized time of day.

6.8 Psychological and Emotional Aspects of the San Jiao

Regarding the issue of the Triple Heater regulating consciousness, in the article entitled 'Triple Burner: Fire-Energy Yang Organ', the unknown author (22) states:

On a psychological level, the Triple Burners mobilise Qi and lift depression which results due to stagnation of Liver Qi. When the Triple Heaters are full, the consciousness becomes unwavering and the Mind's intent is benevolent and kind-hearted. Through the 5 Elements the Triple Heaters are also linked with the Heart and Pericardium and are affected by the emotion of joy. When the Shen of the heart is strong and pure, the desires and thoughts of an individual will be at peace, and the individual will have a peaceful sense of well-being and will be fulfilled.

Summary of Chapter 6

Chapter 38 of the *Nan Jing* says, 'The Triple Burner stems from the Yuan Qi; it governs all Qi in the body.' This movement of Qi by the Triple Burner throughout the entire body is called the *Qi Mechanism*. A major function of the Triple Burner is to ensure that Qi passes through all the cavities and all the organs via the Qi Mechanism. This harmonized Qi Transformation by the Triple Burner results in the appropriate production of Nutritive Qi (Ying Qi), Defensive Qi (Wei Qi), and Blood and Body Fluids as required where required.

Rochat de la Vallée (11) advises that even regarding the process of digestion, the Triple Heater controls reception, transformation, diffusion, and then elimination. She then makes an interesting observation in that the Triple Energizer governs the absorption of elements and nutrients from consumed foods that become 'me' and likewise cause the elimination of any components that we do not wish to have assimilated into our life and being. This regulating and controlling aspect of the Triple Burner at numerous physical and psychological levels must have a powerful impact on our personality and belief system.

These facts explain why the Triple Burner is said to control all kinds of Qi. In the commentaries on the 38th Difficult Issue of Unschuld's (1) translation of the *Nan Ching*, regarding the controlling influence that the Triple Burner has throughout our body, Li Chiung says, 'Hence, the Triple Burner controls the influences of the entire body.'

CHAPTER 7

The Triple Burner as a Threefold Division of the Body

Introduction
In the commentaries on the 31st Difficult Issue on page 355 of Unschuld's (1) translation of the *Nan Ching*, Huang Wei-san states, 'The Triple Burner encloses all the depots and palaces externally. It is a fatty membrane covering the entire physical body from the inside. Although it has no definite form and shape, it represents a great palace among the six palaces. Hence, the final section [of Difficult Issue 31] states that 'it has *three ruling centers* and, in addition, that it has a *specific location where it accumulates its influences*' (emphasis is mine). Regarding the Triple Burner, note that the final sentence states, 'It has three ruling centers.'

7.1 The Upper Heater Gathers Together the Heart, the Lung, and the Sea of Qi
On page 78 of the book *Heart Master Triple Heater*, Rochat de la Vallée (11) discusses the functions of the Upper Heater. She notes that the Upper Heater is responsible for assembling the Heart, the Lung, and the Sea of Qi together so that they ensure the regularity of the heartbeat and the rhythm of respiration. Further, the Upper Heater ensures that the Sea of Qi correctly governs the movement of nutritive qi and the defensive qi throughout the body.

7.2 The Upper Heater as the Starting Place for the Diffusion of All Qi

On page 80 of the book *Heart Master Triple Heater*, Rochat de la Vallée (11) discusses the functions of the Upper Heater. She explains that the Upper Heater is the established starting place for the distribution of all Qi and that the Ying Qi and the Wei Qi proceed to circulate from the Upper Heater because that is where they congregate before flowing along their own defined pathways. Similar to the movement that pulsates the nutritive energy, the defensive energies are also pulsated so that they circulate throughout the body through the meridians and in the *luo* vessels. However, the defensive energy is governed by circadian equilibrium that fluctuates between daytime and night-time.

7.3 The Upper Heater Produces a Protective Defensive Mist that Pervades the Whole Body

In the book *Heart Master Triple Heater*, regarding the nature of the Upper Heater, Elisabeth Rochat de la Vallée (11), in pages 92–93, states:

> What is a mist? It is not the mist or fog that makes you cough or wrap up well. This mist suggests everything that rises up from the earth in the form of vapour or steam which is capable of permeating everywhere. This means that at the level of the upper heater, above the diaphragm which is some kind of filter, there are never water or currents of water, but only that which can rise up as misty vapour. *Everything which has managed to cross the filter of the diaphragm from the middle heater is then able to pervade the whole body like some kind of mist.* So it is everything we have seen in the circulation of qi, the propagation and diffusion from the upper heater reaching the most external parts of the body right up to the skin and body hair, but which also penetrates the bones and flesh just like a vapour. (Emphasis is mine)

She notes that, in Chinese thought, *mist* is seen as delicate and pure enough to ascend from the earth and encounter the qi of heaven and engender life. In a like manner, the mist embodies the nature of Defensive Qi, saturating the skin and ensuring that the flesh remains firm, the integrity of the body remains sound, and functioning results are smooth.

7.4 The Middle Heater Extracts from Food That Which Will Constitute the Body

On pages 81–82 of the book *Heart Master Triple Heater*, Rochat de la Vallée (11) discusses her belief regarding the functions of the Middle Heater. She explains that the Middle Heater relates to the middle of the Stomach, and she notes that the Stomach is one of the 'four seas' constituting the 'sea of liquids and cereals'. She suggests that because *the Stomach has the role of extracting **all** the essential elements from food that proceed to constitute the body*, it possesses one of the most important roles in the body. The author then describes a second major role of the Stomach in that it causes the *jin ye* to vaporize. The delicate, moist jin ye vapors that are beaming with elements essential for life arise from the Stomach and then engender highly purified life-sustaining liquids, which diffuse, and both irrigate and rinse out the entire body. Another fractionated derivative emanating from the Stomach is the *jing wei*, where the *jing* are the essences and the *wei* are very subtle derivatives that are almost imperceptible. Jing wei transformation involves extracting the vital elements and the delicate essences and flavors of foods into a rarefied form that can be readily absorbed and become a literal part of our very being.

While what Rochat de la Vallée says above about the function of the 'middle of the Stomach' is very beautiful and poetic, from a biochemical and scientific point of view, the ethos of her comment (that I have emphasized) does not accord with known biology and physiology. This is exemplified by her further comments as discussed below, where she notes that foodstuffs continue their descent from the Middle Heater because 'all the usefulness' has not been extracted from the food and further extraction is required. I believe that the intent of what Rochat de la Vallée states is correct, but her specific belief of what was originally meant therein is not correct. Regarding the processing of foods to supposedly allow nutrient absorption through the Stomach wall, she suggests that *the Stomach has the role of extracting **all** the essential elements from food that proceed to constitute the body*. I do not agree with that belief, and I propose that it is primarily the bulk water and the biophotons present in the food that this applies to and not to food nutrients, including minerals, proteins, fats, and carbohydrates, as she infers.

Rochat de la Vallée believes that many nutritional food elements are absorbed through the Stomach wall. As I have discussed elsewhere in this book, only a small number of low-molecular-weight chemical substances that have been swallowed are actually absorbed through the Stomach wall. Let me explain. When she refers to the heart of the stomach which is the 'sea of liquids and cereals', she hits the nail on the head because the consumed 'liquids' are predominantly bulk water, H_2O. As I have elucidated elsewhere, the 'cereals' too are composed predominantly of about 80–90% bulk water themselves. I believe that the 'maceration' process discussed below releases the bulk water and the biophotons from the 'cereals' and that these two constituents easily penetrate the Stomach wall and become the 'extract from food that which will constitute the body'. Rochat de la Vallée notes that a very humid emission of vaporous steam ascends and generates very pure life-giving liquids that circulate throughout the entire body to irrigate it and to rinse it out. This exactly describes the life-giving liquid-crystalline-water that Professor Pollack describes in detail, which is derived from the two stated constituents I believe pass through the Stomach wall, namely biophotons and bulk water. When these two constituents come into contact with hydrophilic membranes (which are omnipresent throughout the body), life-giving liquid-crystalline EZ water is generated. Amazingly, she says that the water that 'is full of life' is 'capable of passing throughout the whole organism to irrigate it, and also to rinse it out'. Professor Pollack's living EZ water is 'Exclusion Zone water' and, by its very nature, 'rinses' out the body. At KL 801–811 of the book entitled *The Fourth Phase of Water: Beyond Solid, Liquid, and Vapor*, Professor Pollack (27) states:

> What does the EZ exclude? These experiments showed that the EZ rather broadly excludes substances of many sizes,

> from very small to very large. . . . we could conclude that the exclusion phenomenon was general: almost any hydrophilic surface can generate an EZ, and the EZ excludes almost anything suspended or dissolved in the water.

Pollack (KL 814–817) continued, 'This demonstrably vast exclusionary power implied yet again that we might be dealing with some kind of crystal-like substance, for crystals exclude massively. . . . As the ordered zone grew, it would push out solutes in the same way that a growing glacier pushes out rocks.' I discuss Pollack's living EZ (Exclusion Zone) water in great detail later in the book.

So while Rochat de la Vallée described the actual process perfectly, she was herself unaware of the exact nature of the substances that were truly involved, and she incorrectly supposed that other food components (micro-refined minerals, proteins, fats, and carbohydrates) were actually passing through the Stomach wall.

7.5 The Lower Heater Generates Urine and Feces While the Fire Therein Distils Useful Fluids

On page 85 of the book *Heart Master Triple Heater*, Rochat de la Vallée (11) discusses the functions of the Lower Heater. She notes that *Ling shu*'s Chapter 18 describes the Lower Heater as a drainage canal that causes residues and waste to separate and descend. She notes that the ancient Chinese were well aware that no specific anatomical outlet existed that directly connected fluids from the Small Intestine directly to the Bladder, and she suggested that the fluid movement was due to some form of 'osmosis or filtration'. She notes that all liquids in the body are eventually processed by the treatment plant of the Bladder so that purification and recycling can occur. This processing by the Bladder yields three products. The beneficial fraction of the unclear liquids is purified and recycled and then ascends in the form of constructive mists and vapours. The rinsate derived from the drainage and rinsing fraction is stored in the Bladder and then eliminated from the system as urine, while the most degraded fraction continues to be dehydrated and is eliminated from the body via the Large Intestine as excrement.

Rochat de la Vallée then notes that the Lower Heater constitutes the effective regulation and management of all that which must be eliminated and that which must be recovered. She further notes that, along with the Water in the Lower Heater, Fire is also present to cause that which is still useful to be evaporated to form the *mist* that ascends for reuse.

The concept of Fire in the Lower Heater causing 'purification and recovery of that which can still be useful' is a prevalent thought throughout the TCM Classics. She explains that once these unclear liquids have been purified and regenerated, they are once again able to ascend in the phase of vapours and mists. Pollack's liquid-crystalline-water theory would readily allow the hydrophilic membranes in the Lower Heater to convert the 'unclear liquids' into EZ water, whereby the dissolved and suspended impurities would be 'excluded' and 'finally end up as excrement' and be 'directed toward the large intestine.' EZ water is the catalyst that allows one phase of water to easily transform into the adjacent phase of water. For example, bulk water via EZ catalysis would easily change phase to 'mists and vapours' ready to be directed upwards. I am at a loss to suggest which structure in the Lower Heater is responsible for this procedure of the San Jiao Cycle. I propose that there is a fatty membranous structure present in close proximity to the Bladder, (perhaps the external Bladder membranes) that actually performs this role that has so far not been elucidated.

7.6 The Middle Heater Breaks Down and Separates (Macerates) the Components of Foods

Regarding the term *maceration* in connection to the Middle Heater, on pages 94–95 of the book *Heart Master Triple Heater*, Elisabeth Rochat de la Vallée (11) makes the comment that the term *maceration* isn't very clear and asks if a more appropriate word exists. She then describes the process of maceration that would happen in the obviously warm Middle Heater, and she presents the analogy of placing some food into a receptacle or a bowl and adding water so that the action of the ambient heat would cause the soaked food to soften and finally rot and decompose. I believe that the term *maceration* is a very appropriate term for the food-tenderizing process that occurs in the stomach. Synonyms for *maceration* include *soaking, softening, steeping, marinating, drenching, saturation, infusion,* and *deliquescence*.

7.7 The Defensive Qi Comes from the Lower Heater

In answer to the question about why the Defensive Qi comes from the Lower Heater, on pages 86–87 of the book *Heart Master Triple Heater*, Rochat de la Vallée (11) answers that it is because not all the useful components have been extracted at the level of the Middle Heater. As the partially processed foods and drinks descend, a continuance of the removal and refining process persists, and especially, there is a transformation within the Lower Heater. Subsequently, the Nutritive Qi is derived from the clear component, and the Defensive Qi is derived from the unclear component. Pertaining to the free circulation and the cooperation of both the Nutritive Qi and the Defensive Qi, she then explains that this exemplifies 'the collaboration of all the organs which is in fact called the triple heater'.

When she states above 'the collaboration of all the organs which is in fact called the triple heater', she reconfirms her belief that the Triple Heater is not a dedicated stand-alone organ. She believes that the Triple Energizer is the combined function associated with all the various organs and their individual processing and distribution of the various synthesized components of food and fluids throughout the body. I disagree with this construct completely and absolutely believe the Triple Heater (Triple-Energizer Metasystem) is a stand-alone organ complex and has a defined morphology and location that permeates the entire body in the form of collagenous fatty membranes in affiliation with the Primo Vascular System.

7.8 The Lower Heater as a Canal for the Separation of the Clear and the Unclear Substrates

In the book *Heart Master Triple Heater*, regarding the properties of the Lower Heater, Elisabeth Rochat de la Vallée (11) in pages 95–96 shows that the Lower Heater involves evacuation and irrigation functions and notes that nearly all the zang fu participate in this role. She further states, 'So the triple heater is *never an isolated function*, it is a *gathering of all the physiological functions* rooted in the original qi' (emphasis is mine).

Once again, she shows that she believes that 'the triple heater is never an isolated function, it is a gathering of all the physiological functions rooted in the original qi'. As stated above, I do not agree with this concept of the Triple Heater being considered as only a function. I believe it is an omnipresent organ system in its own right.

Summary of Chapter 7
In the commentaries on the 31st Difficult Issue, regarding properties of the Triple Burner, Huang Wei-san plainly states 'it has three ruling centers and, in addition, that it has a specific location where it accumulates its influences.' Each of the three Burners has different functions, and each burner is essential for the maintenance of life for different reasons.

In Rochat de la Vallée's (11) discussion of the functions of the Upper Heater, she says that a major function is the cooperation of the heart, the lung, and the sea of qi to initiate the beating of the heart, the life-sustaining rhythm of respiration, and the energizing sea of qi. These three provisions engender everything required by the body.

The Stomach in the Middle Heater has one of the most important functions in the body. The Stomach extracts from drinks and food the two major ingredients (namely bulk water and biophotons) required to orchestrate healing and regeneration and maintaining life. Thanks to the Stomach, the jin ye is vaporized. This volatile emission of very humid vapour rises and forms very pure liquids which are full of life and capable of permeating the entire body to irrigate all the cells and also to rinse it out. Thus, the jin ye must be beaming with all the substances essential for life. Pollack's theory on living liquid crystalline EZ water aptly fits this description, whereby the bulk water and biophotons that are macerated from drinks and foods pass through the Stomach wall.

Also in the stomach, the jing wei is produced. The jing are the essences, while the wei are composed of something very subtle and almost imperceptible. This transformation related to the jing wei is associated with the stomach extracting the essences and flavor and vital elements from the ingested foods and converting the components so they can penetrate and permeate into your life.

Ling shu (chapter 18) describes the lower heater as a canal for drainage and for separating and descending the residue and waste. The lower burner directs liquids toward the field of action of the bladder. Liquids derived from the drainage and the rinsing generate urine, which is stored in the bladder. Turbid substances that continue to dry out and finally form excrement are routed toward the large intestine.

There is also fire present in the lower heater, and recoverable fluids that are worthy of recycling ascend in the form of mists and vapours. The defensive qi comes from the Lower Heater. The fire of the Lower Heater endows the defensive qi with the very lively aspect which allows the Wei Qi to fulfill its function of defense throughout the body.

CHAPTER 8

Correlation Between the Triple Burner and the Pericardium (Palaces/Depots and Their Internal/External Relationship)

Introduction
What exactly are the 'palaces' and the 'depots'? Let's define the 'palaces' and the 'depots'. In the 39th Difficult Issue, the scripture states: 'There are five palaces and six depots. What [does that mean]?'

In the commentaries on the 39th Difficult Issue on page 399 of Unschuld's (1) translation of *Nan Ching*, Li Chiung states, 'Gall, stomach, large intestine, small intestine, and bladder [are the five palaces]; liver, heart, spleen, lung, kidney, and gate of life [are the six depots].' So they are the hollow organs (Fu) and solid organs (*Tsang*) respectively. So now, let's discuss the relationship between the 'palaces' and the 'depots'.

8.1 The Cosmological View of Man and the Organs Relative to Natural Phenomena
In the commentaries on the 39th Difficult Issue on page 401 of Unschuld's (1) translation of *Nan Ching*, Yü Shu states, 'Heaven manages [the world] below with the six [climatic] influences; the earth presents the Five Phases [to the world] above. Heaven and

earth exchange [their influences] liberally. [That which reflects their] numbers "five" and "six" is perfect. Man reflects the three powers. Therefore, the depots and the palaces reflect the numbers five and six, respectively.' Yü Shu continues, 'That is to say, the human head is round; it reflects heaven. The feet are rectangular; they reflect the earth. The depots and the palaces, numbering five and six respectively, reflect man. Thus, the three powers are [reflected] completely.'

8.2 Defined Specific Features of the Triple Burner Confirm Its Actual Existence

The 38th Difficult Issue states, 'The depots are but five; only the palaces are six. Why is that so?' The answer is 'It is like this. The [existence of the] Triple Burner is to be named as the reason for the fact that there are six palaces. The Triple Burner represents an additional [source] of original influences; it governs all the influences [circulating in the body]. It has a name but no form. It is associated with the hand-minor-yang conduit. This is an external palace. Hence, one speaks of the existence of six palaces.'

There is a lot of information included in this answer. The statement 'The *existence* of the Triple Burner is to be named as the reason for the fact that there are six palaces' (emphasis is mine) shows categorically that the Triple Burner is a real tangible Fu organ, making six tangible Fu organs in the body. The quote continues that the Triple Burner is a secondary or additional source of 'original influences' (Yuan Qi) along with the kidneys, which are the proper source or original source of the original influences. Note that after just stating that the Triple Burner represents the sixth literal tangible Fu, the comment that 'it has a name but no form' is made. As stated in several locations throughout this book, the possible confusion these two apparently contradictory comments could cause is easily resolved when the following information is borne in mind. Regarding the Connective-Tissue Metasystem, in an article in the 2011 *IDEA Fitness Journal*, Thomas Myers (28) reports that fascia is 'so variable among individuals that its actual architecture is hard to delineate'. Myers is stating that because of the extreme variability of the connective-tissue fascial system within individuals, it essentially has 'no form'. So the Connective-Tissue Metasystem definitely exists, and it has a name but no form. There is no doubt in my mind that the Triple-Energizer Metasystem is intimately associated with the Connective-Tissue Metasystem. Further confirming its literal physical authenticity, it is stated that the Triple Burner governs all the influences, *is associated with the hand-minor-yang conduit,* and is *an external palace.* It is an external palace or external Fu because it is composed of omnipresent connective tissue that surrounds and wraps around all the organs in the body *like an external wall.* Of course, the Triple Burner exists. However, it has 'no form' in that it is does not possess a defined, recognizable morphology as is the case for every other TCM organ.

8.3 Further Confirmation about the Literal Physical Existence of the Triple Burner

In the commentaries on the 38th Difficult Issue on page 396 of Unschuld's (1) translation of *Nan Ching,* Ting Chin states:

> 'It has a name but no form' because it encloses [everything else like a cover] on the outside. Hence, [the text] speaks of an 'external palace.' In the twenty-fifth difficult issue, my commentary stated that *the Triple Burner holds all the depots and palaces like a large bag.* If one compares that [difficult issue] with the meaning of the present paragraph, one could become confused. People of later times have said that the Triple Burner has a form and that the *Nan-ching* is wrong. Well, they failed to penetrate the entire corpus of the *Nan-ching.* (Emphasis is mine)

Here Ting Chin re-emphasizes that while *Nan Ching* states that the Triple Burner has no form, it does not mean that it is not a literal physical organ. Rather, the Triple Burner does not have a specific and unique shape and definable morphology, like a stomach or a heart or kidney, for example. Obviously, if the Connective-Tissue Metasystem of a grossly obese male person was able to be dissected out of their body, there would be no comparison in shape, morphology, and weight to the anatomically removed Connective-Tissue Metasystem of a young female with anorexia, for example. Or for that matter, if the Connective-Tissue Metasystem of a male from one of the Sudanese tribes (with an average height of 1.9 m) was compared to a male pygmy (with an average height of 1.5 m), they would be extremely dissimilar. So it can be seen that the Connective-Tissue Metasystem essentially has no form because of its extreme variability between individuals. While it has no form which is definable and describable, it still exists as a physical, tangible entity. I believe that the Connective-Tissue Metasystem is closely associated with the Triple-Energizer Metasystem, which *Nan Ching* also describes as having 'no form' for the reasons discussed above.

When Ting Chin stated, 'People of later times have said that the Triple Burner has a form and that the *Nan-ching* is wrong. Well, they failed to penetrate the entire corpus of the *Nan-ching*,' he is stating that *Nan Ching* taught that the Triple Burner was an actual physical organ but that it was somewhat amorphous—that is, existing as a solid but lacking long-range order—much like the substances amorphous silicon and amorphous carbon.

8.4 The Triple Burner and the Pericardium Are External/Internal Partners

What does it mean when the quote states that 'it is associated with the hand-minor-yang conduit'?

The Triple Burner and the Pericardium are external/internal partners. Explaining the TCM concept of the relationship between the 12 body organs, the article *Zang-fu* (29) states:

> To understand the zàng-fǔ it is important to realize that their concept did not primarily develop out of anatomical considerations. The need to describe and systematize the bodily functions was more significant to ancient Chinese physicians than opening up a dead body and seeing what morphological structures there actually were. Thus, the zàng-fǔ are functional entities first and foremost, and only loosely tied to (rudimentary) anatomical assumptions.

8.5 Why Are Hollow Fu Organs Called Palaces and the Solid Tsang Organs Called Depots?

In the commentaries on the 25th Difficult Issue on page 310 of Unschuld's (1) translation of the *Nan Ching*, Hsü Ta-ch'un states, 'The [treatise] "Chiu chen lun" of the *Ling*[-*shu* states concerning] the five depots: "The heart stores the spirit; the lung stores the p'o; the liver stores the hun; the spleen stores one's imagination; the kidneys store the essence and the mind."' The six palaces include the small intestine, the large intestine, the stomach, the gall, the bladder, and the Triple Burner. They are responsible for emission and intake of water and grains, resembling a palace treasury that oversees expenditure and income. Hence, they are called palaces.

Summary of Chapter 8

In the Notes for the 25th Difficult Issue (page 316), Unschuld (1) personally stated, 'Each depot [yin, internal] is linked to a specific palace [yang, external]—namely, lung to large intestine, spleen to stomach, heart to small intestine, kidneys to bladder, heart-master/ heart enclosing network [pericardium] to Triple Burner, and liver to gall.'

This fire element Pericardium–Three Heaters couple have a lot in common. The pericardium (from the Greek περί, 'around', and κάρδιον, 'heart') is a double-walled sac containing the heart and the roots of the great vessels (30). The Triple Heater is the largest of the 12 organs and, like a bag or a sack or a wall, contains all the other organs just as the Pericardium contains the Heart. The Pericardium contains pericardial fluid, which protects and lubricates the Heart from shocks and infections. Likewise, the Triple Heater is filled with circulating fluids that lubricate, protect, and defend all the many organs that it contains.

CHAPTER 9

Relationship of the Triple Burner With the Stomach

Introduction
The Triple Burner completely encloses the stomach externally. In the commentaries on the 25th Difficult Issue on page 314 of Unschuld's (1) translation of *Nan Ching*, Ting Chin states, 'All these [transformations occur] because of the [activities of the] stomach. But how could the stomach be stimulated to perform these transformations if it were not *for the fact that the Triple Burner externally completely encloses* [the stomach]?' (Emphasis is mine)

Exactly the same as all the other organs, the Triple Burner 'externally completely encloses the stomach'. This allows the Stomach to perform all its attributed functions thanks to the assistance of the Triple Burner. However, it is a two-way street as the Stomach empowers the Triple Burner. So now, I will discuss the relationship between the Triple Burner and the Stomach.

9.1 Ming Men, and the Influences Sent Out by the Stomach Are What Empower the Triple Burner
In the commentaries on the 31st Difficult Issue on page 348 of Unschuld's (1) translation of *Nan Ching*, Hua Shou states:

The depots and palaces of the human body have form and shape; they are supplied [with influences] and they are generated. For instance, the liver receives influences from the [phase of] wood and is generated by the [phase of] water. The heart receives influences from the [phase of] fire and is generated by the [phase of] wood. There is no exception. Only the Triple Burner has no form and shape, and it is supplied and generated by nothing but the original influences and the influences [sent out] by the stomach. That is why [the text] states: 'It encompasses the passageways of water and grain [in the organism]; it represents conclusion and start of [the course of] the influences.'

Interestingly, in this scripture, Hua Shou states that the original influences—that is, *Ming men*—and the influences sent out by the Stomach are what empower the Triple Burner. I believe the influences sent out by the Stomach that empower the Triple Burner are predominantly the bulk water consumed for hydration along with the 80% water present in most foods (grains) that are transmuted into energized, living liquid-crystalline EZ water along with the biophotons encapsulated in the foods. Professor Pollack has shown that Bulk Water + Biophotons + Hydrophilic Connective-Tissue Membranes = Energized, Living Liquid Crystalline EZ Water. Pollack proved that this subsequent energized EZ water is equivalent to biological batteries and pumps, which I believe is the dominant force used to power the Triple Burner. Water is the main substrate that the Triple Burner is responsible for. The Stomach granary permits the consumed drinks containing generally more than 90% water and foods (grain) containing generally more than 80% water to be macerated in the Stomach so that the solar-energized biophotons present in the foods can be liberated and the vast amount of H_2O molecules present in the drinks and foods can be converted into the energized EZ water, $[H_3O_2]^-$ most likely just after diffusing from the stomach on the way to the Spleen.

9.2 The Influences of the Grains in the Stomach Generates the Triple Burner

In the commentaries on the 31st Difficult Issue on pages 348–349 of Unschuld's (1) translation of *Nan Ching*, Chang Shih-hsien stated:

> The Triple Burner is supplied with influences moving in the supervisor [vessel] as provision of its beginning; *it relies on the influences of the grains in the stomach as provision for its [continuing] generation.* It is the official responsible for the maintenance of the ditches; the waterways originate from it. Water and grains enter [the organism] through the upper [section of the Triple] Burner; they leave through the lower [section of the Triple] Burner. Hence, one knows that [the Triple Burner] represents the conclusion and the start of the [course of the] influences. (Emphasis is mine)

This is further confirmation about what was just stated immediately above.

9.3 Ch'i-Chieh (Stomach 30) Is the Place Where the Triple Burner Collects Its Influences

In the commentaries on the 31st Difficult Issue on pages 351–352 of Unschuld's (1) translation of *Nan Ching*, Yang states, 'The ch'i-chieh ['street of influences'] is a passage way of the influences. The Triple Burner masters the passage of the influences. Hence, [the text] states: 'It collects [its influences] at the ch'i-chieh.' He continues, 'Chieh ['street'] stands for ch'ü ['crossing']. Ch'ü is a place where four roads reach [into different directions]. It has been said that the Triple Burner masters the influences of the three originals, and that it collects [its influences] at the ch'i-chieh.'

This point is believed to be Stomach 30. Subsequently, it makes sense that the influences of the Triple Burner are circulated throughout the body via such a major point on the Stomach, namely Stomach 30 (*Qichong*, or 'Surging Qi'), which is known as the Sea of Water and Grain Point and is the Intersection Point of the Stomach meridian and Chong Mai meridian.

9.4 Stomach 30 Is Where the Triple Burner Affects Membranes and Fascia throughout the Body

As can be seen by the entry directly above, Stomach 30 is where the Triple Burner collects its influences. Regarding the powerful influence that the Triple Burner has on membranes and fascia throughout the entire body, in the article titled 'Stomach 30: St30 QiChong: Rushing Qi', the author, Jonathan Clogstoun-Willmott (31), states:

> The Chong Mo vessel is considered to have been the prime mover when the fertilized cell that grew to be you first implanted itself in the wall of your mother's womb. *Chong Mo's first action was to lay down foundations. In your body as it began to grow these were in the form of supportive fascia or membranes and fatty tissues*, that enabled your organs to grow, be *supported and wrapped protectively. So any problems with these fascia may be treated here.* (Emphasis is mine)

Clogstoun-Willmott (31) further shows that medical conditions throughout the body that involve fascia can be effectively treated using Stomach 30. Such conditions include pleurisy and retained placenta. He further notes that Qi stagnation in the abdomen causing colic-type pain and distension may be due to Qi stagnation within the membranes, which can also cause pain in the thighs and low back. As Stomach 30 is located at the junction of the body and the legs, it can also be used in cases of paralysis of the legs.

As can be seen from directly above, Clogstoun-Willmott (31) confirms that Stomach 30, which is 'the place where the Triple Burner collects its influences', is also an acupoint that has a powerful effect on membrane and fascial dysfunction throughout the body. I thought it very insightful of the author to also associate the very earliest phase of embryology—that is, the implantation of the blastocyst within the womb—with the formation of the 'supportive fascia or membranes and fatty tissues'. These facts are very suggestive that the Connective-Tissue Metasystem and the Triple-Energizer Metasystem have an intimate association with each other.

9.5 The Large Intestine and Small Intestine Belong to Stomach
Regarding the relationship between the Stomach and the San Jiao, Yongping Jiang (8), the author of the *Journal of Chinese Medicine* article titled 'The San Jiao: Returning to the Nei Jing (A Modern Explanation of Original Theory)', states, 'The existence of this system is evidenced in Chapter 2 of the *Nei Jing Ling Shu*, where the Large Intestine and Small Intestine are said to "belong" to the Stomach. . . . This explains why the lower He-sea points of the Large Intestine (Shangjuxu ST-37) and Small Intestine (Xiajuxu ST-39) are located on the Stomach channel.' This concept of the Small Intestine and Large Intestine belonging to the Stomach is an interesting notion but is really not that difficult to comprehend from an anatomical point of view. In reality, the Small Intestine and Large Intestine are a continuous muscular tube which extends from the lower end of the Stomach to the anus.

Summary of Chapter 9
The Triple Burner 'externally completely encloses the stomach'. This is also true of all the other organs. Because the Triple Burner surrounds the Stomach like an outer wall or like a large wrapping bag, there is an intimately close relationship that allows the Stomach to perform all its attributed functions thanks to the proximity and mutual assistance of the Triple Burner. However, due to the macerated mixture of drinks and grains, the Stomach empowers the Triple Burner. I believe the influences sent out by the Stomach that empower the Triple Burner are predominantly the bulk water consumed for hydration along with the 80% water present in most foods (grains) that are macerated within the Stomach and transmuted into energized, living liquid crystalline EZ water along with the biophotons encapsulated in the foods.

Professor Pollack has shown that Bulk Water + Biophotons + Hydrophilic Connective-Tissue Membranes = Energized, Living Liquid Crystalline EZ Water. Pollack proved that this subsequent energized EZ water is equivalent to biological batteries and pumps, which I believe is the dominant force used to power the Triple Burner. Remember, water is the main consumable that the Triple Burner is responsible for throughout the body. The Stomach granary permits the consumed drinks containing generally more than 90% water, and foods (grain), containing generally more than 80% water to be macerated in the Stomach, so that the solar-energized biophotons present in the foods can be liberated, and the vast amount of H_2O molecules present in the drinks and foods can be converted into the energized, life-sustaining liquid crystalline EZ water $[H_3O_2]^-$, most likely just after diffusing from the stomach, on the way to the Spleen. A future Chapter discusses in depth the recent scientific findings of Professor Pollack regarding energized liquid crystalline EZ water $[H_3O_2]^-$.

CHAPTER 10

The Manufacture of Blood by the Middle Heater

Introduction
With regard to the relationship of Blood and Ying (Nutritive) Qi within the body, from the article 'Blood (Xue): Vital Substances in Chinese Medicine' (32), it can be seen that in TCM, Blood is a denser form of Qi and is inseparable from Qi. The relationship exists such that Qi gives life and movement to Blood, while simultaneously Blood is the mother of Qi because Blood nourishes the very Organs that produce Qi. Blood and Ying (Nutritive) Qi possess a close association and are intimately connected, and they both flow together in the vessels throughout the body.

10.1 The Synthesis of Blood within the Body
The Stomach macerates grains and then subsequently the Food Qi produced by the Spleen is conducted upward to the Lungs, after which the Lungs direct it to the Heart, where it is transformed into Blood. This transformation process requires the catalytic assistance of the Original Qi stored in the Kidneys. The Kidney Essence produces Marrow, which generates the bone marrow, which generates dedicated cells that are added to the composition of Blood. Subsequently, it can be seen that Blood is generated from the interaction of the Postnatal Jing (derived from food and drink after being processed and refined by

the Stomach and Spleen) and the Prenatal Jing (stored in the Kidneys). The article (32) notes, 'Chinese theory of blood-forming function of the bone marrow predated the arrival of Western Medicine.'

10.2 The Function of Blood throughout the Body

Blood is part of Yin and is fluid-like and moistening. Blood nourishes the entire body and complements the nourishing action of Ying Qi. Blood is a denser form of Qi, and it flows along with the Ying Qi in the blood vessels and acupuncture channels all over the body. Blood moistens all body tissues throughout the body, ensuring that they do not dry out.

10.3 The Relationship between Blood and the Shen

The Shen resides in the Heart, and its field of effect continuously circumnavigates the entire body through the extension of the Heart—the vascular system. Blood nourishes and supports the Shen, giving it integrity and foundation. Should Blood become vacuous or deficient, the harmony and tranquility of the Shen can be disturbed, the individual can develop uneasiness, and adverse symptoms can develop, including a feeling of vague anxiety, irritability, unease, and difficulty in sleeping. In serious cases of Blood vacuity, the person can feel unreal and disconnected as if standing outside of themselves.

10.4 The Relationship and Connection between Blood and the Internal Organs (32)

10.4.1 The Relationship and Connection Between Blood and the Heart

In TCM, it is said that the Heart governs the Blood. The Blood Vessels are the established conduit system through which the Blood circulates. The Blood Vessel system is intimately associated with the Heart and constitutes part of the entire system of the TCM Heart. In TCM, the manufacture of Blood is completed in the Heart via the yang Heart Fire. On the other hand, Blood cools the Fire and prevents it from flaring up (32).

10.4.2 The Relationship and Connection between Blood and the Spleen

In TCM, the Spleen produces Food Qi, which is the initial stage in the formation of Blood. Spleen Qi is responsible for keeping the Blood in the Vessels so that it does not extravasate (exude from or pass out of a vessel into the tissues). If Spleen Qi is Deficient, the weak Qi is unable to hold and secure the Blood, resulting in hemorrhages (32).

10.4.3 The Relationship and Connection between Blood and the Liver

In TCM, it is said that the Liver stores the Blood. While a person is up and about and active, the Blood flows to the muscles and tendons (governed by the Liver). However, when a person lies down, Blood flows back to Liver for storage. Liver Blood also moistens the eyes, maintaining good eyesight, and it also moistens and nourishes the sinews, promoting flexibility of the joints. Liver Blood nourishes the uterus with Blood, together with the Penetrating Vessel (Chong Mai, one of the eight Extraordinary or Ancestral Vessels), with which it is closely related. Consequently, Liver Blood is very important for regular and healthy menstruation and fertility (32).

10.4.4 The Relationship between the Liver, the Kidney, Blood, and Gynecology

The Kidneys store Jing, and the Liver stores Blood. As Water feeds Wood, the Kidneys are the mother of the Liver in TCM's five-element theory. Thus, Kidney Jing creates Liver Blood. Jing and Blood mutually support each other, whereby Jing is indirectly transformed into Blood and Blood nourishes and replenishes Jing. Kidney Jing controls reproductive function and influences Blood. The status of Liver Blood is very important regarding menstruation. That is why women's physiology is more dependent on Blood than that of men. Subsequently, if Liver Blood is deficient, this can cause amenorrhea or scanty menstruation, while if Liver Blood is stagnant, dysmenorrhea may result (32).

10.4.5 The Relationship and Connection between Blood and the Lungs

The Lungs direct Food Qi from the Spleen to the Heart to form Blood. The Lungs also supply Qi to the acupuncture channels and Blood Vessels to assist the Heart's pushing action to ensure nourishing oxygenated Blood is circulated throughout the entire body (32).

10.4.6 The Relationship and Connection between Blood and the Kidneys

Original Qi, which is stored in the Kidneys, is required to transform Food Qi from the Spleen into Blood. The Kidney stores Jing, which produces Marrow. The Marrow generates bone marrow, which contributes to the formation of Blood (32).

Thus, to nourish Blood in TCM, we must therefore tonify the Spleen and Kidneys. However, the Heart, Spleen, and Liver have the most direct relationship with Blood. The Heart governs Blood, Spleen holds Blood in the Vessels so it does not extravasate, and the Liver stores Blood. (32)

10.5 Stomach Cells Produce Intrinsic Factor, Necessary for Red Blood Cell Manufacture

In the commentaries on the 25th Difficult Issue on page 313 of Unschuld's (1) translation of the *Nan Ching*, Ting Chin says, 'The treatise "Chüeh-ch'i" states: "The central burner receives influences. It absorbs the juices, transforms them, and turns them red. That is called the blood."'

A critically important substance required for the manufacture of oxygen-carrying red blood cells (erythrocytes) is Vitamin B_{12}. I believe it is a very interesting fact that 'the parietal cells of the stomach are responsible for producing Intrinsic Factor, which is necessary for the absorption of Vitamin B_{12}. Vitamin B_{12} is used in cellular metabolism and is necessary for the production of red blood cells, and the functioning of the nervous system' (33). So here is the stomach producing a substance called Intrinsic Factor, which directly allows the absorption of the very substance (Vitamin B_{12})

which produces the red component of blood, the erythrocytes. Here I propose that the 'juices' include water and fluids consumed plus the greater than 80% water present in most foods.

10.6 Blood Is Derived of Clear and Turbid Portions Flowing Inside/Outside Blood Vessels

In the commentaries on the 30th Difficult Issue on page 342 of Unschuld's (1) translation of the *Nan-ching*, regarding the processing of water and grains (food) in the Stomach, Ting Te-yung says:

> Once man is endowed with life through the true influences of heaven, the water he drinks and the grains he eats enter the stomach. From there they are transmitted to the five depots and six palaces where they are transformed into essence and blood. Both essence and blood have clear and turbid [portions]. The clear [portion] of the essence turns to the lung where it supports the true influences of heaven. Its turbid [portion] strengthens the bones and the marrow. Hence, the clear [portion] in the blood turns to the heart where it nourishes the spirit. The turbid [portion] of the blood provides external splendor to the flesh. The clear [portion] proceeds inside the vessels; the turbid [portion] proceeds outside the vessels.

Here, Ting Te-yung confirms that a living man stays alive by eating food and drinking water and fluids and that, once in the stomach, essential nutritive substances within this water and grain are transmitted from the stomach 'to the five depots and six palaces where they are transformed into essence and blood'. Note that Ting Te-yung states, 'The clear portion proceeds inside the vessels; the turbid portion proceeds outside the vessels.' This dual circuitry may pertain to the recently discovered Primo Vascular System (PVS), whereby hollow thread-like PVS ducts containing diverse biochemical substances have been shown to exist inside of blood vessels, while a coexisting similar structure exists outside the blood vessels. Regarding 'the clear portion in the blood turns to the heart where it nourishes the spirit', ducts of this Primo Vascular System have been found as a floating structure inside the heart and is believed to lubricate the heart valves (34).

10.7 When Fire Meets Water, a Transformation into Influences Takes Place

In the commentaries on the 31st Difficult Issue on pages 353–355 of Unschuld's (1) translation of the *Nan Ching*, Yeh Lin relates:

> The [treatise] 'Ling lan mi-tien lun' of the Su-wen states: 'The Triple Burner is the official responsible for the maintenance of the ditches. The water-ways originate from there.' . . . The Triple Burner is associated with the residence of the minister-fire. The nature of fire is to ascend from below. Hence, the [treatise] 'Ching-mai pieh-lun' of the Su-wen states: 'Drinks enter [the organism] through the stomach where their essential influences float off, moving upward to the spleen.' That is a reference to the central [section of the Triple] Burner. 'The influences of the spleen distribute the essence which ascends [further] and turns to the lung.' That is a reference to the upper [section of the Triple] Burner. 'From there they penetrate into and regulate the passageways of water, moving downward to the bladder.' That is a reference to the lower [section of the Triple] Burner.

This scripture is more definitive and advises that 'drinks' diffuse directly from the stomach and then 'their essential influences float off, moving upward to the spleen'. From the spleen, these essential influences ascend further to the lung. Notice, if you will, the next intoxicating question that Yeh Lin asks: 'But why are *only drinks* emphasized in this discussion of the *influences of the upper, central, and lower [section of the Triple] Burner?*' (emphasis is mine). Yeh Lin continues and thereby discusses the very nature and importance of EZ water. He states:

> [Anybody posing such a question] does not know that *the influences are transformed from water.* Through the

inhalation of the heavenly yang, the water of the bladder follows the fire of the heart downward to the lower [section of the Triple] Burner. There it evaporates like steam and is transformed into influences moving up again, where they become the chin [liquids], the yeh [liquids], and the sweat. All of that rests on the principle that *when fire meets water, a transformation into influences takes place.* The meaning is that heavenly yang [i.e., the influences of the sun] enters earthly yin [i.e., the water in the soil]. The [latter], following the movement of the yang influences ascends and become clouds and rain. (Emphasis is mine)

The energic influences that occur throughout the body result from what happens to normal water and fluids after the Triple-Energizer Metasystem has energized the water with fire from the associated minister fire. Yeh Lin said that 'the influences are transformed from water'. How much clearer could it be? Through the Triple-Heater cycle, the 'inhalation of the heavenly yang' (absorption of biophotons from the sun into the water and foods we consume) occurs, and the EZ water is generated directly from the Stomach, perhaps with the catalytic assistance of the very acidic hydrochloric acid imparting the low pH. Many chemical transformations are optimized at very low pH. Note also that the influences move 'up again, where they become the chin [liquids], the yeh [liquids], and the sweat'. All biological solutions in our body must be alkaline for optimal health. In the process of the generation of EZ water, higher alkalinity occurs at the hydrophilic regions along the connective tissues. This is the purest and most valuable component, and the most essential for life processes and nutritional functionality. The normal bulk water is subdivided in the Stomach into the alkaline fraction, whereby 'their essential influences float off' to the Spleen. From the Spleen, fluids are directed upwards for further processing by the Lung and the 'five depots and six palaces', where it becomes the 'chin liquids and the yeh liquids'. An acid fraction is also produced, and what better place to store it than in the stomach as hydrochloric acid? Interestingly, Pollack stated that the acidic by-products produced in the EZ water cycle must be eliminated. Guess what the acidic waste products are in the human body? Sweat, urine, and hydrochloric acid.

In the information source titled 'The Fourth Phase of Water: Central Role in Biology', Professor Pollack (35) advised that the human body is generally negatively charged—60% of the body is cells, which are negatively charged, and the extracellular space of the body makes up a further 20%. Thus, 80% of our body is composed of negative charges. So where are the low-pH positive charges? The positively charged components of our body include hydrochloric acid, urine, sweat, and our exhaled breaths (which contain positively charged CO_2 gas in water as carbonic acid). In good health, the body strives for the state of highest electronegativity.

10.8 Transformation, Transportation, and Excretion Products of the Triple Burner

Regarding the fluid products generated by the Triple Burner, in the 2011 article titled 'The Triple Burner (2)', Giovanni Maciocia (10) states:

> The end result of the complex process of transformation, transportation and excretion of fluids leads to the formation of various body fluids in each of the three Burners. The fluids of the Upper Burner are primarily sweat which flows in the space between skin and muscles; those of the Middle Burner are the fluids produced by the Stomach which moisten the body and integrate Blood; those of the Lower Burner are primarily urine and the small amount of fluids in the stools.

It can be readily seen that Maciocia's statement above agrees with the fluid mechanics of Pollack's theory.

10.9 The Triple Burner Is Where Blood Is Manufactured

In the commentaries on the 25th Difficult Issue on pages 313–314 of Unschuld's (1) translation of the *Nan Ching*, Ting Chin states, 'It was said further that "the upper burner disperses the taste [-influences] of the grains; the central burner receives influences, absorbs the juices, transforms them, and turns them red. That is called the blood.'" He further states:

> It was said further that 'the constructive [influences] emerge from the central burner; the protective [influences] emerge from the lower burner.' The constructive [influences] become the blood because they are [generated from] the essence of the taste[-influences] of the grains. The protective [influences] are [volatile] influences [because they are] generated from the [volatile] influences of the grains. All these [transformations occur] because of the [activities of the] stomach. But how could the stomach be stimulated to perform these transformations if it were not for the fact that the Triple Burner externally completely encloses [the stomach] and manages the movement of the influences?

Ting Chin advises that Blood is manufactured in the Middle Burner. In the commentaries on the 45th Difficult Issue on page 436 of Unschuld's (1) translation of the *Nan Ching*, Hsü Ta-ch'un states, 'The ke-shu [holes] [Bl 17] belong to the foot-great-yang [vessel]. . . . They are locations in the central section of the [Triple] Burner where the essence is transformed into blood. Hence they are the gathering-points of the blood.' The 'Sea of Blood' point is Bl 17 (*ke-shu*).

10.10 Nutrients Extracted from Food in the Stomach Are Sent to the Spleen to Produce Blood

In the 30th Difficult Issue in the commentaries on page 343 of Unschuld's (1) translation of the *Nan Ching*, Chang Shih-hsien states: 'Man's root and basis are his drinks and his food; they maintain his existence. Hence, man receives his influences from the grains. The grains enter the stomach, from which their essential influences flow out to be transported upward to the spleen.' From the Spleen, Chang Shih-hsien reports, 'The influences of the spleen distribute the essential [influences] further to the five depots and to the six palaces. They all [are supplied with] the influences of the grains in the stomach.'

10.11 The Triple Burner Fire Causes Digestion and Biochemical Biosynthesis from Food

In the commentaries on the 31st Difficult Issue on page 350 of Unschuld's (1) translation of the *Nan Ching*, Hua Shao states, 'The Triple Burner represents the minister-fire. Fire is capable of spoiling and processing the ten thousand things. [The character] chiao ["burner"] is derived from "fire"; it, too, [refers to] influences which spoil things. The meaning is to be taken from the terms.' Thus, it can be seen that the Triple Burner's fire spoils foodstuffs by decomposing whole foods and processing the chemical substances therein into ten thousand constituents—nutrients, vitamins, minerals, trace elements, amino acids, monosaccharides, fats, etc.—in all their diversity.

10.12 The Spleen Transforms Water and Grains to Generate Blood

In the 14th Difficult Issue in the commentaries on page 194 of Unschuld's (1) translation of the *Nan Ching*, it is stated by Yü Shu that 'the spleen transforms water and grains to generate (protective) influences and blood'.

10.13 Essential Nutrients from the Stomach Ascend to the Lung and Are Transformed into Blood

In the commentaries on the 31st Difficult Issue on pages 352–353 of Unschuld's (1) translation of the *Nan Ching*, Hsü Ta-ch'un states:

> The central [section of the Triple] Burner, too, is associated with the center of the stomach. It emits [its influences] upward after the upper [section of the Triple] Burner [has done so]. The influences received by the [central section of the Triple] Burner are gushing dregs and steaming liquids. The essential and subtle [portions] of these [dregs and liquids] are transformed and flow upward into the vessel [associated with the] lung. There they are transformed into blood. Nothing is more valuable concerning the maintenance of life in one's body than the [blood]. Hence,

it alone may move through hidden conduits. It is named 'constructive influences'.

It is most probable that the 'constructive influences' moving through the 'hidden conduits' referred to here are part of the structure made up by newly discovered Primo Vascular System conduits, which will be elucidated as more research is performed into the PVS circuitry.

10.14 The Heart Generates the Blood and the Liver Stores the Blood

In the commentaries on the 45th Difficult Issue on page 436 of Unschuld's (1) translation of the *Nan Ching*, Chang Shih-hsien stated:

> 'The gathering-point of the blood' refers to the blood of the entire body. Ke-shu is the name of holes which are located at a distance of one inch and five fen on both sides of the spine below the seventh vertebra. The blood of all the conduits moves from the diaphragm (ke-mo) upward and downward. The heart generates the blood and the liver stores the blood. The heart is located above the diaphragm; the liver is located below the diaphragm. [The Blood] travels through the diaphragm. Hence the 'diaphragm transportation' [hole—i.e., the ke-shu] is the gathering-point of the blood.

This refers to acupoint Bl 17, which is the Sea of Blood.

10.15 The Blood Is Responsible for Moisturizing and Nourishing the Entire Body's Organs

In the 22nd Difficult Issue in the commentaries on page 280 of Unschuld's (1) translation of the *Nan Ching*, Hua Shou states:

> Hsü stands for hsü ('warm'). 'The influences are responsible for providing the [body] with a warm flow' means that man's true influences arrive like a genial breeze, steaming through the interspace between skin and

flesh. 'The blood is responsible for providing the [body] with moisture' means that man's blood vessels moisten his muscles and bones, soften his joints, and nourish his body's depots and palaces.

10.16 Constructive Influences and Transformations in the Stomach Produce the Blood

In the commentaries on the 25 Issue on page 314 of Unschuld's (1) translation of the *Nan Ching*, Ting Chin states, 'It was said further that "the constructive [influences] emerge from the central burner ... the constructive [influences] become the blood because they are [generated from] the essence of the taste [-influences] of the grains".' I discuss this issue in another Chapter. Ting Chin continues:

> The protective [influences] are [volatile] influences [because they are] generated from the [volatile] influences of the grains. All these [transformations occur] because of the [activities of the] stomach. But how could the stomach be stimulated to perform these transformations if it were not for the fact that the Triple Burner externally completely encloses [the stomach] and manages the movement of the influences?

10.17 Blood Is the Constructive Influence and Lymph Is the Protective Influence in the Body

In the 22nd Difficult Issue in the commentaries on page 280 of Unschuld's (1) translation of the *Nan Ching*, Chang Shih-hsien states:

> In man's entire body, the blood constitutes the constructive [influences] while the [proper] influences constitute the protective [influences]. The constructive [influences] move inside the vessels; the protective [influences] move outside of the vessels. Evil [influences] enter from outside; they affect the [protective] influences first and are then transmitted to the blood. Here again, this scripture refers to dual fluid flows, where the 'constructive influences

move inside the vessels' and 'the protective influences move outside of the vessels.'

Recent research into the Primo Vascular System (PVS) has revealed that hollow thread-like PVS ducts are located floating inside blood vessels and an exact-looking structure is also present outside of blood vessels. I am sure that future research will show that the inner fluid likely contains more P-microcell stem cells with the capacity to heal due to their constructive influences, while the fluid in the external-vessel system contains more immunologically active components so that they constitute the protective influences to combat the evil influences before they can penetrate deeper into the bloodstream and then into the organs. I suspect that the fluid in the external-vessel system is not straight Lymph but may be a derivative of it. The dual fluid flow also occurs inside/outside lymphatic vessels too. Prior to knowledge of the PVS, this mechanism was hard to comprehend because no western anatomical equivalents had been determined. PVS knowledge now makes the whole process pretty common sense.

10.18 Blood Is the Constructive Influence and Governs Our Existence and Continued Good Health

In the 22nd Difficult Issue in the commentaries on page 282 of Unschuld's (1) translation of the *Nan Ching*, Yang states:

> Man has been endowed with the [protective] influences and with the blood to govern his existence. The influences are yang. Yang [influences] are the protective [influences]. The blood is yin. Yin [influences] are the constructive [influences]. Normally, these two kinds of influences [i.e., the protective influences and the blood] flow [through the organism]; under this condition no illness is present. Evil [influences] hit the yang [influences]. The yang [influences] are the [protective] influences. Hence, these influences are the first to be affected by an illness. That is because the yang influences are located in the external [parts of the organism]. If no cure is achieved while [the illness is still] in the yang [section], it will enter then the yin [section,

i.e., the yin influences]. The yin [influences] are the blood. Hence, the blood is affected by an illness afterward. That is because the blood is located in the internal [section of the organism].

10.19 The 'Waterways' Are Likened to the Blood Vessels of the Human Body

In the 27th Difficult Issue in the commentaries on page 325 of Unschuld's (1) translation of the *Nan Ching*, Hsü Ta-ch'un states, 'The waterways are used here as a metaphor for the blood vessels of the human body. When the blood vessels are filled completely, the twelve [main] conduits do not suffice to accept the [surplus from the blood vessels]. Consequently, there is an overflow into the single-conduit [vessels]. Hence, the single-conduit [vessels] are separate vessels [branching off from] the twelve [main] conduits.'

10.20 Blood Nourishes and Builds Up and Fortifies the Entire Body

In the 30th Difficult Issue in the commentaries on page 342 of Unschuld's (1) translation of the *Nan Ching*, Yang states:

> Ying ('constructive') is written here as jung ('brightness'). Jung has the meaning of jung-hua ['splendor'], that is to say, man's hundred bones and the nine orifices receive their splendour from these blood-influences. Ying ('constructive') stands for ching-ying ('to build up'). That is to say, the movement in the conduit-vessels continues without stop; it links the [entire] human body and provides it with long life. The two meanings [of ying and of Jung] are identical here. Wei ('protective') stands for hu ('to guard'). That is, man has aggressive influences proceeding outside the conduit-vessels. At day they proceed through the body, and at night they proceed through the depots to protect the human body. Hence, they are called 'protective' influences. Man's yin and yang influences meet in the head, in the hands, and in the feet.

> Their flow revolves [through the organism] without end. Hence, [the text] states: 'Like a ring without end.' The heart [is associated with] the constructive [influences, i.e.,] the blood. The lung [is associated with] the protective influences. The flow of the blood relies on the [movement of the protective] influences. The movement of the [protective] influences follows the blood. They proceed [through the organism] depending on each other. Hence, one knows that 'constructive and protective [influences] follow each other.'

Where Qi goes, blood follows. Where Blood, flows Lymph is generated in the tissue spaces. The Lymph is then collected in lymphatic vessels and returned to the heart so it can be sent out again; hence, the blood circulation is 'like a ring without end'.

10.21 The Middle Heater Is the Foundation of Blood Manufacture

On pages 82–84 of Heart Master Triple Heater, Rochat de la Vallée (11) discusses the production of Blood in the Middle Heater. She states:

> The essences in oneself are the model of composition and recomposition of what makes your life, and in this lies the importance of the middle heater which perpetually recomposes your own vitality from exterior elements. This vitality will appear again under the aspect of blood which is formed thanks to the work of the middle heater from the essences, *the juices and the liquids which become your own* and which are charged with the vitality that you assimilate and what passes from the middle heater to the upper heater. (Emphasis is mine)

Rochat de la Vallée then explains that in Chinese medicine, Blood is unique and differs from other liquids in that it is the only liquid derived from the middle heater that requires participation from the upper heater to empower it. Blood supplies more that nutrients to the body. Blood engenders life. She concludes, 'The *root and foundation of the blood* are in the *middle heater*' (emphasis is mine). From this information, it can be seen that the quality of the food and drink that enters the Stomach in the Middle Heater determines the integrity and vitality of the Blood and the subsequent health and well-being of the individual.

10.22 The Stomach Transforms the Water and Transmits It Upward into the Heart. The Heart Then Generates the Blood

In the commentaries on the 30th Difficult Issue on pages 342–343 of Unschuld's (1) translation of the *Nan Ching*, Yü Shu says:

> The scripture states: 'Man receives his influences from the grains. The grains enter the stomach, from which they are transmitted to the five depots and six palaces.' That is to say, water and grains enter the mouth and move down into the stomach. The stomach transforms the grains into influences. [These influences] are transmitted upward into the lung. The lung masters the influences. [These] influences are the protective [influences]. The stomach transforms the water and transmits it upward into the heart. The heart generates the blood. The blood constitutes the constructive [influences]. The [protective] *influences represent the exterior; they proceed outside of the vessels. The blood represents the interior; it proceeds inside the vessels.* Both depend on each other in their movement. (Emphasis is mine)

Here Yü Shu is stating that once they have entered the stomach, the grains and the water have different paths. The influences of the transformed grains are transmitted to the lungs, which form the protective influences, which represent the exterior, and proceed outside the vessels, while the influences of the transformed water are transmitted to the heart, which form the blood, which constitutes the constructive influences, and proceed inside the vessels. Grains, which are representative of foods, are extremely complex and diverse. It is understandable how they must be transformed, broken down with enzymes, carbohydrates into simple sugars, and proteins into simple amino acids which are small enough to be absorbed into the bloodstream ready to be synthesized into the thousands of biochemical compounds that are necessary for life.

But how exactly can water be transformed? It is a very basic molecule, H_2O. I propose that at this stage, thanks to the hydrophilic nature of the omnipresent connective tissues along with the presence of biophotons in the grains, bulk water (H_2O) is transformed into EZ water, $[H_3O_2]^-$ and that this living water is transformed into the living liquid organ, i.e. blood. Influences also include the biophotons derived from sunshine that has been required in the production of every single food type. Once again, the dual fluid movements which flow internally/externally to the blood vessels are mentioned. As I have discussed elsewhere, this circuitry pertains to the PVS. Interestingly, Yü Shu states, 'Both depend on each other in their movement.' So there is codependence between the established circuitry of the internal PVS and external PVS liquid flows. Future research will get to the bottom of the relationship.

Figure 3. Splash of water crown on blue surface. Photo by trans961. © Depositphotos.com.

Summary of Chapter 10

I set out to find from the TCM medical classics the step-by-step pathway for the manufacture of Blood. So while it is often cited that Blood is manufactured in the Stomach, that is not physiologically so. Poetically, Blood is manufactured in the Stomach in that the Stomach is the origin of Blood, mainly because the liquid portion of Blood is certainly derived from the food and drink that humans consume.

And all the nutrients and components in Blood (and also every single cell throughout the body) also came from food and drink that was macerated and digested inside the Stomach walls. I propose that most of the pathways described in the TCM medical classics will be shown to be existing ducts and conduits that comprise the Primo Vascular System network, which has only recently been elucidated. I propose that hollow collagenous conduits associated with the Connective-Tissue Metasystem are also involved in the distribution pathways. Incredibly, the PVS system has always been there. The ancients obviously knew of its existence. Modern science and anatomy is only now becoming aware of its presence and has a massive amount of PVS infrastructure and fluid composition contained therein to explore and define. I trust that future research will specifically elucidate the chemical and cellular composition of Wei Qi, Yuan Qi, Zong Qi, etc. and define their circulation through hollow ducts and conduits that comprise the Connective-Tissue Metasystem and the Primo Vascular System networks.

CHAPTER 11

Formation of Nutritive Qi in the Middle and Upper Heaters

Introduction
In the article 'Gross Conception of Anatomical Structure of the Triple Burner in Huangdi Neijing', regarding the formation of nutritive qi within the Triple Burner, the authors (9) state:

> That which moves in the Stomach is food, and that which moves in the pharynx is air, but according to Chapter 18, Lingshu, the Generation and the Engagement of the Nutritive and Defensive Qi, 'the human body receives qi from the grains, which enters the stomach and passes itself onto the lungs and later the five Zang-organs and the six Fu-organs, with the clear becoming the nutritive (ying) qi that runs inside the meridian and with the turbidness becoming the defensive (wei) qi that runs outside the meridian. 'The nutritive qi cycles without stop . . . penetrating across yin and yang as if it were a complete ring'.' The Upper Burner runs with nutritive qi at all times, Nutritive qi comes from the pectoral qi and just like the pectoral qi, it is produced by combining the fresh air inhaled into the lung with the food nutrients absorbed and transported by the spleen. Nutritive qi is the 'most essential part' of the food nutrients. The function

of nutritive qi is the transformation of blood to nourish the whole body, especially nourishing the internal organs, maintaining their physiological functions. *Nutritive qi goes in the blood vessel. Only is nutritive qi to go in the meridians or the channels.* (Emphasis is mine)

Once again, the issue of fluid flow differentiation occurring either inside or outside of blood vessels, the meridians, or the channels is discussed. This concept always used to confound me. That was until about a year ago when I became aware of the Primo Vascular System. I believe that this dual circuitry pertains to the recently discovered Primo Vascular System (PVS), whereby hollow thread-like PVS ducts containing diverse biochemical substances have been shown to exist inside of blood vessels, while a coexisting similar structure exists outside the blood vessels (34).

11.1 Absorption of Many Small Molecules from the Stomach Does Occur

It is a well-known fact that larger molecules—including carbohydrates, proteins, and fats—are not absorbed directly through the stomach and that your body absorbs most of the essential nutrients required from the small intestine. However, the large intestine does absorb some nutrients too. The main job of the large intestine is actually to remove water from undigested matter and to form solid waste for your body to excrete. With reference to absorption of nutrients and substances directly from the stomach, an article in the *Encyclopaedia Britannica* (36) states:

> Although the stomach absorbs few of the products of digestion, it can absorb many other substances, including glucose and other simple sugars, amino acids, and some fat-soluble substances. The pH of the gastric contents determines whether some substances are absorbed. At a low pH, for example, the environment is acidic and aspirin is absorbed from the stomach almost as rapidly as water, but, as the pH of the stomach rises and the environment becomes more basic, aspirin is absorbed more slowly.

> Water moves freely from the gastric contents across the gastric mucosa into the blood. . . . The absorption of water and alcohol can be slowed if the stomach contains foodstuffs and especially fats, probably because gastric emptying is delayed by fats.

The article, entitled 'Stomach' (33), further advises that although the absorption of nutrients from food is predominantly the function of the small intestine, some absorption of certain small molecules does occur through the lining of the stomach, and this includes water, medications (including aspirin), amino acids, 10–20% of ingested ethanol (e.g. from alcoholic beverages), and caffeine. According to Vince Calder (37), 'Water and ethanol are rapidly absorbed, even in the mouth and oesophagus to some extent, and are absorbed almost completely in the stomach'.

Regarding the fact that some vitamins and minerals are absorbed directly through the wall of the stomach, in the article titled 'Do You Know Where in the Digestive System Vitamins and Minerals Enter the Bloodstream?', Mateljan (38) states, 'This amount appears to be minor and has not traditionally been considered to be part of our vitamin and mineral absorption process. Exceptions here would be the minerals, copper, iodine, fluoride, and molybdenum, which may be significantly absorbed directly from the stomach.'

So the consensus is that while larger nutrient molecules of carbohydrates, proteins, and fats are not absorbed through the Stomach mucosa, a large variety of smaller molecules can readily be absorbed into the bloodstream through the Stomach wall, and they include water, alcohol, glucose and other simple sugars, amino acids, some fat-soluble substances, medications (including aspirin), caffeine, some vitamins, and several minerals, including copper, iodine, fluoride, and molybdenum. I would suggest that extremely small biophotons could also easily pass through the Stomach mucosa.

11.2 The Lining of the Stomach Wall Is Strongly Hydrophobic Due to DPPC

In 1996, an article entitled 'Gastric Surfactant and the Hydrophobic Mucosal Barrier' was published in the medical journal *Gut*. The author, B. A. Hills (39), stated:

> A striking property of the stomach wall is its hydrophobicity as witnessed when the present writer placed a droplet of water on the luminal surface of canine mucosa rinsed free of gastric contents to find that it 'beaded up', as though placed on Polythene, rather than spontaneously wetting the surface as anticipated on a hydrophilic mucoid layer. . . . This raises the question of what substance might be transforming the hydrophilic mucoid layer into a surface so hydrophobic that the contact angle in many normal stomachs can exceed 90° that is, approaching that of Polythene (93°) or Teflon (108°).

The author realized that many of the industrial synthetic surfactants used as corrosion inhibitors to protect cars and machinery had similar chemical structures to phospholipid dipalmitoyl phosphatidylcholine (DPPC), a known surface-active component that also covers the lung surface. It is the only surface-active component of lung surfactant capable of lowering surface tension to near zero levels. DPPC is also a lecithin.

11.3 Surface-Active Phospholipid (SAPL) Protects the Stomach Lining from Highly Corrosive HCl

The most exciting finding to this writer (39) was the multiple layers of surface-active phospholipid (SAPL) coating the epithelial surface of oxyntic ducts as these surfaces are mucus-free yet exposed to a highly corrosive environment in which proteolytic pepsinogen has been secreted from chief cells to mix with HCl at a pH of 1. The source of the SAPL seems to be lamellar bodies found in parietal cells and mucus neck cells with some reported in chief cells. Hills advised, 'Lamellar bodies are unequivocally the form in which SAPL is produced in the lung.' The author (39) reported that many classic research studies have also established the cytoprotective effect of phospholipid in the surface of the stomach lining and the reduced permeability to hydrogen ions that the phospholipid can impart to mucus. Hills advised that a single monolayer of SAPL adsorbed into a filter paper decreased its permeability to hydrogen ions by order of magnitude.

11.4 Several Organisms that Survive Corrosive Stomach Acid Have Protective SAPL Coats

The author (39) advised that parasites that can survive in the highly corrosive acidic environment of the stomach, including the Barber's Pole worm, which thrives in the abomasum of Australian sheep, and the bacteria that causes stomach ulcers, *H. pylori*, are coated with protective coatings similar to SAPL. Regarding the coating, Hills stated that it

> is similar to SAPL found on the cuticles of many creatures, including the cockroach—which is difficult to kill because droplets of any aqueous insecticide simply 'bounce off' those hydrophobic creatures unless a wetting agent is incorporated. Gastric mucus is also a wetting agent and this could stabilise the surfactant layer on the gastric mucosa by reducing the surface energy of what is otherwise a high energy interface with aqueous gastric contents.

Hills (39) further advised:

> An alternative approach in enhancing a gastric mucosal barrier is to bind the indigenous (or exogenous) SAPL more effectively, improving phosphate to phosphate ionic bonds by interspersing highly charged cations. Tighter binding of phosphate groups—as in 'phosphating' 'pulls' the outwardly orientated fatty acid chains together and, since these are straight (saturated), they will pack tightly into a very cohesive hydrocarbon barrier to water and, hence, to the hydrogen ion. In any aqueous environment these (that is, hydrogen ions) are never H^+ (a proton) but $H(H_2O)_3^+$.

> This compact barrier structure would be dislocated if the surfactant molecules were to lose one alkyl chain, explaining why lysolecithin is such an effective barrier breaker.

This comment by Hills (39) cited directly above reaffirms that, in biological systems, water molecules aren't always simple old H_2O molecules. Instead, in living systems, water molecules do assume various forms depending on the pH and other physical and chemical parameters—in this case, $H(H_2O)_3^+$.

11.5 Does the Surface Active Phospholipid Present in the Lungs Allow the 'Mist' to Form?

In the commentaries on the 31st Difficult Issue on pages 349–350 of Unschuld's (1) translation of the *Nan Ching*, regarding the location and the nature of the Triple Burner, Yü Shu states:

> The upper [section of the Triple] Burner is responsible for the entry of water and grains [into the organism]. It takes in but it does not discharge. The Ling-shu ching states: 'The upper [section of the Triple] Burner resembles fog.' That is to say, when it passes the influences, that resembles mist gently flowing into all the conduits. In other words, the influences of the stomach and the influences distributed by the tan-chung are poured downward by the lung into all the depots. The [Nei-]ching states: 'The lung passes the influences of heaven.' That is the meaning implied here.

The 2014 article entitled 'Pulmonary Surfactant' (40) states:

> Pulmonary surfactant is a surface-active lipoprotein complex (phospholipoprotein) formed by type II alveolar cells. The proteins and lipids that make up the surfactant have both hydrophilic and hydrophobic regions. By adsorbing to the air-water interface of alveoli hydrophilic head groups in the water and the hydrophobic tails facing

towards the air, the main lipid component of surfactant, dipalmitoylphosphatidylcholine (DPPC), reduces surface tension.

In the book *Heart Master Triple Heater*, regarding the mist emanating from the Upper Heater, Elisabeth Rochat de la Vallée (11) on page 92 states:

> This mist suggests everything that rises up from the earth in the form of vapour or steam which is capable of permeating everywhere. This means that at the level of the upper heater, above the diaphragm which is some kind of filter, there are never water or currents of water, but only that which can rise up as misty vapour. . . . So it is everything we have seen in the circulation of qi, the propagation and diffusion from the upper heater reaching the most external parts of the body right up to the skin and body hair, but which also penetrates the bones and flesh just like a vapour.

Note that it is interesting that both the stomach and the lungs are coated with surface-active phospholipid (SAPL) produced by Lamellar bodies. I propose that this surface coating on the Lung lining to reduce surface tension to allow for faster oxygen absorption also may be allowing the surface covering of water in the Lungs to easily change phases from liquid to vapour via the EZ water intermediate phase (as discussed elsewhere) and readily allow the mist to emanate from the lungs, as referred to in the 31st Difficult Issue discussed below. If we breathe onto a cold surface—for example, glass or a mirror—the mist on our breath condenses and is made visible. Similarly, on a very cold day, our exhaled breath can be readily seen as the cold air condenses the moisture in our breath. Likewise, holding our hand or any other part of our body on a cold surface will yield water condensate. We are emitting a *mist* throughout our entire body, especially our feet, which is one of our wettest parts, with each foot producing as much as one pint of sweat per day. So it can certainly be shown that water (H_2O) from the Lungs in the form of vapour or steam or *mist* is capable of permeating everywhere throughout our body.

11.6 TCM Teaches 'Influences' from 'Drinks and Grains' Are Literally Absorbed from the Stomach

In the article 'Gross Conception of Anatomical Structure of the Triple Burner in Huangdi Neijing', regarding the function of the Middle Burner, the authors (9) state, 'The place directly above the duodenal papilla is the boundary that separates the different functions of the digestive tract. Above here, nutrients are generally not absorbed. Only below here do the digestive enzymes, excreted by the pancreas, and the bile, excreted by the liver, enter food, allowing digestion to proceed, even though TCM thinks nutrients are absorbed in the stomach.'

Regarding the TCM concept of nutrients from grains being absorbed by the Stomach in the Middle Burner, the authors further state, 'Here, TCM thinks that the stomach absorbs the essence of water and grains, different from what we know through modern science. However, if we treat history and historical scientific development with respect, this kind of misconception is not a major mistake to be criticized.'

I don't believe it is a misconception. What I do believe is a misconception is that modern TCM practitioners assume that what is absorbed from the stomach are nutrients like carbohydrates, proteins, fats, minerals, and so on and that this was never the intention of the original authors. I believe the main essence of water and grains that is absorbed directly from the Stomach include predominantly bulk water molecules (H_2O) from the drinks and also from the grains, along with biophotons derived from the grains. It is known that other small molecules do pass through the Stomach wall. I propose that what is essentially absorbed from the grains is actually the 80% water that makes up those grains, along with greater than 90% water that makes up the drinks. Additionally, I propose that those grains are bounding with biophotons and that it is predominantly the bulk water present in the drinks and foods plus the biophotons which are absorbed through the stomach lining directly for the manufacture of EZ water. The bulk water plus the biophotons in the presence of the copious hydrophilic material of the membranous Triple-Energizer Metasystem yield living liquid crystalline EZ water, which powers the Triple Burner. At KL 331–332 of his book, Professor Gerald

Pollack (27) states, 'The water molecule is so small that if you were to count every single molecule in your body, 99% of them would be water molecules.' Note too that the San Jiao is not master over carbohydrates, proteins, fats, and minerals so that they should be given preference to be absorbed from the Stomach. The San Jiao is master over water. Author Glenn Grossman (5) stressed this fact regarding the legendary personified irrigation official named Yu. Highly-respected Yu personified the 'San Jiao as "the office of the sluices; it manifests as the waterways," or "the official in charge of irrigation and it controls the water passages"'.

11.7 The Nature and Composition of Nutrient Qi

Nutritive Qi is now called Nutrient Qi by WHO (3). Nutrient Qi is defined as 'the qi that moves within the vessels and nourishes all the organs and tissues, the same as nutritive qi'. Is Nutrient Qi predominantly composed of biophotons and EZ water (energized living water) that is manufactured either within the Stomach or shortly after passing through the Stomach? This is *not* the nutrients derived from the food that TCM practitioners in general assume is meant, which we know is not correct physiologically. I propose that what is absorbed from grains in the stomach is *not* the vast number of molecular nutrients as that occurs predominantly in the small intestine. All foods exist because the sun has irradiated them, and they subsequently have accumulated biophotons. Note too that all foods generally contain greater than 80% water so that when we eat grains (in whatever type of food that is), we are accessing the 80% of available water in sun-irradiated biophoton-enriched food. This solar-energized bulk water is quickly absorbed through the stomach wall. I believe that the necessary nutrients extracted from the food by the Small Intestine are mixed with the transmuted EZ water–biophoton mixture, whereby becoming the circulating Nutrient Qi. I suggest that at different locations—for example, in the Lower Burner—the degraded and contaminated less-pure EZ water forms the Wei qi, which is still energized EZ water, but due to its different composition, it is highly suitable for its defensive role. Note that good old Wei Qi is not even mentioned once in WHO (3). It is now called Defense Qi and is defined as 'the qi that moves outside the vessels, protecting the body surface and warding off external pathogens, the same as defensive qi'.

The solar-energized EZ water plus food-derived nutrients from the Small Intestines (Nutrient Qi) is the ideal initial building block to produce blood plasma, which is the relatively clear, yellow-tinted liquid composed of more than 92% water (along with sugars, fats, proteins, and salt solution) and carries the red cells, white cells, and platelets (41). I suspect that most TCM practitioners believe that the Triple Burner pathway that they are personally familiar with suggests that the red bloodstream commences and flows from directly outside the stomach wall. Nowhere in the TCM medical classics is that directly stated. It is intriguing that the initial building block of blood, the plasma, which the Earth-element Stomach and Spleen organs produce, is yellow.

Note that for babies, their 'grains' initially constitute only liquid breast milk right up until solid foods are introduced. Breast milk is composed of 90% water. Note that even so-called solid foods are still about 80% water or more.

Summary of Chapter 11
Regarding the occurrence of Exclusion Zones and the formation of EZ water alongside hydrophilic surfaces, Pollack (27) (KL 782–788) and his team of researchers found that numerous hydrophilic-material surfaces exhibited the property of forming negatively charged exclusion zones close to their surfaces. Such materials include all water-containing (hydro) gels, gels made of biological molecules and artificial polymers, and natural biological surfaces, including vascular endothelia (the insides of blood vessels), regions of plant roots, and muscle. Interestingly, Pollack even observed substantial EZs adjacent to single molecular layers, which meant that material depth was not consequential. Pollack also reports (KL 1317) that positively charged exclusion zones do exist, although they are less common than the negative ones. He located them next to certain polymers and metals. Note that Pollack has found that the majority of surfaces that form negatively charged exclusion zones close to their surface

are hydrophilic in nature and generally not hydrophobic. Pollack has determined some positively charged exclusion zones located next to certain polymers and metals. The lining of the Stomach wall is strongly hydrophobic due to DPPC. Subsequently, at this early stage of research, it is not known whether the EZ water is manufactured directly inside the Stomach at the hydrophobic Stomach lining or once the bulk water from the drinks and the foods (80% water) has passed through the Stomach lining.

Whichever is the case, I propose that the water derived from drinks and foods is transmuted into EZ water shortly after leaving the Stomach, if not therein. Likewise, I propose that the biophotons present in the foods are released in the Stomach and, being so miniscule, leave the Stomach (likely in the water) to be utilized by the abundant hydrophilic Connective-Tissue Metasystem that pervade the body as a major component of the Triple-Energizer Metasystem. Professor Pollack has proven that only three components are required to convert bulk water to life-sustaining, polymerized liquid crystalline EZ water, namely bulk water (from drinks and foods) + light energy (from biophotons in foods) + hydrophilic surfaces (connective tissues, especially collagen in connective tissues and membranes). As water is rapidly absorbed from the Stomach (37), the transmutation of bulk water to EZ water likely occurs after the water has left the Stomach on its way to the Spleen or is actually performed by the Spleen itself. In the 14th Difficult Issue in the commentaries on page 194 of Unschuld's (1) translation of the *Nan Ching*, it is stated by Yü Shu that 'the spleen transforms water and grains to generate (protective) influences and blood'.

In the commentaries on the 31st Difficult Issue on pages 353–355 of Unschuld's (1) translation of the *Nan Ching*, Yeh Lin relates, 'The [treatise] "Ching-mai pieh-lun" of the *Su-wen* states: "Drinks enter [the organism] through the stomach where their essential influences float off, moving upward to the spleen." That is a reference to the central [section of the Triple] Burner. "The influences of the spleen distribute the essence which ascends [further] and turns to the lung."' Note that this scripture only discusses fluid consumption, such that drinks diffuse directly from the stomach and then 'their essential influences float off, moving upward to the spleen'. Notice that the next revealing question that Yeh Lin asks is *'But why are only drinks emphasized in this discussion of the influences of the upper, central, and lower [section of the Triple] Burner?'* Yeh Lin continues and thereby discusses the very nature and importance of water. He states:

> [Anybody posing such a question] does not know that the influences are transformed from water. Through the inhalation of the heavenly yang, the water of the bladder follows the fire of the heart downward to the lower [section of the Triple] Burner. There it evaporates like steam and is transformed into influences moving up again, where they become the chin [liquids], the yeh [liquids], and the sweat. All of that rests on the principle that when fire meets water, a transformation into influences takes place. The meaning is that heavenly yang [i.e., the influences of the sun] enters earthly yin [i.e., the water in the soil].

Yeh Lin said, 'The influences are transformed from water.' How much clearer could it be? Through the Triple-Heater cycle, the 'inhalation of the heavenly yang' (absorption of biophotons from the sun into the foods we consume) occurs, and the EZ water is generated either inside the Stomach or shortly after it diffuses from the Stomach. Note also that the influences transform into 'the chin [liquids], the yeh [liquids], and the sweat'. It can be seen that water is the major building block of all the diverse numerous biological solutions that are manufactured in our body by the Triple-Energizer Metasystem. As Yeh Lin said, *'All of that rests on the principle that when fire meets water, a transformation into influences takes place.* The meaning is that heavenly yang [i.e. the influences of the sun] enters earthly yin [i.e., the water in the soil].' So here I believe that when Yeh Lin mentions the 'heavenly yang [i.e., the influences of the sun]', he is referring to what we now know are biophotons. If so, he is stating that 'a transformation into influences takes place' when biophotons 'enter earthly yin [i.e., the water in the soil]'. Here earthly yin exactly describes the yin Earth element, the Spleen. Otherwise, 'the water in the soil' could be the bulk water in the soil (Earth) elements, namely the Stomach or the Spleen.

CHAPTER 12

The Triple Burner is Associated with the Residence of the Minister Fire

Introduction
Regarding the confusion associated with the 'minister fire' in traditional Chinese medicine today, the pericardium is occasionally equated with the minister fire but generally only in the context of acupuncture and channel theory. Regarding the confusion regarding this minister fire in the *Australian Journal of Acupuncture and Chinese Medicine* article, the author, Mary Garvey (42), stated:

> The occasional mention of minister fire's various connections and interpretations in English language sources are difficult to reconcile for today's TCM students and practitioners. Wang Bing's (王冰 c. 710–805 CE) version of the Inner Canon defines minister fire as the heart-kidney and sanjiao-gall bladder channels (the shaoyin and shaoyang). The Nanjing (c. 100 CE) commentaries link it with the heart ruler-liver and sanjiao gall bladder channels (the jueyin and shaoyang). Zhu Danxi (朱丹溪 1280–1358 CE) identified minister fire with lifegate fire, and said that minister fire is stored in the kidneys and liver and connected to the heart. According to Li Shizhen (李时珍 1518–1593 CE), minister fire inhabits the liver and gall bladder. Zhang Jiebin (张介宾 1563–1640 CE)

identified it with the kidney, liver, sanjiao, gall bladder and pericardium. TCM has reconciled minister fire's various representations by reassigning its physiological contributions to the kidney yang, and identifying its pathogenic influences with liver and gall bladder yang repletion patterns. Small wonder many contemporary authors have little to say about the minister fire.

It would appear that there is as much confusion about the make-up of minister fire as there is about the nature of the San Jiao, which is being regarded as having a name but no form. Due to the complexity of this issue regarding minister fire, I will only cover the topic briefly.

12.1 The Triple Burner Represents the Minister Fire

In the commentaries on the 38th Difficult Issue on page 397 of Unschuld's (1) translation of *Nan Ching*, Ting Te-yung states, '[The text] speaks of "five depots" and "six palaces." That is to say, the five depots correspond to the Five Phases on earth, while the six palaces correspond to the six [climatic] influences of heaven. In reference to the six [climatic] influences of heaven, the Triple Burner represents the minister-fire. It is associated with the hand-minor-yang [conduit]. Hence, [the text] states: "Only the palaces are six."'

12.2 Triple Burner Related to the Minister Fire because the Pericardium Was Related Too

In the commentaries on the 38th Difficult Issue on pages 397–398 of Unschuld's (1) translation of *Nan Ching*, Katō Bankei said:

> The twenty-fifth treatise of the old version [of the *Nan-ching*] *speaks of the Triple Burner and of the heart-enclosing [network] as outside and inside.* The present paragraph calls it an 'additional source] of original influences.' There, [in difficult issue 25,] the Triple Burner and the heart-enclosing [network] were discussed together as depot and palace, and [the Triple Burner was

associated with] the minister-fire because [the heart-enclosing network was associated with it, too]. Here, the gate of life and the Triple Burner are referred to as beginning and end. The meaning implied is different. (Emphasis is mine)

12.3 When Fire Meets Water, a Transformation into Influences Takes Place

In the commentaries on the 31st Difficult Issue on pages 353–355 of Unschuld's (1) translation of *Nan Ching*, Yeh Lin relates:

> The [treatise] 'Ling lan mi-tien lun' of the Su-wen states: 'The Triple Burner is the official responsible for the maintenance of the ditches. The water-ways originate from there.' . . . *The Triple Burner is associated with the residence of the minister-fire*. The nature of fire is to ascend from below. Hence, the [treatise] 'Ching-mai pieh-lun' of the Su-wen states: 'Drinks enter [the organism] through the stomach where their essential influences float off, moving upward to the spleen.' That is a reference to the central [section of the Triple] Burner. 'The influences of the spleen distribute the essence which ascends [further] and turns to the lung.' That is a reference to the upper [section of the Triple] Burner. 'From there they penetrate into and regulate the passageways of water, moving downward to the bladder.' That is a reference to the lower [section of the Triple] Burner. (Emphasis is mine)

This scripture advises that drinks diffuse directly from the stomach and then 'their essential influences float off, moving upward to the spleen'. From the spleen, these essential influences ascend further to the lung. Notice, if you will, the next intoxicating question that Yeh Lin asks: 'But why are *only drinks* emphasized in this discussion of the *influences of the upper, central, and lower [section of the Triple] Burner?*' (emphasis is mine). Yeh Lin continues, and I believe, thereby discusses the very nature and importance of EZ water. Yeh Lin states:

[Anybody posing such a question] *does not know that the influences are transformed from water.* Through the inhalation of the heavenly yang, the water of the bladder follows the fire of the heart downward to the lower [section of the Triple] Burner. There it evaporates like steam and is transformed into influences moving up again, where they become the chin [liquids], the yeh [liquids], and the sweat. All of that rests on the principle that when fire meets water, a transformation into influences takes place. The meaning is that heavenly yang [i.e., the influences of the sun] enters earthly yin [i.e., the water in the soil]. The [latter], following the movement of the yang influences ascends and become clouds and rain.' (Emphasis is mine)

Summary of Chapter 12

The former citations from *Nan Ching* indicate that the fire of the Pericardium, or the minister fire, is directly associated with the Triple Burner. This should be obvious as the Pericardium and Triple Burner are internal–external partners. Note that Yeh Lin relates, 'The Triple Burner is associated with the residence of the minister-fire. The nature of fire is to ascend from below.' So it is the minister fire within the Triple Burner that causes the energic influences to be transformed from water. The energic influences that occur throughout the body result from what happens to normal water and fluids after the Triple-Energizer Metasystem has energized the water with fire from the associated minister fire. Yeh Lin said, 'The influences are transformed from water.' How much clearer could it be? Through the Triple-Heater cycle, the 'inhalation of the heavenly yang' (absorption of biophotons from the sun into the water and foods we consume) occurs, and the EZ water is generated either directly from the Stomach or shortly after the water diffuses through the Stomach wall on its way to the Spleen.

I ponder over whether minister fire is in some way associated with biophotons derived from life-sustaining foods that we must consume daily to survive and prosper. Mary Garvey (42) commented about the apparent confusion regarding the nature of minister fire at the Introduction of this Chapter, where she stated, 'Minister fire inhabits the liver and gall bladder. Zhang Jiebin . . . identified it with the kidney, liver, sanjiao, gall bladder and pericardium.' Biophotons have certainly been shown scientifically to be associated with some of these organs, and I discuss this issue in detail in another Chapter of this book.

CHAPTER 13

The Triple Burner Encompasses the Whole Body

Introduction

In the commentaries on the 25th Difficult Issue on pages 312 to 315 of Unschuld's (1) translation of *Nan Ching*, regarding the nature and the many aspects of the Triple Burner, Ting Chin makes an extremely wordy comment when he states:

> This paragraph states that the heart-master and the Triple Burner constitute outside and inside, and that both have a name but no form. Because of the two words 'no form', people in later times who did not check the meaning of the [*Nei-*]*ching* have engaged themselves in highly confused argumentations. They not only criticized the [alleged] mistakes of Yüeh-jen but also criticized [what they considered to be] erroneous interpretations forced [on this passage] by [Wang] Shu-ho. Over the past three thousand years, this has never been settled finally. I always think that the *Nan-ching* was not yet distant from antiquity. Of all the authors who appeared [in later times to comment on the ancient scriptures, Yüeh-jen] was the very first. Also, one must base [one's understanding of the *Nan-ching*] word for word on the *Nei-ching*. Why should misunderstandings and a deception of mankind be created just for the two key

[concepts] of the [heart-]enclosing network and the Triple Burner? There is no other way to elucidate [their meaning] except by comparing the meaning in the *Nei-ching* with that in the *Nan-ching*. Thus, the *Nei-ching* states that all the five depots have form and color, that the five palaces, too, can be measured in chang and feet, and that the water and the grains with which they are filled can be recorded in amounts of pints and pecks. If the [heart-] enclosing network and the Triple Burner had a form, why would they be the only ones with colors, sizes, and capacities that are not clearly recorded? Well, one should look at what the *Nan-ching* says about the [heart-]enclosing network and pick its meaning from the term 'enclosing,' and [one should look] also at what [the *Nan-ching*] says about the Triple Burner and pick its meaning from the term 'triple.' Thus, in the *Ling*[-*shu*] and in the *Su*[-*wen*], the treatise 'Pen-shu' states: 'The Triple Burner is a palace [acting as] central ditch; the passage-ways of water emerge from it. It is associated with the bladder and it constitutes the palace of uniqueness.' The treatise 'Pen-tsang' states: 'When the pores are sealed tightly and when the skin is thick, the Triple Burner and the bladder are thick [too].' The treatise 'Chüeh-ch'i' states: 'The upper burner [is responsible for] emissions; it disperses the taste[-influences] of the five grains, [a process] resembling the gentle flow of mist. [What is distributed] is called "the influences". The central burner receives influences. It absorbs the juices, transforms them, and turns them red. That is called the blood.' In the treatise 'Ying wei sheng hui', it is stated: 'The constructive [influences] emerge from the central burner; the protective [influences] emerge from the lower burner.' It states further: 'The upper burner resembles fog; the central burner resembles foam the lower burner resembles a ditch.' The discussion in the [treatise] 'Wu lung chin-yeh pieh-lun' states: 'The Triple Burner emits the influences in order to warm the flesh and to fill the skin.' The treatise 'Hsieh-k'o' of the *Ling-shu* states: 'The heart is the great master of the five depots and six palaces. As a depot it is strong and firm.

Evil [influences] cannot settle in it. If they do settle in it, the heart will be injured; the spirit will leave and [the respective person] will die. Hence, it is stated that if evil [influences] are present in the heart, they are always in the network enclosing the heart.' All the lines quoted above from the *Ling[-shu]* and from the *Su[-wen]* describe the Triple Burner as completely enclosing the five depots and six palaces. The [heart-]enclosing network has the meaning of enclosing only the heart. The 'palace [acting as] central ditch' is the 'palace of uniqueness.' If it were not for the fact that the Triple Burner enclosed the [organism] externally, how could [this Palace] have this singularly honorable designation? It was said further that 'when the pores are sealed tightly, and when the skin is thick, the Triple Burner is thick [too].' Now, if the inside of the skin and the flesh of the entire body were not supported by the Triple Burner, how could their thicknesses correspond to each other? It was said further that 'the upper burner disperses the taste [-influences] of the grains; the central burner receives influences, absorbs the juices, transforms them, and turns them red.' If the Triple Burner did not enclose all the body's depots and palaces, how could all the body's depots and palaces share in the influences of the Triple Burner in order to [further] diffuse and transform them? It was said further that 'the constructive [influences] emerge from the central burner; the protective [influences] emerge from the lower burner.' The constructive [influences] become the blood because they are [generated from] the essence of the taste[-influences] of the grains. The protective [influences] are [volatile] influences [because they are] generated from the [volatile] influences of the grains. All these [transformations occur] because of the [activities of the] stomach. But how could the stomach be stimulated to perform these transformations if it were not for the fact that the Triple Burner externally completely encloses [the stomach] and manages the movement of the influences? It was stated further: '[The upper burner] resembles fog; [the central burner] resembles foam; [the lower burner] resembles a ditch.' Above, [the upper burner] gives orders concerning emissions; below, [the lower burner] manages the passageways of water. How could this be if it were not for the fact that the Triple Burner externally encloses all the body's depots and palaces, exerting complete control over them? It was stated further: '[The Triple Burner] emits the influences in order to warm the flesh and fill the skin.' That is a clear indication of the fact that the Triple Burner constitutes a layer supporting the skin and the flesh from inside. It was stated further: 'Whenever evil [influences] are present in the heart, they are always in the network enclosing the heart.' That is a clear indication of the fact that the enclosing network constitutes a layer holding the heart from outside. Later readers of these texts were to say, if the Triple Burner has no form, how can passageways of water emerge from it? How can it be thick or thin? How can it be like mist or fog or foam or a ditch? How can it emit influences in order to supply warmth to the flesh? And if the enclosing network [of the heart] has no form, how can all the evil [influences] settle in this network enclosing the heart? Why is it the only [entity] that cannot be seen? Why does it lack color, width, and length? They obviously did not know that the [heart-]enclosing network is a small bag providing a network internally and an enclosure externally. Thus, the name already states that it is an 'enclosing network.' Its form does not have to be described in terms of big or small, feet or inches. The Triple Burner is a large bag supporting [the organism] from outside and holding it inside. The uniqueness of its holding [function] is described fully by nothing but the term 'triple.' The term 'burner' fully describes the provision of the entire [body] with influences. Hence, the name already states that it is a Triple Burner. Again, its form does not have to be described in terms of large or small, chang or feet. Anybody who hitherto has harbored some doubts can have them resolved now if he follows this [argumentation]. Also, if one matches this small bag resembling a depot and [therefore] constituting a separate depot with that large bag resembling a palace and [therefore] constituting a separate palace, that is the principle of heavenly creation and earthly organization. Yüeh-jen stated the two words 'no form' here, and again in the thirty-fourth Difficult Issue. An examination reveals that they are highly appropriate; an analysis shows that they are quite correct. How could the people of later times grasp but the hair on the skin of the *Nei-ching* and then criticize exemplary men of former times? Often [enough, their statements] reveal only the dimensions of their ignorance.

Many of the observations that I make about the Triple Burner in my book are based on the comments made by Ting Chin in the quote above. Many of these quotations are repeated several times throughout the book and are based on the context of the topic under investigation.

13.1 The Triple Burner Does *not* Have a Distinct Location and a Distinct Form

In the commentaries on the 25th Difficult Issue on page 315 of Unschuld's (1) translation of *Nan Ching*, Liao P'ing states, 'The [Triple Burner] is spread out [to cover, internally,] the *entire chest and back*. It is unlike the other depots and palaces, which have a distinct location and a distinct form and which can be pointed out as concrete [entities]. If it is said [here] that if one assumes that "it has no form," that was *not even followed by the authors of the apocryphal writings*. It was a *mistake of the one who said that*' (emphasis is mine).

Wow! That is self-explanatory when Liao P'ing states, 'The Triple Burner is spread out to cover, *internally*, the entire chest and back. *It is unlike the other depots and palaces, which have a distinct location and a distinct form and which can be pointed out as concrete entities*' (emphasis is mine). If I need to explain this, there is something very wrong with your mentation processes. What I do want to emphasize is what Liao P'ing says when he states, 'The Triple Burner is spread out.' It is actually 'spread out' to encompass all the Connective-Tissue

Metasystem because I believe that elements of the Triple-Energizer Metasystem in association with elements of the PVS combine in a union and are essentially one and the same as the location of the Connective-Tissue Metasystem.

13.2 Fu's Functions of Emitting, Intake, Revolution, and Transformation Due to the Triple Burner

In the commentaries on the 38th Difficult Issue on pages 397–398 of Unschuld's (1) translation of *Nan Ching*, Katō Bankei said, 'Now, the Triple Burner is not a proper palace. However, without its influences all the other palaces could not fulfill their functions of emitting, intake, revolution, and transformation. It is not a proper palace; hence, *it steams inside of the membrane; it moves in between the palaces and depots. It resembles an external wall. Hence, it is called "external palace"'* (emphasis is mine).

The Triple Burner is an external palace because it is a whole-body fu organ complex that takes up all the space outside all the palaces and depots (hollow fu organs and solid *Tang* organs) of the body. Note that this external palace actually resides inside of the membrane external to all the organs in the body. It thus obviously means that the external palace, which is the Triple Burner, is also the connective tissue/membrane complex of the body, which I define as the Connective-Tissue Metasystem. The text could not be more obvious.

13.3 The Triple Burner Allows the Original Influences and Liquids to Permeate the Entire Body

Regarding the nature of the Triple Burner in allowing the passage of the original influences throughout the entire body and the passage of the liquids through the entire body, in the Notes for the 31st Difficult Issue on page 356, Unschuld (1) personally states:

> The Chung-i-hsüeh kai-lun of 1978 [Shanghai. p. 23] has found an interesting compromise. It distinguishes between the Triple Burner on the one hand and the three sections of the Triple Burner on the other. In the Chung-i-hsüeh

kai-lun, the Triple Burner is considered to be a palace responsible, first, for the passage of the original influences through the entire body—thus stimulating the remaining depots and palaces in their functions—and, second, for the passage of the liquids through the body. The upper, central, and lower sections of the Triple Burner are assumed to refer to the heart and lung, the spleen and stomach, and the liver, kidneys, and bladder, respectively.

13.4 The Ubiquity and Omnipresence of the Triple Energizer throughout the Entire Body

In the book *Heart Master Triple Heater* (11), on page 115, Elisabeth Rochat de la Vallée explains *Nan Jing* Difficulty 66. She notes that the triple heater originates from the original qi of *ming men* and that subsequently the triple heater is responsible for distributing the original qi, yuan qi. This renders the triple heater the intercessor between ming men and all bodily functions, which require the constant presence of the original qi to perform correctly. She continues:

> Yuan qi means the original qi. . . . So this is what the triple heater represents for the original qi, it remains attached to the source but goes everywhere to activate things like a servant. It acts to produce all the effects that this qi commands in every place in the organism. This being done, all free communication and circulation goes well, and in particular the circulation of what we call the three types of qi, which we can consider as being the ancestral, nutritive and defensive—each of them attached to one of the three heaters. But we can also think of it as the expression of the original, unique and single breath in all its possible diversification.

13.5 The Ubiquity and Omnipresence of the Triple Heater throughout the Entire Body

In the book *Heart Master Triple Heater* (11), on page 118, the authors elaborate on the ubiquity and omnipresence of the triple heater throughout the body and the vast array of vital and necessary biological outcomes it produces. They state:

> So there is no limit to this kind of total encompassing quality of the triple heater. *It crosses throughout the body making different circuits, and this means that the triple heater has no distinct form. One can feel it, and have a certain knowledge of it, but you cannot see it.* It creates the harmony and proper functioning of the essences and qi. It opens the passages and ensures free communication. Through blood and qi it allows life to continue—thanks to the spirits. (Emphasis is mine)

It is interesting that regarding the nature and form of the Triple Heater, the authors state, 'It crosses throughout the body making different circuits, and this means that the triple heater has no distinct form. One can feel it, and have a certain knowledge of it, but you cannot see it.' Reading this statement at face value exactly describes the total-body location and the very nature of the Triple Heater. And yet strangely, the authors actually do *not* believe that the Triple Heater is a real organ. So while the authors were 'waxing lyrical' and being poetic in their description, they have in reality exactly described the very nature of the Triple Heater.

13.6 There is Nothing that the Triple Heater Does Not Envelop or Surround

In the book *Heart Master Triple Heater* (11), on page 59, regarding how the Triple Heater ensures free communication and circulation universally throughout the entire body, the authors quote Zhang Jiebin, a great doctor from the beginning of the seventeenth century, when he stated, 'The triple heater, its upper limit and its extreme point below, make it similar to the six reunions or junctions, liu he, of the universe, and *there is nothing that it does not envelop or surround*' (emphasis is mine).

The immensely multitudinous and diverse microbiological life forms emanating from the soil covering the four corners of the Earth cover every surface of every object on the Earth with complex biological coatings and biofilms. These diverse life forms include bacteria, viruses, fungi, yeasts, moulds, paramecium, protozoans, prions, and phages. Due to our intimate contact with the Earth, these same organisms find their way to cover all our personal external and internal surfaces. The eventuality is that, in many respects, the composition and synergisms of the microflora ubiquitous in the soil biosphere of the Earth mimic the human biosphere and as such are then essentially replicated in the gut, internal organs, and skin and hair covering all living creatures on Earth—from lice, lizards, lions, and even humans. Depending on exactly where these organisms reside will determine their specific role as these diverse organisms are responsible for numerous facets of biological processes on Earth. Interesting too is the fact that every atom in every cell in our body also came from the Earth and its atmosphere. How so? Every food item has its origin from the soil—be it honey, water, salt, milk, grain, vegetables, berries, fruit, nuts, meats (beef, pork, chicken, or fish, etc.), wine, beer, or cider. All food produce that we eat had its origin directly from the Earth and the rain and the Sun, of course, along with some Oxygen and Nitrogen and Carbon from the atmosphere. Subsequently, the atoms from the apple that we ate yesterday will become part of our Spleen or Heart. So *all* our atoms and all our microfloral hitch-hikers were derived from the Earth. We are essentially transmuted Earth under the management and control of the Triple Heater. These nutrients are used in the body to form the connective tissues and membranes that surround and communicate with every single cell in our body and simultaneously provide support and structure to our very being. I believe this Connective-Tissue Metasystem is intimately related to the Triple Heater. This is what Zhang Jiebin stated when he said, 'The triple heater, its upper limit and its extreme point below, make it similar to the six reunions or junctions . . . of the universe, and there is nothing that it does not envelop or surround.'

13.7 The Ancients Regarded the Sanjiao as the Biggest Organ Inside the Body

Regarding the large size of the Sanjiao, in the *Journal of Chinese Medicine* article titled 'The Location and Function of the Sanjiao', the authors Qu Lifang and Mary Garvey (43), state, 'The Sanjiao is one of the six fu. *The ancients regarded it as the biggest of the organs inside the body.* In fact, its location is given as inside the body and outside the other zangfu, in a sense *enclosing or holding the other organs*' (emphasis is mine).

13.8 The Sanjiao Extends to the Skin and Flesh and the Extremities

Regarding the body components of the Sanjiao, in the *Journal of Chinese Medicine* article titled 'The Location and Function of the Sanjiao', the authors Qu Lifang and Mary Garvey (43) explain that Chapter 18 of the *Lingshu* includes the channels of the Lung and Spleen along with their coat and lining, which are the yin-yang channel partners, which are the Large Intestine and Stomach respectively. The implication here is that the superficial regions (including the flesh and skin) of the arms and legs, where these four channels flow, and the extremities of the arms and legs are also included in the location and function of the Sanjiao.

Note that the Sanjiao holds or contains all the other *Zangfu*. This obviously means that it is a very sizable organ, extending from the deepest organs (Kidneys and Spleen) out to the surface of the body, including the skin of the extremities. Chapter 2 of the *Nei Jing Ling Shu* states, 'The San Jiao is a water passage, and it belongs to Bladder.' This explains why the most external lower distal acupoint (Bladder 67) can affect the baby immersed in the vast sea of the product of the Triple Energizer (i.e. the amniotic fluid) and cause a breech baby to turn or to stimulate induction. Further, because Bladder 67 is the connection point of the Bladder meridian and Kidney meridian, it regulates the Kidney and tonifies Qi.

13.9 The Two Distinct Qi and Water Passageways of the Sanjiao are Omnipresent and Coexist

Regarding the omnipresent coexisting Qi and Water passageways of the Sanjiao, in the *Journal of Chinese Medicine* article, the authors Qu Lifang and Mary Garvey (43) explain that while the Water and Qi pathways of the San Jiao assist and support each other, they are stand-alone entities and distinct from each other. The Water pathway facilitates hydration and irrigation of the Zangfu and regulates and maintains body temperature and ensures that body fluid balance is harmonized throughout the entire body. The three Qi (Yuan Qi, Zhen Qi, and Wei Qi) are meticulously transported and distributed via the Qi pathways that permeate the body so that Pathogenic Qi is repelled and good health and vitality are maintained. While the Sanjiao supplies the infrastructure and governs the waterways, various Zangfu specifically perform individual roles associated with body fluid metabolism and circulation. Way beyond the three burning spaces that house the Zangfu, Sanjiao's domain embraces all the diverse cavities and spaces throughout the entire body such that every single cell in the body is benefited due to the omnipresent circulation and transformation of Water.

13.10 Description of the Function, Morphology, and Location of the Sanjiao and Couli

Regarding the morphology and location of the Sanjiao, in the *Journal of Chinese Medicine* article, the authors Qu Lifang and Mary Garvey (43) explain that the infrastructure of the Sanjiao is composed of a microsystem of couli (spaces and textures) that permit the passageway of Qi and Water. This intricate network of cavities, spaces, and textures (couli) connects the upper level to the lower level, deep to superficial, and permits Water and Qi (Yuan Qi, Zhen Qi, Wei Qi, and Zangfu Qi) to converge, circulate, and perfuse the Zangfu and their associated tissues, so every cell is moistened and nourished. Due to the Sanjiao's intricate communication network of couli, the balance of yin and yang is maintained, and consequently, body fluid allocation and balance and body temperature are regulated by the processes of sweating and urination, which is under the control of the Sanjiao. The intricate infrastructure and the associated communication and cooperation of the couli microsystem engenders a powerful global defense system throughout the body that maintains a dynamic internal environment that counterattacks superficial Pathogenic Qi attacks and prevents deeper penetration into the body. This body-wide protection is particularly due to the Sanjiao's intricate substructure of couli, which enable the three Qi to congregate swiftly and easily at any location in the body where the Pathogenic Qi attacks. Subsequently, the Sanjiao provides the passageway and battlefield where the *Zheng* Qi can vanquish the Pathogenic Qi.

Thus, it can be seen that the microcosmic Sanjiao exists beyond the macrocosmic three Burning Spaces. The couli are a body-wide network extension of the three major Sanjiao body cavities. The couli could be likened to the capillaries of the Sanjiao Qi and Water circulatory system, and it is via them that the periphery of the body in its entirety is irrigated, nourished, and moisturized by this circulatory system and, at the same time, detoxified through the sweat glands.

13.11 Proper Irrigation of the Internal Spaces (Couli) Is an Obvious Prerequisite for Life

Regarding the proper irrigation of the internal spaces (couli) to maintain homeostasis throughout the body, in the article titled 'Yangsheng and the Channels', Robertson (44) states, 'In Professor Wáng's estimation, the ability to manifest essence is the very definition of "life" (shēng). In order for life to be nourished, circulation in the channels must be optimised.' Robertson notes that while the principal role of the acupuncture channels is to moisten and nourish living tissue throughout the body, the channels also maintain balance and communication between the organs. He notes further that the term *irrigation* designates a slow-moving, steady process throughout the body that is similar to the distribution of water and nutrients to plant roots to guarantee vigorous growth. Robertson then reasons that a prerequisite for vital life includes thorough 'irrigation' of the internal body spaces. This will ensure optimal functioning of the organs and the muscles and that the joints are thoroughly moistened. Appropriate irrigation not only ensures that nourishment is supplied to all cells throughout the entire body via the matrix of spaces that constitute our meridian system but also ensures that metabolic and environmental wastes are also removed.

13.12 The Triple Burner Guides the Yuan Qi through the Secret Circulation of the Entire Body

In the commentaries on the 38th Difficult Issue on pages 395–396 of Unschuld's (1) translation of *Nan Ching*, Hua Shou states, 'The Triple Burner governs all the influences; it is an additional transmitter of original influences. [That is to say], the original influences depend on the guidance of the [Triple Burner] in their ceaseless *hidden movement and secret circulation* through the entire body.'

Summary of Chapter 13

The analogy of the Triple Energizer is like the available access of electrical power and treated water to every nook and cranny within the city limits. Thanks to the engineers employed by the local authority, the distribution of power through the established infrastructural grid of electrical power lines from the power station and the availability of water via the underground water mains surging with treated water sourced from local reservoirs or bores are assured. Perhaps with the addition of extension leads or hoses, you could supply power and water to every conceivable spot within your property if you needed to. The analogy to 'it acts to produce all the effects that this qi commands in every place in the organism' is this. The electricity delivered to our home can be used to light up our nights, refrigerate our foods, warm us using electrical heaters, provide us with entertainment from sound-producing radio devices and televisions and can be used to heat frying pans and boilers so we can prepare meals. All these different requirements are energized from the same source of Qi (power).

Likewise, thanks to modern mobile phone service providers, you can communicate with every person within the city limits using the electromagnetic bandwidths supplied by your provider. I propose that the Triple-Energizer Metasystem is analogous to this description, except I suggest that apart from the established neural system, the water reticulation system also carries the major component of the energy supply system, namely the EZ water through the 'ceaseless hidden movement and secret circulation through the entire body'. It is the fourth phase of water $[H_3O_2]^-$, which is energized water or living water, and it flows alongside and through the ubiquitous omnipresent Connective-Tissue Metasystem that has unbounded or universal presence throughout the entire body and is also known essentially as the Triple-Energizer Metasystem.

I further propose that all the acupuncture meridians constitute a part of the Connective-Tissue Metasystem, and so the EZ water flows through them also, along with biophotons from the sun, free electrons from the earth, and the resultant piezoelectricity.

CHAPTER 14

The Relationship Between the San Jiao, the Cou Li, and the Bladder

Introduction
Regarding the relationship between the San Jiao and the Bladder, the author of the book *Gua Sha: A Traditional Technique for Modern Practice*, Arya Nielsen (20), on page 36, explains that the San Jiao manages Jin Ye fluids throughout the body and that one control valve over these fluids is the Bladder. Nielsen notes that when humans suddenly become cold, the Cou Li and the muscles close to the skin contract, the pores close, and the Bladder fills with clear light-yellow urine. This exemplifies the close interrelationship between the San Jiao, the Bladder, and urine in the *Su Wen* and *Ling Shu*. Regarding the nature of the Cou Li, on page 32, Nielsen (20) advises that some authorities suggest that Cou Li means 'pores' in much of the *Su Wen* and explains that when the pores are open, exogenous factors can penetrate. The same open pores can also allow the release of exogenous factors via sweating. Thus, closed pores act as a barrier to exogenous factors and also hold in the exogenous factor once they have penetrated. Nielsen advises that some commentators of the *Su Wen* translated the term *Cou Li* as 'between the skin and underlying musculature'. She notes that the *Chinese-English Medical Dictionary* (Ou Ming 1988) agrees, and she describes the Cou Li as striae, the natural parallel lines of the skin and muscles and also the spaces between the skin and muscles.

The Dictionary notes that the Cou Li permit the entry and exit of vital energy and blood flow and that they also constitute one of the portals for body fluid excretion and act as a barrier against the penetration of exogenous evils.

Here the Cou Li are the anatomical expression of the San Jiao. The *Ling Shu*, Chapter 47, confirms that 'the Bladder and Triple Heater (San Jiao) have their correspondence and resonance in the most external structure of the body, the Cou Li'. Nielsen (20), on pages 32–33, further explains that the therapeutic and restorative effect of Gua Sha, acupuncture, and even physiotherapy is predominantly due to the relationship between the San Jiao and the Cou Li. She notes that when the body surface is stimulated superficially (massage, physiotherapy) or penetrated by acupuncture or Gua sha, it is the Cou Li that conduct the therapeutic effect internally. She notes that the Cou Li correspond to 'fascial connective tissue and the potential transduction of chemical and mechanical signalling'. She aptly describes the Cou Li as 'the outermost of the San Jiao interconnecting network of bags'.

Here both the *Chinese-English Medical Dictionary* and Nielsen state very plainly that the Cou Li are the outermost defensive structure in the skin. Nielsen states further that the Cou Li correspond 'to fascial connective tissue and the potential transduction of chemical and mechanical signaling' that allows for communication between the outer body and the inner organs. I propose that the Cou Li are just one more bag of relatively superficial differentiated fascia making up the omnipresent continuum of hydrophilic connective tissue that is an integral part of the San Jiao and that the life-sustaining energized EZ water generated within this hydrophilic structure is responsible for powering the battery that drives the superficial defense mechanism responsible for keeping pathogenic forces from entering the body and causing disease processes. I will now discuss the Cou Li (couli) subsystem of the Triple-Energizer Metasystem in more detail.

14.1 The Superficial Layer of the Human Body Is Still Not Well Understood by Researchers

Regarding the Cou Li, in the 2011 article entitled 'The Fascia: The Forgotten Structure', authors Carla Stecco et al. (45), on page 128, state, 'The superficial fascia (or membranous layer of the hypodermis) is still an object of debate; some authors even admit the existence of a membranous layer separating the subcutaneous tissue into two sublayers; others exclude it; and yet others describe multiple such layers.'

This article was published as recently as 2011. So incredibly, even the most commonly investigated superficial connective-tissue layer of the human body is not well understood by medical researchers. Imagine what is yet to be uncovered by researchers about the deeper and more complicated fascial structures and complexes that make up the Connective-Tissue Metasystem!

14.2 Extreme Variation of the Superficial Fascia throughout the Body Means It Has No Form

With regard to the superficial fascia located throughout the body, in the 2011 article entitled 'The Fascia: The Forgotten Structure', authors Carla Stecco et al. (45), on page 129, state:

> *Its arrangement and thickness vary according to body region, body surface, and gender. It is thicker in the lower than in the upper extremities, on the posterior rather than the anterior aspect of the body, and in females more than in males.* Our studies have also revealed the constant presence of a membranous layer of connective tissue of variable thickness inside the subcutaneous tissue, dividing it into superficial (SAT) and deep adipose tissue (DAT). (Emphasis is mine)

Because of its extreme variability relating to arrangement and thickness depending on body region, body surface, and gender, this fascia could easily be described as having no form. As I propose that the Connective-Tissue Metasystem is intimately related to the

Triple-Energizer Metasystem, it is no wonder that the ancient medical text, the *Nan Ching*, likewise said the San Jiao has no form.

14.3 Connection between the Triple Heater, the Kidneys, and the Bladder

On page 61 of her book, Rochat de la Vallée (11) sums up Chapter 2 of *Ling shu* by stating that the Triple Heater has a relationship with the Kidneys and the Bladder and that because it is an intercessor at every level of life, the Triple Heater ensures that the liquids and qi circulate smoothly and irrigate the entire body. This active role is accomplished due to the fire of the kidneys.

14.4 Relationship between the Kidneys and the Three Heaters

In the Appendix of the book *Heart Master Triple Heater*, while summarizing the intimate relationship between the Kidneys and the Three Heaters regarding the proper management of liquids throughout the entire body, Elisabeth Rochat de la Vallée and M. Macé (11), on page 123, advise that sometimes (*Ling shu*, Chapters 2 and 47) the Triple Heater and the Bladder are considered to be the two fu explicitly linked to the zang of the Kidneys. This intimacy with the water element Kidney empowers the Triple Heater to open passages and irrigate tissues as per *Su wen* (8), thus ensuring the proper distribution of liquids throughout the body. She states, 'The link with the transformation of liquids is constantly confirmed in its pathology, as its relationship with water, in all its forms, is emphasized by the traditional titles attributed to each of the three heaters.' She further notes that *Lingshu* (18) uses the term *Wu* to designate 'mist or humid vapours for the Upper Heater', *Ou* to designate 'maceration for the Middle Heater', and the term *Du* to designate 'canal or conduit for the Lower Heater'.

14.5 Relationship between the San Jiao and the Urinary Bladder

The San Jiao and the Urinary Bladder work together to process water and remove waste from the body as urine. This relationship between the Bladder and the San Jiao is described in Chapter 2 of the *Nei Jing*

Ling Shu. Yongping Jiang (8), the author of the *Journal of Chinese Medicine* article titled 'The San Jiao: Returning to the Nei Jing (A Modern Explanation of Original Theory)' states, 'This explains why the lower He-sea point of the San Jiao (*Weiyang* BL-39) is on the Urinary Bladder channel.'

14.6 Waste Water 'Permeates' from the Small Intestine into the Urinary Bladder

Regarding the relationship between the Small Intestine, the Urinary Bladder, and the San Jiao as related in Chapter 12 of the *Nei Jing Ling Shu*, Yongping Jiang (8), the author of the *Journal of Chinese Medicine* article titled 'The San Jiao: Returning to the Nei Jing (A Modern Explanation of Original Theory)', writes, 'The *Nei Jing* states that the lower jiao connects the Urinary Bladder and the Small Intestine, but because there is no actual physical tract between these organs, the lower jiao is said to allow water to "permeate" into the Bladder.' He further relates, 'This water passage, where water drips from the Small Intestine into the Urinary Bladder, is the lower jiao.'

The majority of authorities believe that this drainage system is figurative. Perhaps it is, or perhaps the ancients were once again aware of a literal drainage system through the lower jiao, which is after all a system of membranes that are known to house and manage the passageway of a large proportion of the water present in the body. Apart from this potentially undiscovered section of the Connective-Tissue Metasystem, the drainage system may also possibly be due to an as yet undefined Primo Vascular System network.

14.7 The 'Cou' Are Important Microcomponents of the Sanjiao

Regarding the smaller cavernous components of the Sanjiao, in the article titled 'The location and function of the Sanjiao', the authors, Qu Lifang and Mary Garvey (43), explain that the ancient Chinese medical classics describe the Sanjiao as the 'unique fu' constituting a hollow-organ complex whose strange and novel cavernous structures are spread far and wide throughout the entire body. The predominant and major cavities constitute the 'three burning spaces' of the Sanjiao, consisting of the upper, middle, and lower jiaos, which reside in the chest cavity, the abdominal cavity, and the pelvic cavity respectively. Other smaller spaces and cavities that are located in the extremities and within the muscles are termed *cou* in the *Neijing*, and their explicit physiology and pathology contribute to the overall comprehension and understanding of the Sanjiao's functionality, location, and unique morphology.

14.8 In Ancient Times, *Jiao* Meant Cavities, Spaces, or Gaps, Both Large and Small

Regarding the original meaning of the Chinese word *jiao*, in the *Journal of Chinese Medicine* article titled 'The location and function of the Sanjiao', the authors, Qu Lifang and Mary Garvey (43), explain that the Sanjiao was originally written as *jiao*, and they note that *jiao* is defined in the *Cihai* as muscles failing to fill the shell. The authors then explain that the term *Jiao* therefore denotes the impression of spaces or gaps occurring because the muscles did not fill the shell of the body. Thus, these spaces or cavities culminate in three large cavities or spaces within the trunk of the body, making up the upper, middle, and lower jiao in the thoracic, abdominal, and pelvic cavities respectively. Furthermore, numerous small spaces or hollows develop throughout the entire body, including the extremities. These lesser spaces were called cou in the *Neijing*. The authors quote, 'Cou is a place of the Sanjiao where there is a passageway of circulation and convergence of yuan qi and zhen qi, filled by blood and qi; li is texture of skin and zangfu.' In the trunk of the body the cou-spaces are associated with the zangfu, while in the extremities the cou-spaces are affiliated with more superficial tissues, including the skin, muscles and flesh. Subsequently, the larger cavities and smaller dispersed cou spaces merge in their functionality to form a cohesive network interconnecting the zangfu and their associated tissues, including muscles, flesh, and the skin.

The small cou spaces may well include tubular conduits within the Connective-Tissue Metasystem that carry the many different fluids (yuan qi, zhen qi, wei qi, etc.) throughout the body. Perhaps the cou spaces involves the Primo Vascular System, which is known to be a highly complex system composed of hollow microtubules that do carry various complex fluids throughout the entire body 'and form a cohesive network interconnecting the zangfu and their associated tissues including muscles, flesh and the skin'. I have discussed this recent scientific finding in great detail elsewhere in this book.

14.9 The Original Meaning of the Chinese Word *Li* Is 'Network'
Regarding the original meaning of the Chinese word *li*, in the article titled 'The location and function of the Sanjiao', the authors, Qu Lifang and Mary Garvey (43), explain that the ancient meaning of *li* signified the intrinsic pattern or markings within entities—for example, the natural markings in the mineral jade or the muscle fibers within muscle tissue. In the author's note in the study of the Mawangdui medical manuscripts by Donald Harper, he interprets the word *li* to mean 'network', which conveys the concept that *li* pertains to an intrinsic pattern or a system. The authors note that when the *Neijing* and later medical classics were written, the term *li* was equated to a network or a prevailing pattern in things, and within the medical context, li pertains to a pattern of textures that relate to the zang, fu, tissues, muscles, and skin areas of the body. This construct endorses the five-phase (*wu xing*) association and affiliation of specific organs, tissues, and body areas in accordance with TCM. Subsequently, the terminology *couli* exactly describes the highly structured network system which is distributed throughout the entire body. The authors state, 'The li-textures, their associations and distribution, thereby connect upper and lower, external and internal, shallow and deep. The cou, the spaces, are located between the body tissues and their textures: their existence is dependent on this association.'

With this in mind, my proposal makes perfect sense as the small cou spaces or hollow tubular conduits within the Connective-Tissue Metasystem would carry the many different fluids throughout the entire body due to the li network of the cou tubules with their highly complex and extremely ordered structure, which would 'thereby connect upper and lower, external and internal, shallow and deep' due to the highly-organized infrastructure of the cou-li.

14.10 The Original Meaning of the Chinese Word Couli Is Space and Texture

Regarding further information pertaining to the meaning of the Chinese word *couli*, in the *Neijing*, the authors (43) further state, 'Couli therefore is an anatomical term associated with the grain or texture (li) of the skin, flesh, muscles, zang and fu, and the interstitial spaces (cou) that exist between these structures/textures. Their relationship with specific structures and textures means that, firstly, the cou vary in size and shape, and secondly, that collectively they form a special system of the body.' They further state:

> This system is what is meant when the *Neijing* refers to the couli, 'space and texture'. According to the *Neijing*, the superficial couli are associated with their respective zangfu. The skin and muscle couli are therefore differentiated according to their zangfu association, and their layer or depth, and named accordingly (as for example 'skin texture', 'skin space', 'muscle texture', 'muscle space').

As an example, the authors (43) reference the *Neijing Suwen*, Chapter 71, which advises that human illnesses may begin in the spaces under the skin when External Cold assaults those spaces and, if successful, may result in skin infections, carbuncles, headache, and fever.

14.11 The Triple Energizer Component of Shao Yang Controls Fluid–Qi Movement in the Cou Li

In 2012, acupuncturist Kimberly Thompson (46) posted an article entitled 'What the Heck Is a Triple Energizer Anyway?' on the miridiatech.com website. She explained that the Shao Yang is composed of the Triple Energizer and Gallbladder. Thompson noted that the Triple Energizer is responsible for the movement of the qi and fluids in the body through the interstitial spaces that mantle all the internal organs, and she advised that the Shao Yang is associated with the movement of the synovial fluid that occurs within the spaces between the sinews and bones. Thompson was pointing out that the Cou Li spaces of the body include 'the spaces that surround the internal organs known as the interstitial spaces' and also 'the spaces between the sinews and bones where the synovial fluid moves'.

14.12 Body Tissue Types and Zangfu Relationships Engender Couli Networks

Regarding the body tissue–zangfu relationship within the Sanjiao, the authors Qu Lifang and Mary Garvey (43) further assert that in TCM the couli of certain superficial body tissues are controlled by a defined zangfu relationship. For example, the couli present in the flesh and muscles belong to and are controlled by the Spleen, while the couli present in the skin and soft body hair at the most external layer of the body belong to and are controlled by the Lung. Due to this relationship, Lung Qi deficiency will culminate in a weakness in the body's exterior defenses present in the couli of the skin, and the compromised resistance to external Pathogens would permit them to more readily violate the weakened exterior couli defense and penetrate internally and attack the Lung itself.

The authors further note that the six-channel pairing of the Lung and Spleen engenders the vertical ascending (Spleen) and descending (Lung) potential of Qi and body fluid transportation within the body that links the upper body with the lower body. But further, the zang–body tissue correspondences explain how the couli generate a network linking the interior of the body to the exterior of the body. This body-wide three-dimensional infrastructure thus engenders the highly energetic ramifications of the Sanjiao, permitting fluid and Qi flow from upper to lower and from interior to exterior, thereby defining the prescribed ascending and descending, taking in and giving off properties of the Sanjiao.

In the article '50 Years of Bong-Han Theory and 10 Years of Primo Vascular System', the authors (34) reported that primo nodes (PNs), formerly called Bonghan corpuscles (BHCs), possess a large amount of cells and granules related to the immune system, which suggests that the PVS is involved in a protecting function of the body. This agrees with what the authors Qu Lifang and Mary Garvey (43) said above, namely, 'At the body surface, the couli help form the main line of defence to resist external attack.'

Using fluorescent nanoparticles, modern researchers have traced the Primo Vascular System conduits from the skin surface to deep internal organs. For example, recent research in 2009 (34) revealed that 'the flow of the primo fluid in a certain path was demonstrated in the study using Alcian blue, from the rat acupoint BL-23 in the dorsal skin to the PVS on the surface of internal organs'. Further research in 2010 (34) discovered that 'when Chrome-hematoxylin and fluorescent nanoparticles were injected into testis they were found in the PVs on the organ surfaces between the abdominal cavity and the abdominal wall'. This data certainly establishes proven defined pathways from the skin surface to deep organs via the Primo Vascular System. The book (7), on page 31, states there is a 'close relation between the PVS and the fascia'. I propose that the PVS throughout the entire body will be found to be intimately associated with and directly or indirectly connected to the omnipresent Connective-Tissue Metasystem, and I propose that this amalgamation is synonymous with the Triple-Energizer Metasystem. I further believe that the PVS will be found to be responsible for body tissue types and zangfu relationships as per the example given by Qu Lifang and Mary Garvey (43) above, where they stated, 'The flesh and muscles belong to the Spleen, whilst the skin and soft body hair, at the most external layer of the body, belong to the Lung.'

14.13 The Triple Energizer–Couli Network Extends to the Ultimate Exterior Extremity

I suggest that the couli extend right down to our pores, where the oily liquid material is constantly oozing from our skin. For example, we can wipe our fingers as much as we want, but we continue to leave a fine film of our individuality on everything we touch, called our fingerprints. Even the residue of our fingerprints is secreted through our most superficial couli, thanks to our omnipresent Triple-Energizer Metasystem. Again with every breath, we exhale moisture, carbon dioxide, some 14% oxygen, alcohol (if we have had some as drink), personal chemicals emitted from our mouth flora, and volatile food components (for example, from garlic or peppermint). Some dogs and cats have been known to detect that individuals have medical conditions, including cancer, from the exhaled breath of the individual. Other cats and dogs are known to detect that a person is close to death from their breath. So chemical markers for disease and death are also present on our breath, thanks to the distribution function of the Triple Energizer. Our constantly slightly moist breath is a representation of the mist from the Lungs within our Upper Heater, thanks to the ubiquitous presence of the Triple Energizer. Due to water diffusing across the moist surfaces of breathing passages and alveoli in the Lungs, exhaled air has a relative humidity of 100%. So along with urine removed through the Bladder and sweat secreted through the couli (pores) in our skin, we also regulate water balance through our exhaled breath and also in our feces.

14.14 Three 'Big Qi' Flow Inside the Spaces (Cou) throughout the Sanjiao Network (Li) of the Body

Regarding the flow of Qi inside the spaces (cou) within the Sanjiao network, in the article titled 'The Location and Function of the Sanjiao', the authors Qu Lifang and Mary Garvey (43) cite Difficulties 31 and 36 of the *Nanjing*, which explains that Qi starts and ends in the Sanjiao, which is in charge of allowing the three Qi to pass through the five Zang and six Fu. The authors then state, '"Big qi" and "three qi" refer to yuan qi, zhen qi, and wei qi. Whilst wei qi is concerned with defence against pathogenic influences, the yuan qi and zhen qi transmit the functional aspect of jing-essence from ming men to the zangfu and body tissues. Yuan, then, and wei qi circulate and converge throughout the body, and the Sanjiao supplies a site and thoroughfare for this.'

Here again, it makes perfect sense that the three big qi—that is yuan qi, zhen qi, and wei qi—all flow through the Primo Vascular System's hollow highly organized conduits (cou) which pervade the body as an organized network (li). From the superficial skin (where the cou are the pores) to the couli network of interstitial spaces that pervade the body, the couli network communicates with every single cell. From the synovial joints of the numerous bones throughout the body to the deepest organs, the different forms of fluid and qi are directed to the required locations as required. This way, the three big qi can perform their respective functions, all under the control of the ubiquitous omnipresent Connective-Tissue Metasystem, aka the Triple-Energizer Metasystem.

14.15 The Protective Properties of Yuan Qi, Zhen Qi, and Wei Qi within the Body

Regarding the protective properties of yuan qi, zhen qi, and wei qi within the body, authors Qu Lifang and Mary Garvey (43) state:

> Yuan qi, zhen qi and wei qi therefore are essential for the body's response to pathogenic influences. Because the Sanjiao's qi passageway ensures the ascending and descending, coming in and going out of qi, the three qi together fill the spaces and cavities of the Sanjiao to maintain normal function and health. When the three qi are strong, the couli are compact, the sheaves of muscle are smooth, the five zang and six fu are in harmony, and evil qi cannot attack and penetrate the body.

However, the authors note that should one of the three big Qi become deficient, the power of zheng qi is weakened. This vacuity loosens the couli and causes dysfunction both at the superficial level of the flesh and muscle and at the deeper level of the Zangfu, and Pathogenic Qi can derive from either externally or internally. Thus, the Sanjiao is a potential gateway for external Pathogenic Qi to enter and attack the body, and further, it constitutes the actual battlefield where warfare is fought between pathogenic Qi and antipathogenic Qi either at the superficial level or in more severe conditions at a deeper level. Importantly, the authors note that a close relationship exists between the opening and the closing of the couli and the resultant symptoms of fever and sweating.

The authors (43) quote *Neijing Suwen*, Chapter 42, which says, 'The spaces and textures are opened to make the body shiver, and they are closed to make the body hot and stuffy.' They proceed to explain,

'Pathogenic cold is one of the six external evils. The nature of cold evil has the effect of contracting and stagnating, and this effect often closes the couli at the level of the skin causing fever without sweating.' In this context, the couli pertain to the pores of the skin, which can be opened or closed and where the body fluids are emitted.

14.16 The Sanjiao's Water Passageway throughout the Body Produce Sweat and Urine

Regarding the beneficial aspects of the various body fluids controlled by the Sanjiao, in the article titled 'The Location and Function of the Sanjiao', the authors Qu Lifang and Mary Garvey (43) explain that the body fluids act as the medium for Qi transportation and for waste removal via sweating and urinating. The Sanjiao's delicate fluid balance mechanism that regulates sweating and urination has the secondary function of cooling the body and regulating the body temperature. The authors (43) go on to cite from Chapter 36 of *Neijing Lingshu*, where it says, 'Cold weather or thin clothes make water transform into urine and qi [yang qi to keep the body warm], hot weather or thick clothes make water transform into sweat.' The authors note that because the skin is the largest organ in the body, being definitely larger than the Bladder, Chinese medicine rightfully observes that sweating consumes more yang qi than urination, and thus TCM stresses and cautions that excessive sweating depletes the yang qi. Then showing the omnipresence of the Triple Energizer and its many influences over fluid metabolism throughout the entire body, the article states, 'The Sanjiao's water passageway co-operates downward with the Bladder's excretory function to balance body fluid, and it co-operates externally with the couli at the level of the skin and flesh to balance body temperature by sweating.'

14.17 The Regulation of Temperature and Body Fluid Distribution throughout the Body

Discussing the regulation of temperature and body fluid distribution throughout the entire body further, the authors (43) quote from Chapter 47 of the *Neijing Lingshu*, where it notes that the Kidney coordinates the Sanjiao and the Bladder and that the Sanjiao and

Bladder react to the spaces, textures, and soft hair present on the body. The article proceeds to explain that, after the couli open, sweating occurs, whereby water seeps out via the skin and soft body hair. While the couli are closed, water is directed to the Bladder and excreted from the body as urine. The former process especially relates to temperature regulation, while the latter process pertains mainly to body fluid regulation. The article stresses that the *qihua* function of Kidney yang is crucial in the entire process. Chapter 47 of the *Lingshu* discusses the close energic relationship of the Sanjiao with the Bladder and the Kidney. Due to this intimate relationship between the Sanjiao, the Bladder, and the Kidney, it makes perfect sense that the lower he-sea points for the Sanjiao (Weiyang BL-39) and the Bladder (*Weizhong* BL-40) should be side by side at the popliteal fossa and that the back-shu points for the Sanjiao (*Sanjiaoshu* BL-22) and the Kidney (*Shenshu* BL-23) should be side by side in the lumbar region.

14.18 Body Toxins Are Collected in the Chin Liquid and the Blood and Eliminated in Sweat

In the *Nan Ching* text, regarding clouds and precipitation in the commentaries on the 31st Difficult Issue on pages 354–355 of Unschuld's (1) translation of the *Nan Ching*, Yeh Lin states, 'There it [i.e. water from the Lower Burner] evaporates like steam and is transformed into influences moving up again, where they become the chin [liquids], the yeh [liquids], and the sweat.'

In Unschuld's (1) translation of the *Nan Ching* text, the 34th difficult issue on page 367 asks, 'Each of the five depots has a [specific] sound, complexion, odor, and taste. Can they be known?' Regarding this question, in the Commentaries, Hsü Ta-ch'un states, 'Sweat is an external sign of blood. The heart masters the blood. Hence, [its liquid] is sweat.'

In the *Nan Ching* text in the commentaries on the 35th Difficult Issue on page 378 of Unschuld's (1) translation of the *Nan Ching*, Li Chiung states, 'As to the chin and yeh liquids, that which is emitted as sweat and leaves through the pores is the chin liquid. Those of the liquids that flow into hollow cavities where they stagnate and do not move are the yeh liquids. Chin and yeh liquids are contained in the bladder.'

In the *Nan Ching* text in the commentaries on the 48th Difficult Issue on page 452 of Unschuld's (1) translation of the *Nan Ching*, Hsü Ta-ch'un states, 'Ch'u ('move toward the outside') means that the essential influences are drained toward the outside as, for instance, through sweating, vomiting, or diarrhea. All instances when some thing moves from inside toward outside are meant here.'

In the first quote above, Yeh Lin advises, 'Water from the Lower Burner evaporates like steam and is transformed . . . [to] become the chin liquids, the yeh liquids, and the sweat.' So three different fluid types are synthesized from water from the Lower Burner—namely, the *chin* liquids, the *yeh* liquids, and the sweat. From the second quote above, Hsü Ta-ch'un shows that because the heart masters the blood and sweat is an external sign of blood, then the Heart's liquid is sweat.

While in the first quote above Yeh Li states that water from the Lower Burner is transformed into three separate liquids (chin liquids, yeh liquids, and sweat), in the third quote above, Li Chiung advises that the sweat that leaves through the pores is the chin liquid. So it appears that sweat is a subcategory of chin liquids. In the fourth quote above regarding depletion/repletion disorders where other scholars discuss evil influences and internal damages within the body, Hsü Ta-ch'un states that such evil and damaging influences are 'drained toward the outside as, for instance, through sweating, vomiting, or diarrhea'.

Obviously, vomiting and diarrhea are ways for the body to eliminate chemical poisons and toxic microorganisms from the body. In western medicine, sweating is not considered to be a means of toxin elimination from the body. I believe that this is the case and that the sweat is designed to eliminate toxic matter through the skin as it is more toxic and more acidic than other chin liquids. Because the ancient Chinese medical scholars believed that sweat is derived from the blood, I trust that the ancient medical scholars were aware that sweating is nature's way of eliminating toxins from the blood and the body safely through the very large surface area of the skin. But is there any scientific proof that this ancient belief is actually true?

14.19 Recent Research Confirms that Sweating Eliminates Toxic Heavy Metals and Petrochemicals

Other than the obvious role of regulating body temperature, in the 2015 article titled 'Research Confirms Sweating Detoxifies Dangerous Metals, Petrochemicals', the author, Sayer Ji (47), reported that recent research has confirmed that sweating facilitates the elimination of accumulated toxic heavy metals and petrochemicals from the body. He reported on a groundbreaking 2011 study published in the *Archives of Environmental and Contamination Toxicology* which explored the effects of bioaccumulated toxic elements within the human body and their method of excretion. The article reported that when blood, urine, and sweat were analyzed from the same individual, test results were very inconsistent, and toxic elements were detected at varying levels. Results showed that sweat was the preferred route for excretion of many toxic elements, and remarkably, some of the toxic elements that were present in the perspiration were not present in their serum. It thus appears that induced sweating is a powerful method for the elimination of many toxic substances from the human body. The conclusion was that biomonitoring blood and/ or urine for many toxic substances may grossly underestimate the actual body burden of such toxic substances.

Researchers from the Children's Hospital of Eastern Ontario Research Institute, Ontario, Canada, performed a meta-analysis of 24 studies on toxicant levels in sweat. The findings were published in the *Journal of Public and Environmental Health* in the article entitled 'Arsenic, Cadmium, Lead, and Mercury in Sweat: A Systematic Review'. Ji (47) reported that the researchers made the following observations:

- In individuals with higher exposure or body burden, sweat generally exceeded plasma or urine concentrations, and dermal could match or surpass urinary daily excretion.
- Arsenic dermal excretion was several-fold higher in arsenic-exposed individuals than in unexposed controls.

- Cadmium was more concentrated in sweat than in blood plasma.
- Sweat lead was associated with high-molecular-weight molecules, and in an interventional study, levels were higher with endurance compared with intensive exercise.
- Mercury levels normalized with repeated saunas in a case report.
- The researchers concluded, 'Sweating deserves consideration for toxic element detoxification.'

This research confirmed that blood and urine analyses fail to reveal the true extent of loading of toxic heavy metals within the body.

14.20 Sweating Also Removes the Insidious Petrochemicals BPA and Phthalates

Heavy metals are not the only toxic substances eliminated through the skin. The article (47) reported that two 2012 studies determined that sweating heightens the elimination of hazardous endocrine-disrupting petrochemicals from body tissues. The first study ($n = 20$) determined that the ubiquitous petrochemical bisphenol A (BPA) was excreted from the body through sweat even though, in some individuals, no BPA was detected in their blood serum or urine samples. This clearly confirmed that sweating is very important in the removal of toxic BPA that has bioaccumulated in body tissues.

The second study ($n = 20$) found that the endocrine-disrupting plasticizer agent phthalate was detected in concentrations twice as high in the sweat of test subjects compared to their urine concentration. Further, in several test subjects, phthalate was found in their sweat but not detected in their blood serum. This is suggestive of the possibility of phthalate retention and bioaccumulation within the body tissues.

This recent (2011, 2012) research proves that sweating instigates more than simply a cooling function for the body. The research proved that sweating positively eliminates heavy metals (arsenic, cadmium, lead, and mercury) and ubiquitous toxic petrochemicals, including Bisphenol A (BPA) and phthalate compounds, both of which are endocrine disruptors. Edgar Allan Poe once said, 'The best things in life make you sweaty.' The article noted that natural medical practitioners have argued for decades that the skin is the biggest organ of elimination and that very often skin problems are the outcome of chronic toxicity throughout the entire body. Modern science and western allopathic medicine is only now catching up with these common-sense observations utilized by complementary medical practitioners for a long time.

Summary of Chapter 14

The Cou Li are the anatomical expression of the small cavities associated with the San Jiao. Some authorities suggest that *Cou Li* means 'pores' in much of the *Su Wen*. Thompson (46) explains that the Cou Li spaces of the body include 'the spaces that surround the internal organs known as the interstitial spaces' and also 'the spaces between the sinews and bones where the synovial fluid moves'. Nielsen (20) believes that 'the correspondence of the San Jiao to the Cou Li is essential to understanding the model for the curative effect of Gua Sha, acupuncture and any hands-on physiotherapy'. She cites the *Ling Shu*, Chapter 47, which says, 'The Bladder and Triple Heater (San Jiao) have their correspondence and resonance in the most external structure of the body, the Cou Li.' The larger cavities of the 'three burning spaces' (San Jiao) and the smaller dispersed cou spaces merge in their functionality to form a cohesive network interconnecting the zangfu and their associated tissues, including muscles, flesh, and the skin.

The Cou Li is the outermost defensive structure in the skin and links the exterior of the body with the internal Organs. I propose that each Cou Li is just one more small 'bag' of relatively superficial differentiated fascia making up the omnipresent continuum of hydrophilic connective tissue that is an integral part of the San Jiao and that the life-sustaining energized EZ water generated within this hydrophilic structure is responsible for powering the battery that drives the superficial defense mechanism responsible for keeping pathogenic forces from entering the body and causing disease processes. Anatomically, even as recently as 2011, the most commonly investigated superficial connective tissue layer of the body is not well understood by medical researchers and is prone to conflicting theories as to its structural elements.

There is a connection between the triple heater, the kidneys, the bladder, and the couli to ensure the appropriate circulation and irrigation of the liquids and qi in the body. While the *Nei Jing* states that the lower jiao connects the Small Intestine to the Urinary Bladder, the majority of authorities believe that this drainage system is a figurative 'permeation'. Perhaps it is or perhaps the ancients were once again aware of a literal drainage system via the lower jiao, which is, after all, a system of membranes that are known to house and manage the passageway of a large proportion of the water present in the body. Also the drainage system may well be due to an as yet undefined Primo Vascular System network. The small cou spaces may well include tubular conduits within the Connective-Tissue Metasystem that carry the many different fluids (yuan qi, zhen qi, wei qi, etc.) throughout the body. Perhaps the cou spaces involve the Primo Vascular System, which is known to be a highly complex system composed of hollow microtubules that do carry various complex fluids throughout the entire body and form a cohesive network interconnecting the zangfu and their associated tissues, including muscles, flesh, and the skin. I believe that my suggestion makes perfect sense as the small cou spaces or hollow tubular conduits within the Connective-Tissue Metasystem would carry the many different fluids throughout the entire body due to the li network of the cou tubules with their highly complex and extremely ordered structure that would 'thereby connect upper and lower, external and internal, shallow and deep' due to the structured organization of the cou-li.

Using fluorescent nanoparticles, modern researchers have traced the Primo Vascular System conduits from the superficial skin surface to deep internal organs. For example, recent research in 2009 (34) revealed 'the flow of the primo fluid in a certain path was demonstrated... [to flow]... from the rat acupoint BL-23 in the dorsal skin to the PVS on the surface of internal organs'. Further research in 2010 (34) discovered that 'when Chrome-hematoxylin and fluorescent nanoparticles were injected into testis they were found in the PVs on the organ surfaces between the abdominal cavity and the abdominal wall'. This data certainly establishes proven defined pathways from the skin surface to deep organs via the Primo Vascular System. The book (7), on page 31, states there is a 'close relation between the PVS and the fascia'. I believe that the PVS throughout the entire body will be found to be intimately associated with and directly or indirectly connected to the omnipresent Connective-Tissue Metasystem. I propose that this amalgamation forms the organ complex which is synonymous with the Triple-Energizer Metasystem. I further believe that the PVS will be found to be responsible for body tissue types and zangfu relationships as per the example given by Qu Lifang and Mary Garvey (43) above, where they stated that 'the flesh and muscles belong to the Spleen, whilst the skin and soft body hair, at the most external layer of the body, belong to the Lung'.

Chapter 47 of the *Neijing Lingshu* says that the 'kidney co-ordinates the Sanjiao and Bladder, the Sanjiao and Bladder respond to the spaces, textures and soft hair on the body'. The authors (43) proceed to explain that 'when the couli open, water comes out via the skin and soft body hair by sweating; when they close, water is excreted from the body via the Bladder'. Here is a classic case of the couli of the superficial skin communication with the Bladder organ to fulfill two roles of the Triple Energizer involving fluid balance and regulating body temperature.

Here again, it makes perfect sense that the three big qi—that is, yuan qi, zhen qi, and wei qi—all flow through the Primo Vascular System's hollow highly organized conduits (cou), which pervade the body as an organized network (li) from the superficial skin (where the cou are the pores) to the deepest organs, allowing the different forms of qi to be directed to the required locations so that they can perform their respective functions all under the control of the ubiquitous omnipresent Connective-Tissue Metasystem, which is essentially the Triple-Energizer Metasystem.

Recent research (2011, 2012) proves that sweating instigates more than simply a cooling function for the body. The research proved that sweating positively eliminates heavy metals (arsenic, cadmium, lead, and mercury) and ubiquitous toxic petrochemicals, including bisphenol A (BPA) and phthalate compounds, both of which are endocrine disruptors. Sweat can concentrate arsenic up to 10 times more than blood, cadmium up to 25 times more than blood, lead up to 300 times more than blood, and mercury somewhat more than blood, leading to effective elimination. Thus, it can be seen that the couli in the form of sweat glands have an important role of keeping the blood clean and protecting the body by eliminating toxic heavy metals and petrochemicals, which are excreted through the skin in the sweat.

CHAPTER 15

The Triple Heater Has a Direct Relationship with the Source Points

Introduction

How are the acupuncture meridian source points (Yuan points) associated with the Triple Heater? In the book *Heart Master Triple Heater*, while discussing *Nan Jing* Difficulty 66, Elisabeth Rochat de la Vallée (11), on page 116, says, 'This *original qi*, through the intermediary of the triple heater, *crosses and impregnates the five zang and six fu*' (emphasis is mine). She explains that the very embodiment of the origin is intimately attached to and associated with the Triple Heater's functionality, and consequently, there are specific places along the five zang and six fu channels where the Triple Heater permits communication with the Original Qi. These places along each of the zang and fu meridians are called the Source Points or Yuan Points, and for this reason, *Nan Jing* Difficulty 66 advises that you needle the source point whenever the five zang and six fu are ill.

In this Chapter, I will discuss in detail how the original Yuan Qi is circulated throughout the body inside the meridians and how stimulation of the Yuan acupoints of the Main meridians will correct imbalances that occur within the corresponding hollow Fu and solid Zang organs.

15.1 The Source Points Have a Very Close Relationship with the Triple Heater

While discussing the connection between the three heaters and the source points, Elisabeth Rochat de la Vallée (11), on page 117, cites *Nan Jing* Difficulty 62, which essentially teaches that the six Fu and the Three Heaters unite in a single breath. She explains that because the omnipresent Triple Heater is the agent of distribution of the numerous forms of qi throughout the body, its intimate connection with the Original Qi allows for the unified representation and manifestation of Original Qi at the sixth acupoint along the yang acupuncture meridians, which are thus named the Source points or Yuan points.

As the omnipresent maze of energized connective tissues known as the Triple Energizer surround and envelop every single cell and organ in the body, obviously the representation of the original Yuan Qi, the Triple Heater, 'crosses and impregnates the five zang and six fu' and, by some programmed infrastructure, makes an extra-large presence of the Yuan Qi congregate at the surface at the Yuan point along every zang and fu meridian. Subsequently, when any of the zang or fu is ill and maladjusted, needling the corresponding Yuan point will introduce a surge of healing Yuan Qi into the ailing meridian that will nurture the ailing organ.

15.2 The San Jiao as a Pathway for Source Qi

A major difference between the representation of the San Jiao in the *Nei Jing* and in the *Nan Jing* is the fact that the *Nan Jing* (in Question 66) highlights the San Jiao as a pathway for source qi to the source point of each meridian. Regarding the distribution of Yuan Qi throughout the body by the San Jiao, Yongping Jiang (8), the author of the *Journal of Chinese Medicine* article titled 'The San Jiao: Returning to the Nei Jing (A Modern Explanation of Original Theory)', states, 'The source qi comes from the Kidney, and is the root of the life; the San Jiao brings source qi from the Kidney to the source points of the channels.'

15.3 Integrity of the Yuan Qi within the Body Can Be Monitored at the 24 Yuan Points

There are over a thousand main meridian acupuncture points and extra-meridian acupuncture points in defined locations according to TCM theory. Collectively, the Yuan points are the most energized acupuncture points in the body and act as an accurate gauge of the energy flowing throughout the entire meridian. There are many analytical devices that can diagnose the status of qi in the body, such as electrodermal screening, which measures your electrical skin resistance to gain information about your qi flow. Yuan points are significant in Japanese Ryodoraku acupuncture theory, and measuring their potential is also the way in which the AcuGraph™ determines the status of each of the 12 Main Meridians on the right- and left-hand sides of the body so that the meridians can be balanced by the appropriate placement of needles in acupoints, which are determined by the AcuGraph™ software. Yuan (Source) points are the most energetically active points on the meridians, with four to six times more energetic activity than other acupuncture points. The Yuan point of a meridian is intimately connected with the corresponding internal organ and with the general energetic state of the meridians. Source point graphs tend to show greater variability than Jing well graphs and therefore tend to illustrate imbalances more readily than Jing well graphs.

For nearly ten years, I have personally been using the Meridia Technology™ AcuGraph™ instrument to measure the ting points or yuan points on all 24 main meridians to determine the flow of qi throughout the body. The chart generated allows my patients to 'see the qi' such that deficient meridians are shown as blue, excess meridians as red, and balanced meridians are shown as green. Where there is a large left–right variation (a split) for the same bilateral acupoint, that outcome is displayed as purple. As the patient improves with subsequent acupuncture treatments and the meridians are more balanced, there are more green bars on the chart.

Meridian Balance Report

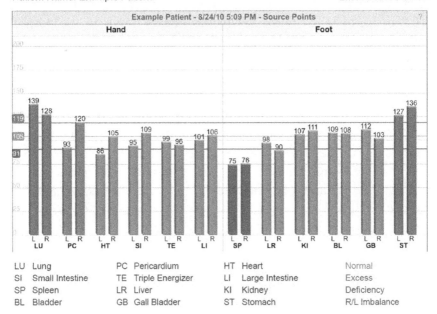

Figure 4. Acugraph meridian balance report chart for source points (permission from Miridia Technology Inc., Meridian, ID 83646).

15.4 The Yuan Qi or P-Microcells Are Indeed 'Persistently Regenerated'

Regarding the nature of the Yuan Qi, Arya Nielsen (20), the author of the book *Gua Sha: A Traditional Technique for Modern Practice*, on pages 34–35, explains that the term *Yuan* means 'original, primordial source' and that it is our primal life force donated by Heaven, engendered through our parents, stored and accumulated in the Kidneys, and circulated through our body by the San Jiao. Nielsen notes that the Chinese ideogram for Yuan embodies three springs gushing forth from a cliff and poetically suggests that an image emerges of the Fire fu organ, the San Jiao, 'gushing' Water within the three body cavities accessed at the three respective acupoints CV 17, CV 12, and CV 6. She remarks that the San Jiao Fire is the original *Yuan Yang* of the Gate of Life (Ming Men) and that the original Yuan

Yang Fire is the original component of all Yang function. Similarly, Water is the Yuan Yin and comprises the original component of all Yin substances and fluids, and alluding to the 31st Difficult Issue of *Nan Ching*, she notes, 'When Fire meets Water in the body, a transformation into Qi takes place.' She cites from Larre and Rochat de la Vallee (1992) when they stated, 'There is no vital Water without Fire to transform it and no Fire of life without Water to fix and express it. . . . *The Yuan not only suffuses the body with motive force but is a catalyst in the formation of Ying, Wei, Qi and Blood. Yuan is itself, in turn, "persistently regenerated" by those very products*' (emphasis is mine). Larre and Rochat de la Vallee explain further that the Yuan Qi congregates in the Kidneys at the Ming Men, and via the San Jiao, Yuan Qi reaches the entire body. They note that Yuan Qi can be directly accessed at the Yuan Source points of the twelve organs at the extremities or at the front Mu acupoints CV 17, CV 12, or CV 6.

So we see here that the original, primordial source, or Yuan Qi, is the life force that we derive from our parents, which is stored in the Kidneys and circulated throughout the body by the San Jiao. The Yuan Qi energizes the body and catalyzes the formation of Ying, Wei, Qi, and Blood as required and where required to maintain optimal biological functioning to maintain life and health. I believe that the Yuan Qi is 'persistently regenerated' along the lines of the 2012 book (7) where, on page 23, Jong-Su Lee explains the general biological concepts behind Bong-Han Theory. He explains that Bong-Han Theory teaches that all old senescent cells get broken down by enzymes, whereby the chromosomes are salvaged and recycled and circulate throughout the body in the Bong-Han system. After receiving some energy, they regenerate new cells, and they regenerate damaged tissues. This Bong-Han system is intimately bound to the ubiquitous Connective-Tissue Metasystem, and I believe that this amalgamation is another name for the Triple-Energizer Metasystem. Thus, the 'original influences' (Yuan Qi), namely the harvested cellular chromosomes consisting of DNA, protein, and RNA are recycled continuously to correct and heal damaged cells throughout the body. Via this continual recycling process, the Yuan Qi or P-microcells are indeed 'persistently regenerated' after they 'receive some energy' from the EZ water battery in intimate proximity within the connective-tissue/Triple-Energizer Metasystem. I have discussed this theory in great detail in another section of this book.

Summary of Chapter 15
The Kidneys store the Yuan Qi. While the Yuan Qi permeates the entire body with motive force, it also catalyzes the synthesis of Ying, Wei, Qi, and Blood. It is no surprise that the original, primordial source of the life force (Yuan Qi) is continually renewed and continuously circulated through the entire body to every cell and structure therein to maintain life and optimal wellness. Obviously, for the extremely exacting circulation of Yuan Qi to be constantly orchestrated throughout the entire body so that omnipresent high fidelity is maintained, it requires a highly skilled conductor. The TCM medical classics repeatedly show that the conductor of the orchestra of life is the San Jiao. Because the San Jiao is not a relatively small defined organ similar to the solid Tsang organs (including the Heart, Liver, Spleen, etc.) or the other hollow Fu organs (including the Small Intestine, Gallbladder, Stomach, etc.) and is actually spread across our entire body mass and surface as discussed elsewhere in this book, I propose that the San Jiao is more appropriately defined as the Triple-Energizer Metasystem.

Subsequently, it is most appropriate that the Triple-Energizer Metasystem is the orchestra conductor that carries and circulates the primordial Yuan Qi throughout the body and especially to the defined Yuan (Source) points of each of the 12 Main meridians on each side of the body, plus the three abdominal Yuan points at CV 17, CV 12, or CV 6 at the front of the body. Because the Main meridian Yuan points connect directly to the associated Tsang and Fu organs, the Yuan points are very beneficial to use to tonify the associated organ directly when they are treated with acupuncture or moxa. I discuss the use of the Yuan points in more detail in a later Chapter of this book.

CHAPTER 16

The Triple Energizer in Relation to the Shao Yang

Introduction
In 2012, acupuncturist Kimberly Thompson (46) posted an article entitled 'What the Heck Is a Triple Energizer Anyway?' on the miridiatech.com website. With regard to the Triple Energizer in relation to the Shao Yang, she explained that the Shao Yang is composed of the Triple Energizer and Gallbladder. She explained that in TCM the Shao Yang division is a pivot between the superficial *Tai Yang* level and the deeper *Yang Ming* level. The Tai Yang level pertains to the outer surface of the body, while the Yang Ming level correlates with digestion. The Yang Ming level controls external movement to the internal organs. Regarding this regulatory pivotal aspect of the Shao Yang level, she advises that if the intermediate Shao Yang level doesn't regulate smoothly, then Tai Yang problems at the exterior surface of the body can arise with immunity and Yang Ming problems can arise internally with digestion.

16.1 When You Think of the Shao Yang, You Should Think Qi and Fluids
With regard to Shao Yang involvement in the body, author Kimberly Thompson (46) explained the following four points:

1. spaces which surround the internal organs
   - The Triple Energizer is specifically in charge of regulating qi and fluids.
   - *It isn't really an organ.* Instead it has to do with the qi and *fluids* which move through the *spaces* that surround the internal organs known as the interstitial spaces.
   - When the Shao Yang is compromised, heat and qi become clumped in the interior of the body.

2. sinews and bones
   - The Shao Yang is related to movement in the spaces between the sinews and bones where the synovial fluid moves.
   - Problems occur in the joints because of poor fluid circulation.

3. poor movement issues
   - Thompson explained that problems within the Shao Yang level included poor movement issues.
   - These include painful constipation due to Stagnant Qi in the Triple Energizer, and high blood pressure, conjunctivitis, dizziness, and tinnitus caused by Stagnant Qi with Heat.

4. channel palpation to diagnose pathological conditions
   - Shao Yang involvement in the body pertained to channel palpation to diagnose pathological conditions.
   - She noted that checking for bumpiness along the TE channel on the arm and along the GB channel on the outer leg is diagnostic. Areas of bumpiness indicate Stagnant Qi where Qi is not flowing smoothly through that region.
   - The presence of tenderness and bumps together indicates that Heat is involved. To treat stagnant Qi, TE 6 and GB 34 are needled to dredge the channel and eliminate clumping along the channel.
   - To treat stagnant Qi with Heat, TE 5 and GB 41 are needled. Where there is Heat and Stagnation, the Heat must be cleared first.

Note that in her article, regarding the Triple Energizer, Thompson categorically states, '*It isn't really an organ*. Instead, it has to do with the qi and fluids which move through the spaces that surround the internal organs known as the interstitial spaces' (emphasis is mine). Thompson further states, 'The Shao Yang is related to movement in the spaces between the sinews and bones where the synovial fluid moves.' I suggest that there are two very important issues discussed here. The first involves the suggestion pertaining to the Triple Energizer that 'it isn't really an organ'. I think it is a tragedy that such an important organ, which I suggest is the very first TCM organ to develop during embryogenesis, isn't even recognized as an organ just because of its extraordinary shape and being enshrouded in mystery. It must feel like a leper!

The second issue is that the Triple Energizer is in charge of regulating qi and fluids throughout the body via the couli, which include all 'the spaces that surround the internal organs known as the interstitial spaces' and also 'the spaces between the sinews and bones where the synovial fluid moves'. Problems in these systems can be treated through the Shao Yang division of the Six Divisions or Six Stages of TCM.

16.2 Mobilizing Nature of Shao Yang Empowers the Triple Energizer to Break Blockages

In the book *Heart Master Triple Heater*, Rochat de la Vallée (11), on page 51, points out that Chapter 8 of the *Su wen* states, 'The triple heater is responsible for the opening up of passages and irrigation. The waterways stem from it.' She then states, 'If we look at Su wen chapter 8, the first thing we see is that the triple heater can *break through passages and pathways*. It has the concentrated strength of shao yang which allows passing through obstacles and a clearing of the ways so that all the irrigation and streams in the body *can circulate freely and keep the qi moving*.' Here she shows that the property of the concentrated and intense strength and motivating force of the Shao Yang division is responsible for empowering the Triple Heater to 'break through passages and pathways' so that obstructions, 'stuckness', coagulations, and blockages are cleared away and free communication can be restored. The Triple Heater also powers fluid flows so that circulating nourishing body fluids can smoothly irrigate all the cells and tissues within the body.

Pollack (27) has provided very convincing evidence that positively charged water molecules inevitably arise from the presence of exclusion zones (EZs) and that those positively charged water molecules have very interesting and diverse properties. He says, 'Their sundry actions include reducing friction, *wedging surfaces apart* . . . running batteries, driving catalysis, and *powering fluid flows*' (KL 3670–3672, emphasis is mine). All these properties of EZ water exactly describe the actions attributed to the Triple Energizer in TCM. Is that a coincidence? I think not!

Summary of Chapter 16
Within the TCM Six Division System, the Shao Yang division is composed of the Three Heaters and the Gallbladder and is positioned halfway between the internal and external level, and subsequently, the symptoms can go backwards and forwards, sometimes for years (e.g. malaria). A Qi obstruction in the Gallbladder channel can produce alternating chills and fever. This division is more superficial than the Yang Ming division. Due to the superficial location of diseases in this division, the body fluids are not severely injured. Because of the five-Phase relationship of the Gallbladders with the Liver, symptoms associated with Liver/Gallbladder obstruction can occur and include blurred vision, frequent sighing, sensation of a lump in the throat and/or trouble swallowing, dry throat, feeling tense and irritable, moodiness, feeling melancholy, abdominal distention, epigastric pain, alternating constipation and diarrhea, irregular menstruation, joint problems, etc. These symptoms involve the stagnation of qi and/or fluids. Heat is often involved, and stagnant qi with heat can cause tinnitus, conjunctivitis, dizziness, and high blood pressure. Lumps and bumps along the Gallbladder meridian indicate blockages in Shao Yang.

The mobilizing property of the reasonably superficial Shao Yang division is responsible for empowering the Triple Heater to remove obstructions and coagulations and open the fluid passages and pathways throughout the entire body so that free communication can be restored and nourishing body fluids can circulate smoothly to irrigate all the cells and tissues within the body.

CHAPTER 17

The Diaphragm (Ke) Membrane Screening off the Turbid Influences

Introduction
In the commentaries (page 350) for the 31st Difficult Issue in Unschuld's (1) translation of the *Nan-ching*, Hsü Ta-ch'un states, 'Ke ("diaphragm") stands for ke ("to screen off"). Below the heart is a membrane screening off the turbid influences. It is called ke.' Here, Hsü Ta-ch'un refers to a membrane screening off the turbid influences below the heart at a location regarded by authorities (3, 21) (in the glossary under *San Jiao*) as a demarcation between the upper energizer and the middle energizer. Hence, the inference is that the membrane constitutes part of the Triple Energizer.

17.1 The Diaphragm Membrane below the Heart
In the commentaries (on page 359) for the 32nd Difficult Issue in Unschuld's (1) translation of the *Nan-ching*, Hua Shou said, 'Everybody has a diaphragm membrane below his heart. It is attached all the way round to the backbone and to the flanks. It provides a barrier screening off the turbid influences and preventing their steaming up to heart and lung.' This comment is similar to the statement above but is more detailed in the distribution of the layer of connective tissue (membrane).

In the commentaries (on page 359) for the 32nd Difficult Issue in Unschuld's (1) translation of the *Nan-ching*, Yeh Lin stated, 'Everybody has a layer of a diaphragm membrane below his heart and lung and above all the [remaining] depots. *It is thin like a fine net*. It ascends and descends following exhalation and inhalation. It provides a barrier for the turbid influences, preventing their steaming up to heart and lung' (emphasis is mine).

The diaphragm is a muscle about 5 mm thick and most likely thicker than a fine net. Is the diaphragm membrane the literal diaphragm or a surrounding fascia, the diaphragmatic peritoneum? In the 'Fascia Research Congress: Evidence from the 100 Year Perspective of Andrew Taylor Still' (48) held in 2012, it was reported that 'fascia is intimately involved with respiration' on page 1. Delegates were advised that projects on respiration and fascia were strongly encouraged to be submitted as abstracts for the Fourth International Fascia Congress to be held in Orlando, Florida, on 15–17 October 2015 on page 6. Whatever the case, according to many experts (3, 21), this location is the separation between the upper energizer and the middle energizer.

17.2 The Diaphragm Membrane below the Heart Most Likely the Diaphragmatic Peritoneum

In commentaries on the 31st Difficult Issue on pages 353–355 of Unschuld's (1) translation of the *Nan Ching*, Yeh Lin relates, '[It is stated that] the upper [section of the Triple] Burner is located below the diaphragm because its upper layer is *attached* to the lower layer of the diaphragm' (emphasis is mine). Yeh Lin is saying that the upper section of the Triple Burner is located below the diaphragm and is not the diaphragm itself but is *attached* to the diaphragm. This layer is most likely the diaphragmatic peritoneum.

Summary of Chapter 17

From the information above, Yeh Lin appears to sum up the meaning of *ke*, the membrane below the heart that screens off the turbid influences, thus preventing them from steaming up to reach the heart and lung. Yeh Lin says that the upper section of the Triple Burner is located below the diaphragm and is not the diaphragm itself but is attached to the diaphragm. This protective layer is most likely the diaphragmatic peritoneum, which is a membrane.

CHAPTER 18

The Triple Burner Is Energized by Fire and Light From the Sun

Introduction

The 37th Difficult Issue asks the question 'The influences of the five depots, where do they originate, where do they pass through. Can that be known?' In commentaries on the 37th Difficult Issue on page 389 of Unschuld's (1) translation of the *Nan Ching*, Hsü Ta-ch'un states:

> Constructive and protective [influences] proceed through [all] depots and palaces. There is absolutely no such doctrine that they proceed through the depots but do not proceed through the palaces. . . . The original text of the [*Nei-*] *ching* states: 'The Yellow Emperor asked: "The influences proceed only through the five depots; they do not circulate through the six palaces. Why is that so?" Ch'i Po replied: "There is no [place] to which the influences do not proceed! Like the flow of water, like the movement of sun and moon, they never rest. Hence [when the influences move through] the yin vessels they circulate through the depots, [when they move through] the yang vessels they circulate through the palaces. It is like a ring without end: nobody knows its break: it ends and begins anew. The influences pour into the depots and palaces internally; they moisten the pores externally".

In the discussion about the influences of the five depots (solid Zang organs) and their origin and where they pass through, the learned commentator Hsü Ta-ch'un likens the continual movement of the sun to the circulation of the Triple Energizer's influences through the Zang organs. How is the movement of the sun involved with the function of the Triple Burner?

18.1 The Sun Is Intimately Involved with the Triple Burner's Influences on Our Life

The question asked in the 23rd Difficult Issue of the *Nan Ching* was 'Can one be instructed on the measurements of the three yin and three yang vessels of the hands and feet?' In the 23rd Difficult Issue in the commentaries on page 292 of Unschuld's (1) translation of the *Nan Ching*, Ting Te-yung states:

> This [refers to] the rise and fall of the yin and *yang influences of heaven and earth in the course of one year, and to the appearance and disappearance of sun and moon, light and darkness, within twenty-four hours.* Similarly, *man's constructive and protective [influences] proceed through twenty-four sections of conduits and network [-vessels]* before they meet once again with the inch-opening and the jen-ying. The so-called inch-opening is the vessel-opening of the hand-great-yin [conduit]. This hole is called t'ai-yüan (Lung 9). Hence, the [movement in the] vessels meets with the t'ai-yüan [hole]. All the twelve conduits and fifteen network [-vessels] are supplied [with influences] by the Triple Burner; [as long as this continues, a person will] live. Hence, [the influences] start from the central burner and flow into the hand-great yin and [hand-] yang-brilliance [conduits]. (Emphasis is mine)

Note that with such a simple question pertaining to the measurements of the 12 main meridians, Ting Te-yung discusses in even more detail how the sun and moon and light exposure over the earth during the 24-hour daytime cause the Triple Burner to supply 'all the twelve conduits and fifteen network-vessels' and how, importantly, 'as long as this continues, a person will live'. Every food that we eat (epitomized as grain) owes its existence to the energy therein derived from the rays and photons from the sun. I will discuss this more in detail in this Chapter.

It is extremely noteworthy that while the word *sun* appears nine times in Unschuld's (1) translation of the *Nan ching*, six times, it is in intimate association with the Triple Burner.

18.2 Biophotons and Infrared Energy from the Sun Affect the Triple Burner via Our Skin

In the text above, the very existence of the Triple Burner is associated with the 'yang influences of heaven'—that is, the sun via annual (in the course of one year) and circadian (within 24 hours) cycles. Being compared to the sun and the light, the quote says, 'Similarly, man's constructive and protective [influences] proceed through twenty-four sections of conduits and network [-vessels].' This may allude to the sun endowing the grains consumed with biophotons, which are processed and absorbed, ready for use throughout the entire body, as stated above: 'Hence, [the influences] start from the central burner.' The Stomach, in the central burner, is where the intrinsic nutrient quality of the food (except for McDonald's, of course) along with its energetic biophotons component are macerated and circulated throughout the entire body via the Triple Energizer. Further too is the probability that the Triple Burner system in humans is directly irradiated and energized by sunshine penetrating our skin. It is known that direct sunlight on the skin causes the synthesis of vitamin D_3. There is also the strong likelihood that sunlight affects us through our skin indirectly by the effect of infrared radiation from all the objects that we encounter—including walls, furniture, and dogs, for example—that have absorbed sunlight during the day and have intrinsic infrared light to emit because of their inherent ambient warmth.

18.3 Most Foods Contain a Large Amount of Water; All Natural Foods Contain Biophotons

Note too that all foods are composed of a considerable amount of water. Fruits and vegetables (49) contain large quantities of water in proportion to their weight. When these foods are eaten, the water therein can be absorbed by the body via the Triple Energizer. For example, apples contain 84% water, tomatoes 93%, carrots 87%, Lettuce (iceberg) 96%. The book entitled *The Composition of Foods* (50) advises that cooked grains ready to eat contain a little less water than fruits and vegetables. For example, boiled barley (page 22) and boiled rice (page 26) both contain approximately 70% water; fresh whole eggs (page 30) contain 73.4%; a raw lean beef steak (page 36) contains 68.3%; roasted chicken flesh (page 36) with basting contains 61.1%; a raw lean mutton chop (page 40) contains 67.1%; stewed rabbit flesh (page 44) contains 63.9%; steamed whole bass excluding guts (page 46) contains 73.3%; and even staple bread (page 22) contains 38–40% moisture.

Figure 5. Water splash and fruits. Photo by arcoss. © Depositphotos.com.

Even foods like nuts (50) (page 82) that appear to be relatively dry, contain a reasonable amount of moisture. For example, shelled peanuts contain 4.5%, shelled almonds contain 4.7% water, shelled Brazil nuts contain 8.5% water, shelled walnuts contain 23.5%. Shelled chestnuts contain a whopping 51.7% water compared to other nuts. Various beans (50) (page 84)—whether broad, butter, and haricot—after soaking and when ready for eating, contain 70–84% water.

As can be seen from the data above, all foods contain various amounts of water in their intrinsic composition so that when grains are mentioned in the *Nan Ching*, large amounts of water and fluids often make up a sizable portion of those grains. All the foods discussed above, and indeed all foods, automatically contain a large quantity of biophotons that are derived from the sun. I discuss how foods store biophotons from the sun in great detail in another Chapter of this book.

18.4 Heavenly Yang (Influences of the Sun or Biophotons/Infrared Radiation) Affect San Jiao

In commentaries on the 31st Difficult Issue on page 355 of Unschuld's (1) translation of the *Nan Ching*, Yeh Lin relates:

> Through the inhalation of the heavenly yang, the water of the bladder follows the fire of the heart downward to the lower [section of the Triple] Burner. There it evaporates like steam and is transformed into influences moving up again, where they become the chin [liquids], the yeh [liquids], and the sweat. All of that rests on the principle that when fire meets water, a transformation into influences takes place. The meaning is that heavenly yang [i.e., the influences of the sun] enters earthly yin [i.e., the water in the soil]. The [latter], following the movement of the yang influences ascends and become clouds and rain.

So once again, in the 31st Difficult Issue, the commentator Yeh Lin references that the Triple Energizer is directly affected by 'the influences of the sun', as Ting Te-yung said in his discussion on Difficult Issue 23.

18.5 Precedent for Human Organs Being Controlled and Energized by the Sun and Moon

Regarding the fact that I propose that the functionality of the Triple Energizer is dependent on light and especially from sunlight, are there any precedents in the human body to support that concept?

Yes, indeed, there is. The pineal gland is known to absorb light and be affected by light from the sun and also from the moon and as such is responsible for orchestrating our circadian rhythms and our circannual rhythms. Are there any other organs that might store the essence of the sun and the moon?

18.6 Kidneys Store Essence of the Sun and Essence of the Moon to be Used by the Triple Burner

Regarding the question of whether other organs might be able to store the essence of the sun and the moon, in the commentaries on the 66th Difficult Issue on pages 567–568 of Unschuld's (1) translation of the *Nan Ching*, Yang states:

> The [two] kidneys are distinguished by [their storing] the essence of the sun and of the moon, [respectively. This essence] consists of immaterial influences; it constitutes man's source and root. . . . Hence it is obvious that the cinnabar field is the basis of life. When the Taoists contemplate the spiritual, when the Buddhist monks meditate, they all cause the influences of their heart to move below the navel. That corresponds exactly to what is meant here. Hence [the text] states '"origin" is an honorable designation for the Triple Burner' because the Triple Burner's influences are accumulated in the kidneys. (Emphasis is mine)

Please read again, carefully, exactly what Yang states above! He states, 'The two kidneys are distinguished by their storing the essence of the sun and of the moon.' Both the sun and the moon are light sources; obviously, the sun is the light source for the day, and the moon is the light source for the nighttime. So Yang shows that the ancients knew that light energy or the essence coming directly from the sun and moon can penetrate our body. Note how this essence or photonic energy is stored by the two kidneys and that this photonic essence 'constitutes man's source and root'. That is very plainly stated, is it not? Note how Yang finishes the quoted paragraph above: 'Hence the text states '"origin' is an honorable designation for the Triple

Burner" because the Triple Burner's influences are accumulated in the kidneys.' So essentially, Yang is saying that immaterial influences, including light energy from the sun (and light from the sun reflected off the moon at nighttime), enter our body and are stored by the two Kidneys to be utilized by the Triple Burner for dispersing nutrients, healing damaged tissues, and maintaining optimal health. The sun's essence also energizes plant and animal produce that we consume so that the biophotons therein can be released into our Middle Heater so that the Stomach can perform its role at extracting nutrients and solar essence and bulk water and convert these into the many forms of Qi as per the Triple Heater fluid and energy production cycle. Some force or feature of these immaterial influences from the sun allows the Kidneys to function as a timekeeper to synchronize the body with the 24-hour circadian cycle and the 12-month circannual cycle, along with the 7-year and 8-year cycles that rule females and males respectively. Some force or feature of this essence from the moon allows the Kidneys to synchronize the lunar-related menstrual cycle in females so that ideally women should ovulate on or near the full moon and menstruate on or near the new moon.

18.7 Melanin May Drive the San Jiao by Converting Solar Irradiation into Metabolic Energy

Melanin is a broad term for a group of natural pigments found in most organisms. It is a complex biopolymer that is derived from the amino acid tyrosine. Melanin is responsible for determining skin and hair color in humans. The process called melanogenesis occurs within the skin after it is exposed to UV radiation, causing the skin to visibly tan. Melanin is known to very effectively absorb light, and the pigment dissipates over 99.9% of the potentially damaging UV radiation absorbed. But is this the only function of melanin present in the melanocytes of human skin?

In the 2016 article titled 'Could Melanin Convert Radiation into Harmless, Even Useful Energy?', the author, Sayer Ji (51), noted that the biological role of melanin in the human body may be much greater than simply protecting us against damaging UV radiation. He stated that 'one recent and highly controversial paper proposes that melanin is responsible for generating the majority of the body's energy, effectively challenging the ATP-focused and glucose-centric view of cellular bioenergetics that has dominated biology for the past half century'. Ji further states, 'Melanin may function in a manner analogous to energy harvesting pigments such as chlorophyll.' If this is indeed the case, then this supports my theory that sunshine exposure on our skin drives the San Jiao that in turn energizes every function and cell in the body. Melanin is a very interesting molecule possessing remarkable biological properties. It was the first organic semiconductor discovered, and being black, melanin absorbs a large range of the broad electromagnetic spectrum. Melanin has numerous physiological roles. For example, it converts and dissipates potentially harmful ultraviolet radiation into thermal energy, it scavenges destructive free radicals, it chelates toxic materials and protects DNA from damage and potentially converts sunlight into metabolic energy.

Remarkably, it appears that melanin can convert highly damaging ionizing gamma radiation into useful energy. In 2001, a Russian report discussed the discovery of a melanin-rich species of fungi that was thriving on the walls at the highly radioactive Chernobyl meltdown reactor site. Three years later, soil fungi were found proliferating at the still highly radioactive Chernobyl site. Melanin-rich fungi have also been found flourishing in the soils of a Nevada nuclear test site in spite of being exposed to radiation 2,000 times greater than the exposure dose lethal to a human. Other microorganisms, including pyomelanin-producing bacteria, have been isolated, thriving in radioactive uranium-contaminated soils. There are obviously some very interesting chemical and biological properties of melanin, and the author (51) further cited a 2007 study published in *PLOS ONE* entitled 'Ionizing Radiation Changes the Electronic Properties of Melanin and Enhances the Growth of Melanised Fungi', where the authors suggested that their findings on irradiated melanin from these fungi raised 'intriguing questions about a potential role for melanin in energy capture and utilisation'.

More recent research (2012) proved that melanin was a promising and safe radioprotector when mice were exposed to lethal levels of gamma radiation, and it appears that 'diets rich in melanin may be beneficial to overcome radiation toxicity in humans'. Why is melanin so different from other molecules that are destroyed by high levels of radiation exposure? While the majority of biomolecules are severely impaired due to oxidative damage when exposed to high levels of radiation, the author (51) reported that an article published in 2011 in *Bioelectrochemistry* stated, 'Melanin remained structurally and functionally intact, appearing capable of producing a continuous electric current. This current, theoretically, could be used to produce chemical/metabolic energy in living systems. This would explain the increased growth rate, even under low nutrient conditions, in certain kinds of gamma irradiated fungi.' The author suggested that a 'good source of supplemental melanin for those interested in its radioprotective and radiotrophic ("radiation eating") properties' is the nutritionally dense mushroom Chaga, which contains a massive amount of melanin. He says it 'was known by the Siberians as the "Gift from God" and the "Mushroom of Immortality," the Japanese as "The Diamond of the Forest," and Chinese as the "King of Plants"'.

Summary of Chapter 18

I thought it was amazing that while the word *sun* appears nine times in Unschuld's (1) translation of the *Nan ching*, six times it is in intimate association with the Triple Burner. To my knowledge, no other author has determined this connection. Ting Te-yung discussed how the sun and moon and light exposure over the earth during the daytime cause the Triple Burner to supply qi and fluid nutrients to 'all the twelve conduits and fifteen network-vessels' and, significantly, 'as long as this continues, a person will live'. This shows that the influence of the sun is vitally important to energize the Triple Burner. Every single food that we eat (epitomized as grain) owes its existence to the solar energy that was derived from the rays and photons from the sun, or immaterial influences, as Yang described them.

I believe that there are three ways in which the sun's energy enters our body and empowers us. Firstly, the Stomach, in the central burner, is where the intrinsic nutrients and stored energy present in all the diverse foods that we consume are extracted. Further, I believe that large amounts of bulk water and biophotons are also extracted from the food and pass through the Stomach wall to be circulated throughout the entire body via the Triple Energizer. Secondly, there is the possibility that the Triple Burner system in humans is driven and energized by sunshine penetrating directly into our skin and activating melanin's energy capture and utilization property to produce the Qi that flows through our channels. Thirdly, there is also the strong likelihood that sunlight affects us through our skin indirectly due to the absorption of ambient infrared radiation from all the objects that we encounter in our daily lives. It is a fact that due to solar radiation of infrared rays, all objects on earth are heated—including walls, furniture, and dogs, for example—after having absorbed sunlight during the day and have intrinsic infrared radiation to emit because of their inherent ambient warmth. Many researchers believe that water and humans absorb this ambient infrared radiation. Even when we consume water that has 'seen' daylight, we are consuming some of the sun-derived immaterial influences. These immaterial influences that Yang described include many sun-derived factors that are essential for our health and well-being and include biophotons, infrared radiation, and ultraviolet B (UVB) rays, which interact with 7-dehydrocholesterol (7-DHC) present in the skin to produce vitamin D_3.

Yang revealed some amazing information when he stated, 'The two kidneys are distinguished by their storing the essence of the sun and of the moon, respectively. This essence consists of immaterial influences; it constitutes man's source and root. . . . Hence, it is obvious that the cinnabar field is the basis of life.' Yang clearly stated that the Kidneys store the essence of the sun and the moon and that this essence constitutes man's source and root, equating the solar-derived essence to the basis of life. Note further that Yang then advised that '"origin" is an honorable designation for the Triple Burner' because the Triple Burner's influences are accumulated in the kidneys'. I propose that the essence of the sun in the form of biophotons enters our body directly through our skin and also in a large degree in the food and water we consume. I believe that the biophotons and bulk water in the consumed food are extracted by maceration in the Stomach in the Middle Heater and passed through the Stomach wall to be processed by the Spleen. When these two components (biophotons and bulk water) come into contact with the omnipresent hydrophilic membranes and connective tissue that pervade our being, they are converted into life-giving liquid crystalline EZ water that powers the pumps, adjusts the pH, and initiates the separation of pure and impure liquids associated with the so-called Triple-Energizer Fluid and Energy Production Cycle, also known as San Jiao.

CHAPTER 19

Introduction to the Nature of Water and the Triple Energizer

Introduction
For years, I have believed that the Triple-Energizer organ system was some kind of omnipresent, ubiquitous electrochemical arrangement that permeated our entire body and was responsible for all the biological and biochemical processes that occur naturally in a healthy body. While the *Nan Ching* suggested that the Triple Burner has a name but no form, I suggest that the former classics indicate that it does indeed have a form but unlike any of the other tangible dissectible organs. The second I heard Professor Pollack's (27) theory on the fourth phase of water, I was convinced that his hypothesis regarding liquid crystalline EZ water exactly fitted and explained the complex underlying mechanism of fluid metamorphosis and energy synthesis associated with the so-called Triple Energizer. So before I plunge in and describe exactly what I believe the Triple Energizer system to be, it is imperative that I describe Professor Pollack's theory regarding the fourth phase of water, liquid crystalline EZ water.

I have worked in Chemical and Biological laboratories for 35 years and have witnessed first-hand many of the bizarre behaviors of water that Pollack has researched. Pollack's answers completely satisfied all the questions that I had pertaining to these bizarre phenomena of water. While his theory does not concur with the numerous accepted and muddled paradigms associated with water chemistry and biology at this time, I propose it is only a matter of time before scientists around the world recognize that his theory is truly groundbreaking and profoundly correct. His theory universally explains the numerous bizarre phenomena associated with water that no other theory has elucidated. While numerous other prestigious scientists have strongly hinted at many of the concepts that Professor Pollack had proposed, they have never researched the subject to the rigorous degree that Pollack has. So what are some of the bizarre water phenomena that defy explanation by the numerous accepted theories of water scientists? Note that the fourth phase of water, liquid crystalline water, EZ water, EZ, and $[H_3O_2]^-$ are all synonymous terms throughout my book. For the next few pages, I have picked out the most exciting scientific findings from his revolutionary thought-provoking book *The Fourth Phase of Water: Beyond Solid, Liquid, and Vapor*. In the commentaries of the 31st Difficult Issue on page 348 of Unschuld's (1) translation of the *Nan Ching*, Li Chiung states, 'The Triple Burner has been compared in the Su-wen with the official responsible for maintaining the ditches; the passage-ways of water originate from there.' As the Triple Energizer is the official in charge of water in TCM, it is essential that we understand the nature of water.

19.1 Only Pollack's Theory Can Explain Water's Bizarre Properties
Pollack's research offers rational scientific explanations for many otherwise unexplainable mysterious phenomena pertaining to water, where the existence and presence of EZ would demystify the enigmas. The presence of EZ in the mix explains why vortexing water cools it by up to 4 °C, why adding salt to a completely filled glass does not cause overflow, why adding different chemicals (e.g. ammonium persulfate) to water cools the water by up to 8 °C and increases its combined volume, why adding water to concentrated sulfuric acid makes the water boil with potentially explosive force, why adding sodium hydroxide pellets to a flask of water reduces its volume, why Russian vodka is 40% alcohol, and why concrete sets hard. The book has taken into account the presence of EZ to rationalize other unexplained phenomena. These include why isolated individual clouds form in a blue sky, why wet sand will make sand castles and

dry sand won't, why tree roots can crack concrete, why ice is slippery, why bubbles eventually gather on the walls of plastic cups filled with water, why a kettle whistles, why underwater dolphins blow round rings and then play with them, why the sea recedes from land just prior to the tsunami's arrival onshore, why warm water freezes faster than colder water when both are placed in a freezer simultaneously. As water cools, its density changes. Other water Scientists still are unable to explain why water's density is highest at 4 °C. What's magical about that number?

19.2 EZ Properties Are the Same as Functions of the Triple Energizer
Pollack has also given very convincing evidence that positively charged water molecules inevitably arise from the presence of exclusion zones (EZs) and that those positively charged water molecules have interesting and diverse properties. He says, 'Their sundry actions include reducing friction, wedging surfaces apart . . . running batteries, driving catalysis, and powering fluid flows' (KL 3670–3672). All these properties involve actions attributed to the Triple Energizer in TCM. Is that a coincidence? I think not!

19.3 Water Molecules in Living Systems Are Much More Reactive Than Biologists Realize
The book (27), at KL 480–523, says that biologists, for example, often consider water as the vast sea that surrounds the more important molecules essential for life. They don't realize that water molecules are actively interacting with everything. Water molecules simply must react and interact with other molecules around them. For example, in a simple water droplet, some of the water molecules that make up the periphery of the droplet must adhere to molecules, for without that resultant cohesion, there could be no individual water droplet. Pollack explains that the theories of water–water interactions are very complex, and even highly educated water scientists experience difficulty in understanding one another regarding the seven different and diverse prevailing theories. However, most of these differing theoretical models share a common feature: multiple states.

19.4 Eminent Scientists Believe Water Organization Is a Major Pillar in the Edifice of Life

Pollack (KL 817–828) reiterates that such molecular ordering is not a new idea, and he relates that the idea of long-range water ordering was advanced by a number of prominent scientists, including Walter Drost-Hansen, James Clegg, and especially Albert Szent-Györgyi and Gilbert Ling. Szent-Györgyi was a seminal thinker who won the Nobel Prize for discovering vitamin C. A cornerstone of his thinking was the long-range ordering of water, which he regarded as a major pillar in the edifice of life. Ling argues that the cell's charged surfaces order nearby water molecules, which in turn exclude most solutes. According to Ling, this ordering is the very reason why most solutes occur in low concentrations inside the cell: the cell's ordered water excludes them.

19.5 The Molecular Nature and Structure of Water Is a Mystery in Spite of Its Abundance

At KL 469–473 of the book entitled *The Fourth Phase of Water: Beyond Solid, Liquid, and Vapor*, Pollack (27) explains that the chemistry and molecular architecture of water is still a mystery and cites Philip Ball, who is a premier science writer of our time and author of the book H_2O: *A Biography of Water*. Ball, who has been a science consultant for the journal *Nature* for a long time, has stated that scientists don't really understand the dynamics of water in spite of it covering two-thirds of our planet.

19.6 On a Molecular Basis, Water Makes Up 99% of the Molecules in the Human Body

At KL 331–332 of the book, Professor Gerald Pollack (27) states, 'Your cells are two-thirds water by volume; however, the water molecule is so small that if you were to count every single molecule in your body, 99% of them would be water molecules. That many water molecules are needed to make up the two-thirds volume.' This is an incredible fact that of all the molecules in our body, 99% of those molecules constitute the one molecule, namely H_2O. On this basis alone, it is inconceivable that water isn't a major player in biological and biochemical processes that we call life.

19.7 Hydrophilic Surfaces Generate Liquid Crystalline Water

Regarding the occurrence of Exclusion Zones, Pollack (27) and his researchers, (KL 782–800), found that numerous material surfaces exhibited the property of forming exclusion zones close to their surface. Such materials include all water-containing (hydro) gels, gels made of biological molecules and artificial polymers, and natural biological surfaces, including vascular endothelia (the insides of blood vessels), regions of plant roots, and muscle. Interestingly, Pollack even observed substantial EZs adjacent to single molecular layers, which meant that material depth was not consequential; it appeared possible that creating an exclusion zone merely required a molecular template. Even charged polymers also produced exclusion zones. An especially potent polymer is Nafion, which possesses a Teflon-like backbone, which contains many negatively charged sulfonic acid groups, which make this polymer one of the more potent EZ excluders. Subsequently, this material is used often in research trials. Pollack advises that the EZ-nucleating materials mentioned above are hydrophilic, or water-loving in nature, and their profoundly intense affinity for water causes the exclusion of all other 'suitors' such that only water is permitted to stay.

19.8 Does Exclusion Zone Water Exclude the Clear Fluids from the Turbid Fluids?

At KL 801–811 of the book entitled *The Fourth Phase of Water: Beyond Solid, Liquid, and Vapor*, Pollack (27) explains that research revealed that the EZ excludes numerous substances extending from small dissolved solutes to large suspended particles. Microspheres fabricated from a diverse range of chemical substances were excluded regardless of whether they were small (0.1 µm) or large (10 µm). The diverse list of substances that were excluded included red blood cells, several strains of bacteria, ordinary dirt particles, the protein albumin, and even various dyes with low molecular weights (100 Daltons), which is only slightly larger than sodium chloride molecules. Remarkably, the width between the smallest to the largest of the excluded substances was one thousand billion times. Pollack stated, 'These experiments showed that the EZ rather broadly excludes substances of many sizes, from very small to very large . . . we could conclude that the exclusion phenomenon was general: almost any hydrophilic surface can generate an EZ, and the EZ excludes almost anything suspended or dissolved in the water.' Pollack (KL 814–817) continued, 'This demonstrably vast exclusionary power implied yet again that we might be dealing with some kind of crystal-like substance, for crystals exclude massively. . . . As the ordered zone grew, it would push out solutes in the same way that a growing glacier pushes out rocks.' Here Pollack is exactly describing the EZ mechanism that the Triple Energizer utilizes in numerous locations throughout the body for excluding (or separating) *clear* fluids from *turbid* fluids. Please note the following four examples.

In the commentaries on the 30th Difficult Issue in Unschuld's (1) translation of the *Nan-ching* (page 342), regarding the processing of water and grains (food) in the Stomach, Ting Te-yung says, 'Both essence and blood have *clear* and *turbid* [portions]. The *clear* [portion] of the essence turns to the lung where it supports the true influences of heaven. Its *turbid* [portion] strengthens the bones and the marrow. Hence, the *clear* [portion] in the blood turns to the heart where it nourishes the spirit. The *turbid* [portion] of the blood provides external splendor to the flesh. The *clear* [portion] proceeds inside the vessels; the *turbid* [portion] proceeds outside the vessels.'

In the commentaries for the 31st Difficult Issue in Unschuld's (1) translation of the *Nan-ching* (page 350), Hsü Ta-ch'un states, 'Ke ("diaphragm") stands for ke ("to screen off"). Below the heart is a membrane screening off the *turbid* influences. It is called ke.' Here, Hsü Ta-ch'un refers to a membrane screening off the *turbid* influences below the heart, at a location regarded by authorities (3, 21) (in glossary under *San Jiao*) as a demarcation between the upper energizer and the middle energizer. Hence, the inference is that the membrane constitutes part of the Triple Energizer.

Also in the commentaries on the 31st Difficult Issue on page 351 of Unschuld's (1) translation of the *Nan-Ching*, Li Chiung states, '[The lower section of the Triple Burner] separates the water and the grains which were taken in through the upper [section of the Triple] Burner. The *clear* [portions] become urine; the *turbid* [portions] become feces. They are then transmitted to the outside.'

Arya Nielsen (20), on page 34, states, 'Jin fluids circulate with the Wei Qi at the body exterior as an agency of protection and nourishment to the skin and muscles. Jin leaves the body as *clear* light fluids: sweat, tears, saliva and mucus. The Lungs and the Upper Jiao of the San Jiao control movement and expression of Jin. Ye fluids are more dense and *turbid*. They circulate with the Ying Qi at the interior, lubricate the joints and orifices (eyes, ears, nose, mouth). Ye are controlled by the Spleen and Kidneys and the Lower Jiao of the San Jiao.'

19.9 Many Physical Properties of EZ Water Differ Vastly from Normal (Bulk) Water

Pollack (27) (KL 913–916) confirmed that the physical properties of EZ water are very different to normal bulk water. Studies showed that EZ water has a higher viscosity and is more stable than normal bulk water. The molecular motions of EZ are more constrained. The light absorption spectra of EZ differ in both the UV visible-light range and the infrared range. EZ has a higher refractive index than bulk water. These numerous physical differences confirm that EZ water hardly resembles liquid water at all. EZ water is very much more ordered, and it has a net negative charge. EZ water also possesses some semiconductor-like features (KL 1307). Pollack proposes that the EZ water is polymerized crystalline water that has reformed to produce a molecular matrix represented by $[H_3O_2]^-$, which to the uninformed may seem a little apostate, chemically speaking, but this is not a new proposal at all.

Numerous by-products of water are known to exist. As soon as water is placed into hydrochloric acid solution, for example, it dissociates based on the following formula: $H_2O + HCl \rightarrow H_3O^+ + Cl^-$. Further, highly complicated and intriguing examples of dissociated water by-products are presented in the article entitled 'HCl Hydrates as Model Systems for Protonated Water' (52), which was published in the *Journal of Physical Chemistry* in 2008, where the authors stated, 'Depending on the composition, the [HCl] hydrates include distinct protonated water forms, which in their equilibrium structures approximate either the Eigen ion $H_3O^+(H_2O)$ (in the hexahydrate) or the Zundel $H_2O \cdots H^+ \cdots OH_2$ ion (in the di- and trihydrate).' Notice that the authors point out that the Eigen ion $H_3O^+(H_2O)$ identified in their research occurs in the hexahydrate form, which Pollack's research shows is the form of temporarily polymerized liquid crystalline EZ water.

More scintillating information can be found on the website of Look Chemical (53), where the variant of water (H_2O) with the Molecular Formula of $[H_3O_2]^+$ or $[HO-OH_2]^+$ is presented. It has an IUPAC name of hydroxyoxidanium and a systematic name of oxidanyloxidanium. Its Traditional name is hydroxyoxonium. But it is also called hydroxyperoxonium. If your son was born a TCM water type, which of these pretty names would you choose? This water variant has a Molecular Weight of 35.02262 and is fairly well known in industry as an intermediate substance during chemical synthesis. Note, however, that this molecule is a cation, having a positive charge.

Another water variant is the negatively charged intermediate, the anionic molecule $[H_3O_2]^-$, which also has several names, including the hydrated hydroxide anion, Hydroxide Monohydrate Anion, or Hydroxide Hydrate Anion $[H_3O_2]^-$. In agreement with Pollack, Martin Chaplin (54) states, 'The hydration of the hydroxide ion (OH^-) is very important for biological and non-biological processes. Unfortunately, it is neither well-known nor simply described.' Pollack believes this ubiquitous structured molecule, $[H_3O_2]^-$, in its polymerized form, is the liquid crystalline complex that is omnipresent in living biological systems and also non-living structures, including fluffy clouds. This is also the molecular structure that I propose powers the omnipresent Triple Energizer complex, which pervades the body in accompaniment with and generated by the omnipresent hydrophilic connective-tissue complex.

19.10 Nature Loves Hexagonal Biochemical and Biological Structures

Pollack (27) (KL 1166–1168) advised, 'You see them throughout the domain of organic chemistry. You also see this structure in graphite, where constituent honeycomb (graphene) sheets can slide easily past one another, resulting in low friction.' Regarding the molecular composition of EZ water, the author concluded that hexagonal sheets seemed a natural option to consider. This low-friction slipperiness of the EZ would further endow the adjacent connective tissues with enhanced gliding capacity, which is a property connective tissue is known to possess.

Regarding the work of other water chemists, Pollack's book notes (KL 1152–1174) that E. R. Lippincott and collaborating chemists from the University of Maryland hypothesized a hexagonal structure for controversial polywater in a 1969 lead article in the respected journal *Science*. In the article, the authors provocatively stated that while the polywater was composed of oxygen and hydrogen atoms, their arrangement in an ordered hexagonal lattice had no resemblance to the water molecule. The research team asserted that the polywater 'should not be considered to be or even called water, any more than the properties of the polymer polyethylene can be directly correlated to the properties of the gas ethylene'. A third distinguishing feature of the polywater was the ratio of hydrogen atoms to oxygen atoms. Water (H_2O) has a ratio of 2:1 respectively, while the polywater had a ratio of 3:2. Pollack (KL 1193–1198) proceeded to explain that scientific research has determined water hexamers adjacent to many diverse surfaces, including protein subunits, quartz, metals, and graphene, which precisely agree with the proposed model.

Pollack (27) (KL 1253–1254) states, 'In sum, pouring water onto a hydrophilic surface triggers EZ growth. Water is the raw material. From this raw material, EZ honeycomb layers build.' Chemists are not able to explain scientifically why gels (for example, gelatin jellies) lock up and contain so much water. Common gels don't leak even when their fractional water content exceeds 99.9% of their total mass. Pollack believes (KL 1356–1360) that because the gel matrix is composed of numerous hydrophilic strands, the strands' surfaces convert bulk water into EZ water, which in the form of the gel resembles the stable structural integrity of a brick wall.

19.11 EZ Water Engenders Biological Processes and Energy Production

Regarding numerous life aspects, Pollack (KL 1394–1397) advises that EZ provokes and stimulates numerous biological and biochemical processes. He says:

Charged entities such as membranes, proteins, and DNA all interface with water; exclusion zones should appear in abundance. Those EZs bear charge, which means they carry electrical potential energy. Since nature rarely discards available potential energy, EZ charge may be used to drive diverse cellular processes ranging from chemical reactions all the way to fluid flows. Opportunities abound.

Notice what Professor Pollack said there, suggesting that water in contact with hydrophilic surfaces, including 'charged . . . membranes' generate 'electrical potential energy' that may 'drive diverse cellular processes ranging from chemical reactions all the way to fluid flows'. It sounds like Pollack is describing exactly what happens inside the Triple Energizer. Recall that the major function of the Triple Energizer involves the processing and metamorphosis of water and the production of Qi (energy). Remember too that, in the commentaries for the 31st Difficult Issue in Unschuld's (1) translation of the *Nan-ching* (page 355), Huang Wei-san said, 'The Triple Burner . . . is a fatty membrane covering the entire physical body from the inside.' And in the commentaries for the 38th Difficult Issue (1), on page 396, Li Chiung states, 'The Triple Burner represents nothing but membranes attached to the upper, central, and lower opening of the stomach.' So here in one place, in one organ, we have the two essential requirements to produce biologically and biochemically diverse fluids (e.g. lymph, blood, interstitial fluid, tears, sweat, urine) and different energy forms. Pollack has scientifically proven that if you put bulk water alongside hydrophilic materials, including membranes, EZ water is readily generated. Pollack is probably oblivious to the existence of the concept of the Triple Energizer, yet his research findings adequately describe the Triple Energizer fluid metamorphosis and distribution and energy production mechanism simply described by the ancient medical classics. As Pollack summed it up, 'opportunities abound'.

19.12 Hydrophilic Gel Generates EZ with Ten-Thousand-Fold Increase of Proton Concentration

The research team put a gel into a beaker of water and positioned a pH probe just outside the gel's EZ. Pollack stated (KL 1408–1412) that the team would have been thrilled to measure a drop of one pH unit, demonstrating a tenfold rise in proton concentration. However, descents of three to four pH units and occasionally even more were measured adjacent to the EZ of the polyacrylic acid gel. Because the pH scale is a logarithmic scale, this indicated that a ten-thousand-fold increase of proton concentration occurred. Pollack and his team were amazed at the results.

19.13 The EZ Phenomenon Leaves inside of Cells Negative and Outside Positive, Which Is Natural

The prevailing view of the function of nerve cells suggests that the nerve cell membrane contains ion pumps and channels that allow nerves to function, leaving the inside of the cell negative and the outside positive. Pollack suggests an alternative view (KL 1496–1501), whereby the charge separation arises from water as any water lying beside a charged or hydrophilic surface becomes EZ water. Likewise, since cells pack charged surfaces so densely inside, most of the water inside cells is EZ water. As EZ water predominates, cell negativity could merely reflect EZ negativity.

19.14 Water-Based Batteries Form Everywhere that Hydrophilic Surfaces Interface with Water

Regarding energy production in the body, Pollack (27) (KL 1526–1546) advises that after immersing a hydrophilic material (including membranes) into water, the EZ builds and charges separately. He explains that the differentiated charges remain separated due to the EZ's dense lattice, and the charge separation preserves the potential difference. He notes that while the magnitude of that difference only reaches 100 to 200 mV, the two zones are nonetheless densely charged and thus able to deliver a substantial amount of energy. Amazingly, similar to n-type semiconductor lattices, when the electrical conductivity is measured parallel to hydrophilic surfaces that readily generate EZs, the electrical conductivity is 100,000 times higher than the conductivity measured through bulk water. He explained that this type of water-based battery exists virtually everywhere that water and hydrophilic surfaces are in intimate contact, and he noted that the numerous hydrophilic surfaces in contact with water inside every cell generate EZs, and subsequently, numerous water-driven nanobatteries form inside every cell.

I believe that the predominant fabric of the omnipresent Triple Energizer (Connective-Tissue Metasystem) is hydrophilic collagen, and when it touches water, the water is electrochemically transformed into EZ, which becomes a water-based battery full of electrical potential to fire and influence all sorts of biochemical reactions. This theory is expounded in the commentaries of the 31st Difficult Issue of Unschuld's (1) translation of the *Nan Ching*, where Yeh Lin relates, 'When fire meets water, a transformation into influences takes place.'

19.15 Composition of the Triple Energizer

Remarkably, EZ water behaves like semiconductors. The book (27) (KL 1546–1557) goes on to say that beyond the EZ resides the positively charged water molecules, hydronium ions (H_3O^+). These hydronium ions carry the positive charge of the battery and are packed with energic potential. Because like-charged molecules repel each other, the hydronium ions relocate to remote locations, and subsequently, a liquid current flow is generated. Further to this, any negatively charged sites that are remotely located strongly attract those positively charged hydronium ions, accentuating liquid-flow dynamics. These counterpoised attractive and repulsive dynamic forces establish a primitive driver of natural fluid movement throughout the body. Subsequently, the additive combination of both the EZ's electrons and the hydronium ions from the bulk water generate considerable potential for doing work. Hydronium ions generate and drive fluid flows and can also strongly drive reactions, requiring positive charge. Subsequently, both of those electrically charged species can provide an abundant amount of work-producing energy.

Due to the foregoing scientific findings, Pollack (27) (KL 1593) believes that 'the EZ battery could be a *versatile supplier* of much of nature's energy' (emphasis is mine). Is Professor Pollack discussing the functional aspect of the Triple-Energizer Metasystem, or is he touting his theory on the nature of Exclusion Zone Water in biological systems? I believe it is one and the same. I propose that the Triple

Energizer is predominantly composed of hydrophilic connective tissues and membranes, which are omnipresent throughout the body in intimate association with bulk water that has been transformed into EZ water. Pollack has proved beyond doubt that bulk water (present in drink and foods) in contact with hydrophilic materials (including connective tissues and membranes) is converted into EZ water. What Professor Pollack just stated could easily describe the machinations of how the Triple Energizer pump and distribute water and biological fluids throughout the entire body and simultaneously supply energy throughout the entire body as the various differentiated forms of Qi. This mechanism of Pollack allows for both features of the Triple Energizer to coexist simultaneously via the same medium, namely, the energized fourth phase of water, which is life-giving liquid crystalline EZ water. Subsequently, I consider I now know where the elusive Triple-Energizer Metasystem has been 'hiding' all this time.

19.16 Hypothetically Possible Structural Components of the Triple Energizer

In the article entitled 'Connective Tissue', Kenneth S. Saladin (55) advises that there are only four basic kinds of tissue that make up the entire human body. These tissue types are nervous tissue, muscular tissue, epithelial tissue, and connective tissue. He advises further that connective tissue is the most abundant tissue type and is widely distributed and very varied. Hypothetically, from a very simplistic point of view, if the Triple Energizer were an actual organ system, they could be constructed from only one or more of these four tissue types. As the classically described properties of the Triple Heater are not in keeping with the nerve fiber system or muscular-tissue qualities, they can immediately be eliminated as the primary component of the Triple Energizer. Epithelial tissue is specialized to form the covering or lining of all internal and external body surfaces, so theoretically, epithelial tissue could likely be included in the development of the Triple Energizer organ complex. However, notice what the article (55) goes on to say about connective tissues. It states:

> Some of the cells of connective tissue are fibroblasts (which produce collagen fibers and are the only cell type in

tendons and ligaments); adipocytes (fat cells); leukocytes (white blood cells, also found outside the bloodstream in fibrous connective tissues); macrophages (large phagocytic cells descended from certain leukocytes); erythrocytes (red blood cells, found only in the blood and bone marrow); chondrocytes (cartilage cells); and osteocytes (bone cells).

Note that some of the cell lines of connective tissue are various blood cells, including leukocytes and erythrocytes. Where is blood produced according to Triple Energizer philosophy? Blood is produced in the Middle Heater and Upper Heater. In the article 'Bone Marrow', the author, Regina Bailey (56), explains that bone marrow is the soft, malleable connective tissue that is found inside bone cavities. Bailey continues:

> The non-vascular sections of the bone marrow are where hematopoiesis or blood cell formation occurs. This area contains immature blood cells, fat cells, white blood cells (macrophages and plasma cells), and thin, branching fibers of reticular connective tissue. While all blood cells are derived from bone marrow, some white blood cells mature in other organs such as the spleen, lymph nodes, and thymus gland.

Note that all blood cells are made in the bone marrow, which is the soft, malleable connective tissue that is found inside bone cavities. So once again, specialized connective tissue is the exact location of a function attributed to the Triple Energizer. This agrees with my hypothesis that the Triple Energizer predominantly constitute various diversified components of the connective-tissue system of the body.

19.17 The San Jiao Is the Actual Membranes or in Close Proximity to Them

Regarding the controversy surrounding the anatomical location and the nature of the San Jiao, in the article entitled 'San Jiao', the highly enlightened author, Linda Barbour (23), states:

Its very existence has been questioned by westerners, as it is not one specific organ that can be found upon dissection. The San Jiao has been said to be a collection of 'greasy membranes' that includes such things as the peritoneum of the abdominal cavity, the serous membranes of the ventral cavity, the mesentery, the diaphragm, the omentum, and possibly even the meninges, the cell and organelle membranes and the endoplasmic reticulum. It is also described as the cavities around these membranes. The membranes are described as 'greasy' because they have high lipid content. Lipids are metabolically very active. They can carry electrical charges and release great amounts of ATP, which fuels all activities. These attributes of 'greasy membranes' correlate with the functions of the San Jiao— whether the San Jiao is the actual membranes or in close proximity to them is inconsequential. (Emphasis is mine)

19.18 The Nature of the Triple Energizer Similar to the Endocrine System or the Immune System

In 2012, acupuncturist Kimberly Thompson (46) posted an article entitled 'What the Heck Is a Triple Energizer Anyway?' on the miridiatech.com website. Regarding the Triple Energizer, she stated in the article, 'It has a name but it has *no form*. The TE is *not an organ which can be removed* from the body or observed on a lab table' (emphasis is mine). This comment is true, but can the endocrine system or the immune system be removed from the body and observed on a lab table? I think not!

19.19 The Triple Heater Does Has a 'Form' but Like No Other Organ

In the *Nan-ching*, many references discuss the subject about the Triple Burner having a name but no form. In Unschuld's (1) translation of the *Nan-ching*, he includes many of the commentaries by learned individuals that dispute that issue. The learned individuals categorically argue that the Triple Burner does indeed have a 'form', but not like other organs of defined shape and defined size and defined location. I believe that the Triple Energizer is a complex connective-tissue metasystem including various structures, which predominantly include the omnipresent connective tissues of the body in all their extremely diversified forms and including, of course, in the middle energizer, the abdominal viscera. In reality, do the normal abdominal viscera have a standardized anatomical form? Note what was stated in *Basic Human Anatomy* (57). It said:

> The positions of the abdominal viscera vary with the individual and with gravity, posture, respiration, and degree of filling. Radiological studies have shown that 'the normal abdominal viscera have no fixed shapes and no fixed positions, and every description of them must be qualified by a statement of the conditions existing at the time of observation. Moreover, profound change may be caused not only by mechanical forces but also by mental influences.

However, the connective-tissue metasystem has a name but no form. The author of *The Economist* article (58) entitled 'The Human Microbiome: Me, Myself, Us' succinctly stated:

> One way to think of the microbiome is as an additional human organ, albeit a rather peculiar one. It weighs as much as many organs (about a kilogram, or a bit more than two pounds). And although it is not a distinct structure in the way that a heart or a liver is distinct, an organ does not have to have form and shape to be real. The immune system, for example, consists of cells scattered all around the body but it has the salient feature of an organ, namely that it is an organised system of cells. The microbiome, too, is organised.

I believe that this present-day understanding of anatomical knowledge about the formless but omnipresent connective-tissue metasystem of membranes and the highly essential but formless microbiome clearly explains why the ancient scholars stated that the Triple Burner had a name but no form. So now that I have proposed that connective tissues and membranes are at the heart of the Triple-Energizer

Metasystem, as Linda Barbour (23) also suggested above, I would like to discuss in depth what modern scientific research has determined about hydrophilic materials, which include connective tissues and membranes.

19.20 Automatic Flow of Fluids Through Hydrophilic Tubules
The most dramatic example of mechanical energy at work via EZ is the flow of water through hydrophilic tubules. To perform this experiment, drop a 1 mm length of Nafion tubing into a small chamber of water, taking care to ensure that the water fills the inside of the Nafion tubing. Make sure that the tube lies flat at the bottom of the chamber. To easily visualize any flow, add some microspheres or a blob of dye to the chamber. After a few minutes of chaotic start-up, you will observe a steady flow of water flowing through the tube, much like the blood running through a vessel. The flow direction is unpredictable from one trial to the next. However, once the flow initiates, it persists with little diminution for as long as an hour. Nafion tubes work, but so will cylindrical tunnels bored within various gels. The results are similar. Thus, rather than this phenomenon being specific to any one material, this flow occurrence happens due to the material's hydrophilic nature (KL 2035–2047). Pollack (27) explains the mechanism by which this simple pump works. He explains that an EZ can be readily observed, building up just inside the tube. As the EZ builds, hydronium ions accumulate inside the core of the tube and can be easily measured. Once the hydronium ion concentration reaches a threshold, those positively charged water molecules spontaneously escape from one end of the Nafion tube, which initiates the flow of internal fluid to the fluid outside. The fluid flow to one end of the tube draws fresh water into the tube at the other end of the tube, the fresh water likewise becomes protonated, and this mechanism perpetuates the flow of fluid. Pollack notes that light energy presumably releases protons and that 'the energy driving this intra-tubular flow evidently comes from light' (KL 2050–2059).

Is this mechanism how various liquids flow through the body within fine conduits within connective tissues and membranes oblivious to scientific and medical practitioners due to their concealed nature and present lack of understanding by the practitioners? How do microorganisms get from the gut to the mammary glands and from infected periodontal region to initiate premature births in pregnant women, and is there a real channel distribution of Blood from the Heart to the Uterus through the *Bao Mai*? Are there concealed passageways within the hydrophilic membranes that have evaded detection due to ignorance and preconceptions and misconceptions by pathologists who believe that they already know it all? This novel concept would most likely not fit into their prejudiced paradigm. Is blood and lymph flow throughout the respective vessels being driven by the EZ energy flow mechanism and not solely by the pumping heart and muscles compressing lymph vessels respectively? I am sure that time will prove this is the case.

19.21 The Chemical Reason EZ Water Collapses Back into Standard Bulk Water

The EZ builds by locking those $(OH)^-$ units into the EZ lattice one at a time. For EZ water to transform back into bulk water, Pollack explains that a portion of the negatively charged lattice-structural unit $(OH)^-$ in the EZ molecule combines with the hydronium ion $(H_3O)^+$ present in the solution to yield two water molecules—i.e. $H_3O^+ + OH^- \rightarrow 2\ H_2O$ (KL 1796–1798). Thus, bulk water present throughout the body transforms into the EZ form required by the functions of the Triple Energizer and back into bulk water once that function has been completed such that the same water can be recycled over and over.

19.22 How Liquid Crystalline Water Interacts with Toxic Damaging Free Radicals

Regarding the effect of damaging Free Radicals on biological systems and EZ formation throughout the body, Pollack (27) explains that because of their high reactivity, free radicals are also called reactive oxygen species (ROS). The superoxide radical, or anion, is the most common ROS and is composed of two oxygen atoms with a single negative charge (O_2^-). Other ROS include the OH radical, which bears no charge, and hydrogen peroxide (H_2O_2). All these ROS contain oxygen atoms, and all of them can potentially form during exclusion zone breakdown. Due to their high chemical reactivity, ROS can cause problems because they can bind instantly to many cellular substances and that chemical binding can potentially modify those substances and make them toxic. As an example, the superoxide anion (O_2^-) can be a powerful destroyer of microorganisms. Subsequently, to prevent cellular damage from occurring, nature ensures that a scavenging enzyme called superoxide dismutase, or SOD, is present in every cell in your body to immediately neutralize these free radicals as they form. Biochemists believe that the omnipresence of this enzyme is an enigma. However, Pollack explains that if free radicals result as natural by-products of exclusion zones that form practically everywhere, then the SOD enzyme's ubiquity becomes reasonable as it should obviously be manufactured practically everywhere as well (KL 1826–1836).

19.23 Charging the EZ Water Battery with Light Energy from the Sun

Pollack (27) and his team initially had difficulty determining the energy source that kept the EZ charged and were unable to find an answer (KL 1609–1612). The query was finally resolved serendipitously by accident when it was realized that the energy supply was radiant electromagnetic energy in the form of not only the visible part of the electromagnetic spectrum but also the ultraviolet and infrared portions too. It was confirmed that bulk water absorbs the electromagnetic energy and uses it for building the EZ and maintaining the attendant charge separation. Photons, especially from sunlight, somehow donated their energy toward EZ growth. This meant that energy from the sun could power nature's biological water battery in much the same way that the sun's energy powers photosynthesis. The book notes that ultraviolet light at 270 nm was the least effective. Visible light was more effective. What was very surprising at first was that infrared—particularly at 3,000 nm—was the most effective. The team later realized that it made perfect sense as the 3,000 nm wavelength of light is the one that is most strongly absorbed by water. Pollack thus proved that 'the *most strongly absorbed wavelength* is the one that *most effectively drives EZ growth*' (emphasis is mine). Pollack noted that because of the massive implications, this finding was an extremely satisfying correlation. He proceeded to explain that because infrared energy is the most effective for building EZs and because infrared energy is omnipresent, 'the *fuel for building EZs* is always available. The fuel comes *free*' (KL 1631–1682, emphasis is mine).

You might ask, why is this energy free? Pollack explains that everything emits infrared radiation and gives the example that, during daytime, all objects absorb the radiant energy from the heat of the sun and subsequently emit plenty of infrared radiation (IR) whether the lights are turned on or off. He says that infrared radiation is always present, and he describes IR as nature's gift which is free for the taking (KL 1689–1691). Pollack (27) further advises, 'Light achieves many wondrous things because photon energy readily converts into other forms of energy.' Examples include incident light of one wavelength converting to another wavelength, producing fluorescence; light energizing the vibrational energy that causes constant movement in solutions called Brownian motion; light releasing electrons in semiconductors to produce the photoelectric effect; light catalyzing chemical reactions; and light separating charge and producing energy in photosynthesis in plants (KL 1693–1697). It was also found that EZ buildup can also occur from other energy sources, including ultrasound at 7.5 MHz, similar to that used for imaging embryos (KL 1702–1704).

19.24 EZ Formation Resembles Photosynthesis-Like Energy Conversion

When comparing the process of EZ formation to photosynthesis, the book explains that the initial stage of photosynthesis involves the splitting of water molecules into positive and negative components that is initiated by light-absorbing chromophores residing next to the water molecules. The photosynthesis scenario of plants very closely resembles the one that Pollack describes for humans except a hydrophilic surface resides next to water molecules in humans instead of light-absorbing chromophores in plants. Remarkably, in both instances of photosynthesis and EZ formation, light induces water molecule splitting (KL 2116–2120).

19.25 EZ Generated Protons Explain Industrial Catalysis and Biological Enzymes

In the book, Pollack (27) proposes that catalysts used in the chemical industry may owe their power and functionality to the EZ effect. He explains that catalysis is the ostensibly mysterious process whereby a catalyst massively accelerates the rate of chemical reactions often by millions of times and over and over without being consumed. The most conventional catalysts are the so-called acid catalysts, which work by mobilizing protons. Other common catalysts greatly accelerate chemical reactions by mobilizing OH^- groups. Pollack wondered whether those charged groups might be derived from EZ formation as EZs very commonly produce free H^+ cations and less commonly produce free OH^- anions. Pollack posits that hydrophilic surfaces should be natural catalysts and hydrophilic surfaces with the highest charge should be the most powerfully catalytic. He noted that Nafion's potent catalytic activity does fit that expectation (KL 3590–3597).

Enzymes are large protein molecules that massively accelerate biological and biochemical reactions and are thus also seen as catalysts. Substrate-specific enzyme catalysis is believed to occur due to the highly specific active site of the protein. However, up until the early twentieth century, it was believed that enzymes triggered changes in the surrounding water, which then massively accelerated chemical reactions in nearby molecules. Pollack advises that like most protein surfaces, enzyme surfaces commonly possess a substantial negative charge, and subsequently, those enzyme surfaces should generate EZ layers. If this is indeed the case, then the biological catalysis of enzymes may closely resemble generic catalysis, encompassing little more than high concentrations of EZ-generated protons (KL 3626–3631).

I propose it is possible that, in a like manner, the known hydrophilic surfaces that predominate on connective tissues and membranes could be acting as biological catalysts to ensure that the necessary molecules required for life are formed in the body when required, thanks to the catalytic property of the membranes dispersed throughout the omnipresent Triple-Energizer membrane matrix.

19.26 Liquid Water, Bubbles, and Mist! Triple Energizer or EZ?

Pollack (27) explained that water bubbles and water droplets can sometimes look similar from the outside of the container, but bubbles contain gas, while water droplets contain liquid. It is known that water droplets can persist on water surfaces for extended times without breaking down and slipping into the liquid water below. Their spherical structure can be readily seen when water is splashed onto a Teflon surface, for example. Pollack even points out that, paradoxically, droplets can even exist within water. Because of the spherical uniformity, scientists accept the presence of some kind of membranous sheath as a given. Less clear is the source of the pressure. Under the microscope, underwater droplets often look like clusters of smaller droplets within the larger droplet. Pollack concluded that an EZ sheath made sense from a functional point of view as an outer shell composed of EZ material could supply protons, which could accumulate inside the water droplet and subsequently provide the repulsive forces that engender the pressure required to explain the roundness of water droplets (KL 3704–3789).

Pollack believes that both water droplets and water bubbles possess EZ shells. The EZ's characteristic absorption at 270 nm is confirmed. He has shown that a light source attracts bubbles just as it draws EZ-shelled particles, and the EZ's sheet-like structure naturally accommodates volume changes. He summarizes by saying, 'So we continue thinking that bubbles and droplets are structurally similar—although the core of the droplet may contain liquid while the core of the bubble may contain gas' (KL 3861–3865).

It is interesting that while discussing the Triple Heater with respect to the proper management of liquids throughout the entire body, Elisabeth Rochat de la Vallée and M. Macé (11), on page 123, state, 'The link with the transformation of liquids is constantly confirmed in its pathology, as its relationship with water, in all its forms, is emphasized by the traditional titles attributed to each of the three heaters (Lingshu 18): Wu, mist or humid vapours for the upper heater, Ou, maceration for the middle heater, [and] Du, canal or conduit for the lower heater.'

19.27 Three Phases of Water Predominate throughout the Body—Mist, Bulk Water, and EZ

It is incredible that Pollack's book discusses the various phases of water and so does the *Nan Ching* when alluding to the Triple Energizer. Apart from emergency cases of frostbite, which I have never treated, fortunately, the solid phase of water (ice) is not represented in the body. However, the other three phases of water predominate throughout the body. When it comes to molecules present in the body, the water molecule is so small that if you were to count every single molecule in your body, 99% of them would be water molecules. Regarding the sensual observation of drifting vapours emanating from a warm cup of coffee, for example, Pollack (27) says warm water in a pot may appear flat and featureless to the naked eye, but infrared photographic images taken from a viewpoint above the warm water reveal ringlike mosaic configurations somewhat similar to the configurations seen in the vapour above the pot (KL 4285–4287).

It is believed that these mosaic configurations in the liquid are where the vapour stream from a cup of coffee emanates from as they rise and swirl magically to enhance the experience. Pollack believes that there is an obvious connection between the water patterns in the vessel beneath where the water exists as a liquid and the vaporous water patterns present in the steam above the vessel (KL 4294–4295).

Pollack explains that these patterns are actually well known to aficionados and are called Rayleigh-Bénard convection cells. These convection cells have been investigated in many liquids and, to a lesser degree,

Figure 6. Coffee cup with coffee beans and cinnamon. Photo by SSilver. © Depositphotos.com.

in water. He notes that while the mosaic patterns appear very distinct in photographs taken with an IR camera, the same mosaic patterns are readily observable in visible light, using the naked eye if you carefully scrutinize the surface of a pot of warm water or even a cup of warm water. He notes that they can even be photographed with an ordinary camera (KL 4311–4331).

Pollack (27) further states that presumably the visible boundaries in all these situations must be made up of something other than bulk water, and he suggests that a good candidate is EZ water. He explains that the optical properties of EZs are different than bulk water in at least two ways. For a start, their light absorption property differs. Secondly, the refractive index of the EZ is about 10% higher than that of bulk water. These differences in optical properties could easily produce the discernible boundary that exists between EZ and non-EZ zones. He notes that warm miso soup exemplifies the phenomenon as it elicits distinct mosaic boundary lines, with the added feature that soup ingredients circulate upward and then downward constantly in a loop constrained within one of the cells while never crossing the peripheral boundary. Further observation reveals that the individual liquid mosaics give rise to vapour mosaics directly above the individual liquid cells (KL 4339–4394).

Along the same lines as the liquid phase of water becoming the gas phase of water over a coffee cup, regarding the formation of rain clouds, Pollack says imagine evaporated vesicles, also called aerosol droplets, ascending high into the atmosphere and ultimately condensing to form clouds. He notes that the mass of water suspended in those clouds is immensely heavy. Pollack commented that one of his colleagues who is an atmospheric scientist does not estimate the mass of clouds in kilograms as that is not a fathomable unit. Rather, he uses the bizarre unit of elephant-equivalents, and advised that the total aerosol-droplet mass in a large cumulonimbus cloud can be equivalent to fifteen million elephants. Exactly how can the massive mass of 15,000,000 elephants remain suspended in one single cumulonimbus cloud in the sky? (KL 4444–4448).

In the commentaries on the 31st Difficult Issue on pages 354–355 of Unschuld's (1) translation of the *Nan Ching*, regarding clouds and precipitation, Yeh Lin relates:

> There it [i.e. water from the Lower Burner] evaporates like steam and is transformed into influences moving up again, where they become the chin [liquids], the yeh [liquids], and the sweat. All of that rests on the principle that when fire meets water, a transformation into influences takes place. The meaning is that heavenly yang [i.e., the influences of the sun] enters earthly yin [i.e., the water in the soil]. The [latter], following the movement of the yang influences ascends and become clouds and rain.

Yeh Lin's description of water being 'transformed into influences' mirrors Pollack's description of bulk water being energized by the sun and transformed into energized crystalline water that influences a multitude of biological functions throughout the body. Pollack's formation of EZ is essentially a case of 'when fire meets water' where the fire is derived from the sun, especially in the form of infrared radiation, which is omnipresent, having its energizing effect on the water that pervades our very being. Where does

Figure 7. Photo by palinchak. © Depositphotos.com.

this transformation into influences occur? Yeh Lin says, 'The water in the soil.' And where is that 'soil' in the human body? Of course, the Stomach and Spleen, the Earth phase of the Zang/Fu organ complex. To my knowledge, Pollack has not singled out the Stomach as a predominant location of EZ formation. But it does happen to be the major location of a large amount of protonated water in the body in the form of hydrochloric acid. In every case where EZ water is generated in vitro, in vivo protons are generated at one end of the EZ cell. Note too that Yeh Lin related that 'water from the Lower Burner evaporates like steam and is transformed into influences moving up again, where they become the chin liquids, the yeh liquids, and the sweat'. The chin liquids and the yeh liquids are the pure liquids, and the sweat (pH 4–6) is the impure liquid, which just happens to be the most acidic liquid formed in the body after gastric juice (pH 1.5–3.5).

19.28 Scientific Culture Bows to the Regality of Prevailing Dogma and Does Not 'Rock the Boat'

In keeping with the innovative advances of the ostracized Dr Ignaz Semmelweis and Dr Barry J. Marshall, for example, Pollack (27) states that rather than being resilient and courageous, present-day scientific culture has become increasingly timorous and diffident. Rarely do scientists question established accepted concepts, especially entrenched concepts, even when they are intrinsically faulty and have outlived their usefulness. Science seeks only petty incremental changes to an established paradigm. The scientific culture has become disciplined and bows complacently to the presumed regality of prevailing dogma. Subsequently, in numerous research fields, modern scientific investigation has produced an abundance of specious data but made little enlightened advancement that progresses our fundamental understanding (KL 5419–5423).

Regarding the field of medical research and advancement, Serge Gracovetsky expressed very similar sentiments more succinctly when he stated, 'Medicine is perhaps the only human activity in which an attractive idea will survive experimental annihilation.' The scientific and medical communities massively resist any novel concepts that differ from the accepted norm, and unfortunately, I tend to feel that this is what has happened to the TCM community as well, where TCM practitioners don't want to buck the system and contradict any established paradigms for fear of rejection and ostracism. Having a scientific background spanning 50 years, I try to fit what I learn daily about new scientific discoveries and revelations into what I have been taught about TCM theory and dogma. While there often appears to be jarring dissimilarity if you look at ancient translation in a certain prejudicial way, I am often astounded at the detailed intricate understanding that ancient Chinese practitioners had thousands of years ago—for example, the Chinese Clock and how that ties in with modern chronopharmacology and circadian rhythms.

Three of the most influential books I have read in my lifetime are the *Nan Ching*, the GAPS book written by Dr Natasha Campbell-McBride (59), and *The Fourth Phase of Water: Beyond Solid, Liquid, and Vapor* written by Professor Gerald Pollack (27). Many of the bizarre natural occurrences he discusses I have personally pondered. Several I had never noticed even though they were under my nose or microscope.

I believe that several orthodoxies regarding the San Jiao (Three Heaters, Triple Burner, Triple Energizer, etc.) are completely misunderstood and that the underlying truth that was told by the ancients has been misunderstood by lecturers and then subsequently by their students. My aim is to relay my understanding of the ancient medical scriptures based on my interpretation of those medical scriptures alongside modern scientific findings, which may possibly appear at first glance to be heretic or apostate. One needs to keep an open mind.

19.29 The In Vivo EZ Apparatus Can 'Separate the Pure Fluids from the Impure Fluids'

Pollack described that the father of modern biochemistry, Albert Szent-Györgyi, believed that the exploitation of electron energy explained biological processes and that because Exclusion Zones generate a ready source of electrons, EZs could easily drive a multitude of biological reactions. Additionally, he noted that the accompanying complementary hydronium ions that form might engender an equally vital role of driving fluid flows as positive-ion concentrations build pressure. Pollack expounded that fluid flows exist practically everywhere throughout nature, including, for example, within primitive cells and highly developed cells throughout our circulatory systems and within the fluid vessels of short plants right through to tall trees. He believes that the pressure resulting from hydronium ions could easily drive many of those fluid flows, and pertinently, he noted that the potential energy of EZs could easily drive practical devices. He announced that a simple and highly effective prototype of an EZ-driven water purifier has already been constructed. Amazingly, he stated, 'Because the EZ excludes solutes, including contaminants, harvesting the EZ amounts to collecting untainted water' (KL 5457–5464, emphasis is mine).

Note here that Pollack states that 'EZ's potential energy can also drive practical devices'. Citing Pollack word for word, he blew me away when he stated in the book, 'Because the EZ excludes solutes, including contaminants, harvesting the EZ amounts to collecting untainted water.' Regarding the function of the Triple Energizer concerning separation of pure fluids from the impure fluids, in the commentaries on the 35th Difficult Issue on pages 376–377 of Unschuld's (1) translation of the *Nan Ching*, Li Chiung states, 'The large intestine transmits the impure [portions] of water and grain. The small intestine is filled with the impure [portions] of water and grains. The stomach takes in and contains the impure portions of water and grains. The bladder stores the impure [portions] of the chin and yeh liquids. Only the gall is clear and pure.' A major function of the Triple Energizer is the separation of pure fluids from the impure fluids so that the pure fluids can be transformed into vital body fluids, including Blood. Here, Pollack is describing a mechanism whereby that exact process occurs in the exact regions abounding with hydrophilic connective tissues, where EZ is automatically produced. It seems highly probable that the EZ mechanism proven to exist by Pollack in hydrophilic connective tissues and membranes could well be the same mechanism behind the function of the Triple Energizer, where one of its roles is to separate the pure fluids from the impure fluids.

19.30 Everyday Water Sitting in a Glass at Ambient Conditions Is Not as Bland as You'd Think

With respect to the nature of everyday, common household water sitting in a glass on the kitchen table at ambient conditions, Pollack (27) further stimulates our gray matter when he says we don't normally think of a glass of water as receiving and absorbing energy but that rather we consider that the water in the glass is in a state of equilibrium with its environment. He stresses that in fact the water in the glass is normally far out of equilibrium with its environment. While this notion may sound somewhat eccentric, the previous chapters have abundantly demonstrated that water does in fact absorb energy from the environment continually and transduce that energy into potential work. The transduction construct is more acceptable once you realize that plants absorb radiant energy from the environment continuously and transduce that energy for doing work and we don't blink an eyelid. As plants are predominantly composed of water, it becomes quite reasonable to accept that a glass of water sitting next to your potted plant has the capacity to absorb and transduce photonic energy similarly to the plant (KL 5476–5481). Pollack explains further and gives the example that when sunshine breaks through the clouds and touches our skin, we feel a slight surge of energy. While that sensation may well affect our deeper psyche, it is possible that we have been physically energized by the incident solar irradiation affecting our skin pigment, melanin, which is known to transduce radiation forms of the electromagnetic spectrum. He notes how the light from a penlight placed on our palm penetrates deeply and can be seen on the dorsal side (KL 5483–5486).

19.31 Why Have the Many Principles about Water Discussed above Remained Secret?

Pollack reasons that there are four major motives why so many of the provable principles about water that he has discussed in depth above have remained secret.

1) The field of water science has had a checkered history, including the famous Russian polywater fiasco that left serious scars which kept curious scientists away from

water research for decades. Then came the 'water memory' debacle associated with Jacques Benveniste, which seemed so improbable that it became the butt of scientific jokes. While the same findings of Benveniste's research has been successfully repeated by other laboratories (which vindicates homeopathy theory), scientific critics are powerful foes, and funding is scarce without backers.

2) Because water is so common, everyone, including scientists, presume that the fundamentals have been resolved. That belief could not be further from the truth. Water is central to so many natural processes that few people can conceive that the basics could remain open to question. Unfortunately, this misperception keeps scientists away. Today science institutions only reward researchers who concentrate their research on trendy concepts and dismiss research that questions the accepted supposedly 'foundational science', which in reality is often more correctly called dogma.

3) Pollack advises that intellectual timidity is responsible for slowing the acceptance of novel fundamental principles by fellow researchers as they do not appreciate disruption to their cherished belief systems that violate the erroneous beliefs they entertain. Pollack suggests that you'd expect scientists to relish dramatic advances that further the understanding of fundamental science but that most scientists would rather only accept minor deviations from the status quo and are passionate defenders of established flawed orthodoxy.

4) Pollack pointed out that the fourth reason why water chemistry and biology are so underresearched is outright fear because disputing the supposed wisdom that is being taught means challenging and stepping on the toes of those very scientists who have gained their success based on that supposed wisdom. He cautions that nasty responses can be expected. Pollack notes that his scientific research has discredited many erroneous scientific beliefs, and he anticipates rebuke, particularly from those scientists whose

reputation, funding grants, and other characteristics of power depend on defending their incorrect scientific findings and established dogma. Senior scientists are not forgiven for such presumed apostasy. Subsequently, many career-oriented scientific researchers don't rock the boat when they smell a rat, knowing that their going along with the big boys keeps bread on their tables (KL 5529–5558). Yes, indeed, even for grown-up Scientists, peer pressure is a powerful force.

Summary of Chapter 19
Long-range water ordering has been advanced by a number of very prominent scientists, including Walter Drost-Hansen, James Clegg, and especially, Albert Szent-Györgyi and Gilbert Ling. Szent-Györgyi was a seminal thinker who won the Nobel Prize for discovering vitamin C. A cornerstone of his thinking was the long-range ordering of water, which he regarded as a major pillar in the edifice of life.

Philip Ball is one of the premier science writers of our time, author of *H_2O: A Biography of Water* and a long-time science consultant for the journal *Nature*. Ball puts it this way: 'No one really understands water. It's embarrassing to admit it, but the stuff that covers two-thirds of our planet is still a mystery. Worse, the more we look, the more the problems accumulate: new techniques probing deeper into the molecular architecture of liquid water are throwing up more puzzles.'

Note that the *fourth phase of water, liquid crystalline water, EZ water, EZ,* and *$[H_3O_2]$* are all synonymous terms that I use throughout my book.

In the commentaries of the 31st Difficult Issue on page 348 of Unschuld's (1) translation of the *Nan Ching*, Li Chiung states, 'The Triple Burner has been compared in the Su-wen with the official responsible for maintaining the ditches; the passage-ways of water originate from there.' As the Triple Energizer is the official in charge of water in TCM, it is essential that we understand the scientific nature of water.

Pollack's research offers rational scientific explanations for many otherwise unexplainable mysterious phenomena pertaining to water, where the existence and presence of EZ would demystify the enigmas. The presence of EZ in the mix explains why vortexing water cools it by up to 4 °C, why adding salt to a completely filled glass does not cause overflow, why adding different chemicals (e.g. ammonium persulfate) to water cools the water by up to 8 °C and increases its combined volume, why adding water to concentrated sulphuric acid makes the water boil with potentially explosive force, why adding sodium hydroxide pellets to a flask of water reduces its volume, why Russian vodka is 40% alcohol, and why concrete sets hard.

Pollack has also given very convincing evidence that positively charged water molecules inevitably arise from the presence of exclusion zones (EZs), and that those positively charged water molecules have interesting and diverse properties. He says, 'Their sundry actions include reducing friction, wedging surfaces apart ... running batteries, driving catalysis, and powering fluid flows.' All these properties involve actions attributed to the Triple Energizer in TCM. Is that a coincidence? I think not! Water molecules in living systems are much more reactive than biologists realize.

In his book, Professor Gerald Pollack (27) states, 'Your cells are two-thirds water by volume; however, the water molecule is so small that if you were to count every single molecule in your body, 99% of them would be water molecules. That many water molecules are needed to make up the two-thirds volume.' This is an incredible fact that of all the molecules in our body, 99% of those molecules constitute the one molecule, namely, H_2O. On this basis alone, it is inconceivable that water isn't a major player in biological and biochemical processes that we call life.

A H_2O variant is the negatively charged intermediate, the anionic molecule $[H_3O_2]^-$, which has several names, including the hydrated hydroxide anion, Hydroxide Monohydrate Anion, and Hydroxide Hydrate Anion $[H_3O_2]^-$. In agreement with Pollack, Martin Chaplin (54) states, 'The hydration of the hydroxide ion (OH^-) is very important for biological and non-biological processes. Unfortunately, it is neither well-known nor simply described.' Pollack believes this ubiquitous structured molecule, $[H_3O_2]^-$, in its polymerized form is the liquid crystalline complex that is omnipresent in living biological systems and also non-living structures, including fluffy clouds. This is also the molecular structure that I propose powers the omnipresent Triple-Energizer complex, which pervades the body in accompaniment with and generated by the omnipresent hydrophilic connective-tissue complex.

Professor Pollack has determined that water in contact with hydrophilic surfaces, including 'charged . . . membranes' generate 'electrical potential energy' that may 'drive diverse cellular processes ranging from chemical reactions all the way to fluid flows'. It sounds like Pollack is describing exactly what happens inside the Triple Energizer. Recall that the major function of the Triple Energizer involves the processing and metamorphosis of water and the production of Qi (energy). Remember too that in the 31st Difficult Issue in Unschuld's (1) translation of the *Nan-ching*, in the commentaries on page 355, Huang Wei-san said, 'The Triple Burner . . . is a fatty membrane covering the entire physical body from the inside.' And in the commentaries on the 38th Difficult Issue on page 396 (1), Li Chiung states, 'The Triple Burner represents nothing but membranes attached to the upper, central, and lower opening of the stomach.' So here in one place, in one organ, we have the two essential requirements to produce biologically and biochemically diverse fluids (e.g. lymph, blood, interstitial fluid, tears, sweat, urine) and different energy forms. Pollack has scientifically proven that if you put bulk water alongside hydrophilic materials, including membranes, EZ water is readily generated. Pollack is probably oblivious to the existence of the concept of the Triple Energizer, yet his research findings adequately describe the Triple-Energizer fluid metamorphosis and distribution and energy production mechanism simply described by the ancient medical classics. As Pollack summed it up, 'opportunities abound'.

Due to the foregoing scientific findings, Pollack believes that 'the EZ battery could be a versatile supplier of much of nature's energy'. Is Professor Pollack discussing the functional aspect of the Triple-Energizer Metasystem, or is he touting his theory on the nature of Exclusion Zone Water in biological systems? I believe it is one and the same. I propose that the Triple Energizer is predominantly composed of hydrophilic connective tissues and membranes which are omnipresent throughout the body in intimate association with water that has been transformed into EZ water. Pollack has proved beyond doubt that bulk water (present in drink and foods) in contact with hydrophilic materials (including connective tissues and membranes) is converted into EZ water. What Professor Pollack just stated could easily describe the machinations of how the Triple Energizer pump and distribute water and biological fluids throughout the entire body and simultaneously supply energy throughout the entire body as the various differentiated forms of Qi. This mechanism of Pollack's allows for both features of the Triple Energizer to coexist simultaneously via the same medium—namely, the energized fourth phase of water, which is life-giving liquid crystalline EZ water. Subsequently, I consider I now know where the elusive Triple-Energizer Metasystem has been 'hiding' all these centuries.

CHAPTER 20

Do Other Scientific Authorities Agree with Pollack's EZ Water Theory?

Introduction
Dr Mercola operates mercola.com, which is renowned as the most popular alternative-health website on the Internet. His knowledge base is remarkable, and he and his team support many dynamic medical researchers and scientists that dare to challenge outdated and dangerous medical paradigms accepted as gospel by orthodox allopathic medicine. The medical clowns from the so-called medical watchdog site, Quackwatch, have criticized Mercola on numerous issues. I have no doubt that these watchdog 'bitches' would also have criticized Dr Marshall, who was ridiculed for years by the medical fraternity for his unbelievable 'radical' and apostate hypothesis that the bacterium *Helicobacter pylori* (*H. pylori*) would dare to survive in hydrochloric acid in the stomach and be the cause of most peptic ulcers. This 'radical' physician reversed decades of incorrect medical doctrine holding that ulcers were caused by stress, spicy foods, and too much acid. Marshall has been quoted as saying in 1998, 'Everyone was against me, but I knew I was right.' Marshall's apostasy is now accepted gospel, and the Quackwatch bitches have swallowed their own vomit and retracted their opposition. I can imagine what the Quackwatch bitches would have to say about the apostate unsubstantiated views of Dr Ignaz Semmelweis when, prior to the findings of Joseph Lister, he had the common sense and audacity

to suggest that doctors should wash their hands prior to delivering babies. Doctors then believed it was perfectly acceptable to deliver babies straight after performing autopsies without washing their hands. Semmelweis's observations conflicted with the established medical opinions, orthodoxy, and dogma of the time, and his ideas were totally rejected by the supposedly educated medical community as absurd. Semmelweis's apostasy is now accepted gospel—but only years after the poor apostate was beaten to death. The blatant medical lack of knowledge that occurred at that time continues and happily lives in ignorance of anything that they did not learn at college 45 years previously. Their retirement villages go by the name of Quackwatch and Fiends of Science in Medicine.

20.1 What Well-Informed Medical Practitioners Say about Pollack's EZ Water Theory

Dr Joseph Mercola (60) interviewed Professor Gerald Pollack in August 2013 and presented the information gleaned on his website in an article entitled 'The Fourth Phase of Water: What You Don't Know About Water, and Really Should'. I have extracted the main points of that interview below.

Dr Gerald Pollack is a professor of bioengineering at the University of Washington and is one of the leading research scientists in the world when it comes to understanding the physics of water and how water impacts your health. He is the founder and editor-in-chief of a scientific journal called *Water* and has published many peer-reviewed scientific papers on the mysterious nature of water molecules. He has even received prestigious awards from the National Institutes of Health. In the web article (60), Dr Mercola states that Pollack believes it is incredible that, in spite of muscle tissue containing 99% (on a molecular basis) of highly structured water molecules in the form of $[H_3O_2]^-$, the currently accepted muscle contraction theories do not involve water. Mercola stated that Pollack has confirmed that water in our cells has special properties. It is more viscous and more alkaline than regular water. It is also more ordered, denser, is negatively charged, and can store and deliver energy, just like a battery does. Amazingly, Pollack advised that light energy in the form of visible light or infrared wavelengths is the key ingredient to generate this highly structured water. Mercola believes Pollack's scientific work is nothing short of groundbreaking.

Mercola advises that Dr Pollack's book touches on several of the most central characteristics of water, many of which are not understood to this day. For example, how does the process of evaporation occur? Why does a tea kettle whistle? Interestingly, in spite of the fact that conventional science informs us that the freezing point of water is supposed to occur at 0 °C, scientific experiments confirm that water can actually freeze at many different temperatures down to as low as −50 °C.

Remarkably, it further appears that the supposed boiling point of water of 100 °C (or 212 °F) does not always hold true either. Having worked in laboratories for 35 years and having calibrated thousands of thermometers over that time, I initially found this revelation remarkable and actually very hard to fathom. Then I recalled an experience that happened when I was working at the Department of Primary Industries Otto Madsen Dairy Research Laboratory in Hamilton, Queensland, about 1978, when I overheated pure water in a brand-new beaker in an industrial microwave oven. When I removed the superheated water from the oven, the vibrations I set up caused the superheated water to undergo violent ebullition before my eyes and seriously burned my hand. That water certainly did not behave as science says it should and boil at 100 °C.

Experimenting at the laboratory at the University of Washington over the last decade, Dr Pollack's team performed many experiments that clearly showed the existence of an additional phase of water called the exclusion zone or EZ. Mercola (60) proceeded to explain that because this fourth phase of water profoundly excludes particulates, including small molecules, it was named exclusion zone water or EZ water. It is abundant throughout the body and is present inside most of our cells. Mercola reported that the optical properties of EZ differ from H_2O because the refractive index of EZ water is about 10% higher and the density is also about 10% higher than ordinary water.

Pollack further believes that the electronegativity of EZ could explain why human cells are negatively charged rather than the currently held textbook theory that 'this negative electrical potential has something to do with the membrane and the ion channels in the membrane'. In support of this revolutionary theory, Pollack advises that laboratory findings confirm that while gels have no membrane, they still have the same potential of between 100 mV and 150 mV negative as cells in the body that possess a membrane. Pollack concludes, 'I think the cells are negatively charged because the water inside the cell is mainly EZ water and not neutral H_2O.'

20.2 How Is EZ Water Formed in Nature?

The article (60) amazingly reports that the major ingredient required to create EZ water is light, i.e. electromagnetic energy, which can include visible light, ultraviolet (UV) wavelengths, or infrared wavelengths, which surround us at all times. Infrared light is the most powerful form, especially at wavelengths of approximately 3 μm, which is all around you. The EZ water can assemble on any hydrophilic (water-loving) surface whenever infrared energy is present. Layer upon layer of EZ water forms. Millions of individual molecular layers can form on a surface. This is how EZ water presents in nature. For example, ice does not form directly from ordinary H_2O. Pollack proposes that regular water transforms into EZ water and then transforms into ice. And when ice melts, it transforms from ice into EZ water and then into regular water. So EZ water is an intermediate state between the three known different phases of the water molecule, H_2O. 'Glacial melt is a perfect way to get EZ water. And a lot of people have known that this water is really good for your health,' Dr Pollack says.

20.3 Professor Pollack Found EZ Water Forms Optimally at a 270 Nanometer Wavelength

With regard to the optimal wavelengths of light that generates EZ water, the article (60) further explains:

> Testing water samples using a UV-visible spectrometer, which measures light absorption at different wavelengths,

Dr. Pollack has discovered that in the UV region of 270 nanometers, just shy of the visible range, the EZ water actually absorbs light. The more of the 270 nanometer light the water absorbs, the more EZ water the sample contains. EZ water appears to be quite stable. This means it can hold the structure, even if you leave it sitting around for some time. Water samples from the river Ganges and from the Lourdes in France have been measured, showing spikes in the 270 nanometer region, suggesting these 'holy waters' contain high amounts of EZ water. According to Dr. Pollack, there's compelling evidence that EZ water is indeed lifesaving.

20.4 Are the Health Benefits of Light and Heat Therapies Due to EZ Water?

Heat therapies simply apply infrared energy, and Dr Pollack has found that when you apply infrared therapy, the EZ water concentration accumulates and doesn't diminish. For example, the health implications of sitting in an infrared sauna are profound. It thus appears that one of the reasons why infrared saunas make you feel so good is because your body's cells are deeply penetrated by infrared energy, which causes the concentration of EZ water to accumulate in the tissues. A similar situation occurs when it comes to light therapy—for example, spending time in the sun and also from laser therapy. 'There are various kinds of light therapy using different wavelengths. We found that all wavelengths—some in particular—of light, even weak light, build EZ. If EZ is critical for the health of your cells, which I think is clear, these therapies have a distinct physical chemical basis,' Dr Pollack explains (60).

20.5 EZ Water May Spontaneously Drive Biological Pumps throughout the Body

The concept of EZ water further provides a scientific mechanism that explains other biological mysteries. For example, Dr Pollack describes another fascinating finding that further bolsters our understanding of the mechanism of action behind the health benefits of something as simple as exposing your body to the light and heat of the sun. In the Mercola (60) interview, Dr Pollack reported that, without applying pressure to the system, researchers observed that placing a simple tube made of hydrophilic material into EZ water caused the water to spontaneously flow through the tube at high speed, and it kept flowing. That was an unprecedented finding. Irradiating the system with light made the flow go even faster. Ultraviolet light proved to be the winner in the stimulatory pumping effect. Pollack suggested this same mechanism could apply to our cardiovascular system in that our capillaries constantly receive radiant energy from metabolic reactions occurring inside our body and ambient light present outside our body. Pollack suggested that 'the flow of blood occurring through your capillaries is automatically enhanced by exposure to light'. He advised that this is an important issue because many of the capillaries of our cardiovascular system are actually smaller in diameter than the red blood cells that pass through the capillaries, which would lead to a lot of resistance. Pollack believes that 'your cardiovascular system is assisted by radiant energy in the same way that the flow in the tubes is assisted by radiant energy', and his team is following up on this hypothesis.

20.6 The Three Elements of Pollack's Pump Constitute the Triple-Energizer Organ Complex

Pollack's findings reveal that infrared energy from light outside the body, along with heat derived from internal exothermic metabolic reactions, transforms simple hydrophilic tubes into the equivalent of mechanical pumps but without a motor. The motor force is generated by the exclusion zone and the protons formed when (1) infrared light energy, (2) water, and (3) hollow hydrophilic tubules come together in space and time. The Triple-Energizer Metasystem is an omnipresent body system responsible for driving the numerous fluids (Jin/Ye body fluids) to every nook and cranny within the body, and such tubes include arteries, veins and capillaries, lymphatic ducts, renal tubules, sweat glands, ureters, tear ducts, Eustachian tubes, respiratory alveolar ducts, bile ducts, etc. All these tubes need pumps to move the associated fluids within. The connective-tissue membrane system happens to be composed almost entirely of hydrophilic material.

While there is no doubt that the heart is the predominant pump for the arteries and veins, the pumping mechanism of most other tubes and pipes in the body is inadequately defined. So we can see that two of the components of Pollack's EZ pump—namely, water and hydrophilic material—are bountiful in the body. The third element of the Pollack Pump is light and/or heat energy, including sunshine. Note that while the word *sun* appears nine times in Unschuld's (1) translation of the Nan-Ching, remarkably, six times it is in intimate association with the Triple Burner. Perchance, did the ancient TCM sages recognize that yang electromagnetic energy from the sun was a driving force of the hydrophilic, membranous Triple-Energizer organ complex?

20.7 Are the Benefits of Infrared Laser Acupuncture Mediated via EZ Water?

Regarding the benefits of infrared laser acupuncture, Professor Pollack describes other fascinating findings when he relates:

> One of the more interesting healing modalities I've been exploring lately is the use of a high-powered laser. The K-Laser also has frequencies in the infra-red range, which can deeply penetrate tissue. This kind of laser therapy has shown to provide profound healing for many painful injuries in a very short amount of time—sometimes just *minutes* of treatment. While the benefits of laser therapy are thought to be due to its action on mitochondrial activity, it may very well be that the benefits are also related to 'recharging' your damaged cells' EZ water, as well as promoting increased capillary blood flow.

20.8 Are the Benefits of Hyperbaric Medicine Mediated via EZ Water?

EZ water in your body also plays a role in hyperbaric medicine, which is also good for injuries. In that case, your tissues are exposed to high oxygen concentrations under pressure. Pollack says:

The results are in. We think we understand the mechanism as to why hyperbaric oxygen is so effective for wound healing ... EZ water has a higher density than bulk water. If you take H_2O and you put it under pressure, it should give you $H_3O_2^-$ because the EZ structure is denser than the H_2O. We did the experiments and we found, indeed, that's the case. If you put H_2O under pressure, you get more EZ water. The same goes for oxygen. EZ also has more oxygen than H_2O, and when you increase oxygen content, you get more EZ water. So, hyperbaric treatment builds EZ water in your body, particularly in injured areas where EZ water is needed.

20.9 Vortexing Water Puts Enormous Energy into the Water

Mercola (60) advised that he personally drinks vortexed water nearly exclusively. Researcher Viktor Schauberger performed much pioneering work about the benefits of drinking vortexed water about a century ago. Dr Pollack has confirmed that creating a vortex in a container of water adds energy to the water, thereby increasing the EZ content. According to Pollack, virtually *any* energy put into the water seems to create or build EZ water, and even acoustic energy seems to affect bulk water.

20.10 Negatively Charged Alkaline EZ Water Is Critical for Healthy Body Function

Dr Mercola's article (60) concluded that to remain healthy and vibrant, drinking alkaline water is essential and that our drinking water can be optimized by adding light energy or physical energy into the water by vortexing the water. He advised that a natural EZ water source includes water from deep springs as it is naturally pressurized by nature. He also suggested Earthing™ or grounding ourselves to allow the transfer of negatively charged electrons from the ground into the soles of your feet. I discuss this topic in detail in a Chapter in this book.

20.11 Accidental Discovery Reveals Coherent Biological Liquid Crystals Make Up Life Forms

Dr Mae-Wan Ho (born in 1941) is a world renowned geneticist and biophysicist known for her criticism of the modern evolutionary synthesis and genetic engineering. She is the Director of the Institute of Science in Society. The 2012 article by Dr Ho (61) titled 'Super-Conducting Liquid Crystalline Water Aligned with Collagen Fibers in the Fascia as Acupuncture Meridians of Traditional Chinese Medicine' reported the following. Serendipitously, Dr Ho and her colleague Michael Lawrence stumbled on a new setting for the polarizing microscope that is especially good for viewing biological liquid crystals. The outcome was that they observed that a worm under the polarizing microscope proved to be completely liquid crystalline and coherent to a high degree, even quantum coherent. Ho went on to call this the rainbow worm. This meant that the entire organism was electrically polarized from head to tail, like a single uniaxial crystal. Not only were the macromolecules in all the tissues and cells perfectly aligned, but so too was the 80% by weight of water. Dr Ho went on to say that 'it is the water that makes the entire organism liquid crystalline because this water is liquid crystalline' and that rather than the disorder associated with bulk water, the individual molecules are more organized and electrically aligned. She advised that life is possible only because 'liquid crystalline living water enables macromolecules to function as quantum molecular machines that transfer and transform energy at close to 100% efficiency'. She notes

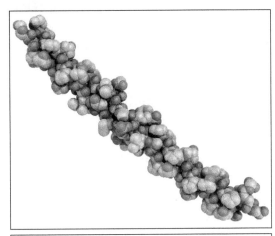

Figure 8. Chemical structure of a collagen model protein. Collagen has a triple-helix structure and is a major component of many tissues, including skin and bone. Photo by Iculig. © Depositphotos.com.

that enzymes are the 'archetypal quantum molecular energy machine' as they are naturally able to reduce the time for chemical reactions to occur in organisms by a factor of 10^{10} to 10^{23}. Dr Ho further notes that the conventional biochemical community has failed to recognize the immense importance of water in the process.

20.12 Liquid Crystalline Continuum from Extracellular Matrix to inside Every Single Cell

Regarding the nature of acupuncture meridians within the body, Dr Mae-Wan Ho (61) profoundly stated:

> The rainbow worm tells us that *a liquid crystalline continuum extends without interruption throughout the extracellular matrix to the interior of every single cell.* Soon after our discovery, I suspected *it might be the key to the rapid intercommunication within the body that enables organisms to function as perfectly coordinated wholes.* ...
> The more specific proposal—that *water aligned with collagen fibers is the anatomical and functional basis of the acupuncture meridians*—is also very much alive. (Emphasis is mine)

20.13 Acupuncture Meridians Have Characteristics Similar to Electrical Transmission Lines

Regarding the nature of acupuncture meridians within the body, Dr Mae-Wan Ho (61) explained that, on an anatomical basis, human acupuncture meridians are generally located alongside connective tissue planes in between neighboring muscles or in between a muscle and bone or a tendon. Acupuncture, moxibustion, and acupressure treatment restores bodily homeostasis by activating specific acupuncture points located at defined anatomical locations close to the surface of the skin. Research has confirmed that acupuncture meridians and individual acupuncture points elicit distinctive electrical properties (e.g. DC skin conductance or impedance) compared to the surrounding skin. Experiments suggest that acupuncture points generally have electrical impedance lower by a factor of 10–100, and acupuncture meridians have been shown to possess characteristics similar to electrical transmission lines. Dr Ho stated, 'As far as I know, *our original proposal is the only one that focuses on the liquid crystalline water structured with the collagen fibers in the connective tissue*, rather than collagen itself, or the connective tissue as such' (emphasis is mine).

20.14 Water Is Essential to Life and Drives the Energy Machine that Powers All Living Processes

Regarding the extreme importance of water, Dr Mae-Wan Ho (61) stated that water was critical for life and is responsible for all living processes, including photosynthesis in green plants, algae, and cyanobacteria, where the intrinsic energy of sunlight causes water to split into hydrogen, oxygen, and electrons, supplying enormous amounts of energy and providing life-sustaining oxygen to the multitudinous life forms on earth. She pointed out that splitting the water that drives 'the living dynamo of photosynthesis and respiration' takes a vast amount of energy, '12.6 eV, to be precise'. Ho related that over half a century ago, Nobel Laureate Albert Szent-Györgyi (the father of biochemistry) 'proposed that water at interfaces such as membranes is in the excited state, and hence requires considerably less energy to split than water in the ground state'. She then advised that the majority of all water present in living organisms is actually interfacial water because it is only a fraction of a micron away from surfaces, such as membranes or macromolecules.

20.15 Dr Ho Cites Professor Pollack's Work Showing EZ Water in Cells Acts as a Battery

Citing Professor Pollack's EZ water research findings, Dr Ho (61) continued,

> A vivid demonstration of interfacial water was achieved by the Gerald Pollack's research team at the University of Washington. Using a hydrophilic gel and a suspension of microspheres just visible to the eye, they created interfacial water apparently tens of microns or even hundreds of

microns thick on the surface of the gel, which excludes the microspheres as well as other solutes such as proteins and dyes, and hence is referred to as an exclusion zone (EZ). EZ water is about 10-fold more viscous than bulk water, it has a peak of light absorption at 270 nm, and emits fluorescence. Del Giudice and colleagues suggest that EZ water is in fact a giant coherent domain stabilized on the surface of the attractive gel. Inside the cell, the EZ would form on surfaces of membranes and macromolecules, as envisaged by Szent-Györgi. Because the coherent domain is excited water with a plasma of almost free electrons, it can easily transfer electrons to molecules on its surface. The interface between fully coherent interfacial water and normal bulk water becomes a 'redox pile.' In line with this proposal, EZ water does indeed act as a battery, as Pollack's research team demonstrated.

20.16 Dr Ho Believes Energy Medicine Relies on Quantum Coherent Liquid Crystalline Water

After discussing hydrophilic Nafion fiber trial results, Dr Ho (61) stated, 'The inverse micelle model of Nafion is especially relevant to the living cell, where interstices between fibers of the cytoskeleton and cytoplasmic membranes effectively form inverse micelle nanospaces and channels that are now known to drastically alter enzyme/substrate relationships and enzyme activity compared to bulk phase thermodynamic models that still dominate conventional cell biology.' Dr Ho further stated:

> The model maybe even more relevant to the connective tissue collagen fibers interwoven with nanotubes of water, most likely the anatomical correlates of the acupuncture meridians of traditional Chinese medicine. It would be a simple matter to measure the proton conductivity of single collagen fibrils for a start. I predict it would be at least as high as the electrospun Nafion fibers. If our hypothesis is right, acupuncture, and all forms of subtle energy medicine, including homeopathy and other traditional

healing practices, may have their mechanism of action mediated via quantum coherent liquid crystalline water.

Dr Ho stressed that their hypothesis agrees completely with Shui-Yin Lo's proposal that stable, clustered water is the foundation of the acupuncture meridian system, which remarkably was developed independently via a completely different line of inquiry.

20.17 Research from a Third World-Renowned Scientist Confirms Fourth State of Water

Dr Shui Yin Lo (62) received his Bachelor of Science in Physics from the University of Illinois in 1962, with highest honors, and his PhD in Physics from the University of Chicago in 1966. Dr Lo's academic career spans the globe as a visiting faculty member and lecturer at leading institutions throughout the world, including California Institute of Technology, Academy of Science (Beijing, China), Stanford Accelerated Center (California), Institute of Theoretical Physics, and State University of New York (Stony Brook, New York). Dr Lo has published over seventy-five scientific papers in internationally recognized physics journals during his 40-year distinguished career as well as authoring over sixty US and world patents in the field of atomic and subatomic particles. In conjunction with Caltech Pasadena and UCLA, Dr Lo was the Senior Scientist that led a team of world-renowned colleagues in the proof of concept of a revolutionary high-energy beam with multiple communications and energy applications. Dr Lo is the inventor and patent holder of the Baser. According to Dr Lo's theory, the extremely cold stable beam of coherent, heavy subatomic particles, known as bosons, may be able to break down molecules or even atoms and their nuclei. The so-called Baser may be able to be used for rock drilling, medical surgery, or precision cutting of metals without distortion. The high-energy beam of enlarged particle clusters is expected to be thousands of times more powerful than the world's most powerful laser.

In between all the above scientific endeavors, over the last 15 years, Dr Lo and his team have uncovered a newly discovered phase of water which he calls Double-Helix Water®. He advised that it is present in ultrapure water without additives or added chemical compounds of any kind. Dr Lo (63) went on to advise that his research team had studied the phenomena for fifteen years and now believe that another phase of water exists. He believes that under a particular set of circumstances, the liquid phase of water condenses into tiny solid particles at room temperature. He further believes that these transmutated water particles are somehow responsible for initiating the self-healing process within the body and that this restructured water phase is a driving force of the body's immune response. He proceeded to say:

> Because of their structure and polar charge, we theorize that these particles are the molecular basis for what Chinese Medicine has suggested for over two thousand years: that an electrical matrix surrounds the body and this electrical matrix is the senior dominating factor in all health issues. This is evidenced by Electron and Atomic Force Microscope photographs, which demonstrate that these particles line up end to end to form circuit-like structures. Therefore it is very feasible that we have found a material basis for the Chinese meridians.

20.18 The Properties of EZ Water Are Identical to the Functions of the Triple Energizer

Pollack's findings suggest that the ambient infrared light which constantly surrounds humans assists the mechanism of fluid circulation and distribution throughout the body. Regarding the scientifically proven effects of light energy–producing EZ water, in the 2014 article 'The 4th Phase of Water: A Key to Understand All Life', Jefferey Jaxen (64) states:

> The mechanical output displayed by EZ water seems to also support the long-held theory that the heart alone is not a sufficient pump to perform the feats it has historically been attributed to. To be sure, the revolutionary properties discovered by Pollack et al. didn't stop there. EZ water acts like a battery by splitting into a negative zone and

a positive zone. When electrodes are placed into each side Pollack comments, 'you can get appreciable electrical energy.' To put things simply he continues, 'the water is essentially an engine or transducer by acting to transform the energy.' What the water uses as fuel, Dr. Pollack explains, is 'external radiant energy' in which it has been demonstrated to transmute into mechanical, electrical, and even optical.

So notice that Jaxen states that EZ water causes 'fluid circulation in the body' and that 'EZ water acts like a battery and produces energy by splitting into a negative zone and a positive zone' and generating different forms of Qi (energy). Jaxen further states, 'The [EZ] water is essentially an engine or transducer by acting to transform the energy.' All these properties of EZ water are established functions of the Triple-Energizer organ complex.

Summary of Chapter 20
A summation of this Chapter shows that at least three highly regarded scientists have performed decades of research between them, confirming that water possesses a fourth phase that has evaded detection until recently. Professor Gerald Pollack is credited with the discovery of the fourth phase of water, which develops in the form of an exclusion zone, or EZ for short, whereby the water separates into a positively charged zone full of solutes and particles plus a negatively charged zone close to the hydrophilic surface that initiated the transformation of normal old H_2O in the presence of light energy. It appears that normal or bulk water absorbs light energy (all wavelengths from ultraviolet to infrared), which causes its molecules to align and resemble a liquid crystal structure. Pollack believes that our body is composed internally of approximately 99% water on a molecule proportion basis, not on a weight basis. Professor Pollack advises that long-range ordering of water molecules is not a new idea and relates that the idea of long-range water ordering was proposed by a number of very prominent scientists, including Sir William Hardy, Walter Drost-Hansen, James Clegg, and especially Albert Szent-Györgyi and Gilbert Ling.

After Professor Pollack, the second advocate for a fourth phase of water is renowned geneticist and biophysicist Dr Mae-Wan Ho. She states above, 'A liquid crystalline continuum extends without interruption throughout the extracellular matrix to the interior of every single cell. Soon after our discovery, I suspected it might be the key to the rapid intercommunication within the body that enables organisms to function as perfectly coordinated wholes.' This liquid crystalline continuum is what Dr Pollack calls EZ water. Notice what Dr Ho states above regarding this EZ water, that it 'extends without interruption throughout the extracellular matrix to the interior of every single cell'. What makes up the 'extracellular matrix to the interior of every single cell'? Is it not the ubiquitous maze of omnipresent membranes or connective tissue that permeates the entire body by wrapping all the numerous organs and structures with connective-tissue fascia like a bag, thereby allowing communication between every cell in the body via the fascia? Li Chiung put it another way when he stated that 'the Triple Burner represents nothing but membranes attached to the upper, central, and lower opening of the stomach' in the commentaries on the 38th Difficult Issue on page 396 of Unschuld's (1) translation of the *Nan Ching*. Note too the title of an article written by Dr Ho (61) in 2012, namely, 'Super-Conducting Liquid Crystalline Water Aligned with Collagen Fibers in the Fascia as Acupuncture Meridians of Traditional Chinese Medicine'. While Dr Ho believes that collagen fibers in the fascia aligned with EZ water constitute the acupuncture meridians, I believe one step further, that the omnipresent connective tissues, collagen fibers, fascia, and membranes make up the Triple-Energizer organ complex. Dr Ho also said, 'It is water that fuels the dynamo of life; water is the basis of the energy metabolism that powers all living processes, the chemistry and the electricity of life.' Again that sounds like the major function of the Triple-Energizer organ complex, which is 'the official in charge of irrigation and it controls the water passages'.

The third world-renowned scientist Dr Shui Yin Lo, a biophysicist, uncovered and has researched for 15 years a newly discovered phase of water, which he calls Double-Helix Water®. He advises that it is ultrapure water, whereby all impurities have been excluded from it. Stone the crows, that sounds a lot like exclusion zone water by another name, Double-Helix Water®. Dr Lo has observed these water molecules can condense into tiny solid particles at room temperature. Pollack and Ho call this a liquid crystal matrix. Lo believes this water phase is involved in the healing process and the body's immune response. So do Pollack and Ho. Like Ho, Lo believes that 'these particles are the molecular basis for what Chinese Medicine has suggested for over two thousand years'. Using Electron and Atomic Force Microscope photographs, Dr Lo has demonstrated 'that these particles line up end to end to form circuit-like structures. Therefore it is very feasible that we have found a material basis for the Chinese meridians.' When I developed my theory about the nondescript, semiamorphous, ubiquitous, omnipresent connective-tissue system constituting the structural corpus of the Triple Energizer several years ago, I never realized so much established scientific research was available to support my theory and was stunned when these scientists even involved TCM and acupuncture meridians in their theories.

CHAPTER 21

Further Information about EZ Water and Biology

Introduction
In the reference entitled 'The Fourth Phase of Water: Central Role in Biology', recorded at the Natural Philosophy Alliance Conference in College Park, Maryland, Professor Gerald H. Pollack (35) discussed the following information pertaining to biology. He said water is absolutely essential to the maintenance of life and that it does much more than carry nutrients to cells. The existence of a fourth phase of water is not entirely unexpected. One century ago, the physical chemist Sir William Hardy argued for its existence, and over the years, many authors have also found evidence for some kind of ordered or structured phase of water. Fresh experimental evidence not only confirms the existence of that ordered, liquid crystalline phase but also details its properties. These properties explain everyday observations and questions ranging from why gelatine desserts hold their water to why teapots whistle. He also believes his theory explains numerous biological happenings that cannot be explained by any other water physics mechanism.

21.1 What Charges the EZ Water Battery When It Is in Its Structured Liquid Crystalline State?
Light, especially infrared (IR), is everywhere, and it is free. Chairs and walls emit IR. The infrared radiation organizes the water and

charges the battery. Just like plants absorb light energy from the sun and grow and form produce, including wood or fruit. On that basis, Pollack formulated the chemical formula $E = H_2O$, where E stands for Energy. Are we absorbing energy from our surroundings—from plants, walls, furniture, ceiling lights, and other people? He points out that water can convert electromagnetic energy into mechanical energy. We do not photosynthesize, *but* we surely use light in some very positive ways.

21.2 Ultralow Friction—Why Don't Our Joints Squeak?
Research has shown that no matter how much load you put on the bones of an intact joint capsule, the bones do not touch, and no one understands why. Pollack believes that because joint capsules are covered by cartilage composed of hydrophilic zones, protons are generated within the resultant exclusion zones in the joint synovial fluid, which repel each other and prevent the bones from touching.

21.3 Why Are Intracellular Fluids Negatively Charged?
Cell components are surrounded by EZ water, so intracellular fluids are negatively charged. The average space between cellular components is seven (7) water molecules. Other water scientists and biologists claim the negative electrical potential inside cells is due to the surrounding membrane. However, Pollack's team has confirmed that the electrical potential of gels is similar to intracellular fluid and that no membrane is present around gels. Pollack believes that the water in cells changing phase from the negatively charged EZ phase (−100 mV) to the neutral charge bulk water liquid phase (0 mV) engenders the action potential that allows muscle cells to exert force, secretory cells to secrete, and nerve cells to allow a nerve impulse through.

21.4 Why Do Local and General Anesthetics Work?
In the reference entitled 'The Fourth Phase of Water: Central Role in Biology', recorded at the Natural Philosophy Alliance Conference in College Park, Maryland, Professor Gerald H. Pollack (PhD) (35)

discussed the following information pertaining to biology. He believes that anesthetics work and eliminate sensations by wiping out the action potential of nerves so they cannot signal pain messages. Pollack actually commenced his career as a bioengineer in the department of anesthesiology. How anesthetics work is still not understood. Experimentation by Pollack showed that as the dosage of two common local anesthetics lidocaine and bupivacaine were increased, the exclusion zone around microspheres was reduced, suggesting that the local anesthetics work by reducing the action potential of nerves. Isoflurane, a common general anesthetic also behaved very similarly, and the exclusion zone around microspheres was reduced as the Isoflurane concentration was increased. Interestingly, an extremely low dose initially gave an excitory stage. Elevating the dosage further then gave the depressive stage. This phenomenon is well known to anesthetists. So it appears most probable that biological action of anesthetics is mediated by EZ water. The famous researcher and scientist Linus Pauling believed that because the noble gases—helium (He), neon (Ne), argon (Ar), krypton (Kr), xenon (Xe), and the radioactive gas radon (Rd)—reacted only with water, the biological mechanism of anesthesia with some of those noble gases involves water. On page 1334 of the book *Linus Pauling: Selected Scientific Papers Volume II*, Pauling (65) states:

> The hydrate-microcrystal theory of anesthesia by non-hydrogen-bonding agents differs from most earlier theories in that it involves primarily the interaction of the molecules of the anesthetic agent with water molecules in the brain, rather than with molecules of lipids. The postulated formation of hydrate microcrystals similar in structure to known hydrate crystals of chloroform, xenon, and other anesthetic agents as well as of the substances related to protein side chains, entrapping ions and electrically charged side chains of protein molecules in such a way as to decrease the energy of electric oscillations in the brain, provides a rational explanation of the effect of the anesthetic agents in causing loss of consciousness.

21.5 How Anesthetic Negatively Affects the Triple Energizer

In 2012, acupuncturist Kimberly Thompson (46) posted an article entitled 'What the Heck Is a Triple Energizer Anyway?' on the miridiatech.com website. She had recently undergone surgery and related how the general anesthetic had adversely affected her general health. She explained that after she had a hysterectomy, her Triple Energizer was greatly affected, so she did some extra study on the 'strange organ' in an effort to speed her recovery. Regarding the outcome of the surgery and especially the effect of the general anesthetic, Thompson stated, 'Anesthesia and the Triple Energizer *don't* get along' (emphasis is mine). She noted that after anesthesia has put the body to sleep, it takes some time for the resumption of normal bodily functions, including bowel function. She advised she had extreme hip pain and was unable to lift her legs, but she made good progress due to having acupuncture performed after the surgery. She reported an extreme lack of energy and believed it was due to her Source Qi being directed toward wound healing rather than the TE circulating Qi elsewhere as it normally did. Subsequently, she believes the body becomes sluggish and slow after surgery to allow the Source Qi to be utilized more efficiently for the healing process. She noted that Source Qi would be quickly depleted if a surgical patient endeavored to resume their normal life activities too soon after their operation. For this reason, she believes that symptoms including hormonal imbalances, low appetite, lack of digestion, and hormonal imbalances all allow the TE to work more efficiently at the healing process. She further noted that the reason it takes three months before a person feels back to normal again is because the TE is responsible for transporting fluids through the lymphatic system so that the interstitial fluid is cleansed.

I thought that Thompson's comments here were very pertinent, especially as Pauling and Pollack, quoted above, both believe that anethetics work due to their effect on water. And more precisely, as Pauling stated, anesthesia worked due to the 'interaction of the molecules of the anesthetic agent with water molecules in the brain'. Subsequently, it makes perfect sense that the organ that is the 'the official in charge of irrigation and it controls the water passages' should be more affected by anesthetics than the other organs.

21.6 What Causes Cellular Pathology?
In the reference entitled 'The Fourth Phase of Water: Central Role in Biology', Professor Pollack (35) discussed the following information pertaining to biology. When there is negative potential in healthy cells, all is well. Around every protein in the cell is EZ water. If there is less EZ water present, then there will be pathology. Pollack believes that Aspirin helps numerous different medical conditions because its innate molecular structure causes an increase in the size of the exclusion zone. In laboratory trials using microspheres, Pollack has confirmed that Aspirin increases the size of exclusion zones to five times the size of controls.

21.7 Why are Living Cells Negatively Charged?
In the human body, 60% is composed of cells, which are negatively charged. The extracellular space of the body makes up a further 20%. Thus, 80% of our body is composed of negative charges. So where are the low-pH positive charges? The positively charged components of our body include urine, sweat, and our exhaled breath (which contains positively charged CO_2 gas in water as carbonic acid). In good health, the body strives for the state of highest electronegativity. All plants too are electronegative as they are connected to the electronegative earth. So Dr Pollack believes that 'maintaining electro-negativity is a fundamental attribute of life'. That is why antioxidants are very beneficial for good health. They prevent the loss of electrons. That is also why negative ions and Earthing™/grounding are so beneficial for maintaining good health. Also, drinking a copious volume of water each day is very healthy. Pollack believes that if we all drank more water, our health would automatically improve. EZ water is like an engine that takes in energy from the environment, transducing it into other forms of energy essential for life and optimum health and well-being.

21.8 Because Pollack's Hypothesis Makes Perfect Sense, He Can Expect Conflict
Pollack's theory makes perfect sense. Pollack's theory explains numerous observations pertaining to water that no other water

scientists or theories can explain. But wow, it is just so novel and different from all the other conflicting and convoluted theories. It is just so very simplistic. His unifying simplistic theory would redefine cellular biology, enzymology, cloud physics, biochemistry, ion pumps that regulate cell membrane potential, and the list goes on and on. How many eminent scientists from vastly diverse fields of research would have their respected theories and research proven wrong? It seems most likely they will all cling together and, as a group, try to oust this apostate with his radical novel hypothesis. They would probably even set out to defame Pollack and his hypothesis and summon the assistance of the flat-earth society groups that dwell in a vacuum, known as Quackwatch and Fiends of Science in Medicine.

21.9 Examples of Other Scientific Researchers Ostracized for Suggesting Novel Ideas

Do you remember Dr Barry James Marshall, Australian physician, Nobel Prize laureate in Physiology, and Professor of Clinical Microbiology at the University of Western Australia? He was ridiculed for years for his unbelievable radical hypothesis that the bacterium *Helicobacter pylori* (*H. pylori*) would dare to survive in hydrochloric acid in the stomach and be the cause of most peptic ulcers. This radical physician reversed decades of incorrect medical dogma holding that ulcers were caused by stress, spicy foods, and too much acid. Marshall has been quoted as saying in 1998 that 'Everyone was against me, but I knew I was right.' Marshall's apostasy is now accepted gospel.

What about Dr Ignaz Semmelweis? *Wikipedia* (66) advises he was a Hungarian physician now known as an early pioneer of antiseptic procedures. He was described as the savior of mothers since Semmelweis discovered that the incidence of puerperal fever could be drastically cut from a mortality rate of 10% to 35% by the use of hand disinfection in obstetrical clinics. Doctors then believed it was perfectly acceptable to deliver babies straight after performing autopsies without washing their hands. Despite various publications of results where handwashing reduced mortality to below 1%, Semmelweis's observations conflicted with the established scientific and medical opinions and paradigm of the time, and his common-sense ideas were rejected by the medical community as absurd. Some doctors were offended at the suggestion that they should wash their hands, and Semmelweis could offer no acceptable scientific explanation for his findings. Semmelweis's proposal earned widespread acceptance only years after his death when Louis Pasteur confirmed the germ theory and Joseph Lister, acting on the French microbiologist's research, practiced and operated using hygienic methods with great success. Semmelweis's supposed apostasy is now accepted gospel. This entrenched medical ignorance reminds me of the quote by Serge Gracovetsky, who concluded, 'Medicine is perhaps the only human activity in which an attractive idea will survive experimental annihilation.'

Hundreds of such *apostates* have fortunately changed the face of science and medicine, and formerly ridiculed proponents of new ideas are now revered. I trust that Pollack's theory will constitute one of those stories. Albert Szent-Györgyi, Nobel laureate (1893–1986), previously proposed many of the same concepts about water that Pollack proposes. He wisely stated, 'Discovery consists of seeing what everybody has seen and thinking what nobody has thought.'

21.10 Western Medicine Hides Behind a Facade of Knowledge and Integrity

It never ceases to amaze me (and annoy me) that western medicine constantly utters their proverbial banter that their deeply entrenched paradigm is based on evidence-based medicine (EBM) and scientifically proven using double-blinded, randomized, controlled trials. They may indeed be pretty words, but all I can say to that claim is 'Poppycock'—and that is being kind.

When I graduated from Acupuncture College, the western medical system incorrectly believed that because plant fiber was not digestible by human intestinal enzymes, it was considered to be an inert component of plant foods and was considered by all western-trained doctors to be of no importance in nutrition. I took great measures to assure patients that was not correct and that the medical system was profoundly wrong. Now the medical system is fully aware of this common-sense issue. I also remember reading medical articles stating that mushrooms likewise were an inert food and of no benefit and that eating 2 eggs a week were bad for your health because they were hypercholesterolemic. Doctors were not aware that the high lecithin level in the eggs counteracted the high cholesterol level. How many patients of the western medical system have been damaged by such continuing faulty medical advice? It seems to me that by using common sense and intuitive wisdom, complementary practitioners have always been decades ahead of the western medical system paradigm in numerous ways when it comes to dispensing sound and correct medical health information. It is apparent that the western medical system dispenses sound, common-sense health information only after it has been proven correct by evidence-based medicine and scientifically proven by using double-blinded, randomized, controlled trials. However, even that statement is a far stretch of the imagination. Numerous western medical myths continue to abound for decades in spite of updated medical findings being published in prestigious medical journals.

21.11 Further Examples of Scientifically Proven Advancements Not Accepted by Doctors

For example, a study by Ponikau et al. (67) published in the *Mayo Clinic Proceedings* in 1999 states that allergic fungal sinusitis (AFS) exists in a significant majority of patients that have been diagnosed with chronic rhinosinusitis (CRS). A massive 96% of the study subjects with chronic rhinosinusitis were found to have fungi (representing 40 different genera) present in their nasal secretions when cultures were prepared from their nasal mucus. Based on their scientific findings, Ponikau et al. conclude that the majority of patients suffering from CRS actually have allergic fungal sinusitis and that allergic fungal sinusitis is significantly underdiagnosed. The fact is that most doctors automatically and incorrectly assume that chronic rhinosinusitis is due to bacterial infection, and unfortunately, they prescribe antibiotics and steroids, probably the worst two pharmaceuticals to use against fungi. Once again their entrenched but incorrect medical stance and treatment paradigm leaves their maltreated patients worse for wear in spite of the correct medical treatment being available for 15 years. I generally get great results when treating chronic rhinosinusitis with acupuncture and lifestyle changes, including nasal irrigation using Jala neti pots.

Another recent example was presented in 2013 when the *Lancet Neurology* journal published the article 'Vasodilation Out of the Picture as a Cause of Migraine Headache', where author Dr Andrew Charles (68) noted that the erroneous medical belief that migraines are a vascular headache has persisted for decades in spite of progressively accumulating evidence to the contrary. He noted that Faisal Amin and colleagues reported their findings in the the *Lancet Neurology* medical journal and that their findings disprove the central precept of the vascular hypothesis for migraine causation. The medical fraternity incorrectly believes that headache is caused by dilation of dural and extracranial vessels. Dr Charles concluded, 'This study might help to finally lay to rest a theory that has been entrenched in the medical consciousness over much the past century.'

Wow! So for over a hundred years, poor migraine sufferers have been maltreated and dosed up on the wrong medication by western doctors, thanks to a totally incorrect paradigm about migraine being a vascular headache. The strange thing is, thanks to my faith in reading alternative medicine journals brimming with good old-fashioned common sense, I have always believed that migraines were due to vasoconstriction and have found the powerful vasodilator magnesium sulfate very beneficial for the treatment of migraines. The problem is, I am still treating patients that are being maltreated by their family doctors that have not read the *Lancet Neurology* article that appeared in the journal cited above. It will probably take another two decades before all doctors have come up to speed with this paradigm shift, albeit very slowly (pun intended). This medical ignorance reminds me of the quote by Serge Gracovetsky, who stated, 'Medicine is perhaps the only human activity in which an attractive idea will survive experimental annihilation.' I love this quote so much because of its honesty. That is why you have seen it before in my book.

21.12 Why 'Science-Based Medicine' Often Assumes the Composition of Bovine Excrement

Trusting the content of new medical findings and practically applying the new information applies only when honest medical scientific research has been progressed with integrity. However, very often this expectation is just not the case. In 2012 Mercola (69) reported, 'The medical and health fields are absolutely riddled with dogmatic beliefs that defy both common sense and scientific truth. And yet they prevail because they are supposedly backed by "science".' Mercola continued:

> But what if that science was not actually trustworthy, but rather a carefully orchestrated product . . . the result of massive conflict of interest, perpetuated by self-interested groups and industries that push unfavorable research findings under the proverbial rug, or even tweak their studies to have the 'right' results?
>
> Would it change the way you feel about your diet, your lifestyle and even your medical decisions if it turned out the research upon which your prior choices were made was actually not science at all, but fraud?

The article proceeded to report that a review 'of over 2,000 biomedical and life-science research papers that have been retracted through May 2012 found that only 21 percent were retracted due to errors. Instead, the most common reason, in over 67 percent of cases, was misconduct, including fraud or suspected fraud (43.4%), duplicate publication (14.2%), and plagiarism (9.8%).' Shamefully, the data showed it was more likely that retractions were attributed to fraud or suspected fraud from the more respected and influential scientific and medical journals. The study confirmed that a disturbing *epidemic of deception is fulminating in the arena of scientific and medical research.* The seething mass of fraudulent science has been proceeding for decades and is estimated to have increased tenfold since 1975. How many newly conceived medical decisions have been made by doctors who have just read an enlightening article which was based on deceptive fraudulent reports so that the pharmaceutical industry could satisfy their greed and be more governed by wealth than by health?

21.13 Just How Scientific Is 'Science-Based Medicine' Really, Honestly?

In the revealing 2012 article titled 'Why Your Doctor's Advice May Be Fatally Flawed', Mercola (70) states:

> I am a big believer in the scientific method, provided it's applied appropriately that is. And that's the key issue here. In order to qualify in the first place, the research must be unbiased, unprejudiced and free from any significant conflicts of interest. Sadly, this is not the case with most of modern medicine—especially not when it comes to drug research, as evidenced by the featured findings.
>
> It's quite shocking that nearly three-quarters of all retracted drug studies are due to pure falsification of data. Especially when you consider that even well-researched drugs can still have significant side effects. Just imagine the potential for tragedy when a drug is based largely on pure fantasy!
>
> Vioxx is perhaps one of the better examples of what can happen when a drug is manufactured and sold under false pretenses. It killed more than 60,000 people in just a few years time, before it was removed from the market. In the case of Vioxx, there are lingering questions about the soundness of the research backing the drug in the first place. Back in 2008, Dr. Joseph S. Ross of New York's Mount Sinai School of Medicine came across ghost-written research studies for Vioxx while reviewing documents related to lawsuits filed against Merck.

21.14 *British Medical Journal* Editor for Thirteen Years Declares Peer Review Is Nonsense

The *British Medical Journal* (BMJ) is one of the most esteemed and prestigious scientific journals in the world. In the 2015 article titled '"Sacred Cow" of Industry Science Cult Should Be Slaughtered for the Good of Humanity, *BMJ* Editor Says', the author (71) noted that a former editor of *BMJ* strongly denounced the current universal acceptance of prepublication peer-reviewed articles being the 'gold standard' of science-based research. The former editor, Richard Smith, was the editor of *BMJ* for about 13 years between 1991 and 2004. Smith has been highly critical of the peer review process for a long time. He believes that for all intents and purposes, the peer review process is definitely not science-based in its approach regarding what research information actually gets published in medical and science journals. Smith poignantly advised that a large proportion of what is accepted for publication is simply 'plain wrong or nonsense'. He warned that no evidence exists to confirm that the peer review process is successful in identifying errors or guaranteeing that only factual scientific discoveries get published in the world's most prestigious medical and scientific journals. Smith commented that if the peer review process was a new pharmaceutical, it would never be released onto the marketplace. He believes this is the case because there is a vast amount of evidence showing its adverse effects, while there is no evidence showing the benefits of the peer review process. Smith stated that 'peer review is a failure and, ironically, it's more faith-based than science-based', and he strongly advised that 'it's time to slaughter the sacred cow'.

21.15 The Proof that Peer Review' Is a Faith-Based Process Rather Than Evidence-Based Process

Richard Smith, the former editor of the *British Medical Journal* (BMJ), presented a speech at an event for the Royal Society. Smith described an experiment that he helped conduct during his 13 years at BMJ. The experiment blatantly revealed that a striking number of intentional errors added to a test paper made it through the peer review process without even being spotted. Huff (71) described Smith's clandestine experiment.

A short paper was written that intentionally contained eight deliberate errors. The paper was sent to 300 individuals for review. Amazingly, Smith reported that only a handful of these chosen reviewers detected any of the intentional mistakes. Not a single reviewer found all eight obvious errors. Shockingly, sixty of the reviewers (20%) did not find a single mistake. Disgracefully, Smith reported, 'No one found more than five, the median was two, and 20 percent didn't spot any.' This bodes very poorly for the supposedly scientific peer review process. What is even worse is that the peer review process breeds discriminatory sentiment against honest pioneering research findings that confront the scientific status quo. There is major discrimination against research papers that reveal the hazards and unnecessary risks associated with issues regarding vaccine safety or the efficacy of orthodox cancer treatments, like chemotherapy. In a 2011 posting at *BMJ* Blogs, Smith wrote that the evidence regarding the prepublication peer review process confirms it is ineffective while, at the same time, it is slow, is very expensive, and is largely a 'lottery'. Further to these demerit points, Smith noted that the peer review process is dismal at spotting error, highly biased, anti-innovatory, highly subject to abuse, and completely impotent in detecting fraudulent research. Smith reported that the global cost of the defective peer review behemoth is $1.9 billion. Not surprisingly, Smith stated, 'It's a *faith-based rather than evidence-based process*, which is hugely ironic when it's at the heart of science' (emphasis is mine).

21.16 What Passes as 'Science' Today Is Really a Cult of Pet Opinions Pushing an Agenda

Regarding the endemicity of fraud and fanciful fabrication that is published in supposedly respected scientific and medical journals, Huff (71) related what the editor of the UK's other top-leading medical journal, *The Lancet*, had to say concerning the peer review process. Editor of *The Lancet* Dr Richard Horton stated, 'Much of the scientific literature, perhaps *half,* may simply be *untrue*' (emphasis is mine). Dr Horton believes this is indeed the case due to many factors, including statistically small sample sizes, researchers with corrupting conflicts of interests, and an obsession that scientists possess to 'pursue fashionable trends of dubious importance'. He observed that unscrupulous research behavior is endemic amongst the science community and noted that, in their yearning to tell a captivating story, scientists all too often mold their data to accommodate their preferred ideal notion of the world. A powerful vested interest that is undeniably anti-science is disguised as factual science, and is promulgating antiscience concepts, such as the absolute safety of vaccines and Frankenfood GMOs and other pseudoscientific misinformation, as fact. For example, in 2014, 120 papers were withdrawn from multiple science journals after a computer scientist admitted that they were all entirely falsified and fraudulent. It is no wonder that the central paradigm of the western medical profession is so immensely corrupted; scientific fraud is so rampant to ensure that the funding and grants for poppycock proposals are allowed to continue due to the propensity of researchers to perpetuate the myths that are flourishing throughout the medical journals. Unfortunately, it generally takes decades (and thousands of corpses) for honest lone-voice researchers without vested interests to finally set the record straight.

Summary of Chapter 21
It can be seen from his 2013 presentation (35) and numerous other published research findings, some of which I have presented in my book (27, 60, 72), that Professor Pollack has considered very diverse phenomena that occur throughout the numerous areas of chemistry, physics, and biology. His EZ theory certainly holds water (no pun intended—well, actually it was), and everything he says makes perfect sense and is scientifically verifiable by repeatable scientific experimentation that is honestly performed without prejudice. The only problem Pollack faces is incorrect but deeply entrenched beliefs and concepts about water chemistry that are well and truly worth reconsidering. The major difficulty is the established scientific and medical opinions, which are adhered to by die-hard scientists that don't like to think that what they have formerly believed for years could be incorrect. That would mean that work from grants they received and information published in their books are wrong. How embarrassing and belittling would that be? They would lose their grants and funding! The truth can certainly hurt and no less so when faulty scientific endeavor is involved and recorded for posterity.

CHAPTER 22

Further Information About EZ Water and Fascia

Introduction

The following is extracted from an editorial written by Gerard Pollack, PhD, from the University of Washington and published in the 2013 *Journal of Bodywork and Movement Therapies*. In the article entitled 'The Fourth Phase of Water: A Role in Fascia?' (72), Pollack advises that because water is so ubiquitous and has such a simple molecular structure, it is generally assumed that the nature of water is completely understood. But nothing could be further from the truth as precious little is known about the molecular structure and alignment of water molecules. However, recently many researchers have determined that water possesses a fourth phase other than the three known phases—namely, solid, liquid, and vapour. At his University of Washington laboratory, Dr Pollack has discovered and extensively researched a surprisingly widespread fourth phase of water that develops and exists next to water-loving (hydrophilic) surfaces. This previously undiscovered phase of water is ubiquitous throughout nature and projects outward from hydrophilic surfaces, forming up to millions of molecular layers away from the substrate.

This information sets the pace for the following Chapter, and I will discuss how water responds when it comes into contact with hydrophilic surfaces. Connective tissues, fascia, and membranes all happen to be hydrophilic surfaces. Imagine that! The omnipresent Connective-Tissue Metasystem pervades the entire body, and so does water. Add to that the fact that infrared radiation also pervades our body so that throughout our entire body at any given time, water, the Connective-Tissue Metasystem, and infrared radiation all coexist. Exactly what do these Three Amigos get up to together?

22.1 Scientists Have Previously Proposed a Fourth Phase of Water

In the article entitled 'The Fourth Phase of Water: A Role in Fascia?' (72), Pollack explains that, about one century ago, the noted physical chemist Sir William Hardy contended that a fourth phase of water existed. Since then, numerous other renowned scientists have uncovered evidence regarding a highly ordered or structured phase of water. Recent experimental trials confirm the existence of the fourth phase of water, which is ordered and liquid crystalline, and have also determined the many properties of the newly discovered phase. These newly determined properties explain why gelatine desserts hold their water without leakage and why teapots whistle when the water boils, for example. Pollack believed that the energy that powers the fourth phase formation of water is the radiant energy from the sun, which converts normal bulk water into a structured polymeric water zone. He explains that water spontaneously absorbs infrared energy that is ever present in the ambient environment, and thus, the buildup of the fourth phase of water occurs effortlessly and spontaneously. Significantly, the fourth phase of water is predominantly negatively charged. The solar-derived energy splits water molecules apart, and the negative end creates the building block used to construct the highly ordered zone, while the positive end of the water molecule reacts with bulk water molecules to generate free hydronium ions (H_3O^+). If more light is added to the system, greater charge separation occurs. Intriguingly, the chemical reaction is closely related to the initial step of photosynthesis in plants.

22.2 How the EZ Water Theory Explains Hydrochloric Acid Production in the Stomach

I propose that this process is what occurs in the Stomach, where water molecules split and the 'positive moiety binds with bulk water molecules [in the Stomach] to form free hydronium ions' in the form of hydrochloric acid for digestion of grains. The negative moiety forms the building block of the ordered zone, which is transmitted from the stomach to the spleen from where it is transformed into influences to produce the numerous fluid and energy forms synthesized by the Triple Energizer.

In the article entitled 'Parietal Cell' (73), the text outlines the formation of hydrochloric acid by parietal cells (also known as oxyntic or delomorphous cells) in the stomach. The article notes that water molecules dissociate to form Hydrogen ions. Subsequently, the following reactions occur:

> The enzyme carbonic anhydrase converts one molecule of carbon dioxide and one molecule of water indirectly into a bicarbonate ion (HCO_3^-) and a hydrogen ion (H^+).
> The bicarbonate ion (HCO_3^-) is exchanged for a chloride ion (Cl^-) on the basal side of the cell and the bicarbonate diffuses into the venous blood, leading to an alkaline tide phenomenon.
> Potassium (K^+) and chloride (Cl^-) ions diffuse into the canaliculi.
> Hydrogen ions are pumped out of the cell into the canaliculi in exchange for potassium ions, via the H+/K+ ATPase.

The article then advises that the gastric lumen is subsequently highly acidic due to the cellular export of hydrogen ions. Note in Pollack's theory this is exactly what occurs in the same chemical pathway, with protonated water (HCl solution) moving in one direction within the parietal cell (in this case, into the lumen of the stomach) while the electronegative moiety (in this case, the bicarbonate ion (HCO_3^-)) diffuses through the basolateral membranes of the parietal cells in the stomach wall into the venous bloodstream. Pollack believes this biochemical pathway is driven by the exclusion zone water consequence and not by the prevailing view of membrane ion pumps and channels. His theory fits all the steps and events that are known to occur and can be reproduced in the laboratory. The only difference is his model does not require membrane ion pumps and channels to perform the work. Pollack noted that EZ water acts as a battery and a pump and is powered by infrared light, which is bountiful in our body and our surroundings.

22.3 How Radiant Energy from the Sun Drives EZ Water Production

Regarding the ubiquity of liquid crystalline water throughout the entire body, in the 2013 article, Pollack (72) explains that during the EZ generation process, the charges that are separated by the influence of light resemble a battery which can deliver energy as required, analogous to the way that separated charges in plants deliver energy as required. He notes that because plants are mostly composed of water, it is likely that a similar energy conversion process takes place actually within the water itself. Pollack stated succinctly, 'The fourth phase can appear nearly everywhere. All that's needed is water, radiant energy, and a hydrophilic surface.' He notes that the hydrophilic surface required could consist of a large piece of polymer or tissue or could be as small as a molecule dissolved in solution. He believes that the liquid crystalline fourth phase predictably builds, and then the behavior of the system is determined by the EZ presence. Consequently, he believes that the liquid crystalline fourth phase of water encloses every single macromolecule throughout the body, especially those of the hydrophilic fascia. He further believes that liquid crystalline water permeates the body and that the majority of water present in the body is liquid crystalline water.

Our entire body, including our hands, emits radiant energy constantly. Regarding massage, Pollack believes this radiant energy emitted from our hands, along with the local pressure resulting from the massage technique, may create the efficacious factors during massage, whereby such radiant energy and pressure affects the ubiquitous fascia, causing it to generate liquid crystalline water. He noted that the presence of liquid crystalline water is necessary for both healthy tissue and for normal physiological function.

22.4 The Relationship between Fire and the Three Heaters

In the Appendix of the book *Heart Master Triple Heater* (page 123–124), while summarizing the intimate relationship between the fire of life and the triple heater, Elisabeth Rochat de la Vallée and M. Macé (11) explain that the Triple Heater regulates the distribution of all the body fluids and is involved in all aspects of digestion and absorption. Its intrinsic fire energizes the respective activity of each of its three components. The authors note that in his Ming dynasty work, the *Yixue Shengshuan*, Yu Tuan considered that the Triple Heater was linked to the minister fire, which engendered all life. The Kidneys are the organ of choice to process and control water and fluids in the body. But the Triple Energizer is also assigned that role, which is an obvious choice due to the water content of the omnipresent hydrophilic (water-loving) maze of connective tissues that pervade the body and occur only where something other than connective tissue wishes to reside. Also obvious is the further connection with the Bladder, which holds the wastewater of the body. Interestingly, all these organs—the yang right Kidney, the Triple-Energizer Metasystem and the Bladder—are resplendent with fire and yang. The highly energized and longest meridian is the Bladder meridian with its 67 acupoints. As these three organs have a major role in water and fluid circulation, they are in constant contact with highly charged EZ Water $[H_3O_2]^-$, which by its very nature is greatly energized and performs as a battery that is contained within the Triple-Energizer Metasystem network, which permeates the entire body and is only micrometers away from every cell in the body that requires an energy top-up. The sagacious Yu Tuan was most enlightened when he considered that the Triple Heater was the fire that circulated throughout the body to develop all life.

I consider that modern science, through Professor Pollack, has confirmed that this fire is contained within the energized water that the Triple Energizer circulates and manages in the form of EZ Water, also known as the hydrated hydroxide ion or hydroxide hydrate anion $[H_3O_2]^-$.

22.5 Organic Heated Living Water Is Circulated by the San Jiao

Regarding the association of fire and the coexisting transportation of the various fluid components of the San Jiao, in the article entitled 'San Jiao', the author, Linda Barbour (23), notes that Dr Tran believes that on orders from the Heart, the Triple Burner meridian transports heated organic water throughout the body. Dr Tran believes that the heated water originates in the lower burner.

Summary of Chapter 22

Summarizing this Chapter, it can be seen that for over one hundred years, many scientists, including the physical chemist Sir William Hardy, argued for the existence of a fourth state of water. Professor Pollack has made the most headway in scientific research to confirm Hardy's hypothesis. Pollack has proven that 'this phase occurs next to water loving (hydrophilic) surfaces. It is surprisingly extensive, projecting out from surfaces by up to millions of molecular layers; and it exists almost everywhere.' The fascial system of the body is omnipresent. The fascial system of the body is hydrophilic. The fascial system of the body is bathed in body-generated infrared energy. The fascial system of the body lies beside *formerly* unstructured bulk water throughout the entire body. Why do I say *formerly*? As Pollack has shown, when unstructured bulk water + hydrophilic surfaces + infrared energy interact, structured, living liquid crystalline water (i.e. EZ water) results. Pollack has proven that EZ water exists almost everywhere within the body and that EZ water produces energy spontaneously, acts like a battery, causes pumping to occur, manifests catalysis, and generates fire. As the omnipresent fascial connective-tissue system makes up all tissue in the body other than connective tissue, I believe that when Pollack describes EZ water permeating the entire body, he is describing the modus operandi of what TCM calls the Triple-Energizer organ complex. Remember, because water molecules are so small, with a Molecular Weight of only 18.01528 g/mol molecule for molecule, H_2O constitutes 99% of the molecules in the body. Remember too, grasshopper, that the *Su Wen* says, 'The San Jiao is the official in charge of irrigation and it controls the water passages.' In the commentaries on the 38th Difficult Issue of Unschuld's (1) translation the *Nan Ching*, Li Chiung states, 'The Triple Burner represents nothing but membranes.'

We have learnt a lot about the scientific properties of water and especially structured EZ water. We need to know more about the scientific properties of membranes or connective tissue and about light. So I am going to discuss the biological functions of light in the next Chapter.

CHAPTER 23

Biophotons and Biological Functions

Introduction
In the article titled 'The Fourth Phase of Water: A Role in Fascia?', Pollack (72) states, 'The fourth phase can appear nearly everywhere. All that's needed is water, radiant energy, and a hydrophilic surface.' I have discussed water in detail. So now let's consider the second requirement for the formation of EZ water in the body, namely, radiant energy.

23.1 What Are Biophotons?
In the book called *Biophotons: The Light in Our Cells*, the author, Marco Bischof (74), states, 'Biophotons, or ultraweak photon emissions of biological systems, are weak electromagnetic waves in the optical range of the spectrum—in other words: light. All living cells of plants, animals and human beings emit biophotons which cannot be seen by the naked eye but can be measured by special equipment developed by German researchers.' Bischof's (74) book continues:

> This light emission is an expression of the functional state of the living organism and its measurement therefore can be used to assess this state. Cancer cells and healthy cells of the same type, for instance, can be discriminated by typical differences in biophoton emission. After an initial decade and a half of basic research on this discovery,

biophysicists of various European and Asian countries are now exploring the many interesting applications which range across such diverse fields as cancer research, non-invasive early medical diagnosis, food and water quality testing, chemical and electromagnetic contamination testing, cell communication, and various applications in biotechnology.

According to the biophoton theory developed on the base of these discoveries the biophoton light is stored in the cells of the organism—more precisely, in the DNA molecules of their nuclei - and a dynamic web of light constantly released and absorbed by the DNA may connect cell organelles, cells, tissues, and organs within the body and serve as the organism's main communication network and as the principal regulating instance for all life processes. The processes of morphogenesis, growth, differentiation and regeneration are also explained by the structuring and regulating activity of the coherent biophoton field. The holographic biophoton field of the brain and the nervous system, and maybe even that of the whole organism, may also be basis of memory and other phenomena of consciousness, as postulated by neurophysiologist Karl Pribram and others. The consciousness-like coherence properties of the biophoton field are closely related to its base in the properties of the physical vacuum and indicate its possible role as an interface to the non-physical realms of mind, psyche and consciousness.

In the book, the author (74) advises that biophoton emission lends credence to some supposedly unconventional methods of healing based on concepts of homeostasis (self-regulation of the organism), such as various somatic therapies, homeopathy, and acupuncture. According to Traditional Chinese Medicine, the Qi energy flowing through the energy channels (meridians) of our body regulates our body functions. This effect may be related to node lines of the organism's biophoton field, which has a basis in weak, coherent electromagnetic biofields.

23.2 Biophoton Communication Demonstrated in Plants, Bacteria, and Human Cells

The Abstract of the article entitled 'Biophotons as Neural Communication Signals Demonstrated by In Situ Biophoton Autography' (75) states, 'Cell to cell communication by biophotons has been demonstrated in plants, bacteria, animal neutrophil granulocytes and kidney cells.' The authors noted that light stimulation at one end of rat spinal sensory or motor nerve roots caused a significant increase in the biophotonic activity at the other end of the nerves. Spectral-light stimulation from infrared, red, yellow, blue, green, and white light all yielded an effect. It was observed that classic metabolic inhibitors and procaine significantly inhibited the effect. Procaine is a local anesthetic known to block neural conduction. The results suggest that light stimulation generates biophotons that conduct along the nerve fibers possibly due to protein–protein biophotonic interactions.

The article (75) further discussed how, under various physiological and pathological situations, biophotonic activity changes. The biophotonic changes result from mechanical, thermal, and chemical stresses; the cell cycle; and mitochondrial respiration; even cancer growth leads to biophotonic changes. Regarding consciousness, meditation, and the status of acupuncture meridians, there is evidence that suggests that biophotons play a central role in the function of the nervous system in humans.

23.3 All Living Cells Emit Detectable Biophotons that Control Cellular Function

Based on experimentation conducted at the Department of Molecular Cell Biology, Utrecht University, Netherlands, and at the International Institute of Biophysics, Neuss, Germany, the author VanWijk (76) believes that biophotons are a form of radiation that possess a bioinformation-carrying capacity. Thanks to the greater sensitivity of modern photomultiplier tubes, it is now clear to every scientist working in this field that ultraweak photon emission in the visible and UV range can be detected from nearly all living cells. A growing body of researchers believe that below the biochemical level of control, very weak electromagnetic interactions play a major regulatory role for all living organisms, including biophotonic information controlling cell division, and other physiological processes occurring in every cell. This theory was originally investigated and suggested by Gurwitsch as early as 1911.

In the research article 'Bio-Photons and Bio-Communication', VanWijk (76) explains, 'The generation of photons requires two phases: (1) the energy pumping that promotes an electron to the excited level, and (2) radiative relaxation that creates a photon. Living organisms can utilize a variety of energy forms and transform a fraction of them into an electronic or vibrational excitation.' He continued:

> Most presently known efficient chemiluminescent systems involve large molecules with easily polarizable and thus excitable π-electron systems, such as flavins, indoles, porphyrins, carbonyl derivatives of aromatic compounds, heterocyclic rings like purines and pyrimidines and species-specific compounds evolved in bioluminescent organisms, the so-called luciferins. These compounds have a relatively high quantum yield due to the short lifetime of the singlet excited state. Good candidates for direct emitters, especially in the near UV range, are tryptophan as well as nucleic acids.

23.4 How Cells May Communicate with Flashes of Light (Qi) within the Human Body

In an interview in 'Science & Spirituality: Living Light: Biophotons and the Human Body—Part 1', Professor Popp (77) discussed the following. In every single cell, there are about 100,000 chemical reactions occurring every second. They all occur at the right time at the right position that is required for life to continue. Each of these reactions requires a photon to make the chemical reaction occur. Our body is bathed in an ocean of light. It is impossible to have a living system that is devoid of light. It is the high degree of coherence of the photons and not the intensity that makes the complex biosystem run so efficiently. They have an extremely high degree of order. The photons can be superpositioned so that the messages from the biophotons are very distinct signals. They are submitters of information. The high degree of coherence of biophotons in living organisms is related to the hundreds of thousands of processes that take place in such a marvellous way in every cell during every second of life. Biophotons send encoded information, communicating between biological cells at the speed of light. Popp (77) believes that biomolecules and chemical messengers are too slow to guarantee the integrity of the organism and that some mechanism operating at the speed of light must be responsible for the vast array of chemical reactions happening every second to sustain life. It is very noteworthy that, regarding radiant energy in the form of light, Professor Popp advises that biophotons in the human body have an extremely high degree of order, while Professor Pollack has proven that EZ water in the body also has an extremely high degree of order and refers to EZ water as ordered water or structured water. Interesting!

The 100,000 reactions per second are controlled and regulated by photons. Each reaction takes a nanosecond to be completed. A spatial dynamic pattern produced by an electromagnetic field is required, which involves the correlation of space and time to tell the cell exactly what to do at exactly the right location at exactly the right time. A photon is supplied for a reaction to occur. The photon is not thermalized and thus converted to heat and lost during the reaction. After the specific chemical reaction occurs, the photon is given back to be used again for the next reaction. So you don't need 100,000 photons in each cell to trigger the reactions in a controlled manner. One photon could potentially be recycled over and over to trigger the 100,000 reactions per second.

23.5 Relationship between Biophotonic Activity and the Bioelectronic Activity of Nerves

In the Discussion section of the article, the authors (75) state, 'It is well accepted that the processing of information by neurons is mediated by their electrical activity, which is generated by the movement of particles, such as ions carrying electrical charges. Our finding in the present study that biophotonic activity may serve as a neural signal raises the question as to whether there is a relationship between biophotonic activity and bioelectronics activity.' The authors (75) further state:

> In vivo imaging of the spontaneous biophotonic activity in a rat's brain correlates with EEG activity. In addition, pulsed electric excitation of frog sciatic nerve has been reported to cause photon emission. In a recent study, it was found that focused mid-infrared light alters membrane potential and activates individual neural processes. These research data suggest that biophotonic or bioelectronic activities in the nervous system are not independent biological phenomena. In our view, the interactions between bioelectronic and biophotonic activities might be an important way for neural information exchange to take place, in which bioelectronic activities may be only the basis for biophotonic activities, providing new perspectives for better understanding the functions of the nervous system, such as vision, learning and memory, as well as the mechanisms of human neurological diseases. In addition, whether biophotons could serve as fundamental signals that can carry and transfer information from one place to another in the nervous system is an interesting possibility for investigation in the future.

23.6 The Puzzling Role of Biophotons in the Brain

In the journal article titled 'Emission of Mitochondrial Biophotons and Their Effect on Electrical Activity of Membrane via Microtubules', Rahnama et al. (78) advised that various research findings suggest that neurons do emit and even conduct photons. Biophotons may help to synchronize the brain. There is growing evidence which shows that photons play a dynamic role in the basic biological functioning of cells. It appears that many cells, perhaps even most cells, emit light as they perform their functions. It appears that many cells use light as a form of intercellular communication. There is evidence that bacteria, plants, and even kidney cells communicate using biophotons. Research has confirmed that rats' brains are literally alight due to the photons formed by neurons as they work, and one group of researchers determined that spinal neurons in rats can actually conduct light.

23.7 Does the Supporting Microtubule Cytoskeleton Also Function Like Fiber Optic Cables?

Majid Rahnama (78) at Shahid Bahonar University of Kerman in Iran and a group of researchers rationalize that, due to the fact that neurons contain many light-sensitive molecules (such as porphyrin rings, flavinic and pyridinic rings, lipid chromophores, and aromatic amino acids), it is hard to imagine that neurons are not influenced by biophotons. For example, mitochondria, the organelles inside cells which produce energy, contain several prominent chromophores. Rahnama et al. hypothesize that microtubules within cells may act as wave guides, channelling light from one part of a cell to another.

The primary role of the microtubule cytoskeleton is mechanical, providing structural support and integrity. However, in addition to structural support, microtubules also take part in many other processes. A microtubule is capable of growing and shrinking in order to generate force, and there are also motor proteins that allow organelles and other cellular factors to be carried along a microtubule. This combination of roles makes microtubules important for organizing cell layout. Could it be that they also work like optical fibers? Rahnama et al. (78) propose that the light channelled by microtubules might help coordinate activities in the diverse regions of the brain. It is a fact that electrical activity in the brain is synchronized over distances that cannot be easily explained. Electrical signals transmitted through nerves are not fast enough to allow for such swift synchronization. Regarding the role that microtubules may play in the functioning of the brain, Rahnama et al. (78) stated:

> Other researchers have previously suggested that microtubules play a major role in the functioning of the brain. Roger Penrose, a mathematical physicist suggested that consciousness is essentially a phenomenon of quantum mechanics and that microtubules were the medium in which quantum mechanics takes place.

23.8 Is Cooperative Coherence of Biophotons an Expression of Coherence of All Living Systems?

In the article entitled 'Quantum Coherence of Biophotons and Living Systems', the author (79) stated:

> Coherence is a property of the description of the system in the classical framework in which the subunits of a system act in a cooperative manner. Coherence becomes classical if the agent causing cooperation is discernible otherwise it is quantum coherence. Both stimulated and spontaneous biophoton signals show properties that can be attributed to the cooperative actions of many photon-emitting units. But the *agents responsible* for the cooperative actions of units have not been discovered so far. The stimulated signal decays with non-exponential character. It is system and situation specific and sensitive to many physiological and environmental factors. Its measurable holistic parameters are strength, shape, relative strengths of spectral components, and excitation curve. The spontaneous signal is non-decaying with the probabilities of detecting various number of photons to be neither normal nor Poisson. The detected probabilities in a signal of Parmelia tinctorum match with probabilities expected in a squeezed state of photons. It is speculated that an in vivo nucleic acid molecule is an assembly of intermittent quantum patches that emit biophotons in quantum transitions. The distributions of quantum patches and their lifetimes determine the holistic features of biophoton signals, so that the coherence of biophotons is merely a manifestation of the coherence of living systems. (Emphasis is mine)

Regarding the statement 'but the *agents responsible* for the cooperative actions of units have not been discovered so far' (emphasis is mine), I propose that the agents responsible are the components of the Triple Energizer. In the commentaries on the 38th Difficult Issue of Unschuld's (1) translation of the *Nan Ching*, Li Chiung states, 'The Triple Burner represents nothing but membranes.' Elsewhere in the book, I discuss the many fascinating and amazing electrical properties of the membranes or connective tissues, especially collagen, that would allow the conduction of biophotons to reach every cell in the body.

23.9 Increasing Research into Endogenous Radiation in the Form of Photon Emission

Regarding the machinations of biophotons in the body, in the research article 'Bio-Photons and Bio-Communication', VanWijk (76) explains, 'There are still a large number of biological phenomena and events that cannot be adequately explained, or even simply described, such as regulation of cell division and cellular differentiation. How is the program of growth controlled, and how does the ordered growth come to be disturbed?' Regarding the increased involvement of researchers from various fields, including physics and biology, VanWijk explains that many researchers have investigated the likelihood that endogenous radiation in the form of photon emission orchestrates the regulation of cellular division. Up until recently, biochemical research has dominated the field, and an abundant number of metabolic and enzymatic reactions have been elucidated. Some of these reactions could be responsible for the liberation of photons by organisms and cells. A third generation of researchers in the field is exploring to see if long-range electromagnetic waves and fields with a bio-informative character are responsible for the precise biological organization required to sustain life. Unlike most Biologists, Physicists aim to simplify complicated phenomena in a unified way, incorporating simple principles, and believe that coherence fits these parameters. The international cooperation of this interdisciplinary cohort of third-generation scientists from all over the world have a major meeting place at the International Institute of Biophysics in Neuss, Germany. VanWijk (76) emphasizes that the field is quickly expanding as can be witnessed by the ever-increasing growth of literature in this field being presented in numerous reviews and books.

The massive interest in this field of research is further exemplified in the journal article titled 'Emission of Mitochondrial Biophotons and Their Effect on Electrical Activity of Membrane via Microtubules', where Roger Penrose (78), a mathematical physicist, said, 'There is no doubt that the field of biophotonics is one of the fastest moving and exciting fields in science today.'

Dr Mercola (80) points out that currently there are about 40 scientific groups worldwide working on many aspects of biophotons. The largest association of scientists is the International Institute of Biophysics based in Neuss, Germany. Mercola noted that, with consideration to both biology and physics, the institute was originally formed to investigate and comprehend living systems. There are 14 Governmental Research Institutes and Universities dedicated in researching various facets, including biophoton coherence in biology, long-range interactions and communication in living organisms, and the general field of biophotonics.

23.10 What Is the Biophoton Field?

Regarding the nature of the biophoton field, in the article entitled 'Five Principles that Can Heal Virtually Any Illness, Part 2', Dr Joseph Mercola (81) noted that your mental body can be likened to a powerful computer where information is processed so that the physical body is regulated at all times. Within every cell in your body, a sophisticated dance is constantly underway such that over 100,000 perfectly timed and sequenced biochemical reactions occur every second. Dr Fritz-Albert Popp proved the existence of the biophotonic field in 1974. He also proved that biophotons are laser-like in nature and originate from your DNA. This laser-like precision and speed helps to control the multitudinous biochemical processes that happen every second.

23.11 So How Does the Biophoton Field Work?

The article by Dr Joseph Mercola (81) discusses how Dr Dietrich Klinghardt believes the biophoton field works. DNA inside each cell in your body vibrates at a frequency of several billion hertz. The vibrations are generated by the coil-like contraction/extension cycles of your DNA. Every time the DNA contracts, it ejects one single light particle called a biophoton, which contains the vast amount of information pertaining to everything occurring within your DNA at that moment in time. It has been estimated that one single biophoton can transmit more than four megabytes of information and transmit this data to other biophotons it crosses in the biophoton field outside of your body. All of the biophotons emitted from your body communicate with one another in this extremely organized and regulated light field that surrounds your body. Klinghardt proposes that this biophoton field is the storehouse of your long-term memory. He also believes this biophoton field also orchestrates the control over your metabolic enzymes. Mercola states, 'The information transfer on biophotons is bidirectional, which means your DNA sends information out on a photon, and on the same photon the information of all the biophotons from your body is broadcast back to your cells, and to your tubulin, which are light conductive molecules in your connective tissue.' Subsequent to that, Mercola continues, 'The tubulin, in turn, receives the information-carrying light impulse and conducts it at the speed of light throughout your body, where it is translated inside each cell into activating or inactivating certain metabolic enzymes.' This information addresses why it is so important to consume high-quality, sun-bathed, vital raw foods to supply adequate biophotons.

23.12 Do You Know Where Your Long-Term Memories Are Stored?

A five-level healing model has been developed by Dr Dietrich Klinghardt. It is influenced by Tibetan medicine, traditional Chinese medicine (TCM), as well as Ayurvedic medicine and is based on 12,000 years of research. In the article entitled 'Five Principles that Can Heal Virtually Any Illness, Part 2', Dr Joseph Mercola (81) discusses the principles of the model. In the article, he advises that the biophotonic field surrounding your body is your 'mental body'. This biophotonic field is where your long-term memory is stored. Dr Klinghardt advises that while the field of neuroscience can explain the brain circuitry of short-term memory, there is no known circuitry for long-term memory storage. He believes that your biophoton field is the storehouse of long-term memory.

23.13 How Your Vital Biophoton Field Is Adversely Affected by Environmental Toxins

Regarding the negative impact of environmental toxins on the vibrant biophoton field, in the article, Mercola (81) reports that tubulins are the light-conductive molecules that are present throughout your connective tissues and are responsible for critical light transmissions coming from your biophoton field to organize and synchronize your cells and your metabolic enzymes. Even extremely low concentrations of highly toxic Mercury can destroy your tubulins, which then disengage the intelligent biochemical synchronization throughout your body. Mercola states that this is 'all the more reason to avoid those yearly flu shots'!

Summary of Chapter 23

Who would have thought that the good old San Jiao could be so much at the centre of modern scientific research in biophysics and water biochemistry? Ironically, I don't believe that any of the researchers discussed even know what the Triple Energizer is. Recall in the commentaries on the 38th Difficult Issue of Unschuld's (1) translation of the *Nan Ching*, Li Chiung stated, 'The Triple Burner represents nothing but membranes.' The 31st Difficult Issue of the *Nanjing* states that the triple burner's predominant responsibility is that 'it represents the conclusion and the start of the course of qi'. This shows that the Triple-Energizer organ complex was believed to be a ubiquitous system of membranes that initiated all energy and communication requirements of the entire body. This has to be some sort of electrical system that is distinct from the established nervous system. I will next talk more about this highly ordered, omnipresent communication system.

CHAPTER 24

Biophotons and the Highly Ordered Nature of Cellular Communication

Introduction
Regarding the informational character of biophotons, VanWijk (76) explains, 'The search for evidence of the "informational character" of ultra-weak photon emission from biological systems was stimulated by Popp in the 1970s. He introduced the term "bio-photons" in 1976.' The author continued to explain that the intimate relationship between biophotonic emission and biological quantum phenomena is most evident in the complicated process of cellular division. A single cell is only one unit in a massively larger biological structure, and the never-ending cell death rate is compensated for exceedingly precisely by the never-ending cell division rate. If this was not permanently the case, serious diseases would result, including abnormal swelling, where excessive cells formed, or shrinking and atrophy of tissues, where the cell division rate was retarded. Individual tissue organization houses much more information than is present in the individual cells that make up that tissue. It has been reasoned that if the information to regulate new cell growth originated from the death of cells, it would not reasonably explain this regulation by messenger molecules from individual dying cells as they would be faulty and corrupted. However, all cell-controlling information for transferring the necessary messages could easily be regulated by

electromagnetic interactions. VanWijk stated, 'Consequently we expect some correlation between growth and bio-photon emission.'

As the Triple Energizer is mainly concerned with nutrient derivation from grains and drink, it is the main organ system to ensure all the necessary requirements for growth and the continuation of life are adequately circulated to the cells requiring the necessary sustenance. When VanWijk states above that 'consequently we expect some correlation between growth and bio-photon emission', it becomes obvious that the maintainer of growth in our body—that is, the Triple Energizer—must also correlate with the biophoton emissions to ensure that sound communication is given by needy cells and that nutrients are subsequently delivered expediently and correctly. So I will now discuss in detail how I propose that the Triple-Energizer Metasystem accomplishes its many tasks in light of recent scientific findings pertaining to biophotons and the highly ordered nature of cellular communication.

24.1 Electromagnetic Information Transfer between Various Cells and Tissues

In the research article 'Bio-Photons and Bio-Communication', VanWijk (76) explains, 'Whereas normal cells show decreasing emission with an increasing number of cells, the photon emission of tumor cells increases in a nonlinear way to higher and higher values, displaying thus a qualitative, not only a quantitative, difference. Photon emission was also cell-type dependent, multipotent fibroblastic cells showing the strongest emission.' VanWijk reported that there are numerous examples of research findings concluding that cell-to-cell communication (bio-information) exists in the absence of chemical mediators or dedicated messenger molecules, including hormones, neurotransmitters, and growth factors.

Regarding cells being able to detect the orientation of neighboring cells, in 1992, researchers reported that the adjoining cells appeared to detect electromagnetic signals that penetrated glass but not thin metallic films. Researchers in 1993 reported that when mammary tissues of lactating mice were stimulated with various hormones, other murine mammary tissues responded similarly even when they were separated by quartz glass. In 1994 researchers proved that neutrophils that were stimulated to experience a respiratory burst activated a second population of neutrophils that were optically coupled in spite of being chemically separated. In 1995 researchers reported that after Raphanus sativus seeds were exposed to low-dose irradiation, they attained a new property. Amazingly, a few hours after being irradiated, the irradiated seeds exerted the influence of accelerated germination on the control seeds that were distant from the irradiated seeds. The authors noted, 'More recently, in 2000 researchers extended their experiments to the question of whether the chemiluminescence burst of the neutrophil cells stimulated by phorbol myristate acetate (PMA) or zymosan could be modulated by the presence of a separated neutrophil cell population in close vicinity.' VanWijk (76) explained, 'The results of all 12 independent tests made during 1996–1997 demonstrated 9 tests with a significant enhancement of the respiratory burst of the PMA-stimulated neutrophil cells by the presence of a separated but optically coupled neutrophil cell suspension, but 3 tests yielded negative outputs.' VanWijk noted that the identical experiment was repeated in 1998, whereby nine independent tests were reproduced. The same positive outcome occurred in eight out of nine tests; however, the negative outcome was statistically insignificant. These results consistently prove that cells and tissues do communicate via electromagnetic means even when light-transparent barriers separate the cells or tissues.

24.2 Models for Explaining Photon Emission from Collective Molecular Interactions

Regarding the formation of biophotons, in the research article 'Bio-Photons and Bio-Communication', VanWijk (76) explains:

> The hypothesis that photon emission originates from the relaxation of superhelical DNA is based on the possibility of excimer formation of polynucleotides at room temperature within the lowest long-living triplet states of DNA. The excimer complex is relatively stable and forms a photon trap since its free energy is lower than that of the molecular

fragments. The natural tendency of exciplexes and excimers to absorb photons and to create excited states associated with ordered and more compact biostructures, e.g., condensed chromatin in the nucleus, fits well with the idea that the relaxation of DNA superstructures releases photons has been substantiated by the findings of at least 7 research projects. In this respect the biophoton emission of fractionated mammalian cells are of interest. However, emission was not detected in purified DNA, neither was it in cell fractions containing cytosol, mitochondria, or ribosomes.

24.3 The Highly Ordered Nature of Biophotons

Regarding the highly ordered nature of biophotons, in the article entitled 'McDonald's and Biophoton Deficiency', Dr Mercola (80) explained that, due to their extremely high degree of order, biophotons behave similarly to biological laser light, and subsequently, their interference can produce many effects beyond the ability of ordinary incoherent light. Due to the extremely high coherence of biophotonic waves, biophotons create order and transmit information. Experiments confirm that biophotons have an essential regulating role within individual cells and also between cells.

24.4 The Omnipresence of the Biophoton Field throughout the Entire Body

Regarding the ubiquity of the biophotonic field effect throughout the entire body, in the article entitled 'McDonald's and Biophoton Deficiency', Mercola (80) explains that the coherent biophoton field pervades the entire body, causing superlative organization throughout the body. This omnipresent biophotonic field coherence allows a highly structured field of very high-density information and regulation to perfuse the entire body, thus causing every single cell to communicate with every other cell so there is intercellular coordination throughout the entire organism at the speed of light in a holographic manner. Recent research strongly suggests that the biophotonic wave emanates from the chromatin of the cell nucleus. This is in keeping with calculations that confirm that the helix structure of the DNA molecule represents the perfect geometric configuration of a hollow resonator, which would allow it to store light very successfully.

24.5 Omnipresent Collagens and Intercommunication throughout the Body

Regarding the nature of collagens, in the 1998 article entitled 'The Acupuncture System and the Liquid Crystalline Collagen Fibers of the Connective Tissues: Liquid Crystalline Meridians', the authors, Ho and Knight (82), explain that most researchers still consider the connective tissues throughout the body in purely biomechanical terms and believe that connective tissue has functions, including maintaining body shape; acting as packing material between the major organs and tissues; strengthening the walls of arteries, veins, intestines, air passages, etc.; and providing the rigid infrastructure for muscle attachments. The authors believe that a more open-minded interpretation is that of a global tensegrity system, whereby bones act as the compression elements and are interconnected with muscles, tendons, and ligaments, which act as the tension elements, and local actions provoke reorganization throughout the entire body. More importantly, they believe that the connective-tissue system may be essentially responsible for the rapid information transfer throughout the body to enable the body to function as an intelligible entity and therefore is fundamental for our optimal health and general well-being. They state, *'The clue to the intercommunication function of connective tissues lies in the properties of collagen, which makes up 70% or more of all the protein of the connective tissues.* Connective tissues, in turn form the bulk of the body of most multicellular animals. Collagen is therefore the most abundant protein in the animal kingdom' (emphasis is mine).

24.6 The Remarkable Physical and Electrical Properties of Collagens throughout the Body

Regarding the physical and electrical properties of collagens, in the 1998 article entitled 'The Acupuncture System and the Liquid Crystalline Collagen Fibers of the Connective Tissues: Liquid Crystalline Meridians', the authors, Ho and Knight (82), state, 'Recent studies reveal that collagens are not just materials with mechanical properties. Instead, they *have dielectric and electrical conductive properties that make them very sensitive to mechanical pressures, pH, and ionic composition, and to electromagnetic fields.*' They explain that, to a large extent, the electrical properties of connective tissues are determined by the amount of bound water molecules present in and around the collagen triple helix. It is known that up to 50–60% of the water present in the cell is contained within the vast microtrabecular lattice that constitutes the entire cell and endows the cell with its solid-state nature. The conductivity of collagen fibers increases most at around 36.9 °C, which happens to be the normal operating temperature of the human body. Remarkably, at the slightly elevated temperature of 40 °C, the triple-helix structure of collagens in dilute solutions melts, which most likely allows the collagen fibers to straighten and subsequently increases their conductivity. It is postulated that this melting and straightening of the ubiquitous collagen fibers may be responsible for the known health benefits of thermogenic physical exercise.

Why would a building material supposedly meant merely for structural stability be teeming with coexisting dielectric and electrical conductive properties with solid-state characteristics? It appears obvious that is because it has both a structural and a coexisting electrical function. Regarding the profoundly incredible coexisting electrical properties, the authors Ho and Knight (82) continue, 'The *collagenous liquid crystalline mesophases* in the connective tissues, with their associated structured water, therefore, constitutes a *semi-conducting, highly responsive network* that extends throughout the organism' (emphasis is mine). They explain that via proteins that penetrate the cell membrane, the external connective tissues network communicates directly to inside individual cells. They conclude by stating, 'The connective tissues and intracellular matrices, together, form a global tensegrity system, as well as an excitable electrical continuum for rapid intercommunication throughout the body'.

I was absolutely stunned to find that many of my initially proposed hypotheses have already been tested and confirmed by recent scientific research. However, to my knowledge, nobody has connected these scientific findings to the probable functionality of the Triple Energizer. So here is Dr Ho, *not* Professor Pollack, discussing 'collagenous liquid crystalline mesophases in the connective tissues, with their associated structured water' (that is, Pollack's EZ water), acting as a 'highly responsive network that extends throughout the organism' that engenders 'an excitable electrical continuum for rapid intercommunication throughout the body'. Dr Ho's description presented here exactly describes the machination of what my hypothesis proposes for the Triple-Energizer organ complex.

24.7 Acupuncture Channels Allow Communication Internally and with External Environment

In the article titled 'Yangsheng and the Channels', the author, Jason D. Robertson (44), explains that modern TCM practitioners accept that traditional Chinese medicine is a holistic system but occasionally fail to appreciate precisely how early TCM physicians actually perceived that unity. He notes that the original Han construct of a holistic system taught that the channel system not only orchestrated communication amongst all internal organs and structures but also coordinated communication with the external environs. Robertson explained that the early practitioners believed in the concept of resonance (*yìng*), whereby one characteristic of the natural world resonated or oscillated in synchrony with another due to an interconnected affinity and predilection. He presented as an example the TCM pattern of the Liver, which resonates with the element Wood, the season spring, the emotion anger, the blue-green color, the direction east, and the *Juéyīn* channel.

Regarding the proper resonance of the internal and external environments, Robertson (44) states, 'If one is resonating properly due to having healthy channels—then it is easier to live in a harmonious way and avoid being out-of-sync with the symphony of qi which surrounds us all.' Regarding the inability of qi to correctly move through the system, he further states, 'Whether arising from excess or deficiency, the net result is a lack of communication within the body and/or an inability to properly "metabolize" the external environment. In the classical model of Chinese medicine, this communication is seen to take place through the medium of the channels.'

24.8 Communication throughout the Entire Body and Beyond Mediated by the Triple Heater

In the book *Heart Master Triple Heater* (page 118), regarding how the Triple Heater ensures free communication and circulation throughout the entire body, the authors (11) finish with a translation attributed to Hua Tuo, who said, 'When the triple heater ensures free communication, then there is free communication internally and externally, left and right, above and below. The whole body is irrigated, harmonized internally and regulated externally, nourished by the left, and maintained by the right, directed from above, propagated from below. There is nothing greater!'

Contrast this comment attributed to Hua Tuo, who lived in the late Eastern Han Dynasty about one thousand eight hundred years ago, with what a recent (2015) physician stated. Dr Joseph Mercola (83) stated:

> Just as your gut houses a microbiome that's critical for your health, our soils contain a microbiome critical to the health of the planet. Your health is directly related to the quality of the food you eat, and the quality of your food depends on the health of the soil in which it's grown. As above, so below. In the rhizosphere, microorganisms form nodules on the roots of plants to help them take up minerals and other nutrients from the soil. Unfortunately, modern industrial agriculture (monoculture) destroys much of the soil's microbiome. Heavy agrichemical use and GE crops rapidly destroy once-fertile soils that took centuries to develop. . . . Although arrogant molecular biologists believe they know, the reality is that we really don't know what effects tampering with the genetics of our food supply will have on your health and your children's health over the long run.

Certainly, the health and vitality of the soil below our feet determines the health and vitality of the food that grows in that soil and then obviously the health and vitality of our body once we eat that produce. The Primo Vascular System (PVS) exists within plants as well as humans. As I believe that the PVS is intimately associated with the Triple-Energizer Metasystem, there is an intimate connection between humans and the earth below us. Note that when the Triple Heater ensures free communication, Hua Tuo said, 'Then there is free communication internally and *externally*, left and right, above and *below*. The whole body is irrigated, harmonised internally and regulated *externally*, nourished by the left, and maintained by the right, directed from above, *propagated from below*. There is nothing greater!' (emphasis is mine). The health of the soil dictates the health of our being. With the exception of the chemical elements in our body that were derived from the air we breathe (predominantly oxygen) and marine-derived food we eat, every other chemical element present in our body came directly from the soil. I suggest that future scientific research will show that sick diseased plants will have sick diseased PVSs and that the damaged plant PVS will adversely affect our human PVS. But this is only conjecture. I further suggest that free electrons from the earth under our bare feet flow abundantly into our electrically conductive body and, along with biophotons, also energize the Triple-Energizer Metasystem. Shoes will prevent this effect. When our body is thoroughly grounded or Earthed™, then the San Jiao will be *propagated from below*.

24.9 The PVS Allows Communication between Living Organisms and the Environment

Regarding the communication of information within the body and from the external environment, in the 2013 *Journal of Acupuncture and Meridian Studies* article titled 'The Primo Vascular System as a New Anatomical System', the authors (84) concluded that the PVS is responsible for the communication of information between living organisms and the environment. They further believe that, during the very early stage of embryological development, the PVS is duplicated by the vascular system and the nervous system, and subsequently, the PVS possesses features associated with the immune, nervous, vascular, and hormonal systems of the body. The PVS has been shown to be omnipresent throughout the body, and because the PVS receives and processes many forms of external and internal information, the PVS regulates and coordinates the multitude of biological life processes. The external information and signals emanate from the ambient environment as electromagnetic waves, while the internal signals are generated by metabolic processes and involve bioelectrical, bioluminous, and acoustical fields. The PVS then processes all this information and then orchestrates the necessary vital bioprocesses to maintain the required high degree or order within the body.

Summary of Chapter 24

I found it fascinating that in the three sections directly above, the different authors presented analogous evidence and concluded in Section 24.7 that the acupuncture channels allow communication internally and with the external environment (44), in Section 24.8 that communication throughout the entire body and beyond is mediated by the Triple Heater (11, 83), and in Section 24.9 that the PVS allows communication between living organisms and the environment (84). From these analogous comments, it appears that acupuncture channels, the Triple Heater, and the Primo Vascular System are all responsible for communication within the body and with the external environment. I propose that these three systems are intimately related and interconnected in that the primordial Primo Vascular System actually initially engendered the Connective-Tissue Metasystem as a support structure, and thereafter the Connective-Tissue Metasystem, in association with the Primo Vascular System, came to constitute the Triple-Energizer Metasystem. I believe further that the Acupuncture Meridian System is a combined derivative of the collagenous Connective-Tissue Metasystem in intimate association with subbranches of the Primo Vascular System. I propose that, due to the hydrophilicity of the omnipresent Connective-Tissue Metasystem, the omnipresent water in the body is transformed into liquid crystalline EZ and this derived ubiquitous material permits biophotonic information transfer through both the Triple-Energizer Metasystem and the Acupuncture Meridian Metasystem to perform their delineated roles.

For the membranous Triple Heater to 'ensure free communication internally and externally, left and right, above and below', there must be a medium for this communication. I suggest that the medium for the communication throughout the membrane complex is biophotons, which flow through the liquid crystalline EZ that coexists alongside the membranes of the Connective-Tissue Metasystem, as I have discussed above. What is the origin of biophotons? The sun. We absorb biophotons from the foods that have grown under the sun and have been irradiated by its light. Thus, there are three requirements for man to be created and sustain life: (1) suitable food, drink, and air containing atoms and molecules derived from the Earth and its atmosphere to act as the building blocks for our human frame; (2) the microbiological biome that makes up 90% of our cells on a cell-per-cell basis and allows the health-sustaining synergy with our material elements to occur (discussed elsewhere); and (3) the solar-derived biophotons, which allow the transit of messages and 'free communication internally and externally, left and right, above and below'. It is very noteworthy that collagen makes up 70% or more of all the protein within our membranes (i.e. the connective-tissue system in the body) and that collagen is endowed with such remarkable electrical properties. All these features of collagen make it the most likely contender for being the agent for the information and intercommunication dissemination throughout the entire electrical continuum, which, I believe, is what the ancient medical classics called San Jiao and WHO calls the Triple Energizer and I call the Triple-Energizer Metasystem.

In the case of noncoherent photons, any interference causes them to collapse within seconds, and they would fail as a communication system. For any optical communication system to perform optimally, high fidelity is essential. The highest fidelity possible must come from a structured state where light waves can form a coherent and communicating field with laser-like coherence. The next Chapter discusses in detail how biophotons are emitted from DNA within the body with laser-like coherence.

CHAPTER 25

Biophotons Emitted from DNA With Laser-Like Coherence

Introduction

Regarding DNA and RNA absorbing photons and emitting laser-like, coherent light within cells throughout the entire body, in the article entitled 'McDonald's and Biophoton Deficiency', Dr Mercola (80) explained that DNA is composed of a double helix, which can turn to the right or to the left. DNA is the central depository for light storage in our body. A recent discovery is that cells both absorb light and also emit light coherently. Both DNA and RNA molecules behave as lasers and can generate an optical hologram that communicates with the resonance of geocosmic fields. The highly coherent light generates an exceedingly organized state where waves can produce a coherent and communicating field. Any noncoherent photons that are out of sync would collapse immediately. It has also been found that the phenomenon of ultra-weak luminescent cell radiation does not radiate chaotically but actually works in a very stable phased fashion, similar to phased coherent laser light. It is noteworthy that recent research has confirmed that highly structured light and highly structured EZ water are associated with the anatomical components that I propose constitute the structured membrane system, which I believe is the Triple-Energizer organ complex.

25.1 Laser-Like Coherent Light from DNA and RNA Allows Bodily Communication

Regarding the function of emitted laser-like coherent light from DNA and RNA within cells throughout the entire body, in the article entitled 'McDonald's and Biophoton Deficiency', Dr Mercola (80) states, 'Communication turns out to be one of our most basic properties-communication within the system as well as communication with the outside. The aim is to counteract entropy, loss of structure, chaos, a state of high disorder, so as to create and maintain a state of excitement'. He continued, 'A high level of order within the body enables an undisturbed flow of information and communication. This, in turn, maintains the metabolism as well as all other life processes'. Mercola explains that all biological processes occur extremely rapidly and require equally rapid communication to ensure biological-system integrity. If this process fails, disease results. Mercola notes that any disease can be construed as an information and communication breakdown within the body.

Please reconsider the quote I previously mentioned above that is attributed to Hua Tuo. He said, 'When the triple heater ensures free communication, then there is *free communication internally and externally, left and right, above and below. The whole body is irrigated, harmonised internally and regulated externally, nourished by the left, and maintained by the right, directed from above, propagated from below.* There is nothing greater!' (emphasis is mine). In the quote in the prior paragraph, Mercola states that 'communication turns out to be one of our most basic properties—communication within the system as well as communication with the outside', almost exactly what Hua Tuo said, 'There is free communication internally and externally'. Regarding the functions of the laser-like coherent light emitted from DNA and RNA within cells throughout the entire body, Mercola further states, 'This, in turn, maintains the metabolism as well as all other life processes.' This last statement sums up a major role of the Triple-Energizer organ complex. So here we can see that radiant energy in the form of laser-like coherent light emitted from DNA and RNA within cells is responsible for communication throughout the entire body to ensure that 'metabolism as well as all other life processes' are maintained in perfect harmony. This harmony would ensure that the cells of the entire body are suitable irrigated, harmonized and regulated, nourished, maintained, directed (to perform their ascribed function), and propagated (to proliferate and not die out). To my knowledge, Hua Tuo did not have a degree in laser physics or cellular biochemistry. He simply described the functions of the Triple Heater in basic terms. Modern practitioners of laser physics and cellular biochemistry are just catching up with Hua Tuo.

25.2 Genetic Information Transfer within Cells Mediated by Coherent Biophotons from DNA

Regarding the known functions of DNA, including coding of genetic information (which is passed on to the next generation in the germ cell) and the storing of information to build all cell components within the body, Dr Mercola (80) states, 'Life and all particles of a system relate to each other coherently and where they communicate with each other to achieve a sensible cooperation in order to produce the optimal condition for the entire system.' He continued, 'Light emission is strongest whenever DNA is reproduced. The DNA emits about 90% of the biophotons in the cell nucleus. The DNA is an excellent storage medium for light and thus also for oxygen because of its form, the double helix. Perhaps this is why DNA is the basis for all processes occurring in the body-and thus also participates in metabolism.'

25.3 DNA and RNA Have Hydrophilic Backbones, and so Produce EZ and Release Biophotons

Regarding why DNA is formed in the shape of a double helix, the 2013 article entitled 'DNAmazing!', Molecular Biologist, Ben Stutchbury (85) advises that the hydrophobic bases hate water and subsequently states:

> *The sugar-phosphate backbone, on the other hand, loves water; it can't get enough of it! So it is termed 'hydrophilic'.* There is a huge amount of water cursing through our body at any one time, so it appears that the hydrophobic bases are

in somewhat of a predicament. To solve this conundrum, two strands of DNA squash themselves together, with the hydrophobic bases facing in and the *hydrophilic sugar-phosphate backbone facing out*, allowing the bases to hide from the water. But this isn't quite enough, if they stayed in this linear position, small water molecules can squeeze through the gaps and get to the bases, so the two strands of DNA tilt themselves and twist around one another, in an anti-clockwise direction, closing the gaps, meaning the bases stay nice and dry. (Emphasis is mine)

Summary of Chapter 25

It is known that each cell contains all the organism's genetic information and instructions stored as DNA. Nevertheless, each cell uses only the instructions from the DNA that specifically pertain to itself. For example, a muscle cell only uses the DNA that pertains to muscle functionality, whereas a nerve cell uses the portion of DNA that pertains to operation of the nervous system. Amazingly, each cell reads only the part of the DNA 'book of instructions' that it requires to operate effectively. The sugar-phosphate backbone is negatively charged and hydrophilic, which allows the DNA backbone to form bonds with water. Based on Pollack's hypothesis, which is discussed in detail above, we can see that the omnipresent DNA existing in each cell has an external hydrophilic (water-loving) structure present, which can convert bulk water into living EZ water. From Mercola's (80) article above, we see that 'the DNA is an excellent storage medium for light and thus also for oxygen because of its form, the double helix'. According to Professor Pollack, both light and oxygen potentiate the formation of EZ. So this one DNA structure brings hydrophilicity, biophotons, and oxygen together in constant contact with water within every cell.

Eureka! In close proximity to the omnipresent hydrophilic connective tissue, this omnipresent hydrophilic DNA structure is able to generate the organizational life-sustaining coherent biophotons to supply the biological commands for all cells. At the same time, the omnipresent hydrophilic connective tissues are generating the omnipresent liquid crystalline EZ water throughout the entire body. I believe that this highly structured biological and biophysical organization is under the control of the omnipresent Triple-Energizer organ complex, which is composed of 'nothing but membranes', connective tissues predominantly (70%) composed of collagen, which is known to possess remarkable electrical properties.

Much research is being undertaken to explore the remarkable properties of light and especially biophotons with regard to their effect on the health of living creatures. This is the topic of my next Chapter.

CHAPTER 26

Biophotons and the Power of Light to Heal

Introduction
With regard to the power of light to heal diseases, in the 2010 article by Dr Joseph Mercola (86) entitled 'The Power of Biological Light in Healing', Mercola notes that the phenomenon known as photorepair is commonly known, whereby if 99% of a cell (including its DNA) is destroyed by UV light, the damage can be repaired within a single day by illuminating the damaged cell with the same UV wavelength but at a greatly reduced intensity. Based on the scientific research of Dr Fritz-Albert Popp, many scientists now consider that the communication system throughout the body may well be a multifaceted network of resonance and frequency. The article (86) continues, explaining that Dr Popp proposed that the herbal plant mistletoe appears to reorganize and harmonize faulty photon emissions of tumour cells back to normal healthy cells. Mercola continued to explain that the field of biophoton emission is an expanding area of research by numerous scientists and physicists around the world. Many scientists believe that biophoton emissions may energize the processes that drive and likely even govern life itself.

26.1 Your Body Is Constantly Emitting Light and Glowing
With regard to the surface of your body constantly emitting light, in the 2010 article by Dr Joseph Mercola (86) entitled 'The Power of

Biological Light in Healing', Mercola notes that humans emit light constantly with an intensity that fluctuates relative to our circadian rhythm. The light is not observable to most people due to it being about 1,000 times less intense than light levels that are visible with the naked eye. Mercola states 'Dr. Fritz-Albert Popp was the first to suggest that this light must come, at least in part, from the foods we eat. . . . The purpose of these biophotons is much more important than many have realized. It turns out they may very well be in control of virtually every biochemical reaction that occurs in your body—including supporting your body's ability to heal' (emphasis is mine).

26.2 Light Controls the Biological Processes within Our Cells

With regard to the complicated biological and biochemical processes that happen inside every single cell, in the 2010 article entitled 'The Power of Biological Light in Healing', Dr Joseph Mercola (86) advises that biologists are aware that there are over 100,000 biochemical reactions per second in every cell in your body. Obviously, these reactions must be carefully timed and sequenced with one another so mishaps don't occur. It was formerly proposed that this happened through a 'mechanical' concept whereby molecules bumped erratically into one another by chance and came together like a lock and key or even by slightly changing shape to come together and form chemical reactions. Common sense has prevailed, and now researchers believe that this intricate and complicated biological and biochemical dance is not random at all, but rather, they believe the convoluted processes and reactions are intimately orchestrated by biophotons. Mercola noted that Dr Popp, who initially confirmed the existence of the biophoton field in 1974, believes that biophoton emissions cause information transfer throughout the entire body.

26.3 The Precise Biochemical 'Dance' Occurring within Every Cell

With regard to the nature of the biochemical 'dance' occurring within cells throughout the body, in the 2010 article by Dr Joseph Mercola (86), he quotes from the article entitled 'The Real Bioinformatics Revolution: Proteins and Nucleic Acids Singing to One Another?' The article was written by Dan Eden for viewzone.com. Eden explains just how this 'dance' takes place. He explained that researchers Veljkovic and Cosic theorized that molecular reactions are electrical in nature and suggested that they take place over large distances compared with the size of the molecules themselves. Cosic, Professor of Biomedical Engineering, RMIT University, Melbourne, Australia, later introduced the concept that molecules distinguish their dedicated targets due to electromagnetic resonance via what became termed as the dynamic electromagnetic field interactions. Cosic believes that molecules produce dedicated frequencies of electromagnetic waves which permit them to perceive each other. This is because molecules essentially '"see" and "hear" each other, as both photon and phonon modes exist for electromagnetic waves'. Consequently, molecules can influence each other from a distance and become inescapably attracted to each other. Remember that there are around 100,000 biochemical reactions occurring in every single cell every single second. For a chemical reaction to proceed, the reacting molecule must be excited by a photon. The participating photon returns to the electromagnetic field after a successful reaction and is then available for further reactions to take place. With all this happening constantly, our body is literally bathed in an ocean of light.

26.4 Where Does Your Body's Light Come From?
With regard to the source of light in your body, in the 2010 article entitled 'The Power of Biological Light in Healing', Dr Joseph Mercola (86) reports that Dr Popp confirmed that light present throughout your body is stored by and emitted from your DNA. Via the coil-like contraction and extension of the DNA molecule, the DNA inside each cell vibrates at a frequency of several billion cycles per second, and one single biophoton or light particle is squeezed out with each contraction of the DNA molecule. Remarkably, the ejected photon stores all the information pertaining to all the chemical reactions going on in your DNA at that instant. Each photon can store more than four megabytes of information and communicates this information to other biophotons encountered in the biophotonic electromagnetic field outside your body. It is believed that photons emitted from your body intercommunicate within this high-fidelity

light field that envelops your body, and it is believed that this ambient electromagnetic field constitutes the reservoir of your long-term memory. Biophotonic information transfer is a two-way street such that your DNA transmits information on a photon, and then that same photon carries the collated information from all the biophotons throughout your body back into your cells and also to your tubulin molecules. Tubulins are spherical protein building blocks that come in many forms and play vital roles in generating microtubule structures with organelle-specific properties and functions. Tubulins make up light-conductive molecules that occur in your omnipresent connective tissues. The tubulin receives the information-carrying light impulse and transmits the information throughout your body at the speed of light to ensure life processes proceed in an orderly manner.

26.5 Illness Occurs when Biophoton Emissions Are 'Out of Sync'

With regard to human connectivity to the earth and to seasonal influences, in the 2010 article entitled 'The Power of Biological Light in Healing', Dr Joseph Mercola (86) notes that research by Dr Popp confirmed that the light emissions of healthy people exhibit a defined biological rhythm for day and night and also by week and month, suggesting they are synchronized to intrinsic biorhythms of the earth. Remarkably, Popp confirmed that the light emissions emanating from cancer patients were devoid of such synchronized rhythms and were scrambled and disharmonious, suggesting that their cells were not communicating effectively with high fidelity. Dr Popp further found that individuals with multiple sclerosis absorb too much light, which similarly leads to confusion on a cellular level. He found that stress also influenced people's biophoton emissions, causing an increased reaction to the stress. Mercola further reported that currently biophotonic- based therapies are at the forefront of medicine. For example, therapies are emerging to reduce pain and stimulate the healing process using specific quantities and frequencies of light. A lead researcher in the field, Dr Dietrich Klinghardt, has been transmitting the electromagnetic information of essential nutrients in the form of light into the human biophotonic field and found that the biological outcome is the same as eating the nutrient.

26.6 An Important Tip for Gathering Valuable Light Energy

With regard to the importance of eating living raw foods, the 2010 article by Dr Joseph Mercola (86) reports that fresh raw foods abounded with biophotonic light energy so essential for good health. It has been established that all living organisms emit biophotons. Research indicates that the greater the amount of light energy that a cell emits, the greater its vitality. Subsequently, there is increased potential for that stored solar energy to be transferred to the individual who consumes it. It naturally follows that the more light that a food can store, the more nutritious it is. He continues, 'The greater your store of light energy from healthy raw foods . . . the greater the power of your overall electromagnetic field, and consequently the more energy is available for healing and maintenance of optimal health. . . . Until then, remember that your body is not only made up of tissue, blood vessels and organs. It's also composed of light.'

Interestingly, in the holy scriptures at John 8:12 in the *English Standard Version* (*ESV*), Jesus says, 'I am the light of the world. Whoever follows me will not walk in darkness, but will have the light of life.' Here, the omnipotent and omniscient Son of God, Jesus, is acknowledging that light engenders life. Thus, it should come as no surprise that light is a nutritive component that is essential for our continued optimal health and well-being.

26.7 Your Entire Body Literally Glows with Light

In the 2009 article by Dr Mercola (87), the author advises modern scientific research is playing catch-up when it comes to how various natural forces and energies rule diverse and intricate functions of our body. He says, 'Eastern medical traditions have operated on this premise for thousands of years, but Western medicine in its myopic focus on dissected parts has been slow to adapt these ancient truths.' Being an acupuncturist, I can certainly concur with that. When I graduated 31 years ago, I remember trying to convince patients that plant fiber was essential for good health, while the myopic medical stance was that it was inert and was not necessary in the diet. Western medical practitioners apparently are not permitted to think for themselves and use common sense. Graduation and the Hippocratic oath seal them to believe their university-entrenched doctrine as immutable. As I recall, the sagacious Hippocrates opined, 'Let your food be your medicine and let your medicine be your food.' I am sure that modern western medical students would reject him as weird should they encounter him today.

26.8 Dynamic Approach to Healing Based on Physics Rather Than Just Chemistry

Mercola (87) advises that a trend is emerging whereby clinicians recognize that a powerful approach to natural healing involves physics more than chemistry. Numerous enlightened natural-health practitioners understand that your body is made up of more than physical tissue types, blood vessels, and organs. They appreciate that the body is composed of energy—or Qi, as it's called in traditional Chinese medicine (TCM). Mercola proceeds to explain that this Qi circulates throughout your body through specifically defined meridian pathways. When acupoints on the surface of your skin are manipulated, the energy circulation within your internal organs is affected, and healing occurs.

In the article, Mercola (87) advises that research confirms that your body is actually surrounded by light or energy at a level about 1,000 times less intense than what your naked eyes can perceive. Interestingly, these light emissions seem to be interconnected with your circadian body clock and the associated rhythmic fluctuations of your metabolism during the course of the day. Mercola states, 'This suggests that you could detect medical conditions with the use of highly light-sensitive cameras.' Hitoshi Okamura is a researcher that investigates circadian biology at Kyoto University in Japan. Regarding testing for medical disorders, he says, 'If you can see the glimmer from the body's surface, you could see the whole body condition.'

Summary of Chapter 26
The purpose of biophotons in the body is much more important than many have realized before now. It turns out that biophotons may very well be in control of virtually every biochemical reaction that occurs throughout your body, including supporting your body's ability to heal. Biologists are aware that over 100,000 biochemical reactions per second occur within every single cell in your body. Understandably, these reactions must be precisely timed and sequenced so that major chemical accidents do not result and mutations occur. When it comes to how various natural forces and energies (e.g. biophotons and EZ water) rule the numerous diverse and intricate functions of our body, modern scientific research is currently playing catch-up. Eastern medical systems have operated on this premise for thousands of years. So far, I have discussed the nature of biophotons inside the body. But how do the biophotons enter the body? In the next Chapter, I will discuss how vital energy radiating from the sun finds its way into every cell in your body via the food you eat in the form of these biophotons.

CHAPTER 27

The Effect of Biophotons on Foods

Introduction
In the article by Dr Joseph Mercola (88) titled 'Eat Your Food Uncooked? Here's the Really Raw Truth', he states, 'The primary reason for making sure you get plenty of raw food in your diet is due to what's called "biophotons". It's a term you may not have heard of before, but in Europe, Germany in particular, there's a lot of research in this area.' Sun-activated foods trap biophotons, which energize our body. Mercola (88) discusses biophotons, which are the smallest physical unit of light. Vital energy radiating from the sun finds its way into every cell in your body via the food you eat in the form of these biophotons. Biophotons from ingested food are stored in and used by all biological organisms, including the human body. He says that biophotons 'contain important bio-information, which controls complex vital processes in your body. The biophotons have the power to order and regulate, and, in doing so, to elevate the organism—in this case, your physical body—to a higher oscillation or order. This is manifested as a feeling of vitality and well-being.' Regarding the vitality and health-imbuing effects of foods, Mercola (88) notes that all living organisms emit biophotons or low-level luminescence possessing a wavelength between 200 and 800 nm. It is generally held that the greater the stored light within the food, the higher the vitality of the food, and consequently, the food is more nutritious and helps in healing. For example, sun-ripened fruits picked at their prime are the most healthy and energizing.

27.1 Unhealthy Interference from Modern Technology—Cell Phone Communication

Mercola (88) discusses how an otherwise-healthy biophoton operation can be disturbed by cell phone communication systems which operate at the same frequency. He explained that DNA molecules present inside each cell of the body vibrate at a frequency of several billion cycles per second. Unfortunately, modern cell phones communicate at that same frequency. The DNA molecules vibrate due to a coil-like contraction and extension of the molecule occurring at the rate of several billion times per second. At each contraction of the DNA molecule, a single particle of light, a biophoton, is released, and thus a harmonizing communication light field occurs throughout the body, which synchronizes all biological activity, including the regulation of your metabolic enzymes.

27.2 Does Food Have 'Vitality', and Can We Absorb Its Life Force?

Regarding the vitality and life energy endowed by natural foods and how their life force permeates throughout the entire body, in the article entitled 'McDonald's and Biophoton Deficiency', Dr Mercola (80) explains that all living organisms emit biophotons or low-level luminescence. Such luminescence is derived from light possessing a wavelength between 200 and 800 nm. This light energy is transmitted continuously by the cell and is believed to be stockpiled in the DNA during the photosynthesis process within plants. It is believed that the greater the amount of light energy that a cell emits, the more dynamic its energy and vitality, and subsequently, the greater the transfer of energy to the individual consuming the produce.

27.3 Transmutation of Sunrays via Foods Is Essential for Optimal Health

There is no doubt that sunlight is vital for life to continue on earth. Without the sun, there would be no life. We readily notice what a revitalizing effect sunlight has on our body and our spirits when, after a long cold dark winter, we enjoy the first invigorating rays of spring sun. But interestingly, we can absorb the sun's dynamic and vibrant energy via our food as well as absorbing sunrays through our skin. Regarding the light energy absorbed by plant foods, which transmutes to life energy that infuses all cells throughout the entire body, Dr Mercola (80) reported on a 2002 research by Professor F. A. Popp and Dr H. Niggli, which showed that apart from the obvious nutrients we derive from plant foods, another vitally important aspect of food quality is the amount of light energy in the form of biophotons that the food possesses. High-quality naturally grown fresh fruits and vegetables are abundant in sun-derived light energy. The capacity of vegetables and fruit to store biophotons determines the vital quality of our food such that the more solar-derived photons that a food is able to accumulate, the more nutritious the produce is. The smallest units of light derived from the sun are called photons. This vibrant life-giving energy from the sun is stored in food as biophotons, and once we eat the energized food, the biophotons can enter our cells as small parcels of light. Popp and Niggli assert that these energized biophotons hold essential bio-information, which regulates complex biological processes within our body and manifests as a sensation of well-being and vitality.

27.4 Biophoton Essence in the Foods We Eat Energizes the Triple Energizer

Regarding the extreme importance of bountiful biophotons being present in our food supply, Dr Mercola (80) reports that eating raw foods greatly energizes our body. He noted that Dr Johanna Budwig from Germany believes that natural live foods are abundant in electrons and, when consumed, behave as highly energized electron donors and as solar resonance fields within the body to accumulate and disperse the sun's radiant energy throughout our entire body. Budwig asserts that when our store of solar energy is abounding then our electromagnetic field will be dynamic and consequently we will be resplendent with healing energy to engender and maintain optimal health.

That is a powerful statement about truly nutritious food that we need to take to heart. Foods really are transmuted sunshine, air, and earth. Obviously, the fresher the food is when we eat it, the more vibrant it is. Vital energized foods certainly store a higher quality of nourishment for our body to absorb and utilize for continuing optimal health and well-being. Consuming highly damaged microwaved food should be something that we keep to an absolute minimum. But regarding what Budwig stated above when he said that 'live foods are electron rich and act as high-powered electron donors and as solar resonance fields in the body to attract, store, and conduct the sun's energy in our body', this is not a new idea. In the commentaries on the 66th Difficult Issue on pages 567–568 of Unschuld's (1) translation of the *Nan Ching*, Yang states:

> The [two] kidneys are distinguished by [their storing] the *essence of the sun and of the moon*, [respectively. This essence] consists of immaterial influences; it constitutes man's source and root. . . . Hence it is obvious that the cinnabar field is the basis of life. When the Taoists contemplate the spiritual, when the Buddhist monks meditate, they all cause the influences of their heart to move below the navel. That corresponds exactly to what is meant here. Hence [the text] states '"origin" is an honorable designation for the Triple Burner' because the Triple Burner's influences are accumulated in the kidneys.

Wow, read again carefully and exactly what Yang states! He states, 'The two kidneys are distinguished by their storing the essence of the sun and of the moon.' Both the sun and the moon are light sources, obviously. The sun is the light source for the day, and the moon is the light source for the nighttime. So the ancients knew that light energy's essence coming directly from the sun and moon can penetrate our body. Note how this essence or photon energy is stored by the two kidneys and that this photonic essence 'constitutes man's source and root.' That is very plainly stated, is it not? Note how Yang finishes the quoted paragraph above: 'Hence the text states '"origin' is an honorable designation for the Triple Burner" because the Triple Burner's influences are accumulated in the kidneys.' So essentially, Yang is saying that light energy from the sun (and light from the sun reflected off the moon at nighttime) enter our body and is stored by the two Kidneys to be utilized by the Triple Burner for dispersing nutrients, healing damaged tissues, and maintaining optimal health.

I believe further that the fact that 'the two kidneys are distinguished by their storing the essence of the sun and of the moon' also is intimately involved with the 24-hour-long circadian rhythm associated with solar days and the Chinese Clock, whereby each of the twelve organs have their maxima for two hours and their minima 12 hours later and that the 29-day Lunar cycle influence directly affects life forms on Earth, as is evidenced by the female menstrual cycle. The kidney is well known for its 7-year life cycles in females and 8-year life cycles in males. Remember too, that the sun's essence also energizes the plant and animal produce that we consume so that the biophotons therein can be released into our Middle Heater and the Stomach can perform its role at extracting nutrients and solar essence and bulk water and convert these into the many forms of Qi as per the Triple Heater fluid and energy production cycle.

27.5 Your Biophoton Field Holds the Key to Your Health

Dr Dietrich Klinghardt explains the existence of light and energy around your body in terms of a biophoton field. In the 2009 article by Dr Mercola (87), he explains, 'The existence of the biophoton field was scientifically proven by Dr. Fritz-Albert Popp in 1974.' Mercola continues, 'Your physical health is dependent not only on what goes on inside of your body, but is also interconnected with and dependent on other non-physical levels of energy, such as the energy surrounding your body, called the biophoton field.' It is well known in biology that every second, over 100,000 biochemical reactions occur within every cell in your body to maintain homeostasis. Mercola explains, 'In highly simplistic terms, your biophoton field can be viewed as a highly sophisticated computer that processes, stores and retrieves information that is then used to regulate your biological processes.'

27.6 Light Energy from the Sun and Moon Enter Our Body and Is Stored by the Two Kidneys

Commenting on the *Nan Ching*, Yang stated that light energy from the sun and moon enter our body and is stored by the two Kidneys to be utilized by the Triple Burner for dispersing nutrients, healing damaged tissues, and maintaining optimal health. Further to that, the sun's essence is absorbed by plant and animal produce that we consume. Subsequently, the biophotons therein can be released into our Stomach within the Middle Heater. So let's consider what is happening in this cauldron called the Stomach. Along with the food nutrients—including minerals absorbed from the earth (the so-called grains), and the bulk water (the so-called drink)—the solar energy in the form of biophotons (the so-called essence) are all present in the one location in the Stomach that just happens to be a strongly hydrophobic containment vessel. In his book, Pollack (72) states categorically, '"Hydrophobic" or water-hating surfaces, such as Teflon, prove inept [at producing EZ water] by contrast; no exclusion zones could be found. It appears that the exclusion phenomenon belongs to hydrophilic surfaces as a class' (KL 800–801). It would appear that the formation of EZ water does not directly occur in the Stomach as the stomach lining is not hydrophilic. In the commentaries on the eighth Difficult Issue on page 133 of Unschuld's (1) translation of the *Nan Ching*, Liao P'ing states, 'The *Nei-ching* considered the stomach as the sea of water and grains. [Accordingly,] all fourteen conduits receive their supplies from the stomach. . . . Furthermore, [the Nei-ching] considered the through-way vessel to be the sea for the twelve conduits.' Note that Liao P'ing states, 'All fourteen conduits receive their *supplies* from the stomach' (emphasis is mine).

I believe that the raw-material supplies for the production of EZ water (the bulk water (the so-called drink) and the solar energy in the form of biophotons (the so-called essence)) are then distributed to appropriate hydrophilic organs and membranes and conduits (acupuncture meridians) that are omnipresent throughout the body, and subsequently, energized EZ water is manufactured as required. As can be seen from the above quote, the *Nei-ching* considered the throughway vessel to be the sea for the twelve conduits. The throughway vessel is more often called the Chong Mai or penetrating vessel and is known as the Sea of Blood. The Chong Mai links the Stomach and Kidney channels. The Chong Mai originates in the ming men and passes through the uterus and then down to CV 1. From CV 1, it emerges at ST 30 and continues upwards along the kidney meridian to KD 21. The Chong Mai then flows up the throat, encircles the mouth, and continues up to the forehead. The Chong Mai has many branches to all the vital organs, and its main purpose is to regulate the Qi and the Blood in the 12 main meridians and their corresponding organs. Subsequently, the Chong

Mai is one possible way that the raw supplies from the Stomach, namely the bulk water and the biophotons, could be circulated throughout the body to the 14 conduits and the omnipresent hydrophilic membranes to be converted into the many forms of Qi as per the Triple Heater fluid and energy production cycle. Note that the Kidney organ and Kidney meridian are both strongly involved with the Chong Mai circuitry. And as I mentioned above, Yang stated, 'The two kidneys are distinguished by their storing the essence of the sun and of the moon.' What does modern scientific research say on the topic? The 2010 article entitled 'Biophotons as Neural Communication Signals Demonstrated by In Situ Biophoton Autography' (75) stated, 'Cell to cell communication by biophotons has been demonstrated in . . . kidney cells.'

Summary of Chapter 27
Life on earth would be impossible without sunshine. For example, the vitamin D_3, which is produced in our skin (thanks to cholesterol sulfate), influences at least 2,000–3,000 genes and probably more. If there is a deficiency of the vitamin D_3, your body is prone to a staggering amount of diseases. Just as we can also absorb vitamin D_3 through our food as well, we can also absorb transmuted sunrays in the form of biophotons from foods. However, the best source of life-giving and sustaining biophotons is ripe, freshly picked, sunbathed fruit and vegetable produce. In the 2009 article by Dr Mercola (87) entitled 'Your Body Literally Glows with Light', he writes, 'Dr. Johanna Budwig from Germany has stated that live foods are electron-rich, and act as high-powered electron donors and "solar resonance fields" in your body to attract, store, and conduct the sun's energy in your body.' Mercola states in the conclusion of the article, 'The greater your store of light energy, the greater the power of your overall electromagnetic field, and consequently the more energy is available for healing and maintenance of optimal health.'

So we have learnt that healthy foods are power-packed with biophotons, which are essential for optimal health and well-being. But is there any scientific evidence for the practical application of biophotons for the treatment of medical conditions? The next Chapter will consider that topic.

CHAPTER 28

Practical Application of Biophotons

Introduction
Regarding the practical and theoretical implications and medical applications for biophotons, in the Abstract of the article 'Properties of Biophotons and Their Theoretical Implications', Professor Popp (89) stated:

> Biological phenomena like intracellular and intercellular communication, cell growth and differentiation, interactions among biological systems (like 'Gestaltbildung' or swarming), and microbial infections can be understood in terms of biophotons. 'Biophotonics', the corresponding field of applications, provide a new powerful tool for assessing the quality of food (like freshness and shelf life), microbial infections, environmental influences and for substantiating medical diagnosis and therapy.

Amazingly, all the cell-regulating and communication phenomena associated with biophotons that Professor Popp discusses are the identical regulatory phenomena attributed to the Triple Energizer. From the scientific research cited above, it is obvious that many prominent physicists propose that biophotons spontaneously and continuously control literally all the biochemical and neurological influences in our bodies and that some external evil forces (physical or emotional traumas) or some energetic blockages from toxins (mercury, fluoride) or electromagnetic pollution can disrupt an otherwise-wholesome body condition defined as wellness.

28.1 Practical Application of Biophotonic Healing Techniques and Devices

Biophotonic healing techniques already exist, as demonstrated in the Inside Out webpage (90), which states:

> To many, words such as biophotons, light therapy and biontology seem very new, even a little mysterious. Surprisingly, this very science can actually be traced back to ancient Egypt. However, it wasn't until much later, the 1970's actually, that Fritz-Albert Popp confirmed the existence of biophotons. It was he, who discovered that light emanates from every living cell at a rate of 100,000 impulses of light per second. These impulses are called biophotons.

28.2 Biophotons Formerly Shown to be Efficacious for Many Medical Conditions

In the 2014 article 'Resurgence of 'Old' Therapy offers New Hope. Ultraviolet Blood Irradiation', Robert A. Eslinger (91) discusses the history of Ultraviolet Blood Irradiation (UBI). He advises that the practice began in the 1920s when a UBI device was developed for the irradiation of the clients' blood outside the body. He reports that by the 1940s, UBI was used to treat medical conditions, including bacterial and viral infections and autoimmune diseases. Researchers believed that the therapy worked due to the secondary emissions of biophotons from blood cells that had been irradiated during the treatment. Eslinger reported that 'enthusiasm over the new antibiotics and vaccines in the 1950's caused the UBI device to be placed on the shelf even though, for certain indications (hepatitis, herpes and viral pneumonia), UBI was demonstrably superior.'

28.3 Recent Renewed Interest in Biophoton Treatment Due to Bacterial Resistance to Antibiotics

The 2014 article (91) informs us that there is a renewed interest in UBI Therapy thanks to the increased resistance of pathogenic bacteria to antibiotics in recent years and also for the search for less-toxic therapies than those offered by mainstream allopathic medicine. Eslinger advises that the records show that 'millions of patients have been successfully treated with UBI and scores of clinical trials have been conducted in Russia, Ukraine and the former East Germany. UBI is also currently used by some physicians in China and the United States.' There is a vast reservoir of documentation regarding the success of UBI treatment, including several controlled studies. The UBI treatment protocol began in 1928 and was practiced by reputable physicians for over 30 years, whereby hundreds of thousands of patients were effectively treated.

In the 2014 article, the author (91) points out that while UBI Therapy was proven to be most effective against many disorders, that it is a significant lapse for American medical science to ignore the procedure, and states, 'It is especially hard to justify this oversight given the intense effort to identify promising approaches to the treatment of HIV and related viral conditions.' It is my personal belief that the allopathic medical system has ceased incorporating UBI Therapy in their preferred medical procedures as they would fail to make the vast profits they have become accustomed to were they to utilize the therapy due to its relatively low cost to perform.

28.4 Scientific Mechanism behind Ultraviolet Blood Irradiation (UBI)

Regarding the theory behind UBI Therapy, Robert A. Eslinger (91) discusses the mechanism of Ultraviolet Blood Irradiation (UBI). He advises that:

> The Ultraviolet treatment of the blood in the treatment chamber destroys or alters bacteria and viruses in the extracted blood in such a way, as to create a kind of vaccination effect when they return to the body. This

provokes a reaction by the immune system, which in turn destroys most or all of the other bacteria or viruses in the body.

Eslinger proceeded to explain that the treatment stimulates normal red blood cells to emit biophotons. Only as little as 5% of the blood volume needs to be treated, and then the efficacious effect spreads throughout the entire body as the treated blood circulates. The generated biophotons destroy viruses, bacteria, and in the case of autoimmune diseases, the activated white blood cells. It appears that in the case of autoimmune disorders, metabolically active T-cells, along with other immune cells, absorb more biophotons than regular body cells and immune cells. Subsequently, the biophotonic influence destroys them, thereafter slowing or stopping the disease process. Regarding the vast array of different medical conditions that UBI Therapy was found to be beneficial for, Eslinger (91) advises that medical research up until the 1940s confirmed that UBI treatments were profoundly beneficial in inactivating toxins and treating numerous types of infections. UBI Therapy was successfully used to treat both viral and bacterial pneumonia, botulism, encephalitis, peritonitis, nonhealing wounds, and medical conditions as dissimilar as polio, asthma, hepatitis, chronic fatigue, fibromyalgia, and rheumatoid arthritis.

Other research findings confirmed that UBI improved oxygen transportation and dilation of blood vessels, thus improving microcirculation throughout the body. Improved blood parameters included activation of white blood cells, stimulation of cellular and humoral immunity, fibrinolysis, decreased platelet aggregation, and the decreased viscosity of blood. The wide diversity of health benefits ranged from favorable effects on infectious, inflammatory, and autoimmune disorders, and even the activation of steroid hormones was observed.

In the 2014 article entitled 'Blood Irradiation Therapy' (92) the commentary states that for therapeutic reasons, the patients' blood is exposed to light, which is often emitted from a laser. Russia and China are the main countries that conduct most of the research. The article states, 'Laser blood irradiation therapy was government-certified in Germany in 2005. In the following two years, this method was established in more than 300 centers in Germany, Austria, Switzerland, Italy and Australia.'

The Russian researchers Meshalkin and Sergievskiy first developed intravenous laser blood irradiation, and the procedure was introduced into clinical practice in 1981 for treatment of cardiovascular abnormalities, and the potential to use intravascular laser irradiation for the treatment of cardiocirculatory pathologies was presented in the *American Heart Journal* in 1982. This procedure is not associated with the gamma irradiation process, which is used in blood transfusion medicine.

The 2014 article (92) described three different applications of the procedure. They include an intravenous route via a vein in the forearm, whereby intravascular blood is irradiated, using laser light emitted from a helium–neon laser (1–3 mW at a wavelength of 632.8 nm). The second method involves intranasal blood irradiation, whereby the nasal cavity is irradiated noninvasively by inserting a small red light diode or a low-intensity laser to illuminate the walls of the nasal cavity. The third method is transcutaneous irradiation, whereby higher-power laser light illuminates unbroken skin over large blood vessels, for example, at the forearm. The three varied procedures assume that the light-stimulatory therapeutic effect will be disseminated throughout the body via the circulatory system.

The fact that the USA Food and Drug Administration (FDA) has approved the concept that ultraviolet light treatment of the blood can convey therapeutic benefit confirms that the principle is legitimate and must irritate the geriatric skeptics at Quackwatch that are paid by Bad Pharma to propagate the purported benefits of pharmaceuticals, radiotherapy, chemotherapy, and GE foods.

Summary of Chapter 28
Ultraviolet Blood Irradiation (UBI) has a long and established history. The practice began in 1928 when a UBI device was developed for the irradiation of the clients' blood outside the body, and the procedure

was successfully used by the established allopathic medical system of the time. UBI was successfully used to treat medical conditions, including bacterial and viral infections and autoimmune diseases along with a plethora of other medical conditions. It was a very inexpensive and simple procedure to perform. That was the 'problem'. Eslinger reported that due to the pharmaceutical development of more profitable antibiotics and vaccines in the 1950s, the UBI device was placed on the shelf, even though, for certain indications (hepatitis, herpes, and viral pneumonia), UBI was demonstrably superior. Typical of their established paradigm, the medical industry opted for higher profitability and then maligned what they knew was a superior medical procedure. The underlying scientific mechanism of the highly successful UBI procedure was not understood until the 1970s, when Fritz-Albert Popp confirmed the existence of biophotons. Subsequent research into biophotons revealed that they are pivotal to all biochemical reactions that occur in the body and that we require a constant supply of biophotons from sunbathed healthy foods, which are power-packed with solar-derived biophotons that are absolutely essential for optimal health and well-being. In 2007, this UBI method was still established and being practiced in more than 300 centers in Germany, Austria, Switzerland, Italy, and Australia.

CHAPTER 29

How Biophotons Energize the Triple Energizer

Introduction
In the commentaries on the 66th Difficult Issue on pages 567–568 of Unschuld's (1) translation of the *Nan Ching*, Yang states:

> The [location] between the two kidneys is called 'great sea' (ta-hai); another name is 'submerged in water' (ni-shui). Inside of it is the spirit-turtle. It exhales and inhales the original influences. When [the original influences] flow [out of the cinnabar field], they penetrate the four extremities as wind and rain; they reach everywhere. The [two] kidneys are distinguished by [their storing] the essence of the sun and of the moon, [respectively. This essence] consists of immaterial influences; it constitutes man's source and root. . . . Hence it is obvious that the cinnabar field is the basis of life. When the Taoists contemplate the spiritual, when the Buddhist monks meditate, they all cause the influences of their heart to move below the navel. That corresponds exactly to what is meant here. Hence [the text] states '"origin" is an honorable designation for the Triple Burner' because the Triple Burner's influences are accumulated in the kidneys.

This is the only location in Unschuld's (1) translation of the *Nan Ching* that the term *spirit-turtle* occurs. In this quote from Yang, the 'cinnabar field' relates to *dan tian*, which is the area around Ren 4, *Guan Yuan* (source gate), which is the location from where the 'moving qi between the kidneys' (Chong Mai) originates. This location between the Kidneys and below the navel is the origin of Yuan Qi, and it is the origin of the Triple-Burner Metasystem. 'Submerged in water' exactly describes the vast majority of the hydrophilic connective-tissue metasystem throughout the body. Professor Pollack advised that any hydrophilic material in contact with bulk water generates life-giving EZ water, and especially so, in the presence of light (biophotons). While the entire omnipresent connective-tissue metasystem throughout the body is indeed 'submerged in water' and in close proximity to a source of biophotons, for some reason, this anatomical region is the master location that initiates the life-sustaining pumping process and 'exhales and inhales the original influences', that is, pumps the Yuan Qi throughout the body. Perhaps it is here that a vast array of PVS microtubules summons up enough pressure via the Pollack-microtubule-pump-apparatus to initiate the Triple-Energizer Metasystem's multiple functions including fluid distribution and energy production throughout the body.

29.1 The San Jiao Regulates Water and Fluid Passageways throughout the Body

Regarding the first half of the scripture cited above, and the nature of the Jin Ye of TCM, the author of the book *Gua Sha: A Traditional Technique for Modern Practice*, Arya Nielsen (20) page 34, explains that the term *hai* describes the internal environment of the body as a 'sea within a sea'. Nielsen notes that while Body Fluid (Jin Ye) is a component of Blood when the Blood is inside the blood vessels, that same Body Fluid when outside the blood vessels 'stays in the slit of the body organs.' She notes that all body organs and tissues are completely surrounded by and in direct contact with the Body Fluids, Jin Ye. The entire body is essentially bathed in an external envelope of fluids. She cites Chapter 36 of the *Ling Shu*, which states, 'The Qi of the Triple Burner goes to the muscles and skin and is transformed into fluids (Jin). Other body fluids do not move and are transformed into liquids (Ye).' She notes that Jin Ye is the Body Fluids.

She then differentiates the two fluid components, such that the lighter Jin fluids circulate through the exterior of the body with the Wei Qi and provide the skin and muscles with protection and nourishment. The Lungs and the Upper Jiao of the San Jiao regulate the movement and expression of Jin throughout the body, and these Jin fluids leave the body as clear light fluids, including sweat, tears, saliva, and mucus. The Ye fluids are more dense than the Jin fluids and are more turbid, and their circulation is more internal and deeper along with the Ying Qi. These denser Ye fluids are thicker fluid discharges that are regulated by the Spleen and Kidneys and the Lower Jiao of the San Jiao, and they lubricate the joints and the orifices including the eyes, ears, nose, and mouth.

I propose that Yuan Qi, the original influences, are derived from the DNA-rich P-microcells that flow continuously throughout the Primo Vascular System. I believe that some form of biophoton pump and synchronizer resides between the kidneys. Maybe the same pump that drives the PVS liquids and P-microcells within the PVS also pumps the biophotons throughout the connective-tissue metasystem. Notice that Yang states that it 'exhales and inhales the original influences'. This seems to be some form of pump. Pollack showed that small hydrophilic tubules (including connective tissues) automatically generate a pump in the presence of water and biophotons. The small hydrophilic PVS tubules exactly fit this description. If this structure stops functioning, then so does life itself for that individual. In such a case, the regulatory function of the biophoton synchronizing Triple-Energizer Metasystem will fail and the human coherence system will crash, and death will result. Note too that this origin 'is an honorable designation for the Triple Burner', which at the beginning of the passage above is also called 'great sea' and 'another name is 'submerged in water'. Because the San Jiao regulates Qi and water passageways throughout the body, I believe that the San Jiao possesses a control station for biophoton synchronization and distribution (Qi function or qihua condition) and for EZ water generation and circulation throughout the body possibly via the omnipresent PVS.

I propose that this control station is the great sea or 'submerged in water' that 'exhales and inhales the original influences' and that it is an as-yet-unidentified anatomical entity which actually resides in the location described as between the kidneys, closer to the right kidney, below the navel, and composed of yellowish fatty fascial membrane.

Note what the kidneys do. Yang states, 'The two kidneys are distinguished by their storing the essence of the sun and of the moon.' What is the essence of the sun and moon? Obviously, light energy in the form of photons. I believe that sunlight energy directly from the sun and indirectly through the infinite varieties of food that we eat supply the biophotons. Every single food we eat is dependent on sunshine and subsequently absorbs photons from the sun into their produce, be it apple, rice, beef, chicken, eggs, nuts, beans, fish, etc. Once consumed, the intrinsic biological influences of the human digestive system and the influences of the individual microbiome present since birth harvests the biological nutrients, the necessary fluids, and the biophotons, which are then distributed throughout the entire body to where they are required for growth and sustaining life.

I found it remarkable that, irrespective of the specific topic of the Difficult Issue under discussion, in six out of the nine times that the sun is mentioned in the *Nan Ching*, it is in connection with the Triple Burner. That is where the fire of the Triple Burner comes from. The sun supplies the photons that all animals and plants convert into biophotons. I propose that these biophotons convert the bulk water H_2O to the energized EZ water $[H_3O_2]^-$ anion that acts like a life-giving battery throughout all the ubiquitous omnipresent hydrophilic continuum of connective tissues that constitute the Triple-Energizer Metasystem throughout our body.

Several types of known communication systems occur concomitantly within the body—chemical, biophotonic, piezoelectric, and neural. I propose that all these communication systems are regulated and synchronized by biophotons that energize the EZ water generation throughout the body via the Triple-Energizer Metasystem.

29.2 I Propose That Pollack's EZ Water Is a Driving Force of the Triple-Energizer Metasystem

Regarding the nature of the Yuan Qi of TCM, Arya Nielsen (20), the author of the book *Gua Sha: A Traditional Technique for Modern Practice* (pages 34–35) explains that the term *Yuan* means 'original, primordial source' and that it is our primal life force donated by Heaven, engendered through our parents, stored and accumulated in the Kidneys, and circulated through our body by the San Jiao. I have mentioned this previously, but I will repeat the information for this context. Nielsen noted that the San Jiao Fire is the original Yuan Yang of the Gate of Life (Ming Men) and that the original Yuan Yang Fire is the original component of all Yang function. Similarly, Water is the Yuan Yin and comprises the original component of all Yin substances and fluids. And alluding to the 31st Difficult Issue of *Nan Ching*, she notes, 'When Fire meets Water in the body, a transformation into Qi takes place.' She cites Larre and Rochat de la Vallee (1992), who state, 'There is *no vital Water without Fire* to transform it and *no Fire of life without Water* to fix and express it. . . . The Yuan not only suffuses the body with motive force but is a *catalyst in the formation of Ying, Wei, Qi and Blood*. Yuan is itself, in turn, "persistently regenerated" by those very products.' Larre and Rochat de la Vallee explain further that the Yuan

Figure 9. Photo by Oleksandrum79. © Depositphotos.com.

Qi congregates in the Kidneys at the Ming Men, and via the San Jiao, Yuan Qi reaches the entire body.

I propose that Fire (biophotons) from the sun reacts with Water (H_2O) from the kidneys to produce EZ water [H_3O_2]⁻, which is the original constituent of all yin and fluids in the body. In former Chapters, I have presented research information, whereby Professor Pollack has shown that this EZ water catalyzes all the life-sustaining biological and biochemical reactions in the body. In the quote above Larre and Rochat de la Vallee state, 'The Yuan not only suffuses the body with motive force but is a catalyst in the formation of Ying, Wei, Qi and Blood.' From the above information, it appears that both the EZ water and the Yuan Qi behave as catalysts in the formation of 'all the life-sustaining biological and biochemical reactions in the body'. That the EZ water is intimately associated with the Connective-Tissue Metasystem and that the Yuan Qi is intimately associated with the Triple-Energizer Metasystem lends credence to the fact that the Connective-Tissue Metasystem and the Triple-Energizer Metasystem are themselves intimately connected.

Summary of Chapter 29
As Larre and Rochat de la Vallée (1992) put it, 'when Fire meets Water in the body, a transformation into Qi takes place'. The EZ water contains the fire that transformed it into living liquid crystalline phase of water that causes all biological functionality essential for life to continue. As Larre and Rochat de la Vallee succinctly state, 'The Yuan not only suffuses the body with motive force but is a catalyst in the formation of Ying, Wei, Qi and Blood.' I have shown that this is exactly what Pollack proves that EZ water does. Thus I propose that as all the functions of EZ water are associated with those of Yuan Qi, the Yuan Qi that circulates through the Triple Energizer is intimately associated with what Professor Pollack calls EZ water.

Quoting from the book *Gua Sha: A Traditional Technique for Modern Practice* above, note that Nielsen (20) states, 'The Yuan not only suffuses the body with motive force but is a catalyst in the formation of Ying, Wei, Qi, and Blood. Yuan is itself, in turn, "persistently regenerated" by those very products'. Notice, in exact harmony with this comment, on page 23 of the 2012 book (7), Jong-Su Lee explains the general biological concepts behind Bong-Han Theory. He states, 'First of all, all living cells of living beings, when they get old, they dissolve to chromosomes of cells, and then they circulate in Bong-Han system. After receiving some energy, they become cells, and they regenerate damaged tissues. All cells are connected to Bong-Han system, Bong-Han canals penetrate to nuclei of cells.' Thus, according to Bong-Han Theory, all old, damaged (senescent) cells are predominantly digested enzymatically, except for their chromosomes, which are recycled and circulated in the fluid within the Bong-Han system microtubules. Notice that after this stage, Jong-Su Lee says, 'After receiving some energy, they become cells, and they regenerate damaged tissues.' These circulating P-microcells (Stem Cells) then regenerate into new cells to heal damaged tissues after being energized by receiving some energy. I propose that this Bong-Han system is intimately bound to the ubiquitous Connective-Tissue Metasystem, whereby the marriage of these two metasystems form the Triple-Energizer Metasystem. Thus, the original influences (Yuan Qi), namely the harvested cellular chromosomes consisting of DNA, protein and RNA are recycled continuously to correct and heal damaged cells throughout the body. Via this continual recycling process, the Yuan Qi or P-microcells are indeed persistently regenerated after they receive some energy from the EZ water battery in intimate proximity within the Connective-Tissue Metasystem/Triple-Energizer Metasystem.

CHAPTER 30

Morphogenetic Fields Bathe the Universe Independent of Time and Space

Introduction
English biochemist and author Dr Alfred Rupert Sheldrake is best known for his controversial hypothesis of formative causation, which proposes that nature itself possesses memory. He asserts that every different system in the universe—including, for example, molecules, cells, crystals, organisms, and societies—reacts in comparable or predetermined patterns in response to invisible fields of influence. He calls this occurrence morphic resonance. Sheldrake believes that the invisible field known as a morphic field is where preestablished patterns accumulate to influence a similar activity that may be occurring contemporaneously. Sheldrake has proposed that memory is inherent to all organically formed structures and systems. Sheldrake introduced the term *morphic field*, which proposes that there is a field within and around a morphic unit which organizes its characteristic structure and pattern of activity. Regarding Sheldrake's theory, in the article titled 'Morphogenetic Field (Body Field)—Rupert Sheldrake, PhD, University of Cambridge', the author, Uyan Dünya (93), states:

> The 'morphic field' underlies the formation and behaviour of 'holons' and 'morphic units', and can be set up by the repetition of similar acts or thoughts. The hypothesis is that a particular form belonging to a certain group, which

has already established its (collective) 'morphic field', will tune into that 'morphic field'.

Dünya thus advises that the specific form will subsequently read the collective information via the process of 'morphic resonance' and then use that preexisting information pool to direct its own specific development.

30.1 Sheldrake's Morphogenetic Fields Are Structures Independent of Time and Space

Sheldrake's theory proposes a literal interpretation of *morphogenetic fields* as structures independent of time and space. All the past morphogenetic fields of a given type are available instantly to or coexist with subsequent similar systems. *Morphic resonance* is the term that Sheldrake calls the influence of one morphogenetic field upon another morphogenetic field, involving a new kind of action at a distance, independent of space and time.

30.2 Embryos Develop By Tuning into Preexisting Morphogenetic Fields

In the website article entitled 'Our Causal Body', the author Robert Najemy (94) notes that Dr Rupert Sheldrake was the former director of studies in biochemistry and cell biology at Cambridge University. In the article, he states:

> Sheldrake claims that embryos and their brains develop by tuning into an entirely new kind of non-material formative force called morphogenetic fields. Just as a magnetic field can orient a pile of iron fillings into a certain pattern, so this mysterious new force can shape body and mind. Rupert Sheldrake's theory is that the changes which occur in the DNA are the results of changes in the morphogenetic field or causal body. These create a corresponding change in the DNA structure, which in turn create corresponding changes in genes, body structure, instinct, behavior and character. The DNA molecules and genes thus are like

the components of the television which simply transfer the changes to the screen. The incoming waves from the station form various invisible wave patterns which create changes in the components of the television, thus forming the various images on the screen.

30.3 The 1 Percent Theory Was Confirmed When the Crime Rate and Divorce Rate Dropped Significantly

Regarding what became called the 1 percent theory, in the website article entitled 'Our Causal Body', the author, Robert Najemy (94), stated:

> The Maharishi Maheesh Yogi, founder of the Transcendental Meditation Society of American, and subsequently a number of universities dedicated to unifying the scientific, social and spiritual, showed in a live experiment that the behavior of the whole population could be changed for the better if one percent of the population would meditate regularly. In the small state of Rhode Island in the U.S.A., this statement was proven. When the number of meditators reached one percent of the population, the crime rate and divorce rate dropped significantly for the total population. Whereas these rates in the surrounding states actually rose during the same period of time. This one percent of the population just might really have had this calming effect on the whole. A basic change might have been made in the morphogenetic field of that area.

30.4 Intelligent Learned Behaviour of Monkeys Adopted by Other Monkeys across the Sea

In the article titled 'Our Causal Body', the author, Robert Najemy (94), reported that, in 1952 on the island of Koshima, scientists were providing monkeys with sweet potatoes dropped in the sand. Sheldrake discussed the learned behaviour of an 18-month-old female monkey named Imo, who found that she could solve the problem of eating gritty sweet potatoes by washing them in a nearby stream. Imo taught this trick to her mother and to her young playmates, who also taught their mothers. This cultural innovation was gradually picked up by various monkeys before the eyes of the scientists. Between 1952 and 1958, all the young monkeys had learned how to wash the sandy sweet potatoes to make them more palatable. It was only the adults who imitated their children that learned this social improvement. The other adult monkeys kept eating the gritty sweet potatoes. Then something startling took place. In the autumn of 1958, a large number of Koshima monkeys were routinely washing their sweet potatoes—the exact number is not known. Then something remarkable happened! One evening, almost all the monkeys were washing sweet potatoes before eating them. This learned behaviour somehow created an ideological breakthrough! The most incredible and surprising thing observed by these scientists was that this newly learned habit of washing sweet potatoes then spontaneously jumped over the sea. Colonies of monkeys on other islands and the mainland troop of monkeys at Takasa Kiyama began washing their sweet potatoes! The concept of Sheldrake's morphogenetic field can easily explain that monkeys' newly learnt behaviour transcended space and then became the norm.

30.5 Is Widespread Antibiotic Resistance Found in Wildlife Due to Morphogenetic Fields?

Regarding the alarming finding of widespread antibiotic-resistant bacteria present in wildlife around the world in locations as diverse as southeast Mexico, Britain, and the Galapagos Islands, in the 2015 article titled 'Scientists Never Expected to Discover Antibiotic-Resistant Bacteria Here', Dr Becker (95) noted that, in a recently published article in the journal *PLOS ONE*, wildlife biologist Jurgi Cristóbal-Azkarate reported his 2014 discovery of superbugs being isolated from the feces of wild howler monkeys near Veracruz, Mexico. What made the research findings surprising was the fact that these wild monkeys are rarely exposed to human contact and have never been exposed to pharmaceutical antibiotics. Consequently, Cristóbal-Azkarate was both astonished and concerned to encounter antibiotic-resistant bacteria deep in the jungle, far removed from medicated civilization. Cristóbal-Azkarate and his team of researchers documented an abundance of antibiotic-resistant superbugs in the feces of seven diversified types of wildlife in the Veracruz region of southeast Mexico, including the howler monkeys, plus spider monkeys, jaguars, a puma, tapirs (large herbivores that resemble pigs), jaguarundis, and a dwarf leopard. Becker further stated, 'The team also learned that monkeys living far from humans were just as apt to harbor drug-resistant bacteria as those living closer to settled areas.' He continued, 'According to Randall Singer, an epidemiologist at the University of Minnesota, "Resistance is everywhere. It is found in places that are 'pristine' and in places that are 'polluted.'" Additional discoveries of superbugs in wildlife include enterobacteria discovered in wild rodents in Britain, and terrestrial iguanas on the Galapagos Islands.'

As most antibiotics in use are produced from soil bacteria and fungi and generally other soil bacteria have been exposed to these microbial antibiotic substances for millions of years, it is quite conceivable that some antibiotic resistance does occur naturally. The author (95) continues, 'However, Cristóbal-Azkarate found isolated bacteria that were resistant even to relatively recently developed synthetic antibiotics such as fluoroquinolones. This suggests the resistance is acquired vs. natural.' Becker continues, 'According to study co-author and microbiologist Carlos Amábile-Cuevas, currently "we are completely in the dark [about] which kind of processes led to this type of resistance."' I suspect that the likelihood of individuals being treated with such recent synthetic fluoroquinolone antibiotics coming into contact with so many different species in so many different locations is highly unlikely. I propose that such widespread global antibiotic resistance found in wildlife in remote locations is due to the effects of morphogenetic fields, as proposed by Dr Alfred Rupert Sheldrake.

Summary of Chapter 30
Sheldrake's theory also permits a new perspective on nonliving phenomena. Take for example, a novel newly synthesized substance that has never before been manufactured and subsequently crystallized

in a laboratory. As the new substance has never existed before, it has never previously had a defined form and does not 'know' what it is supposed to look like as a crystal. A chemist must work out the optimal conditions that will favor the initial crystal formation. Some scientists claim that once a substance has been crystallized many times over, subsequent crystallizations occur more readily. Scientists that object to Sheldrake's theory suggest greater knowledge on the substance, increased skill, and improved technique as the factors conducive to faster crystallization. Some claim that microscopic pieces of the new crystals escape into the air, and subsequently 'seed' newly produced solutions of the chemical and hence speed crystal growth. Sheldrake disagrees and predicts that even when airborne crystal transfer is vigorously prevented, crystallization will become progressively easier as time progresses. He believes this is due to a strengthening of the global morphogenetic field that is responsible for the chemical's initial crystalline pattern.

According to Sheldrake's hypothesis of formative causation, a novel newly synthesized chemical's crystalline form will not be predictable in advance, and no morphogenetic field for this form will yet exist in the universe. But after the substance has been crystallized for the first time, the substance then 'knows' what it should look like, and the established form of its crystals will influence subsequent crystallizations by morphic resonance. The more often it is crystallized, the stronger this influence should be.

Likewise, any phenomenon that has previously occurred can more readily reoccur spontaneously in the future.

CHAPTER 31

The Triple Burner as a System of Waterways

Introduction

Regarding the nature of the Triple Burner, Dharmananda (15) states, 'A model for the triple burner that came to be relied upon, at least as an image to work from, was a fermentation vat, such as used for making rice wine or beer. That fermentation process, developed in ancient times, was utilized in the Han Dynasty period (when the Neijing was written) to make health drinks that combined the alcoholic beverage with herbs.' Dharmananda went on to say:

> The brewers noted that at the top of the vessel, which was where the water and grain and yeast were poured in, there developed a fragrant mist that smelled like the wine or beer essence. At the upper part of the fluid was foam, made from the bubbling mixture that had some impurities contributing to generating the foamy top. Towards the bottom was a mixture in which solids were accumulating, and leaving a relatively clear liquid above them. The clear liquid would then be 'tapped' to provide the drinkable wine or beer. Thus, we will find a description of the upper burner involving 'fog' or 'mist'; the middle burner involving fermentation (rotting), foam, and collections of

bubbles; and the lower burner involving clear liquid and turbid substance separating out as the dregs.

Being a former home brewer, this analogy is close to my heart. The analogy of the middle burner involving fermentation (rotting), foam, and collections of bubbles is exactly what happens in the intestines. But the microflora associated with wine and beer production is yeast, normally *Saccharomyces* spp., while the fermentation in our intestines is orchestrated predominantly by *Bacteroides*, along with *Lactobacillus* spp. and *Bifidobacteria* spp. It should be noted that some *Saccharomyces* spp. are generally present in a healthy human gut. As can be seen from above, the Triple-Energizer Metasystem revolves predominantly around water distribution throughout our body.

31.1 Is the Triple Heater Merely a Summation of the Functional Relationship between the Various Organs That Regulate Water throughout the Body?

Jóhannsdóttir (96) states, 'The Triple Heater can be looked upon as being the functional relationship between various Organs that regulate Water. These are mainly the Lungs, Spleen and Kidneys, but including also the Small Intestine and the Bladder. The Triple Heater does not exist as an entity outside these various Organs, but rather it is the pathway that makes these Organs a complete system.'

I believe that because the author does not yet know about the actual location of the Three Heaters being integrated and predominantly composed of the Connective-Tissue Metasystem that is omnipresent throughout the entire body, she incorrectly suggests, 'The Triple Heater does not exist as an entity outside these various Organs, but rather it is the pathway that makes these Organs a complete system.' This is in keeping with the inappropriate bias of believing that the Three Heaters is said to have a name but no shape, which is primarily based on a misunderstanding derived from the *Nan Ching*.

31.2 The Triple Burner Is Not a Receptacle, but a Passageway
Regarding the 31st Difficult Issue discussed in the *Nan Ching*, namely, 'The triple burner: how is it supplied and what does it generate? Where does it start and where does it end? And where, in general, are its disorders regulated?' Dharmananda (15) states, 'The triple burner encompasses the passageways of water and grain.' He goes on to say, 'The triple burner is indicated in this initial response as *not* being a receptacle, but a *passageway*; that is to say, unlike the other fu organs, it is not retaining material that is later to be passed on.'

I believe that many of the descriptions throughout the ancient texts regarding the operation and functionality of the San Jiao are actually factual, but due to ignorance about the lymphatic system and the Primo Vascular System, to this time, the fluid flows and passageways have not been anatomically observable. Generally though, I believe that the underlying principles and outcomes are sound and factual. Dharmananda (15) concluded his discussion on the triple burner succinctly by stating:

> The ying essence derived from food, a process taking place in the middle burner with passage through the upper burner, will then rain down to the lower body; the yang essence, including wei qi, will rise up from the lower jiao, especially under the influence of the upward action of liver and kidney yang. All these essences circulate throughout the body, so one can only speak of an 'origin' point in a relative manner. The triple burner, subtending the three zones and involving the processing of grains and water, is a fitting marker for the beginning movements of fog and mist, ying and wei, and the yin and yang passageways.

31.3 What Is the Middle Burner Passageway through to the Upper Burner?
I propose that the passageway from the Middle Burner upwards to the Upper Burner is a factual conduit derived of either hollow connective-tissue channels or hollow ducts of the Primo Vascular System or possibly even both systems. Anatomists are finding completely new

structures or previously unknown functions of existing structures on a regular basis in recent years. In 2012, Beck et al. (97) reported, 'Human investigations aimed at dissecting the relationship between the gut microbiome and lung immunity are limited due to experimental and ethical considerations. While important associational relationships are being established, direct hypothesis testing is considerably more difficult. Animal models are providing novel information to test the hypothesis that gut microbiota influence lung immunity.' I propose that future research will confirm that the immunological dynamics of the Lungs will be governed by and potentiated by the gut microbiome. I discuss features of the newly discovered organ, the gut microbiome, in great detail in a later Chapter. Beck's research indicates that there is a biological mechanism whereby gut microorganisms affect the Defensive Qi (immunity) of the Lungs. Elsewhere I have discussed how the gut microorganisms, somehow or other, manage to get from the intestines of the mother into the breast milk. Obviously, bacteria can't time-travel. There is obviously a literal hollow conduit system that permits passageway of the microflora from the Small Intestine to the mammary lacteals. Again I propose the conduit system is a 'passageway from the Middle Burner upwards to the Upper Burner' via the Triple-Energizer Metasystem, also known as the Connective-Tissue Metasystem. This would explain why acupoint SI 1 is good for insufficient lactation, mastitis, breast abscess, and cysts and why SI 11 is so beneficial for mastitis, insufficient lactation, and breast pain.

31.4 One Important Facet of Our Immune System (i.e. TH17 Cells) Located in the Small Intestine Is Absolutely Dependent on the Presence of Microflora

Regarding the necessity of commensal microbes in the small intestines for the development of a competent immune system, in the Drbganimalpharm at Blogspot, the author (98) notes that the major locations of microbial fermentation in the human alimentary tract are the cecum, appendix, and the remainder of the large intestine. The author notes that the proximal portion of the small intestine is nearly sterile and that it is at the distal end of the small intestine called the ileum, where commensal species of bacteria and fungi take up residence. T helper 17 (TH17) cells are one of the important arms of the immune system. Interestingly, the author notes that no TH17 cells appear in the small intestines until commensal microorganisms take up residence in the area, and stresses that laboratory bred germ-free rodents are completely devoid of an entire facet of the immune system until colonization by commensal segmented filamentous bacteria (SFB) occurs. The spore-forming soil bacteria of the genus *Arthromitus* admirably fits the bill since all life forms developed along with the continual presence of dirt for millennia. This shows that at least one important facet of our immune system (i.e. TH17 cells) located in the Small Intestine is absolutely dependent on the presence of microflora. Regarding other essential commensal-organism systems that engender good health, the author (98) continues:

> Initially breastmilk was considered sterile but like many things in science, this was inaccurate. Colostrum contains over 700 live organisms (European Society of Neurogastroenterology and Motility: Gut Microbiota Worldwatch). Are these important? Why? Lysates of strains such as Lactobacillus and Bifidobacter have been shown to tighten up the intestinal tight junctions. Several strains in probiotics have been shown to be associated with reduced mortality in NICU (Neonatal Intensive Care Unit) settings and against often fatal necrotizing enteric colitis. Although babies are born with leaky guts to accommodate mothers' large proteins direct access into blood and lymph circulation (immunoglobulins, IgM), they appear to get a lot of help from natural commensals to switch and develop intestinal impermeability.

Further along in the blog at Blogspot, the author (98) states:

> Where does mammary microflora originate from? Some researchers hypothesize there is a special conduit that transfers gut flora to the mammary glands, an 'entero-mammary pathway'. Lymph circulation? It is unknown and not entirely elucidated. Researchers looked at microbiota in breast milk (at 0 mon/colostrum, 1mon and 6 months), different areas of the mother's body and

compared flora between elective C-section and vaginal/ non-elective C-sections. Breastmilk from C-section moms resembles mouth/oral and skin flora; whereas, breastmilk from moms who went through vaginal birth or some modicum of vaginal delivery that ended in non-elective C-section (that was me) showed flora that matched the gut and feces. Birth does something to make proper milk. The fact that the milk microbiome of mothers who gave birth by nonelective cesarean section had a normal microbial composition that was comparable to that of breast milk from mothers who delivered vaginally suggests that physiologic (eg, hormonal) changes produced in the mother during the labor process may influence the composition of the bacterial community.

Radioisotoped microflora introduced into pregnant animals have been translocated to the breast milk for the immunological benefit of the offspring. Are the microorganisms and their biochemical products and nutrients being transported from the Intestines throughout the body as required? This could well explain how good commensal gut microbiome bacteria make their way to the mammary glands to inoculate the neonate with immune-boosting bacteria of their own. To date this pathway has not been elucidated. I believe that there is an as-yet-undefined pathway through the maze of connective-tissues membranes that are omnipresent throughout our body. Otherwise, the pathway may be due to ducts of the Primo Vascular System.

31.5 The Passageways of Water Originate from The Triple Burner
In the commentaries on the 31st Difficult Issue on page 349 of Unschuld's (1) translation of the *Nan Ching*, Hsü Ta-ch'un states, 'This is a general summary of the meaning of the Triple Burner. It says that [the Triple Burner] is supplied and generated by water and grains and that it constitutes the start and conclusion of the [course of the] influences.' Along the same lines, in the commentaries on the 31st Difficult Issue on page 348 of Unschuld's (1) translation of the *Nan Ching*, Li Chiung states:

The Triple Burner [has been compared in the Su-wen] with the official responsible for maintaining the ditches; the passage-ways of water originate from there. The passage of the water enters [the organism] through the upper [section of the Triple] Burner and leaves it through the lower [section of the Triple] Burner. Thus one knows that [the Triple Burner] represents conclusion and start of the [course of the] influences.

31.6 The Lower-Ditch Controls Elimination of the Urine and Feces

In the commentaries on the 31st Difficult Issue on page 351 of Unschuld's (1) translation of the *Nan Ching*, Yü Shu states, 'The Ling-shu ching states: "The lower [section of the Triple] Burner resembles a ditch." That is to say, the bladder controls the water. The *Su-wen* states: "The Triple Burner represents the official responsible for maintaining the ditches. The waterways originate from there."'

Also in the commentaries on the 31st Difficult Issue on page 351 of Unschuld's (1) translation of the *Nan Ching*, Li Chiung states, '[The lower section of the Triple Burner] separates the water and the grains which were taken in through the upper [section of the Triple] Burner. The clear [portions] become urine; the turbid [portions] become feces. They are then transmitted to the outside.'

31.7 The Triple Burner Is the Palace of the Central Ditch

In the commentaries on the 38th Difficult Issue on page 397 of Unschuld's (1) translation of the *Nan Ching*, Hsü Ta-ch'un states:

> That is to say, it is located outside of all the [remaining] palaces. Hence, it is called an 'external palace.' According to the treatise 'Pen-shu' of the *Ling* [-shu], 'the Triple Burner is the palace of the central ditch. The waterways originate from there. It is associated with the bladder. It is a solitary palace.' it is called 'solitary palace' because it is not attached to a depot. That is the meaning of 'external palace.'

31.8 Triple Energizer Govern the Irrigation of Fluids throughout the Body

In the book *Heart Master Triple Heater*, Rochat de la Vallée (11), on page 51, points out that *Su wen* Chapter 8 states, 'The triple heater is responsible for the opening up of passages and irrigation. The waterways stem from it.' She then states, 'If we look at *Su wen* Chapter 8, the first thing we see is that the Triple Heater can break through passages and pathways. It has the concentrated strength of shao yang which allows passing through obstacles and a clearing of the ways so that all the irrigation and streams in the body can circulate freely and keep the qi moving.' She points out that the Chinese character for irrigation is the ideogram *du* and it represents a small irrigation channel in a field. The term is present in the name of the acupoint TH 9 (si du or Sidu), which means 'Four Rivers'.

She explains that the ideogram *du* reveals how a piece of ground can be rendered fertile by forming an intricate network of channels that allow water to flow from the river and circulate freely through the system to irrigate the field. This network would then permit adequate irrigation and drainage and allow evacuation of wastes. She suggests the analogy of a township, whereby the ideogram *du* would be the municipal drainage and sewage system, and noted that, in both cases, it is essential that free communication occurs and no obstructions or blockages result. This shows that the Triple Heater is likened to the intricate irrigation channels of a field, whereby every small part of the field is moistened and made fertile by the life-giving flowing water diverted from the local river. The defined organ system in the body that is like this is the omnipresent Connective-Tissue Metasystem, which I discuss next.

31.9 Resembling the Triple Energizer, Fascia Makes Up 16% of Total Body Weight and Stores 23 Percent of Total Water Composition, and Fascia Connects All the Tissues of the Human Body

In the article entitled 'Fascia: The Under Appreciated Tissue', the author (99) states that connective tissues play an important role in human function and that fascia makes up 16% of total body weight and stores 23% of total water composition of the human body and that fascia connects all the different tissue types of the human body together, including the muscles, organs, nerves, and vessels of the body. The author continues, 'Fascia is a dynamic connective tissue that changes based on the stresses placed on it.' The author states further that 'embryology helps explain how all the fascial system connects all major systems including the nervous system'.

31.10 Triple Heater and Kidneys Both Control Water Movement in the Body

In the book *Heart Master Triple Heater*, regarding the property of the Triple Heater in controlling the distribution of fluids throughout the entire body, Elisabeth Rochat de la Vallée (11), on page 53, states, 'When *Su wen* Chapter 8 insists on the pathway of liquids and the regulation of fluids it is just to make this action of the triple heater more tangible by means of this image of the movement of liquids inside the body. It makes the links with the kidneys and the water of the kidneys closer.'

Summary of Chapter 31

Regarding the Qi mechanism that allows for the regulated transportation of the various fluid components of the San Jiao to move in and out of cells, tissues, vessels, organs, joints, etc. throughout the entire body, in the article entitled 'San Jiao', the author, Linda Barbour (23), states very eloquently that this Qi mechanism ensures that the metabolism throughout the entire human body maintains a defined equilibrium. Fluid imbalances within the system can cause damp or phlegm conditions, including swollen and/or edematous manifestations throughout the body, and conversely, where there is fluid deficiency, body tissues can become dehydrated. Either of these conditions is not conducive to optimal functioning of bodily systems, and disease states occur. The various body fluids can only move along the pathways that the San Jiao provides. Barbour advises that some body fluids passively diffuse into and out of the cells throughout the body, the interstitial spaces, and the bloodstream. Yet other fluids require an input of energy (Qi) to move according to their function. All these diverse fluid flows are dependent on the directives of the Qi mechanism. She further notes that the San Jiao constitutes the pathway for Original Qi and the numerous body fluids to follow, and the San Jiao ensures that the pathways remain open. Original Qi is the energy form that orchestrates all the diverse fluid flows and allows all the numerous reactions to take place.

Connective tissues play a very important role in human function. That fascia connects all the different tissue types of the human body together—including the muscles, organs, nerves, and vessels of the body—indicates that it has a large role in communication throughout the body just as the San Jiao does too. The fact that fascia makes up 16% of total body weight and stores 23% of total water composition of the human body (99) shows that it has a lot in common with the description of the San Jiao. Grossman (5) stressed that the legendary highly respected irrigation official named Yu personified the San Jiao as 'the official in charge of irrigation and it controls the water passages'.

CHAPTER 32

Acupuncture Meridians Exist Within an Omnipresent Liquid Crystalline Membrane System of Collagen

Introduction
After obtaining her PhD in Biochemistry from Hong Kong University, Mae-Wan Ho embarked on a distinguished research career that included a postdoctoral fellowship in Neurosciences at the University of California at San Diego, and Fellowship of the National Genetics Foundation, USA. She was a Senior Research Fellow in Biochemistry, University of London, then a Lecturer in Genetics, and then a Reader in Biology, Open University. Her research evolved through biochemistry to molecular genetics, non-Darwinian evolution, and since 1988, she has concentrated on the physics of living organization, defining a new field with her present book. She is widely acclaimed by serious scientists across the disciplines and by non-scientists alike. Ho is lead author of the 1998 research article entitled 'The Acupuncture System and the Liquid Crystalline Collagen Fibres of the Connective Tissues: Liquid Crystalline Meridians'. Regarding the nature of acupuncture meridians in the body, in the 2014 article, the authors, Ho and Knight (82), state, 'We propose that the acupuncture system and the DC body field detected by Western scientists both inhere in the *continuum of liquid crystalline collagen fibres that make up the bulk of the connective tissues*' (emphasis is mine).

The authors believe that the layers of liquid crystalline water bound to the collagen fibers throughout the body allow rapid intercommunication throughout the entire body due to the conduction of protons. They believe that this ubiquitous liquid crystalline continuum explains such processes as the immediacy of hyper-reactivity to allergens and the efficacy of subtle energy medicine, including homeopathy. According to TCM, acupuncture meridians allow energy flow throughout the body and constitute a dynamic communication system throughout the entire body. The acupuncture system is a stand-alone system and is not the same as the nervous system. However, there does appear to be functional interconnection with the central and peripheral nervous systems. The authors further note that an electrodynamical field is present in all early embryos and also in plants and animals which are devoid of neural or perineural tissues and that while the DC field is mostly outside the nervous system, there is a functional interconnectivity with the nervous system. Further, it is widely recognized that, in many conditions, the speed of information relay within our body is much faster than optimal nerve conduction can produce, along with the fact that nerves simply do not extend to all parts of our body. Subsequently, Dr Ho believes that 'the acupuncture system and the DC body field detected by Western scientists both inhere in the continuum of liquid crystalline collagen fibres that make up the bulk of the connective tissues'. In this Chapter, I will explore this topic more as it is pivotal to my theory.

32.1 The DC Electrodynamical Field and the Acupuncture System Have a Common Anatomical Basis

Regarding the nature of acupuncture meridians in the body, the 2014 article by authors Ho and Knight (82) states, 'We propose that both the DC electrodynamical field and the *acupuncture system have a common anatomical basis. It is the aligned, collagen liquid crystalline continuum in the connective tissues of the body with its layers of structured water molecules supporting rapid semi-conduction of protons*' (emphasis is mine). The authors explain that this mechanism allows for communication throughout the entire body and coherence is maintained.

32.2 The Entire Human Organism Is Believed to Be a Liquid Crystalline Continuum

With regard to the nature of the entire human organism, in the article entitled 'The Acupuncture System and the Liquid Crystalline Collagen Fibres of the Connective Tissues: Liquid Crystalline Meridians', the authors, Ho and Knight (82), explain that one prerequisite for an intercommunication system is a continuum capable of carrying information, and they suggest that the living human organism is actually such a continuum. They note that it is now accepted that human cells are interconnected both electrically and mechanically in a solid state or tensegrity system such that 'all the cells in the body are in turn *interconnected to one another via the connective tissues*' (emphasis is mine). Ho and Knight state that they recently discovered that the living human continuum is liquid crystalline in nature, thus endowing it with all the properties perfect for communication throughout the entire body.

The authors explain that liquid crystals constitute a phase that exists in between a solid-crystal phase and the liquid phase, hence the term *mesophase*. Liquid phases lack molecular order. While crystal phases are ordered, the liquid crystal phase possesses orientational order. But distinct from solid crystals, liquid crystals are flexible, malleable, and responsive, and characteristically experience rapid changes in orientation when subjected to electric and magnetic fields. This is why modern technology utilizes this property of liquid crystals in display screens. Liquid crystals are sensitive to many varying parameters, including changes in ambient temperature and pressure, hydration rate, and local shear forces. As biological liquid crystals convey static electric charges, they are also affected by the pH, salt concentration, and the dielectric constant of the solvent they are immersed in. The late George William Gray was a Professor of Organic Chemistry at the University of Hull. He was the inventor of long-lasting materials used to make liquid crystal displays so commonplace today. He set up and systematized the science of liquid crystals. Interestingly, he referred to liquid crystals as tunable responsive systems and proposed that they would be the ideal medium for synthesizing organisms.

The authors (82) reported that it is widely acknowledged that all the major components of living organisms may well be liquid crystalline in nature. This includes the lipids present in cellular membranes, DNA, and most likely, all proteins, especially including cytoskeletal proteins, muscle proteins, and the proteins in connective tissues, including collagens and proteoglycans. Remarkably, very few biochemical researchers accept the concept that living organisms may be principally liquid crystalline in nature in spite of the fact that recent nuclear magnetic resonance (NMR) studies have confirmed that muscles in living human subjects are highly suggestive of being 'liquid-crystalline-like' in their composition.

Interestingly, the conception that living organisms are liquid crystalline in nature is not at all a recent construct. In 1927, Hardy suggested that molecular orientation may be a significant parameter for living protoplasm, and in 1929, Peters believed there was a direct connection between molecular orientation and liquid crystals in living systems. Joseph Needham was a member of the Club for Theoretical Biology in Cambridge and was instrumental in the creation of the field of molecular biology. As long ago as 1935, Needham proposed specifically that organisms actually are liquid crystalline in nature. Only recently, Ho (82) and co-workers have provided direct evidence for that assertion. They utilized an interference color technique that magnifies weak birefringences, and the results suggested that the images of live organisms were typical of biological liquid crystals.

32.3 Collagen Fiber Orientation and the Acupuncture System

Concerning the oriented nature of collagen liquid crystalline phases in all connective tissues and the association with acupuncture points, the 2014 article by authors Ho and Knight (82) is very informative. They believe that the highly ordered, oriented structure of collagenous liquid crystalline mesophases present in all connective tissues throughout the entire body concomitantly accounts for the high efficiency of information transfer throughout the entire body. It is accepted that the mechanical stresses and strains required for various tissue types to perform optimally determines the distinguishing orientation and composition of connective tissues. Further to that known function, the authors believe that these same structural parameters may also serve a secondary function of being crucial for body-wide intercommunication. It has been an accepted fact for a long time that the alignment of collagen is significant in the construction of cartilage and bone. The authors note that it is a lesser-known fact that the Langer lines in the skin conform to the principal orientations of collagen fibers, particularly determined by local pressures encountered during development and growth. They state, '*Collagen fibre alignments in connective tissues providing channels for electrical intercommunication may thus be correlated with the acupuncture system of meridians and points in traditional Chinese medicine*, which, as mentioned above, is also related to the DC body field identified by scientists in the West.' The authors note that, due to the water bound within the collagen fibers, they are expected to conduct (positive) electricity along the fibers, and they suggest that these same conduction pathways may correspond to conductive acupuncture meridians. Regarding the fact that acupuncture points generally possess lower electrical resistances than the surrounding skin, they suggest that acupuncture points may occupy the gaps between adjacent collagen fibers or be positioned at locations where the collagen fibers are positioned at right angles to the dermal layer. They stated, 'A number of structures mentioned earlier, which are at or near acupuncture points, have a common feature in that they are *located in local gaps in the fascia or collagen fibres*' (emphasis is mine).

32.4 Oriented Collagens Engender Intercommunication and Body Consciousness

Regarding the topic of an intercommunication system in the body that is faster than the nervous system and an omnipresent body consciousness separate from cranial consciousness, in the 2014 article, the authors, Ho and Knight (82), state, 'Liquid crystallinity will make coherent excitations even more likely to happen. Weak signals of mechanical pressure, heat or electricity, may therefore be readily amplified and propagated by a modulation of the proton currents or coherent polarization waves.' They continued, '*The hydrogen-bonded water network of the connective tissues is actually linked to ordered water dipoles in the ion-channels of the cell membrane* that allow inorganic ions to pass in and out of the cell.' The authors further explained that a direct channel of electrical communication occurs between distant signals and the intracellular matrix, which initiates modifications within various cells, including neurons and glial cells. They believe that this electrical communication channel coexists as the mechanical tensegrity system composed of the external connective-tissue metasystem, which is continuous with the intracellular matrix. The authors stated:

> Any mechanical deformations of the protein-bound water network will automatically result in electrical disturbances and conversely, electrical disturbances will result in mechanical effects. . . . Proton jump-conduction is a form of semi-conduction in condensed matter, and is much faster than conduction of electrical signals by the nerves. Thus *the 'ground substance' of the entire body may provide a much better intercommunication system than the nervous system*. Indeed, it is possible that one of the functions of the nervous system is to slow down intercommunication through the ground substance. Lower animals which do not have a nervous system are nonetheless sensitive. At the other end of the evolutionary scale, note the alarming speed with which a hypersensitive response occurs in human beings. There is no doubt that a body consciousness exists prior to the 'brain' consciousness associated with the nervous system. This body consciousness also has a memory. (Emphasis is mine)

32.5 Collagen Liquid Crystalline Mesophases and Memory

With reference to the intercommunication of liquid crystalline collagens and the intrinsic memory therein, in the 2014 article by Ho and Knight (82), the authors note that researchers have confirmed that subtle changes to the three-dimensional configuration of the collagen triple helix cause the collagen to express altered biological activities. Collagen is known to regulate the growth and movement of cells when contact is made, and numerous cell membrane proteins recognize specific sites of the collagen protein. When subtle errors occur in the amino acid sequencing during the formation of collagen, the outcome can be profound, yielding hereditary disorders, including chondrodysplasias, osteogenesis imperfecta, and Ehler-Danlos syndrome. The authors state, '*As the collagens and bound water form a global network*, there will be a certain degree of stability, or resistance to change' (emphasis is mine). They note that this arrangement constitutes a memory and that further cross-linking and other chemical adaptations of the collagens would stabilize that memory even more so. While the collagenous network does preserve tissue memory of former experiences, it is also able to register new experiences because all connective tissues, including bones, are relentlessly communicating and responding to information as well as experiencing metabolic processes. They state, 'Memory is thus dynamically distributed in the structured network and the associated, self-reinforcing circuits of proton currents, the sum total of which will be expected to make up the DC body field itself.' Note that the authors stated, 'The collagens and bound water form a global network.' Note that as the hydrophilic collagen network is the fixed infrastructure, the attracted water is essentially drawn to the collagen and goes along for the ride throughout the entire body. That sounds remarkably like the major function of the Minister of Dykes and Dredges, the Triple-Energizer organ complex (San Jiao), which is 'the official in charge of irrigation and it controls the water passages'.

32.6 Coupled Collagen-Derived Body Consciousness and Nervous System Brain Consciousness

Regarding the coupled connection between body consciousness derived from liquid crystalline collagen fibers of the connective tissues and brain consciousness associated with the nervous system, in the 2014 article entitled 'The Acupuncture System and the Liquid Crystalline Collagen Fibres of the Connective Tissues: Liquid Crystalline Meridians', the authors, Ho and Knight (82), believe that a body consciousness exists throughout the entire body that has all the features of consciousness, including the ability to feel, perceive, and experience subjectively the ability to communicate and to remember events. They believe that the known consciousness attributed to the nervous system is actually deeply rooted within the body consciousness and is intimately connected with it. Support for this theory exists as exemplified by research findings into the biological mechanism of anesthetics. It is proposed that anesthetics work 'by replacing and releasing bound water from proteins and membrane interfaces, thus destroying the hydrogen-bonded network that can support proton jump-conduction'. This would mean that bound water attached to proteins and membrane interfaces would account for the state of consciousness. Further to support this claim is the fact that Becker showed that general anesthesia results in the 'complete attenuation of the DC body field'. The authors suggested that if their theory is correct, they predict that research should confirm that collagens equilibrated with various solvents and anesthetics would show a decrease in conductivity compared to an equivalently hydrated collagen control.

The authors further note that while brain-and-body consciousness are generally interconnected, decoupling does occasionally occur. For example, while under general anesthesia, some surgical patients have regained brain consciousness of pain but have been unable to move or to advise surgeons about their distress. On the contrary, acupuncture treatment has been successfully utilized to anaesthetize patients who are still fully awake and conscious. Becker produced further evidence of the relative independence of brain-and-body consciousness by confirming that 'during a perceptive event, local changes in the DC field can be measured half a second before sensory signals arrive in the brain'. Evidence from Libet et al. suggests that a readiness potential occurs prior to a subject willfully moving an arm or a leg, suggesting that the directives of the brain may be preconditioned and determined by the local DC field. The authors presented likely test equipment that could differentiate situations where brain and body consciousness had actually decoupled from one another.

32.7 Ho Proposes the Acupuncture Meridian System Exists within the Continuum of Liquid Crystalline Collagen Fibers

After reviewing supporting evidence from biochemistry, cell biology, biophysics, and neurophysiology, authors Ho and Knight (82) state, 'We have proposed that *the acupuncture (meridian) system and the DC body field detected by Western scientists both inhere in the continuum of liquid crystalline collagen fibres and the associated layers of bound water that make up the bulk of the connective tissues of the body*' (emphasis is mine). They suggest that the established TCM acupuncture meridian system may be allied with the layers of bound water that coexist externally to collagen fibres and that this construct provides the conduit system for proton conduction pathways throughout the body to allow for rapid body-wide intercommunication of information. They further suggest that amalgamation of the electrical and electromechanical actions of the ubiquitous liquid crystalline continuum throughout the body institutes a body consciousness that works in cooperative harmony with the brain consciousness associated with the nervous system.

Note that *inhere* means 'to exist permanently and inseparably in something as a quality, attribute, or element' or 'to belong intrinsically to something'. Thus, in the quote above, the authors are plainly stating they believe that the established TCM acupuncture meridian system exists as a permanent functional structure within 'the continuum of liquid crystalline collagen fibres and the associated layers of bound water that make up the bulk of the connective tissues of the body'. Interestingly, Pollack has proven that when bulk water and hydrophilic surfaces come together in the presence of biophotons, highly structured liquid crystalline EZ water is generated with its maze of dynamic electrical properties. This structure would provide 'proton conduction pathways for rapid intercommunication throughout the body'. This is exactly what the acupuncture meridian system does. Note too that Ho and Knight (82) comment that defined acupuncture points have different morphology than surrounding tissues. Has this been substantiated by other researchers? Yes. In the next section, Dr Weber discusses this topic in more detail.

32.8 New CT Scans and Scientific Microsensing Apparatus Reveal Acupuncture Point Locations

In the Panaxea article entitled 'New CT Scans Reveal Acupuncture Points', Daniel Weber (100) discussed major findings from a research article that was published in 2013 in the *Journal of Electron*

Spectroscopy and Related Phenomena. He advised that 'acupuncture points have a higher density of micro-vessels and contain a large amount of involuted microvascular structures', while non-acupuncture points did not have these properties. The acupuncture points (for example, ST-36 (*Zusanli*) and ST-37 (*Shangjuxu*)) displayed very distinct structural differences from surrounding areas. The acupuncture points displayed local high-density vascularization, which was not found in non-acupuncture point areas. Weber advised that the researchers stated that other investigators have determined unique structures associated with acupuncture points and acupuncture meridians using various analytical methodologies as diverse as MRI (magnetic resonance imaging), LCD thermal photography, infrared imaging, and ultrasound along with other CT imaging methods. Weber (100) continued, 'The researchers commented that many studies using these technological approaches have already shown that acupuncture points exist.' They note that 'the high brightness, wide spectrum, high collimation, polarization and pulsed structure of synchrotron radiation' facilitated their discovery. They concluded, 'Our results demonstrated again the existence of acupoints, and also show that the acupoints are special points in mammals.'

32.9 Partial Oxygen Pressure Is Significantly Higher at Acupuncture Point Locations

With regard to modern scientific microsensing apparatus being used to detect anatomical structures, Daniel Weber (100) reported that, using an amperometric oxygen microsensor, researchers detected variations in partial oxygen pressure at various locations on the palmar surface of the wrist. The outcome was that researchers determined that the partial oxygen pressure is significantly higher at the location of acupuncture points. It is important to note that these scientific measurements were not made after the acupuncture points had been stimulated by needling or moxibustion but represented the natural resting states of the acupuncture points. Weber concluded the article by pointing out that the research yielded a truly unique finding, namely that acupuncture points definitely exhibit special oxygen-related characteristics. These results and many other recent research projects confirm beyond doubt that acupuncture points and acupuncture channels have scientifically measurable physical, biochemical, and electrical properties that differ greatly from surrounding nonacupuncture regions of the skin.

32.10 Acupuncture Points and Fascia

The author of the article titled 'Fascia: The Under Appreciated Tissue' (99) states, 'Acupuncture points appeared to correlate with areas of greater amounts of connective tissue. These points are located where nerves, arteries and veins collectively penetrate the fascia. Twisting the needle appears to manipulate the fascia, which help reduce pain. Body work also appears to work in this way.' The author continued, saying, 'The majority (82%) of perforation points are topographically identical with the 361 classical acupuncture points in traditional Chinese acupuncture.'

32.11 Acupuncture Points Behave as Booster Amplifiers along the Energy Transmission Cable

On the tandempoint.com website, Rena K. Margulis (101) presents a 2000 paper entitled 'Possible Scientific Basis for Tandem Point Therapy and Acupuncture for Pain Relief', in which the author noted that Robert Becker, who was the orthopedic surgeon and medical researcher who established the technique for healing broken bones using electricity, theorized that the acupuncture meridians are electrical conductors that carry information to the brain. Margulis cited Becker's work from *The Body Electric*, where he discusses that acupuncture points behave as hundreds of little DC booster amplifiers throughout the body, situated only a few inches apart, to prevent the current flow within the acupuncture meridians from fading out and dying completely before the information is correctly transmitted along the bioelectrical conduit. The principles of electrophysics suggest that if acupuncture points and meridians actually behaved as conductors and amplifiers, the surface skin above the structures would possess specific electrical properties different to the surrounding skin. According to known laws, the resistance should be less and the electrical conductivity should be greater, and a DC power source should be discernible directly over the acupoint.

32.12 Electrical Characteristics, Current Strength and Circadian Rhythm of Acupoints

On the tandempoint.com website, Margulis (101) further reports that biophysicist Maria Reichmanis performed research trials on the Large Intestine and Pericardium acupuncture channels along the arm and obtained the predicted electrical parameters at half of the points tested. Remarkably, the same acupoints were detected on all the subjects tested. The research findings indicated that the acupuncture meridians were conducting current, with the polarity displaying a flow toward the central nervous system. This is comparable to the polarity of the sinew channels showing a flow toward the head. Each acupoint tested positive when compared to the surrounding skin, and each acupoint had a characteristic field surrounding it. TCM theory asserts that the energy (Qi) of the body circulates in a rhythmical 24-hour cycle. Reichmanis detected a superimposed variation in the current strength at the acupoints at fifteen-minute intervals, overlaying the 24-hour circadian rhythm that the team detected a decade earlier in the overall DC system.

32.13 The Properties of Collagen Fibers and Acupuncture Meridian Acupoints

Summarizing Becker's findings pertaining to key research on trigger points, collagen, acupuncture points, and meridians, Rena K. Margulis (101) stated:

- Trigger points are electrically different from the surrounding tissue.
- There is a 71% correspondence between published locations of trigger points and classical acupuncture points for the relief of pain.
- Acupuncture points typically represent local maxima in conductance, elevated by a factor of 10 to 100.
- Acupuncture meridians have the characteristics of electrical transmission lines.
- Collagen in connective tissue, which reaches every cell in the body, can act as a semi-conductor.
- Collagen is liquid crystalline in structure.

- In response to pressure, collagen will produce electricity through the piezoelectric effect.
- Collagen in dilute solution 'melts' at 40 °C; melting may enable the collagen fibers to better realign and hence increase conductivity.
- Bound water surrounding collagen fibers can serve as a vehicle for proton-jump conduction
- Proton-jump conduction is much faster than conduction of electrical signals by the nerves.
- The conductivity of collagen increases strongly with the amount of water absorbed.

32.14 Beneficial Information for Acupuncturists to Incorporate in Their Treatment Protocol

Practical information that could be very beneficial for acupuncturists during their treatment protocol include the following conclusions from the experience of Tandem Point therapy, as discussed by Margulis (101): (1) She stressed that patients must be fully hydrated to get the best outcome. (2) It is best not to treat a person directly after a long plane flight. (3) For optimal trigger point release treatments, the patients must consume water throughout the session. (4) Patients obtain inferior results if they are electrolyte-deficient. (5) Patients should not hold their breath, or trigger points will not release. (6) The amplitude of the pulsation in the trigger point will increase immediately after an appropriate tandem point is pressed. (7) Effective trigger point release often generates heat in trigger points and tandem points.

32.15 Justification for Looking for the Routes of Acupuncture Meridians along Collagen Fibers

In the paper titled 'Possible Scientific Basis for Tandem Point Therapy and Acupuncture for Pain Relief', the author (101) stated, 'Some of the *body's electrical energy* circulates along connective tissue, along fascia, and that *more energy* circulates where fascia is thickest' (emphasis is mine). Rolfing educator and author, Tom Myers established a system involving what he calls anatomy trains or myofascial meridians. These delineated lines of strain spread throughout the body following chains of muscle and fascia and therefore collagen. Margulis prudently observed that Myers's entirely western-analysis derived system determined three major myofascial pathways. Margulis pertinently notes that Myers's myofascial pathways is paralleled quite closely to three main meridian channels of Chinese medicine—namely, the Stomach, Gallbladder, and Bladder channels. She notes further that Tandem Point Therapy utilizes these pathways most frequently and concludes that this affords 'additional justification for looking at the *routes of meridians* along *collagen fibers*' (emphasis is mine).

32.16 Circulation of Energy (Qi) to Every Cell in the Body through Connective-Tissue Channels that Form Acupuncture Meridians
On the website tandempoint.com, Rena K. Margulis (102) presents a 2000 paper entitled 'Acupuncture Theory', in which she explains that acupuncture practitioners believe that a native energy (Qi) circulates throughout the body via many cross-linked channels so that the Qi reaches every single cell in the entire body. Margulis uses a brilliant analogy in which she explains that every single home in the USA can be accessed by the established road system, be it via a large-capacity superhighway or by a small gravel track. They are all interconnected. She noted that ancient Chinese philosophy used a similar analogy that involved waterways rather than roadways and included rivers, tributaries, irrigation channels, and reservoirs. Just as large-capacity superhighways carry more traffic, the larger channels in the body carry more Qi. Similarly, just as the small gravel tracks carry less traffic, the smaller channels carry less Qi. She points out that this analogy includes the notion that there is a prevailing direction of energy flow throughout the meridian channels within the body.

In the article, the author (102) advises that there are many different channels. Some channels are very superficial, and some are very deep. The differing forms of channels include cutaneous channels, minute collateral channels, sinew channels, luo-connecting channels, primary meridian channels, divergent channels, and the eight extraordinary channels. There also exist the deep pathways of the primary channels and the divergent channels. Margulis (102) concluded the article by saying, 'Two months ago you heard James Oschman describe how connective tissue reaches every cell in the body. *I believe that at least some acupuncture channels occur along connective tissue pathways*' (emphasis is mine).

32.17 Fascia Is an Omnipresent Connective-Tissue System that Permeates the Human Body

In the 2014 *Journal of Chinese Medicine* article, authors Steven Finando and Donna Finando (103) quoted Findley et al. when they stated, 'Fascia is defined as the soft tissue component of the connective tissue system that permeates the human body . . . [It includes] aponeuroses, ligaments, tendons, retinaculae, joint capsules, organ and vessel tunics, the epineurium, the meninges, the periosteal and all the endomysial and intramuscular fibers of the myofasciae.' They continue:

> The fascia of the human body is a continuous sheath of tissue that moves, senses and connects every organ, blood vessel, nerve, lymph vessel, muscle and bone. It is a *continuous, three-dimensional, whole-body matrix, a dynamic metasystem that interpenetrates and connects every structure of the human body*, an interconnected network of fibrous collagenous tissues that are part of a whole-body tensional force transmission system. (Emphasis is mine)

32.18 Research Has Demonstrated that the Fascial System Should Be Considered an Organ

With consideration to the fact that fascia contributes to the optimal functioning of all body systems, in the 2014 *Journal of Chinese Medicine* article, authors Steven Finando and Donna Finando (103) note that, up until recently, fascia was incorrectly considered to be simply 'packing material' for encasing and padding all the various tissues within the body and contributing to the body's bodily form. Its two predominant functions were considered to be separation and allowing gliding between structures and the connection and transferring of forces within the body. However, modern research has confirmed that *fascia should actually be considered as an organ that delivers a unified environment that ensures the functioning of all body systems.*

32.19 The Numerous Recently Elucidated Functions of the Fascial Metasystem

Regarding the numerous recently elucidated functions of the fascial metasystem, in the 2014 *Journal of Chinese Medicine* article, authors Steven Finando and Donna Finando (103) state:

> Recent research has shed light on the various functions of the fascia. *Langevin . . . considers fascia to be a metasystem—a complex communication network that both influences and is influenced by every muscle, organ, blood vessel and nerve, and which is intimately connected to every aspect of human physiology.* In addition, the fascial system provides form to the entire body. *It is the ground in which all organs and systems function,* and it connects and influences all physiological systems. Guimberteau . . . *refers to fascia as a single connecting organ that is related to every aspect of human physiology*—a unified whole and the environment for the functioning of all body systems. Stecco et al. . . . have developed the concept of the *'organ-fascial unit', in which the functionality of an organ system is inextricably bound to its associated fascial connections.* Oschman . . . describes fascia as a body-wide communication system. He has coined the term '"living matrix" [which] includes the connective tissue and fascial systems . . . as well as the transmembrane proteins (integrins and adhesion complexes), cytoskeletons, nuclear matrices, and DNA.' Panletti . . . suggests that *to some extent fascia is involved in every type of human pathology. . . .* which connects the fascia to cellular nutrition and metabolism . . . Fascia is directly involved in haemodynamic processes, particularly venous and

lymphatic circulation . . . [Some researchers] . . . link fascia to chronic disease. *Pischinger . . . views the fascia as a link between the external world and the internal environment*, and he connects the fascia to the initial, nonspecific immune reaction of the human body to invading pathogens. (Emphasis is mine)

32.20 Tensions Applied through the Fascial System Affect Biochemical Changes at a Cellular Level

In the 2014 *Journal of Chinese Medicine* article, authors Steven Finando and Donna Finando (103) explain that fascia is a tensional network that is fundamental to physical movement and muscle function throughout the body. Fascia safeguards the body from both external forces and internal stresses and simultaneously supplies lubrication, insulation, and structural integrity. The authors report that Chen et al. demonstrated that 'tensions applied through the fascial system have been *found to affect biochemical changes at a cellular level* via mechanochemical transduction. This means that numerous critical cellular processes, including gene expression, cell differentiation, and growth and ongoing survival can be directly altered by mechanical stresses applied to the fascia on a macro scale.'

32.21 Superficial Fascia Communicates with Deep Viscera and Is Associated with Visceral Pathology

Regarding the transmission of local and distal effects in the form of referred pain, authors Steven Finando and Donna Finando (103) explain that referred pain can be transmitted along fascial planes and that myofascial deformations can result in impaired movement. They note that somatovisceral effects pertain to the association between myofascial surface deformations and dysfunction of the viscera and that viscerosomatic influences pertain to the relationship between gut pathology and myofascial deformations. They further note that conditions involving myofascial tissue deformations have been concomitant with visceral and autonomic dysfunction and the latter resolves when the myofascial integrity is restored.

32.22 Fascia Is a Mechanosensitive Signalling System Analogous to the Nervous System

In the 2014 *Journal of Chinese Medicine* article, authors Steven Finando and Donna Finando (103) state:

> Fascia responds to stimulation. Permeated with four types of sensory receptors, Schleip . . . considers it to be *our richest and most important sensory and perceptual organ*. Langevin . . . describes fascia as a mechanosensitive signalling system that serves to integrate systems in a way that is *analogous to the nervous system. Fascia is involved in the process of proprioception* . . . Schleip et al. . . . discuss the fascial role in *interoception, a largely unconscious sense of the physiological condition of the body* which has affective and motivational dimensions related to the homeostatic needs of the system. (Emphasis is mine)

32.23 The Omnipresent Triple-Energizer Metasystem and the Fascial Metasystem Are the Same

Expanding on the fact that fascia is a metasystem that connects every aspect of human physiology, authors Steven Finando and Donna Finando (103) note that 'fascia, therefore, can be understood as a *metasystem that connects every aspect of human physiology*' (emphasis is mine). They note that fascia is directly connected with circulation, metabolism, immune function, pathology, insulation, protection, and movement. They further note that fascia is integral in proprioception and interoception, which is critical for optimal human functioning. The fact that fascia is 'directly involved in circulation, metabolism, immune function, pathology, insulation, protection and movement' is a remarkable comment. While the authors believe that the fascial Connective-Tissue Metasystem is intimately involved with the Acupuncture Meridian Metasystem, I thoroughly agree with them, but I further believe that the fascial Connective-Tissue Metasystem is also intimately associated with the Triple-Energizer Metasystem. Remember, in the commentaries on the 38th Difficult Issue of Unschuld's (1) translation the *Nan Ching*, Li Chiung declared, 'The Triple Burner represents nothing but membranes.'

Summary of Chapter 32

Compiling this Chapter took my breath away. Where do I start to summarize the research observations and findings of the scientists discussed above? I believe the major point regarding my personal hypothesis is what Margulis (102) said when she quotes James Oschman, who said, 'Connective tissue reaches every cell in the body.' This is by no means an original thought. Ho and Knight (82) state that 'collagens and bound water form a global network' throughout the entire body, and that 'the "ground substance" of the entire body may provide a much better intercommunication system than the nervous system'. In complete harmony with Pollack, they also state that 'the hydrogen-bonded water network of the connective tissues is actually linked to ordered water dipoles in the ion-channels of the cell membrane' and beyond due to the 'connective tissue-intracellular matrix continuum'. Finando and Finando (103) state, 'The fascia of the human body is a continuous sheath of tissue that moves, senses and connects every organ, blood vessel, nerve, lymph vessel, muscle and bone. It is a continuous, three-dimensional, whole-body matrix, a dynamic metasystem that interpenetrates and connects every structure of the human body, an interconnected network of fibrous collagenous tissues that are part of a whole-body tensional force transmission system.' The authors (103) further state that research has demonstrated that fascia should actually be considered as an organ that provides a unified environment contributing to the functioning of all body systems. All these comments show that there is a continuum between all the omnipresent connective-tissue membranes that permeate the entire body right down to the intracellular matrix continuum.

In the book *Heart Master Triple Heater*, on page 118, regarding how the Triple-Energizer Metasystem ensures free communication and circulation throughout the entire body, the authors (11) quote Hua Tuo, who said, 'When the triple heater ensures free communication, then there is free communication internally and externally, left and right, above and below. The whole body is irrigated, harmonized internally and regulated externally, nourished by the left, and maintained by the right, directed from above, propagated from below. There is nothing greater!' So what has been uncovered by recent scientific research regarding the functions and omnipresence of the Connective-Tissue Metasystem throughout the body was perfectly summarized by Hua Tuo, who was describing the Triple Heater. I believe this amazing correlation exists because the Triple-Energizer Metasystem is actually synonymous with the Connective-Tissue Metasystem that is also omnipresent throughout the body.

CHAPTER 33

Relationship Between the Primo Vascular System and the Connective-Tissue Metasystem

Introduction
The Connective-Tissue Metasystem harbors the vast majority of interstitial fluid in the body. The 2007 Fascia Research Congress summary (48) reported a remarkable property of fascia on pages 4–5. The article stated, *'Loose connective tissue harbors the vast majority of the 15 liters of interstitial fluid.* This flows through an extracellular matrix which contains cells such as fibroblasts, tumor cells, immune cells, and adipocytes' (emphasis is mine). The article advised that the flow of interstitial fluid has many important biological impacts throughout the body. These include changes to cell function, morphogenesis of tissues, migration of cells, differentiation, and remodeling of tissues. Fibroblast cells implanted within the extracellular matrix organize themselves so that they are perpendicular to the course of the fluid flow. The biomechanical properties of loose connective tissue can be modified by the alteration of the water content and by a variation of ions and other substances present at that specific location. Further, a slight change in fluid flow at a location can modify the shear stress occurring on a cell surface and modify the biochemical environment of the cell at that site. The article continued, 'Interstitial flow regulates nutrient transport to metabolically active

cells and plays a crucial role in maintaining healthy tissue. It can also give directional clues to cells by guiding lymphocytes and tumor cells to lymph nodes or towards lymphatic capillaries.'

33.1 Direct Mechanical Connection of Connective Tissue Is Responsible for Blood Flow to Muscle

The Fascia Research Congress Summary (48) reported data from 2008 (presented on page 5), and it stated, 'Blood flow to skeletal muscle is tightly regulated by its metabolic demands.' The article noted that a direct mechanical connection is responsible for the rapid dilation of the local arterioles when muscles contract. The mechanism is not controlled by either the skeletal or the autonomic nervous system.

33.2 Connective Tissue Fibers Promote Lymphatic Flow and, if Injured, Cause Edema

Data from 2009 (48) presented on page 5 showed that 'inelastic fascia can promote lymphatic flow. When muscles contract against a thick, resistant fascia layer, it increases the pressure within a compartment, and permits blood and lymphatic fluid pumping against gravity towards the heart.'

Further data from 2010 (48) presented on page 5 showed:

> *Fluid volume is regulated* by interstitial hydrostatic and colloid osmotic pressures, which are constantly readjusting due to alterations in capillary filtration and the lymphatics. Connective tissues can alter transcapillary fluid flux by altering cell tension on dermal fibers which surround the hydrophilic ground substance and prevent its osmotic pressure from drawing fluid out of the capillary. When these fibers relax, this allows glycosaminoglycan ground substance to expand and take up fluid, *resulting in edema formation*. After injury, *fluid flow can increase almost 100 fold within minutes*; most of this is due to the active osmotic pressure of the extracellular matrix rather than to capillary leakage which only increases 2 fold. (Emphasis is mine)

In the commentaries on the 25th Difficult Issue on page 313 of Unschuld's (1) translation of the *Nan Ching*, Ting Chin declares, 'In the Ling-shu and in the Su-wen, the treatise "Pen-shu" states: *'The Triple Burner is a palace acting as central ditch; the passageways of water emerge from it.* It is associated with the bladder and it constitutes the palace of uniqueness' (emphasis is mine). From information discussed in the preceding paragraphs (48), it becomes blatantly obvious that references pertaining to the proven properties of fascia—including 'loose connective tissue harbors the vast majority of the 15 liters of interstitial fluid', 'interstitial flow regulates nutrient transport', 'towards lymphatic capillaries', 'blood flow to skeletal muscle', 'inelastic fascia can promote lymphatic flow', 'fluid volume is regulated', 'resulting in edema formation', and 'fluid flow can increase almost 100 fold within minutes'—all pertain to *the passageways of water*. From the above research findings, it has been shown that the *Connective-Tissue Metasystem* exercises control over interstitial fluid, blood and lymph, the three most voluminous liquids in the body. So here we can see that the recent scientific research findings associated with the connective tissues of the body—acting as a central ditch and regulating *the passageways of water*—are all functions attributed to Three Heaters' control of all the various fluid distribution throughout the body. I propose that there are other fluid pathways that exist within the body that modern science has not yet uncovered.

33.3 Channel Pathways that Exist within the Body that Modern Science Has Not yet Discovered

In a later Chapter, I discuss the newly discovered 'organ', the gut microbiome, in great detail. It has been shown that the DNA from our gut microflora (microbiome) has a very large input as to the daily running of our body and can override our own DNA. In fact, cravings may be coming from our gut hitchhikers. Scary! It is known that our gut microorganisms and their biochemical products and nutrients (lactase, vitamin B_{12}) are being transported from the intestines throughout the body as required. This could well explain how good commensal gut microbiome bacteria make their way to the mammary glands to inoculate the neonate with immune-boosting bacteria of their own. To date, this pathway has not been elucidated. It is known that bacteria associated with periodontal disease (in the mouth) can cause pregnant women to have premature babies. It is known that gut microflora from pregnant women colonize the gut of their baby in utero. How do the bacteria get from the gut of the mother to the gut of the baby in utero? Regarding other channels that I have heard TCM practitioners discuss quizzically, on the webpage titled 'Articles—The Heart Channel: Connection with Uterus', the author, Giovanni Maciocia (104), states:

> The Heart is closely connected with the Uterus through the Uterus Vessel (Bao Mai) and this explains the profound influence of mental-emotional problems affecting the Heart on the Uterus. The Uterus is related to the Kidneys via a channel called the Uterus Channel (Bao Luo). The 'Simple Questions' in chapter 47 says: 'The Uterus Channel extends to the Kidneys'. The Uterus is physiologically related also to the Heart via a channel called Uterus Vessel (Bao Mai). The 'Simple Questions' in chapter 33 says: 'The Uterus Vessel pertains to the Heart and extends to the Uterus' and 'When the period does not come it means the Uterus Vessel is obstructed'.

Just as surely as acupuncture meridian conduits exist, I am convinced that these other channels and vessels actually exist and that they are composed of PVS and/or connective-tissue conduits intimately contained (and hidden) within our omnipresent Connective-Tissue Metasystem. Numerous researchers have determined a very large variation in the physical, chemical, and electrical properties of the tissue in the 'holes' at defined acupuncture points compared to non-acupuncture points. I believe that anatomists and surgeons have not seen the forest for the trees and that it is only a matter of time before the highly specialized and dedicated structural subcomponent of the connective-tissue metasystem that makes up the TCM acupuncture meridian conduits and the other specific channels and vessels mentioned above will be elucidated and readily differentiated possibly by highly selective microscopic and macroscopic histological staining technique. Recent research in 2009 (34) revealed that 'the flow of the primo fluid in a certain path was demonstrated in the study using Alcian blue, from the rat acupoint BL-23 in the dorsal skin to the PVS on the surface of internal organs'. Other research in 2010 (34) discovered that 'when Chrome-hematoxylin and fluorescent nanoparticles were injected into testis they were found in the PVs on the organ surfaces between the abdominal cavity and the abdominal wall'. Cancer researchers between 2009 and 2013 (34) have proposed that primo vessel channels could provide pathways for cancer metastasis throughout the body. From these recent findings, there is overwhelming proof that previously unknown pathways exist within the body. It is for this reason that I have faith in the ancient sages who told us that 'the Heart is closely connected with the Uterus through the Uterus Vessel (Bao Mai)' and that 'the Uterus is related to the Kidneys via a channel called the Uterus Channel (Bao Luo)'. I believe that some 'simple' surgeries have severed these microscopic conduits without even the knowledge of their existence on behalf of the surgeon. Subsequently, due to destruction of these 'unidentified' pathways, real symptoms due to the damage are dismissed by the doctors as not possible and 'all in your head'.

33.4 An Intimate Part of the Omnipresent Circulatory System Only Recently Substantiated

The article (105) titled 'Primo-Vascular System' notes that the Primo Vascular System (PVS) is a major component of the circulatory system, together with systems including blood vessels and lymphatic vessels. While the North Korean medical surgeon and scientist Bong-han Kim (born 1916) first reported superficial primo vessels, also known as Bonghan ducts and Bonghan channels in 1962, it was not until 50 years later in the 2010s that several other researchers confirmed Kim's earlier findings.

33.5 Morphology of the Primo Vascular System

Regarding the characteristics of the primo vascular system (PVS), the article (105) states that the PVS 'comprises a body-wide web of vascular structures called Bonghan ducts (also known as primo-vessels). Histological and immunofluorescent investigations have shown that Bonghan ducts are distinctively different from those of similar-looking lymphatic vessels.' The article continues, 'Bonghan ducts, also known as primo vessels, are surrounded by membrane with high concentrations of hyaluronic acid. The Bonghan ducts contain rod-shaped endothelial nuclei, and electrical excitability resembles that of smooth muscle.' A remarkable feature of the PVS is its very thin walls, with the outermost layer being more porous than blood or lymph capillary vessels. The cross-sectional diameter of these vessels is 20–50 μm. Other researchers (34) found that 'one of the important biochemicals in the primo fluid, which was mentioned in BH Kim's report, was catecholamine (adrenalin and noradrenalin), and suggested that the Bonghan system is a catecholamine-producing novel endocrine organ complex'.

33.6 Overview of the Mechanism of the Primo Vascular System

Regarding the mechanism of the Primo Vascular System, in the 2014 *Integrative Medicine International* journal article titled 'Does the Primo Vascular System Originate from the Polar Body', the authors (106) explain that the recently discovered Primo Vascular System (PVS) is a circulatory system present in many animals and human beings. It is composed of ducts, corpuscles, and Primo Micro Cells (PMCs, sanals, or granules) floating inside those ducts. A sanal contains one sanalsome, which is composed of a large amount of DNA, and also sanalplasm, which contains RNA. A sanal can develop into a cell, and conversely, a cell can develop into a sanal via the 'sanalization' process. Each sanal holds the same amount of DNA that is present in a chromosome. Sanals (PMCs) move into the primo vessels where their nucleus-like configuration can develop, mature, and then migrate to nearby tissues to create new tissues. Astoundingly, the authors state, *'All human tissues and cells have been suggested to be linked to the PVS.* Nucleic acid granules (sanals) have been found to play a crucial role in the physiological functions of the PVS, especially in the regeneration of dead cells. *In the PVS, 70 different proteins have been discovered in the liquid and 270 proteins in the vessels, many of them are not normally found in the blood, or in lymph or blood vessels.'* (Emphasis is mine)

Note that the authors state that 'all human tissues and cells have been suggested to be linked to the PVS'. This also applies for the Connective-Tissue Metasystem. It is thus very interesting that these two structures share the same space around all tissues, which is also said of the Triple Burner.

33.7 Has the Du Mai Channel Been Finally Elucidated by Modern Science?

In their article, the authors (34) described novel threadlike structures and microchannels called '(primo vessels and primo nodes) floating in the venous sinuses of rat brains', and the 'Primo vascular system in the subarachnoid space of the spinal cord of a pig', along with 'novel anatomic structures in the brain and spinal cord of rabbit that may belong to the Bonghan system of potential acupuncture meridians'. So let me recap. Modern-day scientific researchers have found a network of threadlike microchannel structures floating inside the spinal cord and brain of several animal species. In the commentaries of the 23rd Difficult Issue in Unschuld's (1) translation of the *Nan-ching*, on page 291, Yü Shu says, 'The [Nei-] ching states: "The supervisor vessel starts from the bottommost transportation [hole]"; [from there,] *it ascends inside the spine up to the wind palace.* There it enters into and becomes attached to the brain. It is four feet five inches long' (emphasis is mine). So here the Governing Vessel, or Du Mai (or supervisor vessel), is described as commencing midway between the tip of the coccyx and the anus and travelling upwards *inside the spine* to the wind palace, DU-16 (Feng Fu). What is significant here is that Yü Shu says the vessel ascends inside the spine, which is in the cerebrospinal fluid, and then 'it enters into and becomes attached to the brain'. Where did Yü Shu derive his correct understanding of this microscopic complex Primo Vascular System?

33.8 I Believe Bong-Han Kim Is the Father of TCM Acupuncture Point and Meridian Anatomy

About fifty years ago, a North Korean professor from Pyongyang Medical University, Bong-Han Kim, revealed the relationship between the TCM acupuncture meridian system and the Bong-Han

System (BHS) by injecting a blue dye into an acupuncture point and then observing the dye flowing along the meridian. Researchers today call the Bong-Han System the Primo Vascular System (PVS). His so-called Bonghan Corpuscle (BHC) represented anatomical structures of acupuncture meridian-collaterals. Today, researchers refer to them as a Primo Node (PN). In the article '50 Years of Bong-Han Theory and 10 Years of Primo Vascular System', the authors (34) described the advancement of the research initiated by Bong-Han Kim (hereafter B. H. Kim or Kim), who reported his first study results on the anatomical entity of acupuncture meridians in 1962. Kim described the primo vascular system (PVS) via his five research reports. His third report was the most extensive and conclusive in his description of PVS anatomy and physiology, relating to the acupuncture meridians. The authors state that the work and accomplishments of B. H. Kim are truly enormous and that, since 2002, all his trials that have been repeated and have been absolutely confirmed to be accurate.

The primal PVS has been determined in many animal species, including mainly rabbits, rats, and mice. But pigs, dogs, cows, and human placentas have also been studied. Because the PVS components are very small and transparent in nature, identifying the system has been extremely difficult for researchers, and a large amount of time has to be consumed to master the analytical techniques required.

33.9 Extremely Complicated Communication Infrastructure among the Five Subclass PVS Networks

In the article '50 Years of Bong-Han Theory and 10 Years of Primo Vascular System', the authors (34) reported that there are five subclass PVS networks and that the nature of the communication among those five subclass PVS networks is extremely complicated. For example, they state, 'The IV-PVS *floating inside blood vessels* was first identified in the abdominal artery and the caudal vena cava of rabbits, rats, and mice. More importantly, the PVS in the atrium of a bovine heart was *found to form a floating network*.' They further state, 'The IV-PVS *floating inside lymph vessels*, was visualized with help of Janus Green B, fluorescent nanoparticles . . . [and] Alcian blue' (emphasis is mine). They reported that primo nodes (PNs), formerly called Bonghan corpuscles (BHCs), possess a large amount of cells and granules related to the immune system, which suggests that the PVS is involved in a protecting function of the body. They reported another subclass PVS network called the OS-PVS. This network was found *floating on the surfaces* of internal organs and 'was observed in rabbits, rats, mice, dogs, and pigs. The structure of this PVS subclass was characterized by the optical, and electron microscopy.' Yet another subclass PVS network called the N-PVS was detected '*floating in the cerebrospinal fluid* (CSF) in the brain of a rabbit and a rat' (emphasis is mine). It was optically observed by using the cell-staining agent Trypan blue. The authors further reported that the PVS network was also determined 'in the subarachnoidal space of the rabbit and rat brains, and in the spine of the rat and pig.' The authors stated, 'All PVS observed and listed previously were *floating in the body fluid*, such as blood, lymph, abdominal fluid, and CSF.' Where else has the PVS been observed? The authors state, 'The presence of the PVS *entering the adipose tissues around the rat small intestine* was first noticed via optical imaging using Alcian blue, which was injected intravenously at the femoral vein. The Alcian blue entering a *PVS floating inside a blood vessel reached to a PN in an adipose tissue*' (emphasis is mine).

I propose that it may be via the PVS that the bacteria and other microflora that make up the microbiome in the small intestine communicate throughout the body. I discuss this in more detail in a later Chapter. I further believe that the PVS involvement with the adipose tissue of the body allows the adipose tissue to be a part of the Triple-Energizer metasystem. I also discuss this topic in more detail in a later Chapter.

33.10 Morphology of the Bonghan Duct (BHD), Which Is Now Called a Primo Vessel (PV)

What Kim originally called a Bonghan duct (BHD) is now called a Primo Vessel (PV) by modern scientific researchers. One of the most distinguished anatomical features of the PV is its bundle-like structure made of multiple subvessels. The authors (34) note that high-fidelity imaging of a Primo Vessel cross section is now readily repeatable, thanks to greatly improved techniques in the sample preparation procedure and subsequent microscopy. B. H. Kim's initial claims have been recently substantiated by transmission electron microscopy (TEM) of OS-PVS, a subclass of PVS found on the surface of internal organs. The procedure revealed the endothelial cell layer of a PV, while the surrounding extracellular matrices (ECM) were composed of collagen fibers.

33.11 Omnipresent Primo Vascular System Found Inside the Blood and Lymphatic Vessels, in the Cerebrospinal Fluids of the Central Nervous System, and on the Surface of the Various Internal Organs

In their article, regarding the primo vascular system being identified within blood vessels, the authors (34) described an *intravascular* threadlike structure and discussed 'in vivo visualization of Bonghan ducts *inside blood vessels* of mice by using an Alcian blue staining method'. Regarding the primo vascular system being identified within lymphatic vessels, the authors (34) described 'novel threadlike structures (Bonghan ducts) *inside lymphatic vessels* of rabbits visualized with a Janus Green B staining method', and they discussed the 'Alcian blue staining method to visualize Bonghan threads *inside large caliber lymphatic vessels*'. Regarding the primo vascular system being associated with organs, the authors (34) described 'biofluid inside novel threadlike structures on the surfaces of mammalian organs'. In the article '50 Years of Bong-Han Theory and 10 Years of Primo Vascular System', the authors (34) summarized the recent findings, stating, 'Approximately fifty years after BH Kim, the presence of the PVS inside the blood and lymphatic vessels, cerebrospinal fluids of the central nervous system, and on the surface of the various internal organs was indeed confirmed by various techniques developed by the SNU [Seoul National University] team' (emphasis is mine).

33.12 The PVS May Explain the TCM Six Divisions of Penetration of Perverse Influences

Regarding the penetration of cold or wind-cold pathogens deeper into the body through the TCM six divisions, in the 2013 *Journal of Acupuncture and Meridian Studies* article titled 'The Primo Vascular System as a New Anatomical System', the authors (84) report that comparatively independent networks of PVS have been determined at superficial and at deep levels and that even elaborate organ circulation systems exist. In the case of systemic PVS structures for whole-body responses, the interconnections and dormant components of the PVS may be triggered. The authors suggest that in pathological situations throughout the body, the damaged tissues send out signals, and the PVS responds by transmitting the primo fluid to the impaired cells to provide the necessary substances required to repair the cells. They believe that acupunctural needles may act as antennae and influence various electromagnetic fields known to be present throughout the body. I found it fascinating to consider that these recent findings about the composition of the PVS show that there are in fact defined nets of circulation of PVS structures at superficial and deeper levels. This recent finding could well explain the TCM theory of the six divisions. The six divisions tend to be remembered more as a philosophical explanation for deepening and exacerbation of disease states in general as clinical manifestations evolve as the pathogenic qi lodges in deeper energetic layers of the body.

33.13 Does the Primo Vascular System Constitute the Cerebrospinal Fluid (CSF) Pump?

The article (105) notes that the inner ducts lie freely within the flow of blood and they are subdivided with blood vessels. It is within these inner ducts that primo fluid, which contains a high amount of basophilic granules, is conveyed throughout the human body via the PVS. It has been determined that the primo fluid flows at a rate of 0.3 ± 0.1 mm/s. This equates to an average of 1,080 mm per hour. As the primo vascular system has been found inside the cerebrospinal fluid of the central nervous system, I wonder if the primo vascular system constitutes the unidentified CSF pump that ensures that cerebrospinal fluid is constantly renewed to bathe, nourish, and protect the brain and spinal cord. As

I have discussed in other Chapters regarding pumping mechanisms in the body, in the Mercola (60) interview, Dr Pollack reported that without applying pressure to the system, researchers observed that placing a simple tube made of hydrophilic material into EZ water caused the water to spontaneously flow through the tube at high speed and it kept flowing. That was an unprecedented finding. Irradiating the system with light made the flow go even faster. Ultraviolet light proved to be the winner in the stimulatory pumping effect.

33.14 Did Bong-Han Kim Discover What Are Now Known as Pluripotent Adult Stem Cells?

In the article '50 Years of Bong-Han Theory and 10 Years of Primo Vascular System', the authors (34) reported that, 50 years ago, very little was known about the theory of stem cells, especially of the adult stem cell. However, 50 years ago, Bong-Han Kim asserted that the very small (1–5 µm) P-microcell (Sanal) was the main agent responsible for wound healing and regeneration. We now know that those properties are the two fundamental functions of stem cells. The fact that Sanal budding has been previously observed by atomic force microscopy and that several stem cell biomarkers have been detected on Sanals is highly suggestive that Sanals and stem cells are one and the same. Kim also claimed that Sanals proliferated in the presence of light. While this observation has not been confirmed yet, it has been established that the movement of Sanals in liquid does increase when they are illuminated by UV-A light at 360 nm.

In the article about Bonghan channels written by Tima Vlasto (107) in 2009, which referred to a photograph showing the stereomicroscopic image of acupuncture meridians, she noted that small granules of DNA or microcells were also visible. These small granules of DNA or microcells were only about 1–2 µm in diameter, and they contained chromosomal material that was highly reactive to stem cell antibody stains. Concerning these microcells, she continued, 'When these cells were isolated and then induced to differentiate, they grew into cells of all three germ layers. These may be our body's natural source of pluripotent adult stem cells, with the potential to develop into any cell in the body.'

33.15 Modern Science Catches Up with Intuitive Ancient TCM Wisdom

The major unconfirmed aspect of PVS research at this time ironically pertains to the PVS in the skin, which is supposed to be the anatomical acupoint. Kim proposed that the acupuncture meridians extended into the PVS within the body. This still requires verification because the analytical techniques used to successfully determine the PVS inside the body do not work for the PVS in the skin. In the article '50 Years of Bong-Han Theory and 10 Years of Primo Vascular System', the authors (34) strongly believe that when the entire PVS network and its roles in mammalian biology are fully unraveled, the acupuncture therapy system of ancient traditional eastern medicine will be truly accepted as scientific medicine. Furthermore, the authors believe a greater understanding about the PVS will bring about a paradigm shift in the understanding of pain control, stem cell therapy, regenerative medicine, immune deficiency, cancer, and many other important aspects in human health-care management. The authors (34) reported that up until this time, three of the five PVS classes that B. H. Kim classified have been confirmed as factual, and they noted that experimental work strongly suggests that the PVS is implicated in cancer metastasis. Summing up, the authors believe that clarifying the functional relationship between the acupuncture meridian system and the PVS may be instrumental in binding eastern and western medicine.

33.16 Bonghan Channels Contain Substance to Lubricate the Joints, Eyes, Skin, and Even Heart Valves

In a revealing article about Bonghan channels written by Tima Vlasto (107) in 2009, she described a photograph showing the stereomicroscopic image of acupuncture meridians that were composed of tubular structures ranging from 30 to 100 μm wide. In comparison, red blood cells are only 6–8 μm in diameter. She notes that due to their transparency and the fact that they are hardly visible with low-magnification surgical microscopes, these structures have not been discovered previously. To further complicate their prior discovery, these structures resemble fibrin, which coagulates and conceals these structures from ready observation in the presence of bleeding within dissected tissues. Since their rediscovery, medical researchers are scrutinizing their composition and their function. Vlasto further noted that the tubular structures that constitute Bonghan channels contain a *flowing liquid* that includes abundant *hyaluronic acid*, a substance that cushions and *lubricates the joints, eyes, skin and even heart valves*.

In the commentaries on the 31st Difficult Issue on page 354 of Unschuld's (1) translation of the *Nan Ching*, regarding the root and the origin of the Triple Burner, scholar Yeh Lin states, '*It is a fatty membrane* emerging from the tie between the kidneys.' Also in the 31st Difficult Issue on page 355 in the commentaries, Huang Wei-san said, 'The Triple Burner encloses all the depots and palaces externally. *It is a fatty membrane covering the entire physical body from the inside*' (emphasis is mine).

In the commentaries on the eighth Difficult Issue on page 132 of Unschuld's (1) translation of the *Nan Ching*, regarding the origin of the vital influences or the transmutation of bulk water into different biological fluids and liquids, scholar Yeh Lin relates:

> Together with the fire of the heart, these [yang influences] cause the water of the bladder to rise as steam. [Hence, the water] is transformed into [volatile] influences. These ascend via the through-way and 'employer [vessels].' They pass the diaphragm and enter the lung whence, in turn, they leave [the body] through mouth and nose. [Some of the] influences which ascend in order to leave [the body] are transformed into liquids in the mouth and by the tongue and in the depots and palaces. These [liquids] leave [the body] through the skin [and its] hair by way of the 'influence paths' (ch'i-chieh). They serve to steam the skin and to soften the flesh. They constitute the sweat. [All of] this occurs [in accordance with] the principle that fire is transformed into [volatile] influences when brought into water.

In the commentaries on the 31st Difficult Issue on page 355 of Unschuld's (1) translation of the *Nan Ching*, where Yeh Lin is specifically discussing the generation of the chin liquids, the yeh liquids, and the sweat, he relates:

> Through the inhalation of the heavenly yang, the water of the bladder follows the fire of the heart downward to the lower [section of the Triple] Burner. There it evaporates like steam and is transformed into influences moving up again, where they become the chin [liquids], the yeh [liquids], and the sweat. All of that rests on the principle that when fire meets water, a transformation into influences takes place.

Notice the similarity of the two passages even though they refer to comments made about the 8th and the 31st Difficult Issues. 'The ch'i-chieh is regarded as [the place] where [the Triple Burner] collects [its influences].' According to Yang in the commentaries on the 31st Difficult Issue, this point is believed to be Stomach 30, which is called the Sea of Water and Grain. Perhaps these two influence paths stemming from the bilateral Stomach 30 acupoints are major production sites of the many biological fluids and lubricating liquids flowing throughout the body under the control of the Triple-Energizer Metasystem. Tima Vlasto (107) stated, 'Bonghan channels contain a flowing liquid that includes abundant hyaluronic acid, a substance that cushions and lubricates the joints, eyes, skin and even heart valves.' As stated above, Kwang-Sup Soh et al. (34) reported the IV primo vascular system (PVS) has been *detected floating inside* the caudal vena cava of rabbits, rats, and mice and that, 'more importantly, the PVS in the atrium of a bovine heart was *found to form a floating network*' (emphasis is mine). It thus appears that the Bonghan channels have a lot in common with the Triple-Energizer Metasystem in that they are composed of membranes which 'steam' and lubricate numerous body structures, including hair, joints, eyes, skin, and even heart valves.

33.17 Professor Kim Discovered a Novel Circulatory System with Therapeutic Effects

In the 2005 article 'Feulgen Reaction Study of Novel Threadlike Structures (Bonghan Ducts) on the Surfaces of Mammalian Organs',

Hak-soo Shin et al. (108) stated, 'A radical challenge to modern anatomy is to explain recent claims that hitherto unnoticed novel threadlike structures exist on the surfaces of the internal organs. These claims are based on recent reports of three groups.' The article continued, 'In brief, 50–100 μm thick semitransparent threadlike structures have been found on the surfaces of internal organs, such as the stomach, liver, large and small intestines, and bladder, of rabbits and rats. These structures do not adhere to the surface, but move freely and are sparsely and irregularly fixed to the peritonea'. Hak-soo Shin et al. (108) continued:

> Bonghan Kim (1963) sought the anatomical basis of acupuncture meridians in humans and animals and found a new circulatory system that was completely different from the vascular, nervous, and lymphatic systems. The meridians formed an anatomically distinctive system of threadlike ducts that spread under the skin. In addition, by tracing the ducts with a staining dye, he discovered that the ducts continued to spread onto the surfaces of internal organs and that they existed even inside blood vessels. He also found that a liquid flowed through the Bonghan duct system, and that the liquid played a physiological role akin to modern cell therapy by totipotent adult stem cells. The flow of this liquid was correlated with the therapeutic effects resulting from acupuncture treatments of damaged internal organs.

Regarding more recent research information on the movement of fluid and cells within the PVS, a very recent 2014 study published in the journal *Cancer Cell & Microenvironment* (109) showed that 'migration of tumor cells to secondary sites was more efficient in the PVS than in the lymphatic system. In addition, the *PVS is also a conduit for a fluid, which, according to proteomic analyzes contains remarkably high levels of carbohydrate metabolic derivatives that are usually associated with stem cells, cancer cells and differentiated myeloid cells*' (emphasis is mine). The article poignantly continued, 'Thus, the PVS has the potential to transport growth and communication factors between primary and secondary tumor sites, thereby enhancing the oncogenicity of tumor cells at secondary sites'. Thus, the PVS has been proven to transport 'growth and communication factors' throughout the body. Very interestingly, this is what the Triple-Energizer Metasystem also does. [Bolding is mine].

The book titled *The Primo Vascular System: Its Role in Cancer and Regeneration* (7) 'explores the first international opportunity to exchange the research results on PVS among multi-disciplinary experts'. The book discusses the functional characteristics of the PVS, comprising its roles in the areas of regenerative medicine and cancer. Some of the findings include information on the sinus in the Primo Vessel discovered inside bovine cardiac chambers. The Primo Vascular System has been detected in deep internal organ structures, including on the intestinal fascia of dogs, in the mesentery of a mouse, in rodent thoracic lymphatic ducts, in the subarachnoid space of rats, and as superficial as in the rat hypodermis. Primo nodes have been detected in the abdominal membrane and lymph nodes of rats. Small stem-like cells have been identified in the primo vascular system of adult animals. It has been suggested that Primo microcells in a Primo Node may be the origin of adult stem cells. There are also interesting implications regarding the unusual optical properties of collagen in the Primo Vascular System, which may well be involved in information transfer and communication throughout the body.

33.18 The PVS May Control the Cardiovascular System and the Nervous System

While discussing the controlling influences of major systems in the body, in the 2013 article titled 'The Primo Vascular System as a New Anatomical System', the authors (84) explain that because the PVS exists side by side to the vascular system and the nervous system, this duplicated PVS configuration utilizes the infrastructure as if it were a highway and subsequently directly influences body systems and organs by its supplying, draining, and innervating role of tissues and organs. The authors state, 'This duplication may be a *way of controlling* these functions. The PVS controls the *cardiovascular system* (which provides substances and hormones to the organs) and the *nervous system* (which provides impulses to the organs)' (emphasis is mine).

This team of researchers is stating that they believe that the PVS 'controls the cardiovascular system ... and the nervous system'. Remember that these authors, and many others, consider the PVS to be intimately connected with and part of the acupuncture meridian system. This means that the acupuncture meridian system controls the cardiovascular system, the nervous system, and subsequently, all the organs of the body. If this is proved to be true, and I believe that it is only a matter of time, all medical, physiology, and anatomy textbooks will have to be rewritten. The fact that the supposedly nonexistent acupuncture meridian system is in fact a real metasystem and is responsible for producing and controlling all other cells, systems, and organs in the body will be very hard to swallow for all the doubters in the medical arena.

33.19 The Primo Vascular System May Also Be an Endocrine Organ

Regarding one of the properties of the PVS, in the 2013 *Journal of Acupuncture and Meridian Studies* article titled 'The Primo Vascular System as a New Anatomical System', the authors (84) state, 'The PVS may be an endocrine organ because of the presence of chromaffin cells in the acupoints, and the PVS liquid carries adrenalin and noradrenalin.' They state further, 'The PVS in the vitelline membrane of eggs was formed after 16–24 hours of incubation, and *the putative PVS was clearly developed earlier than the extraembryonic vessels, the heart, and the intramembrane vessels*' (emphasis is mine).

These findings show that the PVS forms very early after conception and that the PVS develops prior to the *extraembryonic vessels*, the heart, and the intramembrane vessels. While the heart was formerly considered to be the first organ formed after conception, it has now been shown that the PVS is actually the first organ structure that develops after conception.

33.20 Claim to Fame of the Primo Vascular System (PVS)

The article (105) states, 'In 2004 researchers suggested that Bonghan channels may act like optical fiber cables and transmit DNA related information around the body using biophotons.' The article concluded, 'In 2005, the Bonghan duct and the primo-vascular system were both featured on the cover page of the *Anatomical Record*, an official publication of the American Association of Anatomists.' So finally, anatomists from around the world accept that there is a third circulatory system distinct from blood circulation and lymph circulation.

33.21 The Primo Vascular System Has Been Acknowledged by Numerous Scientific Journals

Regarding the legitimacy and the validity of the Primo Vascular System, in the 2013 *Journal of Acupuncture and Meridian Studies* article titled 'The Primo Vascular System as a New Anatomical System', the authors (84) note that recent research findings pertaining to various aspects of the PVS have been published in numerous prestigious international scientific journals. Scientific journals reporting on physics associated with the PVS include *New Journal of Physics*, *Journal of Biomedical Optics*, *Applied Physics Letters*, and *Current Applied Physics*. Scientific journals reporting on anatomical findings about the PVS include *Microcirculation*, *Microscopy Research and Technique*, *Lymphatic Research and Biology*, *Naturwissenschaften*, and *Anatomical Record Part B: The New Anatomist*. Other diverse fields of PVS research have been published in *PLOS One*, *Lymphology*, *Journal of Health Science*, *Cardiology*, *Biologia*, and *Journal of International Society of Life Information Science*. The authors further noted that modern scientific findings have also been published in specialized scientific journals dedicated to acupuncture research.

33.22 Comparison of the Old TCM Medical Classics with New Scientific Research Findings

So now, let's reflect on some of the wisdom reflected in the *Nan Ching* and the ancient medical classics, in light of these new research findings pertaining to the newly-discovered Primo Vascular System.

33.23 The 'Original Influences' (Yuan Qi) Ascend from the Kidneys through the Chong Mai

In the commentaries on the eighth Difficult Issue on page 132 of Unschuld's (1) translation of the *Nan Ching*, regarding the moving influences between the two kidneys, scholar Li Chiung relates:

> *The moving influences between the two kidneys are the original influences which man has received from father and mother.* Furthermore, the [movement of the] influences in the through-way vessel emerges from between the kidneys. The body's five depots and six palaces have twelve separate conduits. Their root and foundation are, indeed, the kidneys. The [influences of the] kidneys ascend, on both sides of the employer vessel, toward the throat. [The throat is responsible for] passing the breath; it serves as the gate of exhalation and inhalation. The Triple Burner of man is patterned after the three original influences of heaven and earth; the kidneys are the basic origin [of the Triple Burner]. (Emphasis is mine)

Note that Li Chiung later states on page 159 in the commentaries on the 11th Difficult Issue: 'That means that the original influences which were received by the kidneys from father and mother diminish'. So we can see from the two quotes from Li Chiung above that the original influences (Yuan Qi) are received by the kidneys from father and mother. These original influences engender the moving influences between the two kidneys, which then move upwards through the throughway vessel (Chong Mo), which embodies the bilateral Kidney meridians. This is concluded from where Li Chiung states, 'The influences of the kidneys ascend, on both sides of the employer vessel (Conception vessel or Ren Mai), toward the throat.' The original influences (Yuan Qi) involve genetic inheritance received from father and mother. This Yuan Qi is housed between the kidneys but in closer proximity to the right Kidney, below the navel. This Yuan Qi is the basic origin of the Triple Burner. The Triple Burner then distributes Yuan Qi to the 12 conduits and all the organs.

33.24 The 'Original Influences' (Yuan Qi) Engender Vibrant Life

In the commentaries on the eighth Difficult Issue on page 133 of Unschuld's (1) translation of the *Nan Ching*, confirming that the original influences (yuan qi) engenders vibrant life, scholar Hua Shou states, 'The kidneys are, furthermore, the spirit guarding against evil [influences]. When the original influences prevail, evil [influences] cannot enter. When the original influences are cut off, death follows. Similarly, when the roots of a tree are cut off, the stalk and the leaves wither.'

This means that the original influences pertain to the vibrant life force that allows a person to be alive and to stay alive with vitality. This scripture further shows that the original influences (yuan qi) are associated with the immune system and protects the body against evil influences, including pathological microorganisms that could penetrate our body and cause disease. The quote 'When the original influences are cut off, death follows' shows that should our Yuan Qi become depleted, the outcome is death. Stem cells throughout the body act as a repair system, replenishing adult tissues. Obviously, if stem cells are abundant and healthy, we experience wellness, potent healing, and longevity. It appears that Primo Vascular System P-microcells (Sanals) are the main agent for wound healing and regeneration, which are the two fundamental roles of stem cells. Could the pluripotent P-microcells constitute a major component of the Yuan Qi?

33.25 The 'Original Influences' (Yuan Qi) Are Associated with the Right Kidney

In the commentaries on the eighth Difficult Issue on page 134 of Unschuld's (1) translation of the *Nan Ching*, regarding the 'moving influences between the kidneys', scholar Ting Te-yung states:

> The 'moving influences between the kidneys' means the following. To the left is the kidney; to the right is the gate of life. The gate of life is the domicile of the essential spirit; the original influences are tied to it. It is also named the 'spirit guarding against the evil.' When the spirit of

> the gate of life holds guard firmly, evil influences cannot enter at will. If they enter, [the person] will die. In this case, the kidney influences will be cut off internally first. The respective person does not yet appear ill. When he falls ill, he will die.

This scripture shows that the left Kidney pertains to the renal Kidney function according to western physiology, but associated with the right Kidney is the gate of life and its accompanying Yuan Qi. The gate of life is also referred to as the 'spirit guarding against the evil', as discussed above. Could the P-microcells constitute the gate of life in that they are composed of DNA and are pluripotent and able to mend all types of different damaged tissues and organs and imbue longevity. This defensive aspect of P-microcells could easily make them the 'spirit guarding against the evil'.

33.26 The Triple Burner Transmits the 'Original Influences' (Yuan Qi) through the 12 Conduits

In the commentaries on the eighth Difficult Issue on pages 135–136 of Unschuld's (1) translation of the *Nan Ching*, regarding 'the influences on which man's life depends', scholar Tamba Genkan relates:

> The moving influences between the kidneys are identified... as the influences of the gate of life—that is to say, as the influences on which man's life depends. [Such explanations are] not as comprehensive as Lü [Kuang's] commentary. Mr. Lü lived close to antiquity; he must have received [his ideas] from a teacher. If one tests [his commentary] against the text of the scripture, [it becomes evident that] it is completely reliable. Now, between the kidneys is the place from where the through-way vessel emerges. In addition, it is the kuan-yüan section and also the origin for the transformation of influences by the Triple Burner. Why would one speak here of 'moving influences'? That which is at rest is [categorized as] yin; that which moves is [categorized as] yang. 'Moving influences,' then, means 'yang influences.' ... The sixty-sixth difficult

issue states: 'The moving influences below the navel and between the kidneys constitute man's life; they are root and foundation of the twelve conduits. Hence, they are called 'origin'. The Triple Burner is a special envoy [transmitting] the original influences. It is responsible for the passage of the three influences and for their procession through the [body's] five depots and six palaces. 'Origin' is an honorable designation for the Triple Burner.' The meanings [of the sixty-sixth difficult issue] and of the present [paragraph] explain each other. Obviously, the moving influences are the influences controlled by the through-way vessel. They are genuinely yang; they are the origin of the transformation of influences by the Triple Burner. Life is tied to them.

In this quotation, scholar Tamba Genkan teaches that man's continuing life depends upon the influences of the gate of life, which are the moving influences between the kidneys. The throughway vessel, or Chong mai, emerges from the location between the kidneys. Interestingly, this location between the kidneys and closer to the right Kidney 'is the kuan-yüan section and also the origin for the transformation of influences by the Triple Burner'. Then in a nutshell, Tamba Genkan summarizes the situation when he states, 'The moving influences below the navel and between the kidneys constitute man's life; they are root and foundation of the twelve conduits. Hence, they are called 'origin'. The Triple Burner is a special envoy [transmitting] the original influences.' Here, Tamba Genkan clearly shows that 'the Triple Burner is a special envoy transmitting the original influences' (Yuan Qi) throughout the twelve conduits, originating from the *kuan-yüan* section, which is the acupuncture point CV 4 (Origin Pass). This acupoint is also the front Mu point of the Small Intestine. Later in the book, I discuss the bacterial microbiome and how the microorganisms which constitute the microbiome, and their DNA, and their vital metabolic biochemical by-products play a major role in synchronizing biochemistry throughout the body, conveniently directly connecting to CV 4. Amazingly, the book mentioned above (109) showed that the PVS is 'a conduit for a fluid, which, according to proteomic analyzes contains remarkably high levels of carbohydrate metabolic derivatives that are usually associated with stem cells'. The article poignantly continued that 'thus, the PVS has the potential to transport growth and communication factors' throughout the body. These exact functions have been attributed to the Triple-Energizer Metasystem for thousands of years.

33.27 Chong Mai Also Transmits the 'Original Influences' (Yuan Qi) through the 12 Conduits

The classic Nei-ching considered the throughway vessel to be the sea for the twelve conduits. In recent times, the throughway vessel is more often called the Chong Mai, penetrating vessel or thrusting vessel, and is also known as the Sea of Blood. The Chong Mai connects the Stomach and Kidney channels. Some authorities believe the Chong Mai originates in the ming men and passes through the uterus and then down to CV 1. From CV 1, it emerges at ST 30 and then continues its upward path along the kidney meridian to KD 21. Other authorities believe that the throughway vessel originates from the ch'i-ch'ung hole, which is the acupuncture point ST 30 (Qichong). The Chong Mai then flows up the throat, encircles the mouth, and continues up to the forehead. The Chong Mai has many branches to all the vital internal organs, and its main purpose is to regulate the Qi and the Blood in the 12 main meridians and their corresponding organs. Consequently, the Chong Mai is one potential way that the raw materials from the Stomach, namely the essence of drinks and food along with the biophotons derived from the grains, could be readily circulated throughout the body to the 14 conduits and the omnipresent hydrophilic membranes to be converted into the many forms of Qi as per the Triple Heater fluid and energy production cycle. Note that the Kidney organ and Kidney meridian are both strongly involved with the Chong Mai circuitry.

It thus appears that both the Triple Burner and the Chong Mai distribute the Yuan Qi through the twelve conduits and to the internal organs and that, in this case, is under the control of the Chong Mai, as Tamba Genkan stated above when he said, 'The moving influences are the influences controlled by the through-way vessel. They are genuinely yang; they are the origin of the transformation of influences by the Triple Burner. Life is tied to them.' As mentioned in the prior paragraphs, the book (109) mentioned above showed that the PVS is a conduit for a fluid, which contains remarkably high levels of carbohydrate metabolic derivatives that are usually associated with stem cells, and the PVS has the potential to transport growth and communication factors throughout the body. For thousands of years, the omnipresent Triple-Energizer Metasystem has been ascribed these exact functions.

33.28 Is the Throughway Vessel (Chong Mai) a Lymphatic Vessel?

In the commentaries on the 29th Difficult Issue on page 336 of Unschuld's (1) translation of the *Nan Ching*, regarding the throughway vessel (Chong Mai) of the Eight Extra Meridians, scholar Liao P'ing relates:

> The through-way vessel is the sea [in which] the twelve conduits [end]; man's ancestral influences (tsung-ch'i) emerge from it. In particular it rules the reproductive affairs. It is called 'lymphatic vessel' (lin-pa-kuan) by the Westerners. Its main [course proceeds] through the abdomen, but at the same time it proceeds along the back. Hence, Mr. Yang's Tai-su considered the three vessels—the through-way, controller, and supervisor—to constitute one entity.

The final statement about the throughway, controller, and supervisor vessels all constituting one entity is a very interesting comment. A further interesting comment about this quote is that it is the only occurrence of the word *lymphatic* in the text of Unschuld's (1) translation of the *Nan Ching*. If this is correct, then the Chong Mai is considered to be a lymphatic vessel. That is very interesting. Should this indeed be the case, P-microcells have been found inside Lymphatic vessels.

33.29 The Kidneys Harbor in the Yin Region the True and Original influences

In the commentaries on the 34th Difficult Issue on page 372 of Unschuld's (1) translation of the *Nan Ching*, regarding which spirit lodges in each organ, scholar Katō Bankei states, 'Spleen and kidneys both have two spirits because the spleen is the basis for the generation and transformation of the constructive and protective [influences] and because the kidneys harbor, in the yin [region], the true and original influences.' Throughout the TCM medical classics, the original influences are described as being housed below the navel and between the kidneys and closer to the right Kidney. The fact that this quote says that the kidneys harbor the true and original influences in the yin region surely could not mean in the yin-dominant left Kidney. I suggest that this quotation means that the original influences or Yuan Qi is 'harbored' in the yin fluid region of the right Kidney. P-microcells have indeed been found in animal kidneys. I propose that time and future research will show that the right Kidney and the location medial to the right Kidney and below the navel is the major nucleus and control centre for the Primo Vascular System and the Yuan Qi and the gate of life and the spirit guarding against the evil and the Triple-Energizer Metasystem and the Chong mai. That sure is a busy intersection. I sure hope I am right on at least one of them!

33.30 The Triple Burner Originates from the 'Gate of Life'

In the 36th Difficult Issue on page 382 of Unschuld's (1) translation of the *Nan Ching*, the following question is asked: 'Each of the depots is a single [entity], except for the kidneys which represent a twin [entity]. Why is that so?' The question is answered: 'It is like this. The two kidneys are not both kidneys. The one on the left is the kidney; the one on the right is the gate of life. The gate of life is the place where the spirit-essence lodges; it is the place to which the original influences are tied. Hence, in males it stores the essence; in females it holds the womb. Hence, one knows that there is only one kidney.'

In the commentaries pertaining to this question, on page 383, the scholar Yü Shu says, 'The [Nan-] ching states: "The one on the right is the gate of life; it is the place to which the original influences are tied." The Mai-ching says that the [gate of life] is related to the Triple Burner like outside and inside. The Triple Burner, furthermore, masters the influences of the three originals. From this one may infer that the Triple Burner originates from the gate of life'.

In further commentaries pertaining to this question, on page 385, the scholar Hua Shou says, 'There are two kidneys. The one on the left is the kidney; the one on the right is the gate of life. In males, the essence is stored here. The essence [transmitted] from the five depots and six palaces is received and stored here. In females, the womb is tied here. It receives the essence [from the males] and transforms it. The womb is the location where the embryo is conceived.' Regarding this same issue, Yeh Lin states:

> The kidney has two lobes; one on the left and one on the right side. One masters the water; one masters the fire. They correspond to the mechanics of rise and fall. The gate of life is the root of the Triple Burner and the sea of the original influences of the 12 conduits; it is the utensil that stores and transforms the essence, and it is the place to which the womb, which conceives the embryo, is tied. Thus, it is the origin of man's life. Hence, it is called gate of life.

Regarding the location of the life gate or Ming Men, in the *Australian Journal of Acupuncture and Chinese Medicine* article, the author, Mary Garvey (42), on page 20, stated, 'By the end of the sixteenth century however, the lifegate's location was generally considered to be between the left and right kidney zang, that is, level with the lumbar two/three area—the location of the Inner Canon's minor heart, and level with the acupoint GV4.'

33.31 Kidneys Are the Proper Source of Yuan Qi; The Triple Burner Is an Additional Source

In the commentaries on the 38th Difficult Issue on page 395 of Unschuld's (1) translation of the *Nan Ching*, regarding the Triple Burner representing an additional source of original influences, the scholar Li Chiung relates:

> The kidneys are the proper [source] of the original influences; *the Triple Burner represents an additional [source] of original influences*. The tan-chung is a sea of influences; it is located in the upper [section of the Triple] Burner. Also, a sea-of-influences hole exists two inches below the navel; it is located in the lower [section of the Triple] Burner. [Hence, the Triple Burner] controls the influences of the entire body. (Emphasis is mine)

In further commentaries on the 38th Difficult Issue on pages 395–396 of Unschuld's (1) translation of the *Nan Ching*, regarding the Triple Burner representing an additional source of original influences, the scholar Hua Shou relates, 'The Triple Burner governs all the influences; it is an additional transmitter of original influences. [That is to say], the *original influences depend on the guidance of the [Triple Burner] in their ceaseless hidden movement and secret circulation through the entire body*' (emphasis is mine).

I propose that the Primo Vascular System is a structural component of the Connective-Tissue Metasystem (San Jiao or Triple-Energizer Metasystem), and I suggest that the P-microcells flowing within the microtubules do indeed depend on the 'ceaseless hidden movement and secret circulation through the entire body' within hollow ducts of the PVS incorporated within the Triple-Energizer Metasystem. I propose that the P-microcells are an essential component of the Yuan Qi. The outcome of Yuan Qi reactions throughout the body relates to energic characteristics (including thermogenesis and the release of biophotons) and metabolic products derived from the microbiome present in the Small Intestine, causing biochemical and biological transformation and transmutation of foods and drinks consumed. Scientists believe that biophoton light is stored in the DNA within the cells of the organism and likely inside the DNA within the P-microcells flowing inside the Primo Vascular System microtubules. It is theorized that light is constantly released and absorbed by the DNA and that this process may interconnect the organelles, cells, tissues, and organs within the body and serve as the main communication network throughout the entire organism and serve as the principal regulating mechanism for all the diverse life processes that exist within the human body. Note that all these features are attributed to the San Jiao. So I propose that the Triple Burner is an additional source of Yuan Qi in that the Triple Burner is composed of the Connective-Tissue Metasystem that houses the Primo Vascular System, which contains the microtubules filled with fluid that allow the Yuan Qi (P-microcells) to maintain their 'ceaseless hidden movement and secret circulation through the entire body'. The 'nesting' system is much like the principle of the Russian matryoshka doll, also known as Russian nesting doll or babushka doll. This refers to a set of wooden dolls of decreasing sizes, one placed inside the other.

33.32 The Triple Burner Is the Special Envoy Transmitting the Yuan Qi or PVS P-Microcells

In the commentaries on the 66th Difficult Issue on pages 568–569 of Unschuld's (1) translation of the *Nan Ching*, regarding the yuan points (origin or source points) of the 12 meridians, scholar Hua Shou states:

> 'The rapids [holes] where the Triple Burner passes [its influences] are the origin [holes]' means the following. The moving influences below the navel and between the kidneys constitute man's life; they are the source of the twelve conduits. Consequently, the Triple Burner is the special envoy [transmitting] the original influences. It is responsible for the passage of the upper, central, and lower influences through the body's five depots and six palaces. '[The Triple Burner is responsible for] the passage of the three influences' means, according to Mr. Chi, [the following]: The lower [section of the Triple] Burner is endowed with the true primordial influences; these are the original influences. They move upward and reach the central [section of the Triple] Burner. The central [section of the Triple] Burner receives the essential but unrefined influences of water and grains and transforms them into constructive and protective [influences]. The constructive and protective influences proceed upward together with the true primordial influences and reach the upper [section

of the Triple] Burner. 'Origin' represents an honorable designation for the Triple Burner, and all the locations where [its influences] stop represent origin [holes] because [the movement of the influences of the Triple Burner] resembles the arrival of the imperial herald, announcing the places where [the Emperor] will pass by and rest. When any of the five depots or six palaces has an illness, it is always appropriate to remove it from these [holes].

In the commentaries on the 66th Difficult Issue on page 561 of Unschuld's (1) translation of the *Nan Ching*, it states, 'The Triple Burner is the special envoy that transmits the original influences.' The source points of a particular meridian will remediate the injured organ of that meridian. For example, the yuan point (source point) of the Liver, Liver 3, will benefit the Liver organ when there is illness there. So thanks to the Triple-Energizer Metasystem, the healing Yuan Qi arrives at Liver 3, and then the Yuan Qi (healing Primo Vascular System P-microcells (Sanals)) are directed to the ailing Liver. Note that these pluripotent P-microcells (stem cells) are the main agents for wound healing and regeneration. I believe it is only a matter of time before it is confirmed that the many different functions of the different types of acupuncture channels are due to the circuitry of the PVS ducts. The PVS has been confirmed to be present in numerous deep internal organs (7). The PVS has been confirmed to be present in the superficial hypodermis (7). Remember too that research in 2009 (34) revealed that 'the flow of the primo fluid in a certain path was demonstrated in the study using Alcian blue, from the rat acupoint BL-23 in the dorsal skin to the PVS on the surface of internal organs'.

33.33 Nucleic Acids (DNA) Recently Found Circulating in the Primo Vascular System

Regarding the presence of a large quantity of nucleic acids being found in the PVS, in the book, J. Kim et al. (7), on page 10, state:

> A large quantity of nucleic acids, DNA in particular, is contained in the Primo node and primo vessel. DNA in the

primo vessel exists in a peculiar way, outside the nucleus in the homogeneous primo fluid . . . In view of this, we consider that the action of the PVS is closely connected with nucleic acids. And the specific form of the existence of nucleic acids in the PVS also requires the study of the functions and metabolism of nucleic acids from a new viewpoint.

Nucleic acids are extremely important because they contain the genetic information that make living things function effectively. While there are two categories of nucleic acid (DNA and RNA), the DNA houses the basic instructions for living things. It is passed down from parent to offspring and is found in the nucleus of the cell and, as only recently determined, also throughout the PVS. It is remarkable that the presence of DNA in the PVS has only recently been elucidated. Note that the Primo Vascular System was previously called the Kyungrak System.

33.34 Exterior Primo Vessels Covered by a Thick Membrane of Connective Tissues

Regarding the structures for exterior primo vessels, the book (7), on page 11, states, 'These are composed of ducts and corpuscles along the blood vessels and nervous system. They are covered by a thick membrane of connective tissues.'

33.35 Collagen May Facilitate Tuning Photon Emissions throughout the Entire Body

Regarding the properties of the PVS, in the 2013 *Journal of Acupuncture and Meridian Studies* article titled 'The Primo Vascular System as a New Anatomical System', the authors (84) state, 'The cells of the PVS show smooth muscle-like excitability with calcium channels and the subvessels have adventitia that contains connective tissue. *Collagen is the main component of the connective tissue.*' The authors note that research indicates that collagen affects photon emissions originating from biomolecular sources and that collagen may very well be capable of fine-tuning the photon emissions known to be propagated throughout the body. This observation strongly supports the theory that the PVS behaves as an omnipresent optical conduit for biophoton emissions throughout the body. They further suggest that DNA may act as a photon reservoir and perform as a coherent radiator, managing the delivery of synchronizing biophotons throughout the entire organism to ensure that cellular development and differentiation occurs correctly. The authors further state, 'The light propagation function of the PVS may explain the instantaneous effect after needling at acupoints' (emphasis is mine).

This recent (2013) research confirms that the PVS is composed of ubiquitous connective tissue and is predominantly collagen and, as such, is intimately related to the omnipresent Connective-Tissue Metasystem that pervades the body. Note that the authors suggest that the light propagation function of the PVS accounts for the instantaneous effect after needling acupoints, suggesting that the acupuncture meridians are indeed made up from or involved with the PVS.

33.36 Connective Tissue, Fascia, and Membranes Are Omnipresent throughout the Body

Regarding the omnipresence of connective tissue/fascia and the Primo Vascular System, in the 2013 *Journal of Acupuncture and Meridian Studies* article titled 'The Primo Vascular System as a New Anatomical System', the authors (84) state:

> The fascia covers the muscles and enters between them. The serous membranes cover nearly all organs. *Loose connective tissue is the most distributed tissue in the body* and does not exist only in the brain, the penis, and the clitoris. *The PVS is associated with the vessels and the nerves and is abundant in loose connective tissue, fat tissue, serous membranes, and fascias; therefore, it is possible that it is distributed as a web among all body systems, including the tissues of organs.* The ePVS is in the skin's hypodermal layer and superficial fascia. The iPVS and the nPVS follow the fascia, loose connective tissue,

and serous membrane distribution, and then reaches the oPVS. (Emphasis is mine)

Regarding the distribution of the PVS, note what the authors stated above when they said, 'The PVS is *associated with the vessels and the nerves* and is *abundant in loose connective tissue, fat tissue, serous membranes, and fascias*; therefore, it is possible that it is distributed as *a web among all body systems*, including the tissues of organs' (emphasis is mine). This recent (2013) research again confirms that the PVS is composed of ubiquitous connective tissue and that the PVS is intimately associated with and abundant in loose connective tissue, serous membranes, and fascias. As it is likely 'distributed as a web among all body systems, including the tissues of organs,' this puts the PVS/connective-tissue network in the same location that the Triple Energizer occupies. For this reason, I believe there is no doubt that the Connective-Tissue Metasystem includes elements of the Primo Vascular System and is intimately related to the Triple-Energizer Metasystem.

33.37 Modern Scientific Hypothesis for the Nature of Qi
Proposing a scientific hypothesis for the description of Qi, in the 2013 *Journal of Acupuncture and Meridian Studies* article titled 'The Primo Vascular System as a New Anatomical System', the authors (84) offer a novel interpretation regarding the possible nature of Qi. They propose that Qi may be the electromagnetic field that travels throughout the fiber-optic-like channel of the PVS throughout the entire body. Because the PVS may be an optical channel for photon emission and because DNA granules have been found within the PVS and because DNA has been implicated as a photon storehouse, all these features lend support to their hypothesis. They further believe that the structure of DNA is capable of storing ambient environmental information pertaining to physical force fields, including electromagnetic fields, and that these electromagnetic waves are able to be transformed into information necessary for biological and biochemical processes. In complete harmony with this theory, the authors reported that the team of physicists led by Toybe suggested a novel process for converting information into

energy based on Szilard's concept of the correspondence between energy and information. They further noted that the theoretical physicist Stephen Hawking once stated, 'Electromagnetism is the basis for life itself.' They noted that the omnipresent PVS has all the physical characteristics necessary to supply electromagnetic waves to every cell in the body. The authors concluded, 'The ancient vital energy Qi probably is an *electromagnetic wave* that is transported through the PVS, and the information obtained from that electromagnetic wave may be stored in the *DNA of the PVS microcells*' (emphasis is mine).

33.38 The Biochemical Composition of Primo Fluid Circulating throughout the PVS

The book (7), on page 12, advises that the PVS meridians are a multicirculatory system for the primo fluid. The biochemical composition of the primo fluid includes a large quantity of nucleic acid and ribonucleic acid. Total Nitrogen is 3.12–3.40%, nonprotein nitrogen is 0.10–0.17%, Fat is 0.57–1.00%, reducing sugar is 0.10–0.12%, and the hyaluronic acid content is 170.4 mg%. More than 19 free amino acids, including the essential amino acids, are present in the primo fluid, and more than 16 free mononucleotides are present. The book notes that primo fluid pathways differ from blood circulation pathways in that they are interconnected and consist of comparatively independent multicirculation pathways. When tracing dyes and radioisotopes are injected into a Bonghan pathway, they circulate only within a dedicated region. However, the primo fluid within a dedicated pathway can be transferred to other pathways via interconnections between the pathways. The book further explained that Primo vessels elicit bioelectrical activity, excitatory conductivity, and mechanical motility.

33.39 When the 'Original Influences' or Yuan Qi or Primo Vessels Are Cut Off, Death Follows

In the commentaries on the eighth Difficult Issue on page 133 of Unschuld's (1) translation of the *Nan Ching*, confirming that the original influences (yuan qi) engenders vibrant life, scholar Hua Shou states, 'When the original influences prevail, evil [influences] cannot enter. When the original influences are cut off, death follows.' This quote shows that the original influences (Yuan Qi) engenders vibrant life and that 'when the original influences are cut off, death follows'.

The book (7), on page 13, advises that 'changes of the primo fluid circulation affect functions of organic tissues'. For example, 'stimuli to the primo vessels change the number of beats and power of the heart, and the intestinal canal movements. It also affects the fatigue curve for the skeletal muscles.' The book explained that prominent changes to the tissue cells occur when the primo vessels are cut and disconnected. For example, 'if the primo vessels are disconnected, a kind of karyolysis occurs in the attached tissue cells and induces apoptosis'. The book continued to explain that if 'the primo vessels in the peripheral nervous system are disconnected, excitability of nerves is prominently reduced' and that if 'the primo vessels in a motor nerve are disconnected and the motor nerve is stimulated repeatedly, there is no muscle movement'. The implication of these findings on surgical procedures is obviously profound.

So note that Hua Shou stated, 'When the original influences are cut off, death follows.' And the book stated, 'If the primo vessels are disconnected, a kind of karyolysis occurs in the attached tissue cells and induces apoptosis.' Karyolysis is the termination of the cell nucleus by swelling or necrosis, and apoptosis is a form of cell death whereby a programmed sequence of events results in the elimination of cells without adversely affecting the local environment. Both these events result in the cell death. The book (7), on page 16, states, 'The PVS appeared to control and dominate the formation, maintenance, and death of tissue cells. All the fundamental processes of life appear to be based on the sanals' movement. So studying the law of this movement should become a fundamental subject of biology.' I believe there is no doubt that the DNA-based sanals (P-microcells) circulating throughout the entire body within the Primo Vascular System constitute a major component (if not all) of what the ancients called Yuan Qi.

33.40 Primo Vascular System Development and Proliferation Precedes the formation of *all* Other Structures during Embryogenesis

Regarding the critical initial formation of the PVS during embryogenesis, the book (7), on page 13, states:

> *Proliferation of the [PVS] meridians takes place ahead of proliferation of any other organs, such as the blood vessels and the nervous system.* Embryo development follows the following steps: the step for the formation of the primo vessel blast cell occurs 7–8 h after fertilization; the step for primordial primo vessel occurs 10 h after fertilization; the step for the formation of primitive primo lumens occurs 15 h after fertilization; and the final step for the completion of the primo lumens occurs 20–28 h after fertilization. *The fact that the proliferation of the PVS precedes the formation of other structures suggests the PVS plays an important role during development of an organism.* (Emphasis is mine)

The PVS is shown here to be intimately and inextricably involved with embryogenesis from as early as 7–8 hours post fertilization. I suggest that this is the initiation of the gate of life and also of the initiation of the Triple-Energizer Metasystem. This PVS structural conception is obviously tied to 'the original influences which man has received from father and mother', as discussed in the eighth Difficult Issue of the *Nan Ching*. As this omnipresent PVS structure is ubiquitous throughout the adult body, the embryological vestige is truly dispersed throughout the entire body via the omnipresent Connective-Tissue Metasystem that is the domicile of the Triple-Energizer Metasystem, which is intimately entwined with the PVS.

33.41 The Connective Tissue Fascial System Starts about 2 weeks into Embryological Development

On the web article entitled 'Fascia & Tensegrity', Myers (110) states:

> *Our single fascial system starts about 2 weeks into development as a fibrous gel that pervades and surrounds*

all the cells in the developing embryo. It is progressively folded by gastrulation and the rest of the motions of development into the complex layers of fascia we see in the adult. In fact there is no discontinuity in the layers of fascia, so 'layers' is a useful but deceptive concept (like 'muscles'). (Emphasis is mine)

Likely, initially after fertilization, the miniscule Primo Vascular System conduits are able to support themselves inside the miniscule fertilized egg. A human oocyte is only about 100 μm in diameter. As further cellular differentiation and then gastrulation occurs, a more stable and strong supporting and protective infrastructure would be required. Enter the supportive omnipresent connective tissue structure to secure and protect the delicate Primo Vascular System.

33.42 The Primordial System that Engenders All Other Major Body Systems and Organs

Regarding the primordial body system that engenders the major body systems, in the 2013 article titled *The Primo Vascular System as a New Anatomical System*, the authors (84) comment that morphological scientists are at a loss to propose a new biomedical model that clarifies the existence of new, intricate biological systems, such as the Primo Vascular System (PVS). It is absolutely astounding that a ubiquitous, physical biological system that integrates the features of the cardiovascular, nervous, immune, and hormonal systems has remained undiscovered until recently. The authors believe that the PVS constitutes the physical framework in which the acupuncture points and acupuncture meridians reside. Discovery of the sophisticated morphological structure abounding with organization and communication obviously assumes an architectural design at the foundation level of life. This discovery of the integrated functionality of the PVS has profoundly altered the basic tenets of biology and medicine since the PVS is intimately involved in the initiation, development, and the integrated functions of living organisms.

According to Western medicine, acupuncture meridians have no known anatomical foundation, and unknown nervous, circulatory, endocrine, and immune mechanisms mediate the effects of acupuncture. The discovery of this PVS may well explain why acupuncture so powerfully benefits nearly all medical conditions because the acupuncture meridian system integrates and regulates the features of the cardiovascular, nervous, immune, and hormonal systems. Many hardened skeptics of acupuncture will have to eat lots of 'humble pie'.

33.43 Does the Primo Vascular System Originate from the Polar Bodies?

Regarding the role of the Polar Body within the developing embryo, in the 2014 *Integrative Medicine International* journal article titled 'Does the Primo Vascular System Originate from the Polar Body', the authors (106) commence their article by explaining that polar bodies degenerate on the very first day of embryonic life and that, to date, no known role for them in the human embryo has been elucidated. They note that it is illogical to have useless cells.

In the commentaries on the 66th Difficult Issue on page 569 of Unschuld's (1) translation of the *Nan Ching*, while discussing the yuan points (origin or source points) on the 12 meridians, scholar Hua Shou states, '"Origin" represents an honorable designation for the Triple Burner.' If Avijgan and Avijgan (106) are correct in this matter, then the Primo Vascular System would be conceived from the very first day of embryonic life. That certainly is at the origin of life. As I believe that the PVS and the Triple Burner are intimately connected, this absolutely would explain why the term *origin* does represent 'an honorable designation for the Triple Burner'. I propose that the PVS and the acupuncture meridian system and the Triple Burner are all intimately connected.

33.44 The Pivotal Role of the PVS Immediately after Conception

Regarding the role of the Primo Vascular System immediately after conception, in the 2014 *Integrative Medicine International* journal article titled 'Does the Primo Vascular System Originate from the Polar Body', the authors (106) state, 'It could be postulated that the sanal cells (PMCs-VSELSCs [primo microcells—very small embryonic-like stem cells]) in the PVS of the embryo are probably the precursor and/or transformer of the inner cell mass into epiblast cells, which is the origin of the three embryonic layers.' They continue, 'This hypothesis also assumes that, in embryonic life and the developmental period of an embryo, the PVS rules the differentiation of epiblast cells into the three (next) layers ectoderm, mesoderm and endoderm of an embryo.' They further propose that the sanal cells 'are deposited in each organ during gastrulation and organogenesis and may serve as reserve pool of quiescent PSCs [pluripotent stem cells]'.

Intriguingly, the authors further suggest that the PVS may appear immediately prior to or simultaneously to the fertilization of the ovum by the sperm, to prevent the entrance of a second sperm into the already fertilized ovum. If this was the case, then the very first role of the PVS would be to preserve the zygote organism.

33.45 Mysterious Polar Bodies Could Be the Precursors for the Primo Vascular System

Regarding the role of the mysterious Polar Bodies during embryogenesis, in the 2014 *Integrative Medicine International* journal article titled 'Does the Primo Vascular System Originate from the Polar Body', the authors (106) hypothesize that 'polar bodies could be the pioneers for the PVS' in that, after fertilization, 'the polar bodies form the primitive PVS'. They suggest that this transformation 'of pre-PVS to proto-PVS to PVS happens in the space between dividing zygote and zona pellucida, during the first week of life and before the zona hatching occurs'. This certainly is at the origin of life, when the life gate is opened according to TCM.

33.46 Does the Primo Vascular System Originate from the Polar Bodies?

Discussing the matter further, regarding the role of the mysterious Polar Bodies during embryogenesis, in the 2014 article titled 'Does the Primo Vascular System Originate from the Polar Body', the authors (106) explain that their hypothesis is based on the recent finding that, within hours of chicken egg fertilization, the PVS emerges in the vitelline membrane of the chicken egg. They suggest this occurrence is comparable in human zygote formation. Some researchers believe that the zona pellucida has the role of thick albumin which surrounds the yolk sac of the chicken egg. The authors propose that if the PVS is indeed a component of the conception product structure, then the PVS could also be involved with processes including early cell differentiation, junction, and the later cavitation process. At this time, the mechanism of these complicated fetal developmental processes is unknown. The authors note that former research suggested that 'polar bodies are transformed into extraembryonic tissue'. Based on their theory, the authors believe that the PVS may be extraembryonic tissue, which is located between the zygote and the zona pellucida. If this is in fact the case, *it strongly suggests that the PVS originates from the polar bodies.*

33.47 Dominating Functions of the Polar Bodies for All Aspects of Embryonic Development

With regard to the dominating functions of polar bodies for all aspects of embryonic development during embryogenesis and beyond, into adult life, in the 2014 *Integrative Medicine International* journal article titled 'Does the Primo Vascular System Originate from the Polar Body', the authors (106) explain that, after successful fertilization, and within the first week of conception, the resulting three biological constituents include (1) the blastocoelic cavity which develops into the embryo, (2) its membranes (which the authors believe constitutes the PVS), and (3) the surrounding zona pellucida. The zygote has to be set free from the encasing zona pellucida via the process called zona hatching before it can implant into the uterine endometrium and commence further embryological development. The authors hypothesize that *the PVS is present in the zona pellucida as extraembryonic PVS, and that it controls all the developmental structures of conception, including trophoblast, inner cell mass differentiation, zona hatching and further embryonic differentiation.* The authors note that while a previous study has implicated the polar bodies with having the dominating role in all aspects of embryonic development, they suggest the identical role for the PVS.

If the PVS actually is a major component of the original San Jiao, as I postulate, this agrees with TCM theory about San Jiao's function of maintaining harmony throughout all body structures. The authors believe that the PVS 'controls all developmental processes of the products of conception, including trophoblast, inner cell mass differentiation, zona hatching and differentiation'. The San Jiao is believed to be the Fu organ that is responsible for numerous functions of homeostasis throughout the adult body, including energy production, digestion, fluid metabolism and synthesis, immune protection, hormonal balance, temperature regulation, detoxification through urine and feces formation, nutrient synthesis, etc. It would make sense that the PVS/San Jiao's function of maintaining harmony throughout all body structures, systems, and organs would continue from embryogenesis throughout the rest of the creature's life.

33.48 Inner Cell Mass Differentiation during Embryonic Development

With regard to inner cell mass differentiation during embryonic development, in the 2014 article titled 'Does the Primo Vascular System Originate from the Polar Body', the authors (106) state that the transition of the epithelial to mesoderm leads to the generation of new cells, which via the primitive streak will expand to engender a new layer of cells or interact with already existing layers of ectoderm, mesoderm, or endoderm. The ectoderm engenders the skin and the central nervous system. The endoderm engenders the endothelial cells of the respiratory and digestive systems and also some organs related to the digestive system, including the liver and pancreas. The mesoderm engenders the mesenchymal cells, which can migrate easily and are recognized by a matrix containing a loose aggregation of reticular fibular and unspecialized stem cells (SCs). The authors note that *the mesoderm appears to be the most significant layer of the embryo as it engenders all the major organs of the body.*

In agreement with the above description of the connective tissue, in the article entitled 'Bioelectric Responsiveness of Fascia: A Model for Understanding the Effects of Manipulation', the author, Judith A. O'Connell (111), states, *'Connective tissue is a major derivative of the mesodermal germ cell layer in the embryo.* As the name implies, it connects with mesodermal, endodermal, and ectodermal derivatives as the differentiation of organs and systems occurs.' She stresses that 'this special property of connective tissue allows it to act as an intermediary both embryologically and in the fully developed individual'. Then, regarding the special property of connective-tissue types, she states, 'There are four groups of connective tissue: connective tissue proper, cartilage, bone, and blood. Together they form the support structure for homeostasis that allows for normal function in the body. A special property that connective tissue maintains throughout life is embryologic plasticity' (emphasis is mine).

From the above information, it can be seen that the mesoderm is the most important layer of the embryo. Mesoderm cells 'can migrate easily', are predominantly composed of collagen, contain unspecialized stem cells, and importantly, the mesoderm gives rise to all major organs of the body. Connective tissue is a major derivative of the mesodermal germ cell layer in the embryo. The mesoderm connects with 'endodermal and ectodermal derivatives as the differentiation of organs and systems occurs'. By means of the mesoderm generating connective tissues, an intricate omnipresent supporting infrastructure is generated that enmeshes every cell, system, and organ within the body. Remember too that the *mesoderm* 'has a great responsibility by *inducing some other internal organs and structures* like the heart, kidney, spleen and connective tissue'. This supporting structure exists from immediately after conception and continues throughout the lifetime of the individual, maintaining its original embryologic plasticity. The connective-tissue metasystem forms 'the support structure for homeostasis that allows for normal function in the body'. Because of the minuteness of the PVS conduits, I propose that the entire Primo Vascular System requires the connective tissue framework and infrastructure to hold, support, and protect the minuscule conduits. Avijgan and Avijgan (106) report that the 'intraembryonic PVS is *spreading amidst the mesoderm*' (emphasis is mine). Thus, the intraembryonic PVS appears to be *derived from the same cellular structure* that the connective tissue infrastructure is derived from—namely, the *mesoderm*.

33.49 The Critically Important Role of the Polar Bodies during Embryonic Development

With regard to the critically important role of the polar bodies during embryonic development, in the 2014 *Integrative Medicine International* journal article titled *Does the Primo Vascular System Originate from the Polar Body*, the authors (106) state:

> *This report indicates for the first time a transformation of polar bodies into the PVS as an extraembryonic tissue; the PVS having a supportive, dominating and controlling role in all steps of embryonic development.* It seems to have been missed or ignored because it is not visible by usual staining. Trypan blue staining and some other specific stainings can make it visible for a short time. *During the embryonic life (after the third week of gestation), the embryo undergoes major changes like the primitive streak, loss of symmetry, gastrulation and the formation and differentiation of endoderm, mesoderm and ectoderm. The extraembryonic PVS has been suggested to have an important and key role in this.*
>
> *The extraembryonic PVS has been assumed to extend around the embryo and be responsible for the generation of structures like the chorion and yolk sac.* In the gastrulation process, it has been suggested that some part of this PVS will ingress into the embryo and form the intraembryonic PVS. *While the intraembryonic PVS is spreading amidst the mesoderm*, this system will be split into two parts during the delamination of the mesoderm into two layers. (Emphasis is mine)

The authors (106) further explain that this mesoderm delamination yields the inner surfaces of the somatic mesoderm and the splanchnic mesoderm respectively. The somatic mesoderm in liaison with the ectoderm initiates the generation of organs like the amnion during embryonic life and also forms the skin. Meanwhile, the splanchnic mesoderm in liaison with the endoderm, initiates the formation of the mucosal cells of organs including the liver, the lung and the gastrointestinal tract. They note that 'the mesoderm has a great responsibility by *inducing some other internal organs and structures* like the heart, kidney, spleen and connective tissue' (emphasis is mine).

Even at such an early stage in life development, 'the extraembryonic PVS has been assumed to *extend around the embryo*' (emphasis is mine). Here, it surrounds the embryo like an external wall, just as the Three Heaters are described to do to all cells, systems, and organs in the human adult. Note too that, in the early stage of embryogenesis, the intraembryonic PVS is *spreading amidst the mesoderm* and that the mesoderm has a *great responsibility by inducing some other internal organs and structures* like the heart, kidney, spleen and connective tissue. Via differentiation of the mesoderm, the PVS and accompanying supportive Connective-Tissue Metasystem is *omnipresent* throughout the developing embryo and, later in life, throughout the entire adult body. Thus, these two systems are intimately associated with one another.

In the 2013 article titled 'The Primo Vascular System as a New Anatomical System', the authors (84) state, 'The PVS . . . provides a physical substrate for the acupuncture points and meridians.' This agrees with the TCM belief that the Triple Burner (Triple-Energizer Metasystem) is also referred to as the Mother of the entire acupuncture meridian system. As stated elsewhere, the conduits of the Primo Vascular System are minute, and the PVS could not exist and be supported in place without the accompanying Connective-Tissue Metasystem as a supportive and protective framework. The bioelectrical properties of the major component of the Connective-Tissue Metasystem, namely collagen, are believed to play a major part in the conduction of Qi through the acupuncture meridian system.

33.50 The Triple Burner Stems from the Yuan Qi

Chapter 38 of the *Nan Jing* discusses the relationship between the Triple Burner and the Yuan Qi. It says, 'There are 6 Fu *because of the Triple Burner which stems from the Yuan Qi*. The Triple Burner governs all Qi in the body' (emphasis is mine).

Several references above show that proliferation of the Primo Vascular System is the very first phase of embryological development and differentiation, commencing as early as 7–8 hours after fertilization. Soon thereafter, the connective-tissue fascial system originates its formation and differentiation as early as two weeks after fertilization. Thus, the PVS and the Connective-Tissue Metasystem are present at the very beginning of life, actually when the gate of life is opened. As the PVS is composed of connective tissue anyway, it is essentially the Primo Connective-Tissue System. So it appears that the comingled life forces of the Sperm plus Ova engender the Primo Vascular System. Later, the process called gastrulation occurs, whereby the blastocyst (once implanted into the endometrium) develops the three germ layers—ectoderm, endoderm, and the mesoderm—which later gives rise to all the different organs, systems, and structures of the human body. I believe that this is aka the Yuan Qi engendering the Triple Burner, as stated in Chapter 38 of the *Nan Jing*, where it says, 'The Triple Burner stems from the Yuan Qi'. Should the embryo continue to grow and prosper to term, it is via the PVS and the Connective-Tissue Metasystem that the life force is allowed to be molded or woven into the numerous differentiations that become all the individual organs and structures that engender a human being. It is for this reason that the Connective-Tissue Metasystem surrounds every single organ and structure like an outer wall or like a wrapping bag. For these reasons, I propose that the omnipresent Connective-Tissue Metasystem is the same as the omnipresent Triple-Energizer Metasystem.

33.51 The PVS Has Been Found to Exist throughout the Biological World

Regarding the distribution of the Primo Vascular System throughout nature, the book (7), on page 13, explains that the PVS apparently exists throughout the entire biological world and is found in both invertebrates and vertebrates as well as the mammalians and is present in all multicellular living organisms, including plants. Research revealed that the PVS pathway for the primo fluid circulation is as follows: Tissue cells → superficial primo nodes → deep primo nodes → primo nodes for organs → terminal primo nodes → tissue cells. The PVS meridians are a highly organized circulatory infrastructure which enables the flow of primo fluid so that every tissue component in an organism is connected to the meridians. The author suggests that all organisms have PVS meridians.

33.52 The Sanal Cell Cycle Is Responsible for the Continuous Self-Renewal of Organisms

The book notes that living organisms keep themselves alive via regeneration following the sanal cell cycle. Regarding this process, the book (7), on page 15, states:

> All structural elements in living organisms regenerated ceaselessly. If physiological regeneration processes in organisms only depend on cell division, very limited scope in the regeneration processes could he recognized. However, given that the concept, Bonghan sanal-cell cycle, applies to the regeneration process, it was recognized that all tissue cells regenerated continually. This process was also clearly observed in culture conditions. Continuous self-renewal of organisms generally occurred not only in the molecular and individual level but also in the cellular level. In other words, continuous self-renewal processes of structural elements took place with continuous metabolism in organisms.

33.53 Sanals Will Only Become Cells if They Are Cultured in Primo Fluid or an Equivalent Medium

Regarding the extreme importance of the Primo Vascular System throughout nature, the book (7), on page 16, states, 'Primo fluid contained various chemicals that were required to grow sanals. The liquid included plentiful free amino acids, free mononucleotides, hyaluronic acids, and various hormones, along with other proteins, sugars, and lipids.' Research shows that sanals will only become cells if they are cultured in primo fluid or in a suitable nutrient solution with a similar composition to primo fluid. Remarkably, sanals will not grow into a cell when they are cultured in blood, lymph, or tissue fluid. Research shows that all tissue cells are connected to the PVS and that 'tissue cells will die after the dissolution of the nucleus when primo vessels connected to them were cut'. Research has confirmed that primo microcells circulating within the Primo Vascular System are responsible for wound healing and cell therapy for regeneration of damaged tissues.

33.54 The Sanal-Cell Cycle Is Responsible for the Manufacture and Self-Renewal of Blood Cells

Regarding the manufacture and self-renewal of Blood cells, the book (7), on page 17, explains that the sanal cell cycle of blood cells, including their self-renewal process, is directed by the PVS meridian system. Blood cell sanals flow within the intravascular primo vessels and mature into blood cells inside intravascular primo nodes. Hematopoietic organs—including lymphatic nodes, bone marrow, and the spleen—all have well-developed primo vessels possessing a structure and function identical to primo nodes. Blood cells are sanalized within primo nodes inside blood vessels and lymphatic vessels.

Fluid and nutrients from consumed drink and grains leave the Stomach and pass to the hematopoietic Spleen and then continue to be processed on their route to the Lungs and then to the Heart; this can now be readily explained. Knowing that the omnipresent Primo Vascular System conduit system has finally been elucidated, this staged Blood manufacturing process makes perfect sense. This is amplified by the fact that 'blood cell sanals usually *flow inside intravascular primo vessels* and *mature to blood cells in intravascular primo nodes*' (emphasis is mine). So it appears that Blood undergoes the manufacturing procedure within the PVS microvessels, which are themselves inside blood vessels and lymphatic vessels throughout the body. As stated above, Kwang-Sup Soh et al. (34) reported the IV primo vascular system (PVS) has been *detected floating inside* the Heart. The team reported, 'More importantly, the PVS in the atrium of a bovine heart was *found to form a floating network*' (emphasis is mine). Is this primo vascular system (PVS) *floating inside* the Heart responsible for the final step in the manufacture of Blood? I propose that this PVS network somehow is what finally makes the Red Blood Cells functionally active, a feature attributed to the Heart in TCM medical classics.

33.55 Blood Is the Constructive Influence and Lymph Is the Protective Influence in the Body

In the commentaries of the 22nd Difficult Issue on page 280 of Unschuld's (1) translation of the *Nan Ching*, Chang Shih-hsien states, 'In man's entire body, the blood constitutes the constructive [influences] while the [proper] influences constitute the protective [influences]. The constructive [influences] move inside the vessels; the protective [influences] move outside of the vessels. Evil [influences] enter from outside; they affect the [protective] influences first and are then transmitted to the blood.' Here Chang Shih-hsien states, 'The constructive influences move inside the vessels; the protective influences move outside of the vessels.' He is indicating that there are circuits inside and outside the established blood vessels. This was not able to be explained until now that very modern research has discovered how this is possible. Regarding how this is possible, the book (7), on page 9, states, 'The primo vessel has two forms of existence. One of the forms of its existence is that it runs inside the blood vessel or the lymphatic vessel and the other is that it runs outside the vessel. The intravascular primo vessel and the extravascular primo vessel take different directions from each other, but there is no difference between them in structure.' How on earth did the ancient medical sages know this intricate detail without microscopes and technical histological staining procedures? I believe that future research into the components and composition of the two fluids flowing through the Bong-Han intravascular and extravascular primo vessels will confirm that constructive (healing rejuvenating) influences of Blood flows inside the blood vessels while protective (immunologically defensive) influences flow outside the blood vessels.

33.56 The General Biological Concepts behind Bong-Han Theory

Generalizing the aspects of Bong-Han Theory, in the book (7), on page 23, Jong-Su Lee explains in broken English that Prof. Kim Bong-Han uncovered a new circulatory system in our body. He noted that as all living cells get old, they degenerate and the chromosomes of those cells circulate within the Bong-Han circulatory system, where they receive some energy and are rejuvenated to healthy vital cells once again so they can regenerate damaged tissues. All cells are connected to the Bong-Han system, which penetrates to the nuclei of the cells. He believes that the Bong-Han system remediates many medical conditions, including disorders of the hematological and endocrine systems.

Jong-Su Lee further notes that while mitosis is one of the fundamental principles of modern western medicine, it is readily accommodated by Bong-Han theory. He suggests that while plant cells live by photosynthesis, animal cells live by Bong-Han synthesis. He believes that the five organs and six viscera of Chinese acupuncture theory all have their own special Bong-Han system.

33.57 Hypothesis for the Cause of Cancer Related to Bong-Han Theory

Regarding what he believes to be the primary cause of cancer, in the 2012 book (7), on page 24, in broken English, Jong-Su Lee suggests that the cause of cancer is associated with the Bong-Han system, that it is via the Bong-Han system that cancer metastasizes, that hormones are distributed, and that all brain cells are being continuously renewed. He suggested that all doctors should carry acupuncture needles instead of a stethoscope, and he believes that many medical problems can be solved using a shift of paradigm due to concepts associated with the Bong-Han theory. He hypothesizes very strongly that the cause of cancer is due to pathological changes within the Bong-Han system.

The 2014 *Science* article titled 'Variation in Cancer Risk among Tissues Can Be Explained by the Number of Stem Cell Divisions' by Cristian Tomasetti and Bert Vogelstein (112) reported, 'Some tissue types give rise to human cancers millions of times more often than other tissue types. Although this has been recognized for more than a century, it has never been explained.'

From their research, they concluded that the majority of cancers result from sheer 'bad luck' rather than unhealthy lifestyles, diet, or even inherited genes. The authors stated, 'These results suggest that only a third of the variation in cancer risk among tissues is attributable to environmental factors or inherited predispositions. The majority is due to "bad luck," that is, random mutations arising during DNA replication in normal, noncancerous stem cells.' A strong correlation was established between a particular tissue's stem cell division rate and its probability of developing cancer. The more often that cells divide, the more likely that cancer will develop.

Note importantly that stem cells are transported through the Primo Vascular System. I suggest that the compiled findings of both of these two former articles indicate that two-thirds of cancer cases could be due to bad luck, where the stem cells travelling through the Primo Vascular System (or Bong-Han system) undergo 'random mutations arising during DNA replication in normal, noncancerous stem cells'. It makes sense that the Primo Vascular System is where cancers would develop; as previously shown, the Primo Vascular System has been proven to be the greatest cause of cancer metastasis. So maybe a pathological PVS is the one-stop 'cancer shop'. If that is the case, nutraceuticals that nourish and protect both the stem cells and the Bong-Han system would be effective in preventing cancer.

33.58 Therapeutic Substances that May Stimulate the Release of Stem Cells and Fight Cancer

Aphanizomenon flos-aquae (AFA) is a freshwater species of cyanobacteria (blue-green algae). It is believed to support the release of adult stem cells from bone marrow and likely from the PVS. According to Christian Drapeau (113) from Stemtech, the following nutraceuticals are believed to work synergistically with AFA to increase the release of adult stem cells:

1. *Undaria pinnatifida* is a marine alga which supports the immune system. Also known as Wakame, it is a sea vegetable, or edible seaweed. It has a mildly sweet flavor and is most

Figure 10. Cordyceps (a genus of ascomycete fungi). Photo by bedobedo. © Depositphotos.com.

often presented in soups and salads. In Oriental medicine, it has been traditionally used for blood purification, intestinal strength, skin, hair, reproductive organs, and menstrual regularity. The ingredient fucoidan supports the prolonged increase of circulating stem cells.

2. *Polygonum multiflorum* is an herb known to help cellular rejuvenation and documented to support the release of stem cells from bone marrow. *Polygonum multiflorum* is also known by its synonym *Fallopia multiflora* (Chinese Knotweed). It is an herbaceous perennial vine growing to 2–4 m tall from a woody tuber. It is used in traditional Chinese medicine and believed to have antiaging properties. The Chinese name for the herb is *Fo-Ti*.

3. *Cordyceps Sinensis* is associated with stamina and longevity, and Drapeau (113) believes that when it is mixed with the other ingredients, it synergizes the increased release of stem cells. *Cordyceps sinensis* has been described as a medicine in old Chinese medical books and Tibetan medicine. It is a rare combination of a caterpillar and a fungus and found at altitudes above 4,500 m in Sikkim.

33.59 Different Tissue Types Where the Primo Vascular System Has Been Observed

Concerning the various tissues where the PVS has been found, in the book (7), on page 25, Kwang-Sup Soh reports that the primo vascular system (PVS) has been observed in the 'nerve system, cardio-vascular system, lymphatic system, fascia in the abdominal cavity, adipose tissue, generative system (testis), skin and abdominal wall, primo fluid and microcells, egg vitelline membrane, and cancer'. It thus appears that the PVS is omnipresent within animals and humans.

Concerning the confirmation of the Primo Vascular System being associated intimately with the conception vessel acupuncture meridian, in the book, Kwang-Sup Soh (7), on page 29, reported:

In the midline of the abdominal wall of a rat, there is a band of adipose tissues which we named the conception vessel (CV) fat line. Along this CV fat line, we can see a large vein and artery running from the xiphoid through the navel to the bladder. According to the chart of human acupuncture meridians, there is a CV meridian, and the WHO nomenclature named the acupoints on this meridian as CV 14 at the xiphoid and CV 8 at the navel; other points between these two acupoints are located at equal distances. Primo nodes at CV 12, 10, 8 were observed, and basic histological study with H&E and Mason's trichrome revealed that they were different from lymph nodes. By injecting FNP [fluorescent nanoparticles] into the primo nodes, we traced the flow of nanoparticles along the CV line to the ligament wrapping the bladder in the primo vessels. Thus, we established the presence of extravascular primo vessels along the blood vessels just outside the connective tissues of the blood vessel. In this experiment, the PVS ran along the CV fat line to the bladder.

The PVS is intimately associated with the connective tissues. Note that Kwang-Sup Soh reported above that the extravascular primo vessels running along the outside of the blood vessels were located just outside the connective tissues of the blood vessel. The book (7), on page 31, states there is a *'close relation between the PVS and the fascia'* (emphasis is mine). I propose that the PVS throughout the entire body will be found to be intimately associated with and directly or indirectly connected to the omnipresent Connective-Tissue Metasystem. I propose that this marriage of the Primo Vascular System with the Connective-Tissue Metasystem constitutes the Triple-Energizer Metasystem.

33.60 The Origin of the Triple Burner Is a Fatty Membrane Emerging from the Tie between the Kidneys
In the commentaries on the 31st Difficult Issue on page 354 of Unschuld's (1) translation of the *Nan Ching*, regarding the root and the origin of the Triple Burner, scholar Yeh Lin relates, 'It is a fatty

membrane emerging from the tie between the kidneys.' So what could this 'fatty membrane emerging from the tie between the kidneys' be?

The perinephric fat or perirenal fat is also known as the adipose capsule of the kidney. It is a structure positioned in between the renal fascia and renal capsule and is considered to be a part of the renal capsule. The article titled 'Adipose Capsule of Kidney' (114) states, 'A different structure, the pararenal fat, is the adipose tissue superficial to the renal fascia.' What is so important about this fatty membrane of 'adipose tissue superficial to the renal fascia' membrane?

It is an interesting fact that the kidney actually resembles the shape of the ear and the shape of the fetus and that the ear is a homunculus of the inverted fetus and that the ear is controlled by the Kidneys in TCM theory. The right Kidney sits a little lower than the left kidney to accommodate the size of the liver.

33.61 Biochemical Properties of Kidney Fat

The article titled 'Saturated Fats and the Kidneys' (115) advises that recent research indicates that saturated fat, cholesterol and the omega-3 fatty acids all play important roles in maintaining healthy kidney function. Regarding the fact that the kidneys require a stable amount of fats to act as protective cushioning to prevent damage and as a source of readily available energy, the article stated:

> We know that the kidney fat normally has a higher concentration of the important saturated fatty acids than are found in any of the other fat depots. These saturated fatty acids are myristic acid (the 14-carbon saturate), palmitic acid (the 16-carbon saturate), and stearic acid (the 18-carbon saturate). When we consume various polyunsaturated fatty acids in large amounts, they are incorporated into kidney tissues, usually at the expense of oleic acid, because the normal high level of saturated fatty acids in the kidney fat does not change.

33.62 Culinary Use of Kidney Fat Is Prized by Quality Chefs

The very colorful website article (116) presented much of the following information. In cattle and sheep it is interesting how the deep red kidneys are surrounded by a dense layer of fat, to act as a form of protection for them. The layer of fat peels off very easily from the kidneys. This is called kidney fat, internal fat, or leaf fat and is otherwise known as suet. It is the highest-quality fat to be found on these animals. The kidney fat has a thin covering of clear connective tissue surrounding its entirety. You can readily tell the difference between body fat on an animal and leaf or internal kidney fat, because leaf fat is a hard fat and is more crumbly and does not have blood veins running through it. Suet is extremely prized by chefs for use in traditional puddings because of its high quality. Further, its high smoke-point makes it an ideal natural fat for deep frying and pastry production. In recipes calling for suet, there is simply no substitute fat that may be used.

33.63 The Kidneys Owe Their High Regard in the Bible Partly to the Fact That They Are Embedded in Fat

Regarding the fact that kidney fat of livestock is of such high purity, the term 'fat of the kidneys' was chosen by God as a proverbial term for surpassing excellence. Concerning this matter, the Bible encyclopedia titled *Aid to Bible Understanding* (117), on page 991, states:

> As with all the organs of the body, the kidneys were directly designed by Jehovah God the Creator. (Ps. 139:13) In sacrificial animals, the fat around the kidneys was considered especially choice, and was specifically mentioned as something that was to be made to smoke on the altar along with the kidneys in communion sacrifices (Lev. 3: 10, 11; 9:19, 20), sin offerings (Lev. 4:8, 9; 8:14, 16; 9:10) and guilt offerings, (Lev. 7:1, 4) In the installation of the priesthood the kidneys of the ram of installation were first waved and then burned on the altar. (Ex. 29:22, 24, 25; Lev. 8:25, 27, 28) In this significance of choiceness, Moses spoke of Jehovah as feeding his people Israel with the 'kidney fat of wheat' . . . Deut. 32:14.

33.64 In the Bible the Kidneys Symbolize Deep Human Emotions and Motivation

Due to the fact that human kidneys are located very deep in the body and are mostly inaccessible, in the holy scriptures, the Kidneys became a natural symbol for the most hidden parts of a man. Concerning this matter, the Bible encyclopedia titled *Aid to Bible Understanding* (117), on page 991, states:

> The position of the kidneys deep in the body places them as among the most inaccessible organs. The Bible applies the term as relating to the inmost thoughts and deepest emotions. A wound in the kidneys would be a very deep wound, either literally or figuratively considered. (Job 16:13; Ps. 73:21: Lam. 3:13) Several times kidneys are mentioned in close connection with the heart, which is itself intimately associated with human emotions, such as affection, and motivation. (Jer. 11:20; 20:12) The kidneys are, in fact, effected by deep emotions, according to medical authorities, who say that sustained emotional strain can cause such diseases as diabetes insipidous (not 'sugar diabetes'), in which the kidneys fail to function properly. So the Bible usage of the term is not based on imagination or tradition.
>
> Jehovah knows the makeup of man in the most thorough and intimate manner, therefore He is said to search out and to test out the 'kidneys,' even as his Son also searches the 'inmost thoughts literally, 'kidneys' and hearts.' (Ps. 7:9; Rev. 2:23) Jehovah can 'refine' the kidneys or 'deepest emotions' of a person so that they become right before Him, and are made sensitive to that which is right or wrong. - Ps. 26:2; 16:7; Prov. 23:16; Jer. 12:2.

33.65 How the Kidney's Connection with the 'Gate of Life' Is Described in the Holy Scriptures

With consideration to the Bible's viewpoint on the kidneys in association with the TCM's gate of life, *The Holy Scriptures Jubilee Bible 2000* version (118), at Psalms 139:13–16, states:

13 For thou hast possessed my kidneys: thou hast covered me in my mother's womb. 14 I will praise thee; for I am fearfully and wonderfully made; marvellous are thy works, and that, my soul knows right well. 15 My body was not hid from thee, even though I was made in secret and brought together in the lowest parts of the earth. 16 Thine eyes did see my substance yet being imperfect; and in thy book all my members were written, which were then formed, without lacking one of them.

Interestingly, the Holy Scriptures, when discussing the embryological development within the womb, discusses the 'substance' that was to become a human life form, thanks to the 'book' wherein all the 'members were written'. This obviously refers to the genetic code encrypted in the DNA, which constitute 'the original influences which man has received from father and mother,' as discussed in the eighth Difficult Issue of the *Nan Ching*.

33.66 Is the PVS 'Knitting Machine' Responsible for the Gate of Life?

I found even more fascinating the rendering of many translations of Psalms 139:13 of the Holy Scriptures. The *New International Version Bible* (119) stated, 'For you created my inmost being; you knit me together in my mother's womb.' The *New Living Translation* (120) stated, 'You made all the delicate, inner parts of my body and knit me together in my mother's womb.' *The Bible Revised Standard Version* (121) states, 'For thou didst form my inward parts, thou didst knit me together in my mother's womb.' Remember, I previously discussed the critical initial formation of the PVS during embryogenesis (7), where it is discussed that the 'proliferation of the [PVS] meridians takes place ahead of proliferation of any other organs, such as the blood vessels and the nervous system'. The book said, 'Embryo development follows the following steps: the step for the formation of the primo vessel blast cell occurs 7–8 h after fertilization; the step for primordial primo vessel occurs 10 h after fertilization; the step for the formation of primitive primo lumens occurs 15 h after fertilization; and the final step for the completion of the primo lumens occurs

20–28 h after fertilization'. The quote concluded, 'The fact that the proliferation of the PVS precedes the formation of other structures suggests the PVS plays an important role during development of an organism.' Thus, as stated above, the 'proliferation of the [PVS] meridians takes place ahead of proliferation of any other organs'. What are the PVS meridians made from? The PVS meridians are made from very thin hollow threads. Presumably, the PVS hollow threads appear to be differentiated and processed into an intricate and delicately intertwined structure that allows a unique human being to be manufactured from the donated DNA from the two parents. So when the *New Living Translation* (120) version of Psalms 139:13 stated 'You made all the delicate, inner parts of my body and knit me together in my mother's womb' that sounds like what modern science is corroborating.

33.67 Did an Ancient Cook Serendipitously Discover the PVS along with Its Inner Lubricant?

The term *Yangsheng* is often translated as 'nourishing life'. One of the earliest extant examples of the concept of Yangsheng can be found in the Daoist text attributed to Zhuangzi. In the article titled 'Yangsheng and the Channels', author Jason D. Robertson (44) states:

> A chapter from that text titled 'Managing Life Nourishment' includes the famous story of Prince Wén Huì's cook, in which the cook describes the means by which he has been able to use the same knife for nineteen years to butcher several thousand oxen without sharpening his blade once. Prince Wén Huì asks how might this be done, and in answering, the cook mentions that in recent years when using the knife he is able deal with the carcass 'in a spirit-like manner . . . [I] do not look at it with my eyes'. Instead he moves his knife by 'observing the natural lines, [my knife] slips through the great crevices and slides through the great cavities, taking advantage of the facilities thus presented. My art avoids the membranous ligatures, and much more the great bones'. Prince Wén Huì exclaims, 'Excellent! I have heard the words of my cook,

and learned from them the nourishment of life'. Scholars describe this text as advocating a way of following a path of least resistance—a means of living to one's fullest potential by being in line with the movements of nature.

I tend to believe that this story is based on fact and that the observant cook had serendipitously discovered the natural lines associated with the PVS. Note that the PVS conduits are filled with liquid, including hyaluronic acid, which possesses enhanced lubricity. It makes sense that the fluid-filled PVS would be located within the crevices and cavities that are outside of the solid structures within the body.

33.68 Professor Wáng Jūyì Proposes that Acupuncture Channels Are Located Within the Fluids Moving Within and Around the Connective Tissues

Regarding the possible location of the acupuncture meridian system of the body, in the article titled 'Yangsheng and the Channels', Robertson (44) states:

> Where exactly are the channels located? Expanding upon the assertion in the Inner Classic that the channels are 'subsumed in the spaces around the flesh—deep and not visible', Professor Wáng Jūyì proposes that we should conceive of the channels as being found in the spaces surrounding all of the structures of the body (i.e. not just in the areas where acupuncture points are found but even within and around the organs). When reviewing modern research into possible anatomical correspondences to the channels, we see promising leads by those considering the physiological role of fascia and connective tissue. Nevertheless, strictly speaking we might say that it is not the fascia to which the Inner Classic is referring above, but instead to the quality of the fluids moving within and around these connective tissues. The Inner Classic abounds in metaphors likening qi circulation to water systems. Thus as acupuncturists looking to find the channels, our attention should focus more on the nature

of fluids within the connective tissues and less on the surrounding solid structures.

Based on Professor Wáng Jūyì's theory, note that Robertson suggests that the acupuncture meridians occupy 'the spaces surrounding all of the structures of the body'. More specifically, he singles out 'the fluids moving within and around these connective tissues' and confirms that 'the Inner Classic abounds in metaphors likening qi circulation to water systems'. This is very plausible, remembering that Professor Pollack has proven that water in contact with hydrophilic materials (which includes the collagen of connective tissues) is transformed into 4th phase EZ in situ. Thus Pollack's explanation allows the production of energy (Qi) to occur. Thereafter, the Qi could easily flow throughout the continuous acupuncture meridian system.

33.69 Possible Electrical Charge Transmission through the Liquid of the Primo Vascular System

Regarding the possible role of PVS liquids transferring information instantly throughout the entire body, in the article titled 'Can the Primo Vascular System (Bong Han Duct System) Be a Basic Concept for Qi Production?' in the *International Journal of Integrative Medicine*, the authors (122) state:

> *An additional significant role of the liquid is the dispersion of electrical signs throughout the PVS system*, which could provide a structure-based issue for the well-known phenomenon of low electrical obstructions at AP [Acupuncture Points]. A hypothetical function of the liquid is light circulation, which may explain the almost immediate impressions felt throughout the entire body when some needling is conducted at AP. It has been noticed that *more than 80% of the APs and 50% of the meridian intersections of the arm seemed to coincide with inter- or intramuscular connective tissue planes*. (Emphasis is mine)

While some researchers (84) believe that the PVS is the major substrate for the acupuncture meridian system, there appears to be no doubt that the Connective-Tissue Metasystem is also involved in the relationship. Perhaps because the components of the PVS are so very small in themselves, they require the supportive aspect of the Connective-Tissue Metasystem to perform their task. One could also ask, if the connective-tissue matrix is only present for the support of the PVS, why would it be necessary for the predominant collagen fibers to possess such incredible bioelectrical characteristics, including fiber optic and piezoelectric properties?

33.70 DNA in the PVS Liquid as a Source of Biophotons Which Drive the Acupuncture Meridian System

Concerning the possible role of DNA granules commuting within the PVS liquid producing biophotons, which then 'drive' the acupuncture meridian system and transfer information instantly throughout the entire body, in the article titled 'Can the Primo Vascular System (Bong Han Duct System) Be a Basic Concept for Qi Production?', the authors (122) explain that when an electron has captivated a quantum of energy, it subsequently is raised to a higher level of energy, and the excited atom or molecule emits light. In technical jargon, the authors state:

> Charge separation underlies the primary bioenergetic transduction processes associated with biological membranes, and the formation of excitons and their propagation are involved in energy transduction and in biocommunication. A DNA molecule is an exciplex in which photons are stored and can be a source of biophotons. Exciplex formation in DNA has been shown in various conditions, even at room temperature. Noncoding DNA may act as a photon store and a coherent radiator, because of its enormous polymer size and its ability to form exciplexes.
>
> The resulting long-range electromagnetic waves and fields can be seen as the basis of several activities such as self-organization, mitosis and cell differentiation. The biophysical model for inter-and intra-cellular

communication developed by Nagl and Popp postulated that the biophoton is trapped and emitted by a cellular physical resonance device, namely DNA, which results in biophoton emissions with a high degree of coherence.

To explain biologically this postulation regarding DNA and biophotons, we still need a network or channel. Bonghan's theory is similar to a channel and is just the answer. It provides the channels with the DNA granules running inside, and the channels are spread all over the body, linking the internal organs to the acupuncture points in the skin.

Thus, we can understand the coherence of biophotons, and the regulation mechanism of the body as a 'coherent body'. This concept is the scientific foundation and base of acupuncture therapy and could bring about a new quantum communication paradigm based upon its biological role.

The authors suggest that this original concept is novel and leaves thousands of questions about the PVS and acupuncture meridians unanswered. For example, does the reinforcing or reduction technique of acupuncture have a real effect or a placebo effect? They question if a relationship actually exists between the various acupuncture channels and between the upper portion and the lower portion of the body and the three dimensional makeup of the PVS.

My heart skipped a beat when the authors asked 'What and where is San Jiao?' while discussing the Primo Vascular System. This gave me confidence that my proposal is not too far out in cyberspace.

33.71 The Unification Theory of How Numerous Physical Activities Produce Healing

Regarding the possible mechanism that explains how numerous physical activities produce healing throughout the body, in the article titled 'Can the Primo Vascular System (Bong Han Duct System) be a Basic Concept for Qi Production?', the authors (122) state:

DNA included in the PMC [primo micro cell, sanal or granule] could be the source and storing place of biophotons which could be suggested to be called Qi. Based on this concept, any Qi (Biophotons) producing stimulators, for example, light (Laser or Infra Red ray), movement (massage), heat (Moxabustion) and needling, and so on, could stimulate the production of Qi. By having a wide web network, such as PVS, any point of body, which contains the PMC and DNA in it, could produce Qi (Biophotons), and this Qi could be distributed. If we were to have a map of PVS, this distribution could be controlled and Qi could be directed to target organs, which is the main purpose of acupuncture.

Wow! Isn't that beautiful? One force not mentioned by the authors above is sound. There is no doubt that acoustic frequencies, including music, tuning forks, drums, singing, Tibetan singing bowls, etc. have a healing effect in the right setting. The main point is that the authors propose that energic stimuli (light, heat, sound, friction, acupuncture) stimulate the production of biophotons from DNA present in the abundant primo microcells that course through our body within the Primo Vascular System. The authors propose that these energy balancing, healing, calming biophotons represent what TCM calls Qi.

33.72 Is the PVS the Physical Substrate for the Acupuncture Points and Meridians?

Regarding the very nature of acupuncture points and the network of acupuncture meridians, in the 2013 *Journal of Acupuncture and Meridian Studies* article titled 'The Primo Vascular System as a New Anatomical System', the authors (84) poignantly state:

> *The PVS is furthermore the physical substrate for the acupuncture points and meridians and is involved in the development and the functioning of living organisms.* The primordial PVS is like a matrix for the vascular and the nervous systems, which are formed around the PVS. The

PVS is duplicated by the vascular and the nervous systems during the very early stage of body development, which is the reason the PVS combines the features of the vascular, nervous, and hormonal systems. After all embryonic body systems have been developed, *the primordial PVS subsequently remains connected with these systems, but dominates and controls them because it is the oldest morphological functional system.* The PVS, which until now has been a missing body system, can explain many of the mysteries of life. The physical substrate for the meridian system is the missing point that can be used to combine the knowledge of ancient Chinese medicine and that of modern science into one successful unit. (Emphasis is mine)

Incredibly, the authors state that the Primo Vascular System constitutes the 'physical substrate for the acupuncture points and meridians'. But more noteworthy is the fact that the PVS is 'the oldest morphological functional system'. I think this is extremely funny and ironic. Note that many ancient scholars referred to the Triple Heater as *origin*. For example, in the commentaries on the 66th Difficult Issue on page 569 of Unschuld's (1) translation of the *Nan Ching*, while discussing the yuan points (origin or source points) on the 12 meridians, scholar Hua Shou states, '"Origin" represents an honorable designation for the Triple Burner.' The authors quoted above believe that the PVS is 'the oldest morphological functional system'. I believe that the PVS/Triple Burner complex is the primordial, original organ to be formed during embryogenesis and subsequently deserves the honorable designation of 'origin'. Thus, I propose that the primordial PVS and the associated Connective-Tissue Metasystem constitute the Triple Burner or Triple-Energizer Metasystem. I believe this is ironic. Modern scientific researchers conclude that the PVS is 'the oldest morphological functional system', while many modern acupuncturists unfortunately believe that the Triple Burner (Triple Energizer) 'has a name but no form' or no morphology.

Summary of Chapter 33

I do not believe that the Bonghan vessels (primo vessels) constitute the acupuncture meridian system in its entirety. For a start, the ducts are only 50–100 μm thick—that is, only 0.05–0.1 mm wide. Acupuncture meridians are much wider and more forgiving than that. I propose that the acupuncture meridians directly communicate with and are intimately involved with the Bonghan system which is immersed and floating within a subsystem of the Connective-Tissue Metasystem. I propose that this combination of the Primo Vascular System and the Connective-Tissue Metasystem constitutes the San Jiao (Triple-Energizer Metasystem). I suggest that the Bonghan ducts are an integral component of the Triple-Energizer Metasystem, for example, in their tissue lubricating function, their healing capacity, and their communication facility. I believe that due to their presence on the major organs of the body, the Bonghan ducts act as the mediators between the organs and the acupuncture meridian system. The fact that Primo microcells (P-microcells) flowing in primo vessels are hypothesized to function like very small totipotent embryonic-like adult stem cells could account for the healing component of Triple-Energizer Metasystem. I propose that the P-microcells within the microtubules of the Primo Vascular System constitutes the major component of the metasystem that the ancients called Yuan Qi circulation, constituting 'the original influences which man has received from father and mother', as Li Chiung put it.

Further to this argument, note that totipotent cells can form all the cell types in a body—plus the extraembryonic, or placental, cells. It is obvious that Kim found a new circulatory system organ complex unlike any other system. This novel hollow circulatory system connects the tissues just under the skin to the internal organs and allows fluids (including introduced dyes) to flow from the external skin to the internal organs. This description sounds remarkably like what the acupuncture meridians do. Note that these hollow thread-like ducts also exist 'inside the blood and lymphatic vessels, cerebrospinal fluids of the central nervous system, and on the surface of the various internal organs' (34). It appears that these totipotent cells that circulate inside the hollow thread-like ducts of the primo vascular system (PVS) throughout the entire body have therapeutic effects. Researchers injecting dye at superficial acupuncture points have traced the dye to internal organs and structures. It appears that these hollow thread-like ducts of the primo vascular system (PVS) that are distributed throughout the entire body are somehow intimately connected to the acupuncture meridian system. But they are also associated with the vascular system, the lymphatic system, cerebrospinal fluids of the central nervous system, various internal organs, adipose tissues, and even cancer cells. I suggest that these structures are intimately connected with the Triple-Energizer Metasystem embodied within the Connective-Tissue Metasystem (CTM), and it is through this PVS–CTM system that communication throughout the body takes place. Some researchers believe that PVS components differentiate into the vascular system, the lymphatic system, and the central nervous system, along with the major organs of the body. I believe this to be true and that the PVS is the original organ formed within hours of fertilization.

So here is another previously unknown organ complex that has, until recently, defied detection by modern science in spite of all their sophisticated detection equipment. The book *The Primo Vascular System: Its Role in Cancer and Regeneration* (7) state, 'These research results have also suggested the extensive roles of PVS in humans, potentially changing the entire paradigm of medicine.'

Sperm plus ova comingle, and their DNA differentiates on account of the primordial Primo Vascular System (PVS). Embryological differentiation proceeds via the assistance of the physical connective-tissue infrastructure that supports and interweaves with the hollow thread-like fluid-filled Primo Vascular System containing the circulating P-microcells within for cellular repair and rejuvenation. In later life, these primordial threads exist inside and outside of blood and lymph vessels, on organs and inside organs—for example, floating within the heart chamber. The PVS threads enmesh the body from the superficial skin to the deepest organs. They are held in place, often floating, by the Connective-Tissue Metasystem. The primordial qi of the P-microcells (Sanals) flow throughout the body in the PVS. The book (7) says every tissue component is connected to the PVS. Interestingly, the Connective-Tissue Metasystem also possesses this feature.

I propose that hollow conduit-like structures of collagen contain the numerous fluids of the Triple-Energizer Metasystem. Many researchers believe that acupuncture meridians are located within connective tissues and predominantly collagen. Others believe that the PVS is the substrate for the acupuncture meridian system. The PVS interconnects many diverse systems—including blood vessels, lymph vessels, and the CNS—and communicates with every cell in the body. Just as we know that every cell in the body has blood flow to it through the vascular system in vessels as small as capillaries, so too does the PVS have even more proximity to the individual cells. It moisturizes and lubricates every cell in the body, using fluids from within the hollow thread-like omnistructure. While it is presently generally believed that Blood causes the healing of damaged cells, it appears that the finer vascular PVS system with its sanals is more responsible than blood for self-renewal and rejuvenation in the body and its multifactorial healing symphony. All structural elements of our body are regenerated ceaselessly. The Bong-Han sanal cell cycle appears to be responsible for the regeneration process. Only primo fluid or a synthetic equivalent permits sanals to differentiate into cells. Sanals will not mature into cells in blood, lymph, or interstitial fluid.

Every single tissue in the body is either connective tissue or something other than connective tissue. Connective tissue exists throughout the entire body and enmeshes and covers every tissue type and every organ that is NOT connective tissue. Interestingly, Avijgan and Avijgan (2014) (106) suggest that all human tissues and cells are linked to the PVS. Thus, connective tissue and the PVS appear to directly connect to every human cell in the body. Whether these structures contact bacterial cells of our microbiome has not been determined. This omnipresent connective-tissue matrix makes up the Connective-Tissue Metasystem. Wherever the Connective-Tissue Metasystem exists, so does the Triple-Energizer Metasystem, for they are intimately connected. This agrees exactly with the 31st Difficult Issue of *Nan Ching*, where scholar Huang Wei-san (1), on page 355, said, 'The Triple Burner encloses all the depots and palaces externally. It is a fatty membrane covering the entire physical body from the inside.' Concurring with Huang Wei-san was Li Chiung (1) who said on page 396, 'The Triple Burner represents nothing but membranes attached to the upper, central, and lower opening of the stomach.'

Stefanov et al. (84) state, 'A hypothetical light propagation function of the PVS may explain the instantaneous effect that is felt throughout the entire body on needling acupoints.' They further propose that 'the PVS may be an endocrine organ' as 'the PVS liquid carries adrenalin and noradrenalin'. They state further that 'the putative PVS was clearly developed earlier than the extraembryonic vessels, the heart, and the intramembrane vessels'. The authors (84) further note that 'loose connective tissue is the most distributed tissue in the body and does not exist only in the brain, the penis, and the clitoris'. They advise that the 'PVS is associated with the vessels and the nerves and is abundant in loose connective tissue, fat tissue, serous membranes, and fascias'.

This team of researchers (84) believes that the PVS 'controls the cardiovascular system . . . and the nervous system'. Remember that these authors, and many others, consider the PVS to be intimately connected with and part of the acupuncture meridian system. This means that the acupuncture meridian system controls the cardiovascular system, the nervous system, and subsequently, all the organs of the body. If this is proved to be true, and I believe that it is only a matter of time, all medical, physiology, and anatomy textbooks will have to be rewritten.

Stefanov et al. (84) suggest that 'the ancient vital energy Qi probably is an electromagnetic wave that is transported through the PVS, and the information obtained from that electromagnetic wave may be stored in the DNA of the PVS microcells'. They suggest that 'the PVS is furthermore the physical substrate for the acupuncture points and meridians and is involved in the development and the functioning of living organisms'.

The authors (84) posit that the primordial PVS is like a matrix for the vascular and the nervous systems, which are formed around the PVS. They believe that the primordial PVS is 'the oldest morphological functional system' and that it dominates and controls the vascular, nervous, and hormonal systems. I believe that the PVS/Triple-Burner complex is the primordial, original organ to be formed during embryogenesis and subsequently deserves the honorable designation of 'origin'. Thus, I propose that the PVS and the associated supportive Connective-Tissue Metasystem constitute the Triple Burner or Triple-Energizer Metasystem.

Regarding the liquid that flows through the PVS, in agreement with Stefanov et al. (84), Avijgan and Avijgan (2013) (122) believe that a significant role of the liquid is the dispersion of electrical messages throughout the PVS system and that this might explain the almost immediate impressions felt throughout the entire body when some needling is conducted at acupoints. It has been observed that more than 80% of the acupoints and 50% of the meridian intersections of the arm coincides with inter- or intramuscular connective-tissue planes.

Avijgan and Avijgan (122) suggest that DNA included in the PMC (primo micro cell, sanal, or granule) could be the source and storing place of biophotons, which could be suggested to be called Qi.

The authors (106) postulate that the sanal cells (PMCs-VSELSCs [primo microcells, very small embryonic-like stem cells]) in the PVS of the embryo are probably the precursor and/or transformer of the inner cell mass into epiblast cells, which is the origin of the three embryonic layers. This hypothesis also assumes that, in embryonic life and the developmental period of an embryo, the PVS rules the differentiation of epiblast cells into the three (next) layers ectoderm, mesoderm, and endoderm of an embryo. But because epiblast cells are the source or original cells, the sanal cells, which can be the origin of PMCs (VSELSCs in the PVS) or epiblast cells (in the embryo), are deposited in each organ during gastrulation and organogenesis and may serve as a reserve pool of quiescent PSCs [pluripotent stem cells].

In the commentaries on the 66th Difficult Issue on page 569 of Unschuld's (1) translation of the *Nan Ching*, while discussing the yuan points (origin or source points) on the 12 meridians, scholar Hua Shou states, '"Origin" represents an honorable designation for the Triple Burner.' If Avijgan and Avijgan (106) are correct, then the Primo Vascular System would be conceived from the very first day of embryonic life. That certainly is at the 'origin' of life. As I believe that the PVS and the Triple Burner are intimately connected, this absolutely would explain why the term *origin* does represent 'an honorable designation for the Triple Burner'. I propose that the PVS and the acupuncture meridian system and the Triple Burner are all intimately interconnected.

The authors (106) further hypothesize that 'polar bodies could be the pioneers for the PVS' in that, after fertilization, 'the polar bodies form the primitive PVS'. They suggest that this transformation 'of pre-PVS to proto-PVS to PVS happens in the space between dividing zygote and zona pellucida, during the first week of life and before the zona hatching occurs'. This certainly is at the origin of life, when the life gate is opened. According to this hypothesis, the PVS, which is in the zona pellucida (as the extraembryonic PVS), controls all developmental processes of the products of conception, including trophoblast, inner cell mass differentiation, zona hatching and differentiation. The dominating role of polar bodies for all aspects of embryonic development has been reported in a previous study.

If the PVS actually is a major component of the original San Jiao along with connective tissue, as I postulate, this agrees with TCM theory about San Jiao's function of maintaining harmony throughout all body structures. The authors believe that the PVS 'controls all developmental processes of the products of conception, including trophoblast, inner cell mass differentiation, zona hatching and differentiation'. The San Jiao is believed to be the Fu organ that is responsible for numerous functions of homeostasis throughout the adult body, including energy production, digestion, fluid metabolism and synthesis, immune protection, hormonal balance, temperature regulation, detoxification through urine and feces formation and elimination, nutrient synthesis, etc. It would make sense that the

PVS/San Jiao's function of maintaining harmony throughout all body structures, systems, and organs would continue from embryogenesis throughout the rest of the creature's life.

Even at such an early stage in life development, 'the extraembryonic PVS has been assumed to extend around the embryo'. Here, it surrounds the embryo like an external wall, just as the Three Heaters are described to do to all cells, systems, and organs in the human adult.

In agreement with the above description of omnipresent connective tissue, O'Connell (111) states, 'Connective tissue is a major derivative of the mesodermal germ cell layer in the embryo. As the name implies, it connects with mesodermal, endodermal, and ectodermal derivatives as the differentiation of organs and systems occurs.' She stresses that 'this special property of connective tissue allows it to act as an intermediary both embryologically and in the fully developed individual.'

From the above information, it can be seen that the mesoderm is 'the most important layer of the embryo'. Mesoderm cells 'can migrate easily', are predominantly composed of collagen, contain unspecialized stem cells, and more importantly, 'gives rise to all major organs of the body'. Connective tissue is a major derivative of the mesodermal germ cell layer in the embryo. The mesoderm connects with 'endodermal and ectodermal derivatives as the differentiation of organs and systems occurs'. By means of the mesoderm generating connective tissues, an intricate omnipresent supporting infrastructure is generated that enmeshes every cell, system, and organ within the body. Remember too, that the mesoderm 'has a great responsibility by inducing some other internal organs and structures like the heart, kidney, spleen and connective tissue' (106).

CHAPTER 34

Western Medical Description of Body Cavity Membranes

Introduction
According to Western medical anatomy, throughout the body, there are two dorsal body cavities and six ventral body cavities and subcompartments. The following information by J. Thompson (123) shows the breakdown of the body cavities.

34.1 The Two Anatomical Dorsal Body Cavities
The dorsal body cavity, which houses the central nervous system, is arbitrarily subdivided into two cavities, a cranial cavity containing the brain, and a vertebral cavity containing the spinal cord and the roots of the spinal nerves (123).

1. The cranial cavity is the partially closed, membrane-lined, sterile anatomical space containing the brain.
2. The vertebral cavity is the partially closed, membrane-lined, sterile anatomical space that houses the inferior portion of the central nervous system (i.e. the spinal cord).

34.2 The Six Anatomical Ventral Body Cavities

The ventral body cavity houses various internal organs; its linings are various serous membranes. It is located medially on the anterior of the trunk and housed within the confines of the rib cage and trunk musculature. It is subdivided into (1) a thoracic cavity containing the lungs, heart, and the organs of the mediastinum and (2) an abdominopelvic cavity with two partially separated subcompartments: (a) an abdominal cavity containing the stomach, liver, intestines, and spleen and (b) a pelvic cavity containing some of the reproductive organs, the urinary bladder, and the distal colon. This cavity provides a protected space for those organs (123).

1. The thoracic cavity is the closed, partially membrane-lined, sterile anatomical space that houses the lungs, heart, and the organs of the mediastinum. It is located medially on the anterior of the trunk and housed within the confines of the rib cage; it provides a protected space for those organs.
2. The pleural cavity consists of two closed, membrane-lined, sterile anatomical spaces, which house the right and left lungs respectively; it provides a protected space for the lungs.
3. The pericardial cavity is the closed, membrane-lined, sterile anatomical space that houses the heart. It is located within the inferior portion of the mediastinum on the anterior of the trunk and beneath the sternum within the confines of the rib cage. It provides a protected, lubricated space for the heart to contract.
4. The mediastinum is the closed, sterile anatomical space that is not lined by a serous membrane and houses the heart in its pericardial sac, the great vessels, the trachea, the esophagus, and the thymus; it is located medially between the two pleural cavities to the right and left and between the thoracic vertebral column and the sternum.
5. The abdominal cavity is the partially closed, membrane-lined, sterile anatomical space that houses certain internal organs, the stomach, liver, intestines, and spleen. It is located superiorly within the abdominopelvic cavity, bounded superiorly by the diaphragm muscle and inferiorly by the

pelvic cavity, with which it is continuous. It provides a protected space for those organs.
6. The pelvic cavity is the moderately closed, membrane-lined, sterile anatomical space which contains some of the reproductive organs, the urinary bladder, and the distal colon. It is located inferiorly inside the abdominopelvic cavity, and it is bounded superiorly by the abdominal cavity, with which it is continuous. It is bounded inferiorly by the walls of the pelvic girdle and its musculature, and it provides a protected space for those organs.

Note that all the cavities are closed or partially closed and that they are all membrane-lined or partially membrane-lined. Only the mediastinum is not lined by a serous membrane. Note too that the organ contents of the thoracic cavity, abdominal cavity, and the pelvic cavity closely resemble the traditional organ contents of the Triple Energizer.

34.3 Organ Contents of the Body Cavities

The book *Human Biology* (124), under the heading 'Body Cavities', states:

> The human body is divided into two main cavities: the ventral cavity and the dorsal cavity. Called the *coelom* in early development, the *ventral cavity* later becomes the thoracic, abdominal, and pelvic cavities. The thoracic cavity contains the lungs and the heart. The thoracic cavity is separated from the abdominal cavity by a horizontal muscle called the *diaphragm*. The stomach, liver, spleen, pancreas, gallbladder, and most of the small and large intestines are in the abdominal cavity. The pelvic cavity contains the rectum, the urinary bladder, the internal reproductive organs, and the rest of the small and large intestine. Males have an external extension of the abdominal wall called the *scrotum*, which contains the testes.

34.4 Body Cavity Membranes Secrete Fluids

The book *Human Biology* (124), under the heading 'Body Membranes', states:

> Body membranes line cavities and the internal spaces of organs and tubes that open to the outside. The body membranes are of four types: mucous, serous, and synovial membranes and the meninges.
>
> *Mucous membranes* line the tubes of the digestive, respiratory, urinary, and reproductive systems. They are composed of an epithelium overlying a loose fibrous connective tissue layer. The epithelium contains specialized cells that secrete mucus. This mucus ordinarily protects the body from invasion by bacteria and viruses. Hence, more mucus is secreted and expelled when a person has a cold and has to blow her/his nose. In addition, mucus usually protects the walls of the stomach and small intestine from digestive juices. This protection breaks down when a person develops an ulcer.
>
> *Serous membranes* line and support the lungs, the heart, and the abdominal cavity and its internal organs. They secrete a watery fluid that keeps the membranes lubricated. Serous membranes support the internal organs and compartmentalize the large thoracic and abdominal cavities. Serous membranes have specific names according to their location. The pleurae (sing., *pleura*) line the thoracic cavity and cover the lungs. The pericardium forms the pericardial sac and covers the heart. The peritoneum lines the abdominal cavity and covers its organs. A double layer of peritoneum, called mesentery, supports the abdominal organs and attaches them to the abdominal wall.
>
> *Peritonitis* is a life-threatening infection of the peritoneum.
>
> *Synovial membranes* composed only of loose connective tissue line the cavities of freely movable joints. They secrete synovial fluid into the joint cavity. This fluid lubricates the ends of the bones so that they can move freely. In rheumatoid arthritis, the synovial membrane becomes inflamed and grows thicker, restricting movement.

The *meninges* (sing., meninx) are membranes found within the dorsal cavity. They are composed only of connective tissue and serve as a protective covering for the brain and spinal cord. Meningitis is a life-threatening infection of the meninges.

Summary of Chapter 34

Note that all four different types of body cavity membranes are composed of connective tissues and that, except for the meninges, the other three different types of body cavity membranes secrete specialized, protective, lubricating fluid. However, the meninges do provide a space for the flow of cerebrospinal fluid, which is produced in the choroid plexus of the brain. What has traditional Chinese medicine (TCM) had to say about body cavity membranes? This topic is discussed in the next Chapter.

CHAPTER 35

The Connective Tissues (Membranes) of the Body are the Major Component of the Triple Energizer

Introduction
In Unschuld's (1) translation of the *Nan-ching*, the word *membrane* is mentioned at least 17 times, mostly within the commentaries in association with discussions about the Triple Energizer. It should be noted that what was called membranes as mentioned in the *Nan-ching* would be called connective tissue today. The word *membrane* is also used to describe the Pericardium, aka protector of the heart. In the commentaries of the 25th Difficult Issue on page 312, Li Chiung states, 'The heart-enclosing network constitutes a fine *muscular membrane*, located outside of the firm fat [of the heart]. It resembles silk threads and *is linked to the heart and to the lung*' (emphasis is mine).

The article 'Pericardium' (30) substantiates the comment of Li Chiung above when it states that the 'Pericardium is a tough double layered membrane which covers the heart. The space between these two layers is filled with serous fluid which protects the heart from any kind of external jerk or shock.' The article further states, 'The visceral layer (of the pericardium) extends to the beginning of the great vessels, becoming one with the parietal layer of the serous pericardium. This happens at two areas: where the aorta and pulmonary trunk leave the heart and where the superior vena cava, inferior vena cava and pulmonary veins enter the heart.' Note that these membranes 'are linked to the heart and to the lung'. I have not found any recent medical or anatomical reference to the pericardium such that it does 'resemble silk threads'. It would be amazing if Li Chiung's 'silk threads' were actually a part of the Primo Vascular System thread-like matrix.

In further commentaries of the 25th Difficult Issue in Unschuld's (1) translation of the *Nan-ching*, on page 312, Hsü Ta-ch'un says, '[The text] states that the Triple Burner *has no form*. That cannot be. It states [further] that the hand-heart-master (Xin Zhu, Pericardium) has no form, but such a doctrine definitely does not exist. The heart-master is the network enclosing the heart; it consists of a *fatty membrane* protecting the heart. How could it have no form?' (emphasis is mine).

As Hsü Ta-ch'un points out the obvious (in that the pericardium definitely has an anatomical form and a fixed location around the heart), I believe that the Triple Energizer too has a form, but due to their individually diversified composition, their location is not as anatomically fixed as are the other organs—including, for example, the liver, kidney, spleen. I will further discuss what I believe to be the structure and composition of the Triple Energizer later in this text.

35.1 The Triple Burner Encloses All the Depots and Palaces Externally

In the 31st Difficult Issue in Unschuld's (1) translation of the *Nan-ching*, in the commentaries on page 355, Huang Wei-san said:

> The Triple Burner encloses all the depots and palaces externally. It is a *fatty membrane* covering the entire physical body from the inside. Although *it has no definite form and shape*, it represents a great palace among the six palaces. Hence, the final section [of Difficult Issue 31] states that it has three ruling centers and, in addition, that it has a specific location where it accumulates its influences. (Emphasis is mine)

Huang Wei-san could not have made the matter clearer than when he said that the Triple Energizer 'encloses' (or encases or surrounds or enfolds) *all* the 'depots' and 'palaces', meaning solid and hollow organs externally. He goes on to say further that the Triple Burner is a *'fatty membrane covering the entire physical body from the inside'* (emphasis is mine). No organs just levitate inside of the human body. Every organ is held in place by a network of fascia and fatty membranes that surgeons believe are inert structures that can be simply dissected and discarded in the biohazardous waste bin. Modern anatomy confirms that various types of connective tissue cover every organ and structure inside the body right down to membranes, covering individual cells like bags or packaging wrap.

Regarding the membranes (*huang*) of the body, Maciocia (18) sums up the situation succinctly when he noted that the various organs inside the abdominal cavity are not just suspended in a vacuum and interconnected by acupuncture channels. Rather, the membranes wrap, anchor, and connect the various organs and the Triple Burner causes the continuous movement of Qi in and out of the Membranes. I believe that this is because the membranes are the Triple Burner, by another name.

35.2 The Triple Burner Represents Nothing but Membranes

In the commentaries on the 38th Difficult Issue on page 396 of Unschuld's (1) translation of the *Nan Ching*, Li Chiung states, '*The Triple Burner represents nothing but membranes* attached to the upper, central, and lower opening of the stomach. The Triple Burner has a name, but no real form.' In this text, Li Chiung states very clearly that 'the Triple Burner represents nothing but membranes'. This is what I believe to be the case where the Triple-Energizer Metasystem is a ubiquitous maze of omnipresent membranes or connective tissue that permeates the entire body by wrapping all the numerous organs and structures with connective-tissue fascia like a bag, thereby allowing communication between every cell in the body via the fascia and the lymphatic system that is supported and held in place by the structural framework and scaffolding of the connective-tissue fascia that is omnipresent.

This is also the view of Peluffo (125).

> The connective tissue is a versatile organic system, quite widespread and omnipresent which, being both well-known and well analyzed by modern medicine, interconnects all parts of human body in each level, from the macro to microscopic one and can be found even in the simplest of the organelles, in each cell and in each texture or body framework. Of greater importance for Eastern medicine—both theoretical and practical—are the now known properties these tissues show: to generate energy and to conduct it.

Where Li Chiung stated that the Triple Energizer is 'nothing but membranes attached to the upper, central, and lower opening of the stomach', note that the lesser omentum ascends from the stomach, making a connection with the upper energizer, while the greater omentum descends from the stomach, whereby the connective tissues extend downwards and into the area associated with the lower energizer. This is confirmed by modern anatomists. For example, Robert Acland (126), author of the website AclandAnatomy.com, states, 'Two double-sided sheets of peritoneum, the greater omentum and the lesser omentum, extend from the greater curve and lesser curve of the stomach.'

35.3 Gastrosplenic Ligament Connects the Greater Curvature of Stomach with the Hilum of the Spleen

Note too that, regarding membrane connections, the Stomach is connected to the Spleen via a membrane conduit system. In her journal article, Electra Peluffo (125) states:

> *Su Wen* 29 in its final lines describes the connection between spleen and stomach through tissues and membranes that keep the two organs attached so that body fluids jin ye, can circulate through both of them. And thus, the density of the spleenstomach relationship is highlighted thanks to the membranes that gather them together with the same

density of relationships on earth, so intimate they reach physical contact, and that enables the control of that which flows between those two membranes.

When the Emperor asks *Su Wen* 29 - whether the jin ye actually go through this kind of tissue or membrane to reach the entire body and specially until the viscera, Qi Bo answers that zutaiyin (spleen) runs through the stomach and establishes a shu dependency relationship with the spleen (penetrates it) and a luo relationship with the throat (envelopes it). Thus, taiyin makes the qi circulate in the three yin levels: tai, shao and jueyin.

Spleen and stomach interact so as to share the circulatory function of the yin yang meridians because although it is assured in *Su Wen* 29 that both zang and fu get qi through the yangming (stomach makes jin ye body-fluids circulate) this would not happen without the intervention of the spleen, that makes this simultaneous yangyin movement irrigate the whole body.

Here Peluffo (125) is discussing the gastrosplenic ligament, which is made of peritoneum that connects the greater curvature of stomach with the hilum of the spleen. This shows that fluids travel throughout the body and indeed irrigate the whole body through tissues and membranes that attach organs to other organs and attach organs to body walls. Note in this regard that, in western anatomy, it is widely known that the Spleen is the master of the lymphatic system.

The AnatomyExpert.com webpage (127) states:

> The gastrosplenic ligament extends between the fundus of the stomach and the hilum of the spleen, and is continuous below with the greater omentum. It consists of two layers of peritoneum, between which pass the short gastric arteries of the splenic artery, which run to the fundus of the stomach. The structures in the gastrosplenic ligament

are the short gastric vessels, left gastroepiploic vessels, lymph vessels, and sympathetic nerves.

Thus, this ligament carries the infrastructure (arteries for blood and lymph vessels) that permits the fluid movements between the Stomach and Spleen organs.

35.4 *Nan Ching* States that Body Membranes Produce Fluids and Resemble a Wall

Regarding the fact that the Triple Burner exists and moves external to all the other organs, in the commentaries of the 38th Difficult Issue in Unschuld's (1) translation of the *Nan-ching*, on page 398, Katō Bankei states:

> However, *without its influences all the other palaces could not fulfill their functions of emitting, intake, revolution, and transformation.* It is not a proper palace; hence, *it steams inside of the membrane; it moves in between the palaces and depots. It resembles an external wall.* Hence, it is called "external palace." The *Ling-shu* calls it "solitary palace." That is the same meaning. The commentary by Hua [Shou] stated: "The Triple Burner has a conduit externally, *but it has no form internally.* Hence, it is called "external palace." (Emphasis is mine)

35.5 Modern Research Shows Membranous Retinacula Contain Hyaluronic Acid-Secreting Cells

In the article titled 'The Fascia: The Forgotten Structure', the authors, Carla Stecco et al. (45), on page 133, state:

> Wrist and ankle retinacula also form three distinct layers: *an inner sliding layer, with hyaluronic acid-secreting cells*; a thick middle layer containing collagen bundles, fibroblasts, and interspersed elastin fibers; and an outer layer consisting of loose connective tissue containing vascular channels. A recent work has verified that also the

> *deep fasciae of limbs contain hyaluronic acid-secreting cells.* So, they could be considered as a joint capsule, providing both a smooth gliding surface and mechanical resistance to the transmission of force at a distance. (Emphasis is mine)

Interestingly, the Triple Burner is assigned the role of circulating fluids throughout the entire body. In the reference 'it steams inside of the membrane', Katō Bankei is alluding to the fact that most membranes produce fluids. The book *Human Biology* (124), under the heading 'Body Membranes', describes the four types of body membranes that line body cavities and the internal spaces as mucous, serous, and synovial membranes and the meninges. The article notes that mucous membranes line the tubes of the digestive, respiratory, urinary, and reproductive systems and secrete protective fluid mucus. The article says, 'Serous membranes line and support the lungs, the heart, and the abdominal cavity and its internal organs. They secrete a watery fluid that keeps the membranes lubricated.' The article further states, 'Synovial membranes composed only of loose connective tissue line the cavities of freely movable joints. They secrete synovial fluid into the joint cavity. This fluid lubricates the ends of the bones so that they can move freely.' The meninges are membranes found within the dorsal cavity and are composed only of connective tissue and serve as a protective covering for the brain and spinal cord.

Regarding the nature of the Triple Burner, Katō Bankei states on page 398 that 'it (i.e. the Triple Burner) moves in between the palaces and depots' and that 'it resembles an external wall', which ensures privacy and affords protection from negative influences. The numerous connective-tissue structures absolutely do reside in between all the palaces and depots (hollow organs and solid organs) throughout the three major body cavities as a scaffolding support structure. Likewise, the Triple-Energizer Metasystem in the form of Connective-Tissue Metasystem does wall off all the palaces and depots (hollow organs and solid organs, respectively) throughout the three major body cavities by covering them externally for protection and support.

35.6 Like the Triple Burner, the Fascia Is Described by Modern Researchers as an External Wall

In the article entitled 'The Fascia: The Forgotten Structure', the authors, Carla Stecco et al. (45), on page 131, state:

> The deep fascia is a fibrous membrane forming an intricate network which envelops and separates muscles, forms sheaths for nerves and vessels, strengthens ligaments around joints, and binds all the structures together into a firm compact mass. The deep fasciae envelop all the muscles of the body, but have different features according to region.

The deep fascia is described exactly like the Triple-Energizer Metasystem in that it enmeshes the structures (muscles, organs) like an external wall. Further in the article, Stecco et al. (45), on page 131, state, 'Under the deep fascia, the muscles are free to slide because of their epimysium. Loose connective tissue rich in hyaluronic acid lies between the epimysium and the deep fasciae.'

Regarding properties of connective tissue, Chapter 26, 'The Abdominal Viscera and Peritoneum', of the book *Basic Human Anatomy* (57) states, 'The peritoneum minimizes friction, resists infection, and stores fat. It allows free movement of the abdominal viscera. In response to injury or infection (peritonitis), it *exudes fluid* and cells and tends to *wall off or localize infection*' (emphasis is mine).

In the commentaries on the 31st Difficult Issue on page 354 of Unschuld's (1) translation of the *Nan Ching*, regarding the root and the origin of the Triple Burner, scholar Yeh Lin states, '*It is a fatty membrane* emerging from the tie between the kidneys.' Also in the commentaries on page 355 of the 31st Difficult Issue, Huang Wei-san said, 'The Triple Burner encloses all the depots and palaces externally. *It is a fatty membrane covering the entire physical body from the inside*' (emphasis is mine). Once again, the description of the Triple-Energizer Metasystem matches exactly the description and function of the Connective-Tissue Metasystem with regard to being an external wall and with regard to being a 'fatty membrane' in that it is a membrane that exudes lubricating hyaluronic acid.

35.7 The Triple Burner Metasystem Is Composed of an Omnipresent 'Membrane' Metasystem

In the commentaries on page 438 of the 45th Difficult Issue in Unschuld's (1) translation of the *Nan-ching*, Hsü Ta-ch'un plainly states, '"Outside of the Triple Burner" means "outside of the *Burner's membrane* (chiao-mo)"' (emphasis is mine). This shows that Hsü Ta-ch'un believed that the Triple Burner was actually composed of membrane.

In his article 'The Triple Burner as a System of Cavities and a Three-Fold Division of the Body', Maciocia (18) notes that when the Triple Burner is considered to be a system of body cavities, rather than being an organ, it is rather a system of cavities that exists outside the internal organs or in between them. Citing from the 'Classic of Categories' (Lei Jing 1624), Maciocia reported the statement of Zhang Jing Yue: 'Outside the internal organs and inside the body [i.e. between the skin and the internal organs], wrapping the internal organs like a net, there is a cavity that is a Fu. It has the name of a ditch but the shape of a Fu [Yang organ].' Zhang Jing Yue also said, 'The Internal Organs have substance; the cavities are like a bag that contains that substance'.

Every single tissue in the body is either connective tissue or other-than connective tissue. Connective tissue exists throughout the entire body and enmeshes and covers every tissue type and every organ that is *not* connective tissue. This omnipresent connective-tissue matrix makes up the Connective-Tissue Metasystem. It appears that wherever the Connective-Tissue Metasystem exists, so does the Triple-Energizer Metasystem.

35.8 Membranes Defensively Seal Off 'Accumulations' (Infections)

In the commentaries of the 55th Difficult Issue in Unschuld's (1) translation of the *Nan-ching*, on page 497, regarding differentiation of yin accumulation illnesses in the five depots and yang concentration illnesses in the six palaces, Tamba Genkan states:

> The treatise 'Pai ping shih sheng' of the *Ling-shu* states: 'The first [condition leading to the] generation of accumulations

is such that one gets cold and [accumulations] develop. [The influences move] contrary to their proper direction and form accumulations.' It states further: 'When a depletion evil hits a person, it will be transmitted until it rests outside intestines and stomach and *within the membrane field. It sticks to the vessels. It stays and does not move. It stops and forms an accumulation.*' (Emphasis is mine)

Here Tamba Genkan is discussing 'evil' infections causing 'accumulations' of pus in the abdominal cavity 'outside the intestines and stomach' that 'sticks to the vessels'. Note that the greater omentum is located in the area 'outside intestines and stomach', which truly is a field of membranes, diverse membranes. The article 'Greater Omentum' (128) describes the functions of the greater omentum, which include fat deposition, immune contribution, having milky spots of macrophage collections, infection and wound isolation. Examine what Tamba Genkan said: 'It sticks to the vessels. It stays and does not move. It stops and forms an accumulation.' Note what the article (128) says the omentum does, and I quote, 'It may also physically limit the spread of intraperitoneal infections. *The greater omentum can often be found wrapped around areas of infection and trauma*' (emphasis is mine). How did Tamba Genkan know such intricacy without modern diagnostic scanning instrumentation?

35.9 The Many Remarkable Biological Functions of the Omentum
In the brilliant thought-provoking article 'Abdominal Explorations: The Omentum', Richard Gold (129) explains that while the Omentum itself has a 'startling anonymity' within the East Asian medical community, in traditional East Asian medicine, the abdominal organs are considered to be extremely important regarding physical and mental health. Gold advises that TCM practitioners are actually in an energetic interchange with the Omentum every time that we touch, needle, or perform cupping on an abdomen, and he asserts that our treatment outcome will improve if we more fully understand the Western physiological aspects and the potential energetic outcomes when we acknowledge the morphology and biochemical dynamics of

the Omentum. Gold notes that two Omenta actually exist, with the Greater Omentum hanging like an apron over the intestines and lower abdominal region. It is approximately 35 cm long and 25 cm wide and is a highly vascularized fatty tissue. While the smaller Lesser Omentum is only attached between the stomach and the liver, the Greater Omentum also attaches to the stomach and the liver but also attaches to the small intestine, colon, spleen, diaphragm, portal vein, bile duct, and hepatic artery.

The Omentum assumes the form of a thin, membranous bag that is 'quilted' into individual pockets, which act as a storage reservoir for fatty materials and nutritional elements that are extraneous for the metabolic needs at the time. As nutritional requirements change, these stored nutrients can be released back into the bloodstream for immediate use throughout the body. The author states, 'The Omentum distils an oily vapor that lubricates the surfaces of the abdominal viscera and facilitates the perpetual motion of these organs over each other.' This prevents the formation of fibrosis and lessens adhesion formation. As fats are a major substrate for the formation of hormones, the Omentum's fat absorption storage role makes it a dynamic component of the endocrine system. The Omentum also functions as a temperature regulator and a dynamic shock absorber that protects the abdominal viscera from physical forces and potential injury. But wait, there's more! Due to its bactericidal properties, the Omentum protects the viscera by sealing off infections and inflammatory processes that develop within the abdomen, thereby protecting the viscera. Omental regions called milky spots can generate macrophages, which are specialized immune cells that bolster healing. Even stem cells are present in Omental tissues.

Regarding neural aspects associated with the Omentum, Gold (129) further explains that because the Omentum has a bounteous supply of neurotransmitters (including serotonin) and because it possesses abundant nerve growth factors that cause axon regeneration and initiate the growth of new nerve tissues and because it contains neural circuitry that helps control the varied viscera of the gut, the Omentum is considered to be the 'brain' of the gut. Interestingly, the primary white marrow within our bones is composed of 98% fat. With regard to the Omentum's role in fat absorption, storage, and distribution, Gold speculates that the Omentum may function indirectly as a source of marrow, and pertinently he states, 'The Omentum has been described as the brain of the gut, and in TCM we consider the brain as the sea of marrow. Might the Omentum be viewed as a "sea of marrow" located in the gut?' He further suggests that with consideration to the extensive and diverse neurological properties of the Omentum and the numerous connections to the major organs of the TCM middle jiao and the TCM lower jiao, the Omentum might be operating as an energetic information relay station.

Regarding the omental lymphatic system, the author explains that it is able to neutralize toxic substances, fight infections, and remove metabolic wastes. The Omentum abounds with angiogenic factors that initiate the development of new blood vessels.

Gold posits that because the Omentum recognizes nutritional deficiencies throughout the body and is able to respond by directing energy and stored nutrients as required, the Omentum may engender functional aspects associated with the TCM transportation function of the Spleen. Gold further questions whether the Omentum's contribution to the dynamic neurological communication between the organs can be considered an aspect of the TCM's role of the Liver in maintaining a smooth and harmonious flow of Qi in the abdomen?

Gold concluded his article by asserting that while the Omentum is a largely ignored organ in our body, in reality, it is a physiologically dynamic, vital, and multidimensional organ complex that has multiple biological functions that help to maintain and restore optimal health. Gold believes that TCM practitioners would prosper if they focused their attention on the Omentum during their evaluation and treatment of our clients.

This brilliant article is certainly Oriental food for thought, so consider it 'Chinese takeaway'. It is amazing that all the properties of the omenta described above by Gold (129) are also attributed to the Triple Burner in TCM. I believe that these two systems of membranes that 'steam from within' to produce lubricating oils to protect the solid and hollow organs are intimately associated with and representative of the omnipresent Connective-Tissue Metasystem and subsequently the Triple-Energizer Metasystem (San Jiao).

35.10 Revised Definition for the Term *Fascia*

Regarding the definition of fascia, in the 2014 article 'Fascia & Tensegrity', Myers (110) explains that the term *extracellular matrix* (ECM) is not synonymous with Myer's revised definition of *fascia*. This is because Myer's believes that the term *fascia* should absolutely include fibroblasts and mast cells and many other cells, including osteoblasts, chondroblasts, and osteoclasts, for example, as these cells are intimately involved with the creation, maintenance, and break down of the extracellular matrix. If the human body was placed into a vat of solvent and all the cells were dissolved, the ECM in its singular organic unity would be observed such that ECM + connective tissue cells = fascia. He uses the analogy of an orange, where he likens the rind, pith, and the walls between the sections to the fascia of the body, whereby those components organize the 'juice' into distinct organized partitions.

Similarly, the heart, liver, and spleen etc. are sectioned off into interconnected compartments with the membranes of the Triple-Energizer Metasystem maintaining the organization within the body so that all the individual organs are exactly where anatomists and surgeons expect to find them.

35.11 Ancient Description of the Triple-Energizer Metasystem (Triple Burner) Matches a Modern 2014 Description of the Connective-Tissue Metasystem

Regarding the omnipresence of the collagen component of connective tissue throughout the entire body, in the article 'Fascia & Tensegrity', Myers (110) further explains that if only the collagen network of the body could be seen, we would be able to observe that the bones, cartilage, tendons, and ligaments are all composed of a dense leathery mesh. While the breast, the cheek, and the pancreas are composed of a very loose mesh, each muscle endowed with and enclosed by a looser network that is still structurally robust. Bones have a tough wrapping cover surrounding them. Likewise, each organ would be 'invested and then bagged in a *fascial sac*' (emphasis is mine). Myers notes that only body structures with hollow tubes—including the respiratory, digestive, and lymphatic systems—do not possess the fascial net structure across their hollow openings. He further notes that this integrated, omnipresent bag-like fascial network system would not just act locally but would *respond to and distribute influences throughout the entire body.*

Regarding Myers's 2014 description of the ubiquitous connective-tissue system throughout the entire body, in the commentaries on page 315 for the 25th Difficult Issue of Unschuld's (1) translation of the *Nan Ching*, Ting Chin states, 'The Triple Burner is a large bag supporting [the organism] from outside and holding it inside.' Once again the ancient description of the Triple-Energizer Metasystem, that 'the Triple Burner is a large bag supporting the organism from outside and holding it inside', perfectly matches a modern 2014 description of the Connective-Tissue Metasystem, that 'every organ would be invested and then bagged in a fascial sac'.

35.12 Fascia and Interstitial Fluid Allow for Global Connectivity within the Body

I am a member of the United Acupuncturists Coalition Network blog. A fellow member, Richard A. Freiberg, put it elegantly when he said, 'The connectiveness between 30–60 trillion body cells is beyond comprehension. There are REAL global physical connections . . . 1) via the complete oneness of all fascia in the body and 2) via the oneness of all interstitial fluid. Affect those in a local place and simultaneously/instantaneously they are global.' I absolutely agree with his comment and believe that the fascia complex that is omnipresent throughout the body is the major component of the Triple-Energizer organ complex.

35.13 Structure without Function Is a Corpse; Function without Structure Is a Ghost

Concerning the global nature and vitally important functionality of fascia, in the 2014 article 'Fascia & Tensegrity', Myers (110) explains that the omnipresent composite neuromyofascial web operates throughout our entire body and allows our body to perform optimally from moment to moment as gravity and many other surrounding forces impact us. He succinctly and poetically explains, 'Structure without function is a corpse. But function without structure is a ghost.' The currently held belief that joints perform their assigned role solely due to muscles impacting bones does not satisfactorily explain the intricacy and fluidity of human stability and movement. It is extremely obvious that because fascia responds locally as well as systemically throughout the body, the unappreciated and forgotten fascia plays a key role in the graceful processes and balance associated with human movement. Myers emphasizes that we simply must comprehend and appreciate the 'geometry of tensegrity' as it relates intimately to all movement within and throughout the body. Myers shows here that fascia and connective tissue respond systemically as well as locally throughout the entire body. As seen elsewhere throughout this book, this is also the case for the Triple-Energizer Metasystem.

35.14 The Fascia: The Forgotten Structure that Has Numerous Biological Roles

In the journal article entitled 'The Fascia: The Forgotten Structure', the authors, Carla Stecco et al. (45), note on page 127 that a large amount of interest has recently been shown into research on fascia as witnessed by increasing numbers of PubMed papers on the topic. There has also been a surge of interest confirmed by congresses focusing on the fascia, along with many alternative therapies and practical manual applications involving the fascial system in disease and wellness. Stecco et al. continue:

> It is increasingly evident that the fasciae may play important roles in venous return, dissipation of tensional stress concentrated at the sites of entheses, etiology of pain, interactions among limb muscles and movement

> perception and coordination due to their *unique mechanical properties* and rich innervation. . . . *Recent studies have emphasized the continuity of the fascial system between regions*, leading to presume its role as a *body-wide proprioceptive/communicating organ.* (Emphasis is mine)

Here, Stecco et al. describe fascias, connective-tissue fibers, primarily collagen, which are omnipresent throughout the body as a 'unique' 'body-wide proprioceptive/communicating organ'. The word *unique*, as the term *palace of uniqueness* occurs only two times in Unschuld's (1) translation of the *Nan Ching*. In both cases, it refers to the Triple Burner as the 'palace of uniqueness'. I discuss the communication aspect of the Triple-Energizer Metasystem and the synonymous Connective-Tissue Metasystem in detail elsewhere in the book.

35.15 'Stuckness' in the Fascia Leads to Disease States

In the article entitled 'The Fascia: The Forgotten Structure', the authors, Carla Stecco et al. (45), state on page 132, 'The capacity of the various collagen layers to slide over each other may change in cases of overuse syndrome, trauma or surgery, all possible causes of myofascial pathologies.' Stecco et al. continue and note that, depending on the particular movement, specific muscles are brought into play, and selective portions of the deep fascia are stretched by the action of the specific myofascial expansions. The authors explain that this procession of events can be observed along all the limbs, suggesting that the fasciae perform like a transmission belt in the middle of two adjacent joints and also in between synergic muscle groups, assuring continuous sensitive and directional action throughout the process. They believe that the anatomical basis of myokinetic chains is most likely due to this mechanism.

The Triple Energizer is the mother of all the meridians which flow in part 'along all the limbs' and are likely responsible for the defined meridian pathways, which I propose are within the piezoelectric, membranous fiber optic fascia. Further along in the article, Stecco et al. note that several pathological conditions have myofibroblasts in common. These include 'frozen shoulder', scars, congenital fascial dystrophy, and Dupuytren's contracture. The authors also note that chronic compartmental syndrome is likely due to increased fascial basal tension. This typifies the mechanics of how dysfunction and pathology occurring within the fascia is involved in the development of numerous disease states throughout the entire body. This mechanism is in direct accord with the TCM construct that 'stuckness' throughout the body (often in the fascia) leads to disease states.

35.16 Nerve Endings Inside the Deep Fasciae Allows Communication via the Nervous System

In the article entitled 'The Fascia: The Forgotten Structure' (45), on page 133, state:

> In the last few years, *several studies have demonstrated the presence of many free, encapsulated nerve endings, particularly Ruffini and Pacini corpuscles, inside the deep fasciae*, although differences exist according to the different regions; retinacula seem to be the most, highly innervated structures. Analysis of the relationship between these nerve endings and the surrounding fibrous tissue shows that the corpuscle capsules and free nerve endings are closely connected to the surrounding collagen fibers, indicating that these nerve endings may be stretched, and thus activated, every time the surrounding deep fascia is stretched. (Emphasis is mine)

The presence of nerve endings inside the deep fasciae confirms that the fascia can cause communication throughout the body via the nervous system.

35.17 Body Fluids Travel along Mesentery Membranes

Regarding the method of fluid transportation within the San Jiao, in the article entitled 'San Jiao', the author, Linda Barbour (23), makes an intriguing comment when she notes that Dr Tran believes that the Small Intestine is related to the Kidney by means of mesentery and that the fluids associated with the San Jiao actually travel along this mesentery.

Here, Barbour is quoting Dr Tran, who believes that body fluids travel along mesentery. Recent (2013) advances in gastrointestinal anatomy have demonstrated that the mesenteric organ is actually a single continuous structure that reaches from the duodenojejunal flexure to the level of the distal mesorectum. I too believe that fluids are circulated throughout the entire body by means of the membranous Connective-Tissue Metasystem, which includes mesentery.

Summary of Chapter 35
In Unschuld's (1) translation of the *Nan-ching*, the word *membrane* is mentioned at least 17 times and always in connection with the Triple Energizer. What was called membranes in the *Nan-ching* would be called connective tissue today. Huang Wei-san (1), on page 355, said, 'The Triple Burner encloses all the depots and palaces externally. It is a *fatty membrane* covering the entire physical body from the inside. Although it has no definite form and shape, it represents a great palace among the six palaces' (emphasis is mine). Li Chiung (1), on page 396, states, 'The Triple Burner represents *nothing but membranes* attached to the upper, central, and lower opening of the stomach. The Triple Burner has a name, but no real form.' Regarding the Triple Burner, Katō Bankei (1), on page 398, states, 'It steams inside of the membrane; *it moves in between the palaces and depots. It resembles an external wall*' (emphasis is mine). The numerous connective-tissue structures absolutely do reside in between all the palaces and depots (hollow organs and solid organs respectively) throughout the three major body cavities as a scaffolding support structure. Likewise, the Triple-Energizer Metasystem in the form of the Connective-Tissue Metasystem does wall off all the palaces and depots (hollow organs and solid organs respectively) throughout the three major body cavities by covering them externally for protection and support. In response to injury or infection (peritonitis), membranes *exude fluid* and *wall off or localize infection to prevent germs spreading*. The Triple Burner certainly is a fatty membrane covering the entire physical body from the inside. The description of the Triple-Energizer Metasystem matches exactly the description and function of the Connective-Tissue Metasystem with regard to being

an external wall and with regard to being a 'fatty membrane' in that it is a membrane that exudes lubricating hyaluronic acid.

It is obvious that the Triple-Energizer Metasystem is a ubiquitous maze of omnipresent membranes or connective tissue that permeates the entire body by wrapping all the numerous organs and structures with connective-tissue fascia, like a bag. This further allows communication between every cell in the body via the electrophysiological properties of the fascia. Along with the fluid supply and communication systems provided by the Triple-Energizer Metasystem, the Connective-Tissue Metasystem supplies the omnipresent structural framework and scaffolding for every organ and tissue system within the body that is not itself connective tissue.

Every single tissue in the body is either connective tissue or other-than connective tissue. Connective tissue exists throughout the entire body and enmeshes and covers every tissue type and every organ that is *not* connective tissue. This omnipresent connective-tissue matrix makes up the Connective-Tissue Metasystem. Wherever the Connective-Tissue Metasystem exists, so does the Triple-Energizer Metasystem, for they are intimately related.

CHAPTER 36

The Omnipresence of Connective-Tissue Membranes Throughout the Body

Introduction
The Fascia Research Congress from the 100 Year Perspective of Andrew Taylor Still (48), presented in 2012, confirmed that the continuum of fascial membranes penetrating throughout the entire body allows it to serve as a highly effective body-wide mechanosensitive signaling system. Information established that the connective tissue that surrounds every muscle is not an isolated and independent entity, but rather it is a continuous substance throughout the body. Information revealed throughout the paper include that highly innervated fascia allows adjoining tissues to smoothly glide over each other effortlessly. It was stated (page 1) that 'fascia is intimately involved with respiration and with nourishment of all cells of the body, including those of disease and cancer'.

36.1 Definition, Structure and Function of the Connective-Tissue Metasystem
The article titled 'Connective Tissue' (130) defines connective tissue as follows:

> Connective tissue (CT) is one of the four types of biological tissue that support, connect, or separate different types of

tissues and organs in the body. The other three types are epithelial, muscle, and nervous tissue. Connective tissue is found in between other tissues everywhere in the body, including the nervous system.

All connective tissue apart from blood and lymph consists of three main components: fibers (elastic and collagenous fibers), ground substance and cells. (Not all authorities include blood or lymph as connective tissue.) Blood and lymph lack the fiber component. *All are immersed in the body water.* (Emphasis is mine)

The nine (9) various functions of connective tissue mentioned in the article (130) include 'Storage of energy, Protection of organs, Provision of structural framework for the body, Connection of body tissues, Connection of epithelial tissues to muscle fibres, Supply of hormones all over the body, Nutritional support to epithelium, Site of defence reactions, [and] Repair of body tissues'.

As can be seen from the article above, of the four general classes of animal tissues, connective tissue (CT) is the most diverse in the cellular arrangements that it constructs. CT is 'found everywhere', like a bag or a wall; 'it is located in between other tissues'. All CT 'are immersed in the body water'. All these facts about the Connective-Tissue Metasystem apply equally to the definition and the structure of the Triple-Energizer Metasystem. Likewise, the nine functions attributed to the Connective-Tissue Metasystem mentioned above apply equally to the Triple-Energizer Metasystem. I believe that they are intimately related.

36.2 Fascia (Connective Tissue) Is Ubiquitous as It Intimately Permeates Our Entire Body

To show how fascia intimately permeates our entire body from the fibrin in blood to the coral structure of bone, Myers (110) stresses that as time progresses, enlightened researchers and professionals around the world are agreeing that fascia is far more encompassing than previously thought, and the definition of *fascia* now includes the array of collagenous-based soft tissues throughout the body, also comprising the various cells that construct and uphold the complex network of the extracellular matrix (ECM). Myers notes that the broadened definition now includes the ligaments, tendons, and bursae, along with the muscle-associated fascia, including endomysium, perimysium, and epimysium. Further to this, organ-associated fascia is included, namely, coelomic bags that contain the organs in the peritoneum and the mesentery within your abdominal cavity, the mediastinum, pericardium, and the pleura that contain the chest cavity organs. Also, membranes comprising the dura, pia, and perineuria that mantle the brain, spinal cord, and peripheral nerves are included as 'fascia'. In spite of the more enlightened definition of *fascia*, many researchers still exclude bone and cartilage as fascia. Myers believes that because these harder structures are actually also derived from the fiber, gel, and water composition of the ECM (as are all the other 'fascia'), they too rightfully should be defined as *fascia*.

36.3 The Major Building Component of Omnipresent Connective Tissue Is Water

An interesting comment is made by Myers (110), that water makes up a large part of this omnipresent fascial connective-tissue structure, which makes up a vast array of diverse structural components throughout the body. He says (110), 'You would need a large shopping cart to purchase all the materials you would need to make a body, but *connective tissue manages to build all of them—strings, wires, elastics, sheets, sacs, insulating material, bushings, struts, and springs—your connective tissue cells wrestle all of these from three simple elements: water, gels, and fibers*' (emphasis is mine). Myers also states that the extracellular matrix (ECM) + connective tissue cells = fascia.

Myers shows that water is a major component of the Connective-Tissue Metasystem. Remember that the Triple-Energizer Metasystem also masters the water throughout the entire body. The author goes further to state, 'However you define it, *fascia is everywhere—top to toe, birth to death, micro to macro. . . . Fascia is one network, embryologically and anatomically.* All these different names we give elements within it—this tendon or that ligament—can tend to hide the fact that it is all one connected system' (emphasis is mine).

Regarding the expanded definition of *fascia*, there is an amazing connection with the fact that the Triple Energizer is the activator of the Yuan Qi, and here we see that the author states that the fascial system is intrinsic from *birth to death* and encompasses one network *embryologically and anatomically* into adulthood.

Compare and contrast what Myers said above regarding the Connective-Tissue Metasystem with the following quote attributed to Hua Tuo (11), which appears on page 118 of Heart Master Triple Heater, when he describes the distribution of the Triple Heater. Hua Tuo stated, 'When the triple heater ensures free communication, then there is *free communication internally and externally, left and right, above and below. The whole body is irrigated, harmonised internally and regulated externally, nourished by the left, and maintained by the right, directed from above, propagated from below.* There is nothing greater!' (emphasis is mine). I have no doubt that the TCM Triple-Energizer Metasystem is intimately associated with the Connective-Tissue Metasystem.

36.4 The Triple Energizer and Fascia Truly Permeate Our Entire Being

Myers (110) continues that, just as our circulatory system acts as a chemical regulator throughout the entire body and the nervous system manages timing sequences throughout the entire body, similarly, fascia constitutes the Biomechanical Regulation System throughout the entire body and should be given the respect it deserves and should be researched and treated as an omnipresent system rather than being segregated into a series of individual elements. He stresses that the omnipresent Biomechanical Regulation System, which is the fascial metasystem, should be more correctly defined in modern terms associated with Einstein's theory on relativity, synergetic systems theory, fractal mathematics, and tensegrity geometry associated with biological systems.

In a like manner, the Triple Energizer needs to be seen for what it truly is—an integral system that permeates the entire body. The diagram on Myers's website asks the question 'Are there really 600 muscles?' It pertinently answers with the comment 'Or only one muscle in 600 fascial pockets?'

36.5 Fascia Is the Dense Connective Tissue that Envelopes Every Individual Cell in the Body

In the 2010 article entitled 'Fascia: A Missing Link in Our Understanding of the Pathology of Fibromyalgia', the author, Ginevra L. Liptan (131), states:

> Allopathic medicine has historically regarded fascia as relatively inert. According to a recent article in *Science* magazine 'medical books barely mention fascia and anatomical displays remove it'. However in osteopathic medicine, the fascia has long been recognized as a potential cause of pain and soft-tissue dysfunction. As one osteopath writes 'The whole of OMT [osteopathic manipulative treatment] has been concerned, purposefully or not, with manipulation of the fascia'. Fascia is the dense connective tissue that envelopes muscles grossly, and also surrounds every bundle of muscle fibers and each individual muscle cell. This connective tissue is inextricably linked with the muscle, and is continuous with the tendons and periosteum. The fascia is composed of cells including fibroblasts, macrophages and mast cells and extracellular matrix. The extracellular matrix (ECM) is composed of ground substance and collagen and elastin fibers. Fascia is essentially a dense gel (the ground substance) in which cells and fibers are suspended, giving it colloidal properties. Fascia is richly innervated. A histological study found nerve fibers in all specimens of the deep fascia, including a variety of both free and encapsulated nerve endings, especially Ruffini and Pacini corpuscles. In fact muscle innervation is primarily located in the fascia: consisting of 25 percent stretch receptors

of muscle cells, and 75 percent free nerve endings in intramuscular fascia, and in the walls of blood vessels and tendons. The principal cell of the connective tissue is the fibroblast, which produces the extracellular matrix, in addition to its roles in regulation of inflammation and wound repair. Fibroblast activation is induced by various stimuli that occur with tissue injury.

Summary of Chapter 36

What is Connective Tissue? The terminology *connective tissue* denotes the tissues that surround, support, and protect all the other anatomical structures of the body. Connective tissue is the matrix that attaches and fastens together all of the body's organs and systems while simultaneously providing compartmentalization and differentiation between them. The connective tissue that surrounds every muscle, for example, is not an isolated and independent entity, but rather, it is a continuous substance extending throughout the body. As Myers says, 'You would need a large shopping cart to purchase all the materials you would need to make a body, but connective tissue manages to build all of them—strings, wires, elastics, sheets, sacs, insulating material, bushings, struts, and springs—your connective tissue cells wrestle all these from three simple elements: water, gels, and fibers.'

Fascia, a specific kind of connective tissue, is a strong, continuous sheath that provides structural support for the skeleton and the soft tissues, including muscles, tendons, ligaments, etc. Increasingly in scientific and research circles and professionals worldwide, *fascia* has a wider definition and includes all the collagen-based soft tissues in the body, including the cells that create and maintain that network of extracellular matrix (ECM). Fascia is everywhere—top to toe, birth to death, micro to macro. Fascia is one network embryologically and anatomically. All these different names we give elements within it—this tendon or that ligament—can tend to hide the fact that it is all one connected system. Myers states that the extracellular matrix (ECM) + connective-tissue cells = fascia.

Of the four general classes of animal tissues, connective tissue (CT) is the most diverse in the cellular arrangements that it constructs. CT is 'found everywhere', like a bag or a wall; 'it is located in between other tissues'. All CTs 'are immersed in the body fluids'. All these facts about the Connective-Tissue Metasystem apply equally for the definition and the structure of the Triple-Energizer Metasystem. Likewise, the nine functions attributed to the Connective-Tissue Metasystem above apply equally to the Triple-Energizer Metasystem. I absolutely believe that these two metasystems are substantially one and the same.

CHAPTER 37

Incredible Physical Properties of Connective-Tissue Membranes (Fascia) in the Body

Introduction
Regarding the fact that piezoelectric properties are common in numerous biological materials, in the article entitled 'Research & Massage Therapy, Part 2: Why Does Massage Benefit the Body?', Ross Turchaninov (132) reports that, early in the 1960s, two American scientists, Becker and Basset, conducted a series of brilliant electrophysiological experiments which proved that collagen molecules in bone are mostly responsible for the negative potential piezoelectric property of bone during the bone's deformation.

Turchaninov (132) related that, after their results were published, numerous scientists in different countries started to examine and research the piezoelectric properties of other biological materials, and the results were astonishing. It was found that keratin (1982), elastin of the skin (1967), ligaments (1969), collagen in the tendon (1968), actine and myosin in the skeletal muscles (1970), hyaluronic acid (1975), even DNA molecules (1982), and some individual amino acids (1970) also exhibited piezoelectric properties. All this positive research allowed scientists to conclude that piezoelectricity is an innate property of most, if not all, tissues in the plant and animal kingdoms and that this piezoelectric property of fascia changes mechanical forces applied to the fascia into electric energy.

The *Fascia Research Congress from the 100 Year Perspective of Andrew Taylor Still* (48) presented in 2012 confirmed that:

> Fascia is also capable of transmitting electrical signals throughout the body. One of the main components of fascia is collagen. Collagen has been shown to have semiconductive, piezoelectric and photoconductive properties in vitro. Electronic currents can flow over much greater distances than ionically derived potentials. These electronic currents within connective tissue can be altered by external influences, and cause a physiologic response in neighboring structures.

I will now discuss in detail some of the remarkable electrophysiological properties of connective tissues throughout the body.

37.1 Ten Times More Sensory Nerve Endings Are in Your Fascial Tissues Than in Your Muscles

In the 2011 *IDEA Fitness Journal* article titled 'Fascial Fitness: Training in the Neuromyofascial Web', the author, Thomas Myers (28), proposed that, contrary to popular opinion, most injuries are due to connective-tissue (fascial) damage, not to damage to the muscles themselves. Actually, there are ten times more sensory nerve endings in your fascial tissues than in your muscles, and it is more accurately termed the *neuromyofascial web*.

37.2 Major Reasons the Triple-Energizer Metasystem Is Equivalent to the Connective-Tissue Metasystem

Regarding the description of the omnipresent fascial system throughout the body, Thomas Myers (28) notes:

> Traditional anatomy texts of the muscles and fascia are inaccurate, based on a fundamental misunderstanding of

our movement function—so how can we work with fascia as a whole, as the *'organ system of stability'*? Fascia is much more than 'plastic wrap around the muscles.' Fascia is the organ system of stability and mechano-regulation. Fascia is the Cinderella of body tissues—systematically ignored, dissected out and thrown away in bits. However, *fascia forms the biological container and connector for every organ (including muscles).* In dissection, *fascia is literally a greasy mess* (not at all like what the books show you) and *so variable among individuals that its actual architecture is hard to delineate.* For many reasons, fascia has not been seen as a whole system; therefore we have been ignorant of fascia's overall role in biomechanics. (Emphasis is mine)

Note here that Myers calls the fascial system an *organ system* of 'stability and mechano-regulation'. Myers considers that the omnipresent fascial metasystem is a dedicated organ. The omnipresent Triple-Energizer Metasystem is a dedicated organ system too. Myers says that 'fascia forms the biological container and connector for every organ' and that fascia is 'much more than plastic wrap around the muscles' and other organs. Ting Chin (1), stated (page 315), 'The Triple Burner is a large bag supporting the organism from outside and holding it inside.'

Remarkably, Myers further states, 'Fascia is literally a greasy mess.' In Unschuld's (1) translation of the *Nan-ching*, on page 355 in the commentaries, Huang Wei-san said, 'The Triple Burner . . . is a fatty membrane covering the entire physical body from the inside.'

Regarding the Connective-Tissue Metasystem, Myers reports that fascia is 'so variable among individuals that its actual architecture is hard to delineate'. Myers is stating that because of the extreme variability of the connective-tissue fascial system within individuals, it essentially has no form. So the Connective-Tissue Metasystem definitely exists, and 'it has a name but no form'. Regarding the same issue, in the commentaries on the 25th Difficult Issue on page 312 of Unschuld's (1) translation of the *Nan Ching*, Hsü Ta-ch'un says, 'The text states that the Triple Burner has no form. That cannot be. It states further that the hand-heart-master has no form, but such a doctrine definitely does not exist. *The heart-master is the network enclosing the heart, it consists of a fatty membrane protecting the heart. How could it have no form?*' (emphasis is mine). There is no doubt in my mind that the Triple-Energizer Metasystem is essentially one and the same as the Connective-Tissue Metasystem. I will clarify the actual composition of the Triple-Energizer Metasystem later in the book.

37.3 The Integrating Mechano-Biological Nature of the Ubiquitous Fascial Web Is Being Unraveled

In an article in the *IDEA Fitness Journal* published in 2011, Thomas Myers (28) advises that recent research confirms that the integrating mechano-biological nature of the ubiquitous fascial web is being unravelled. You can get in your vehicle and drive from any house in your country to any other house in your country, using the established road system, because every laneway, roadway, and highway is interconnected. Likewise, your fascia really is one all-embracing net with no separation from the top of your head to the tip of your little toe, from your skin to your deepest inner core, or from birth to death. Every single cell in your body is connected to and reacts to the elasticized tensional milieu of the all-encompassing fascia. If you do something to modify your mechanics, the function of cells will change.

37.4 Your Infinitely Communicative Integrated Fascial Matrix Functions as a Unified System

The flawed paradigm of western medical practice is to petition and 'pigeonhole' individual structures of the human body, including the fascial components, and we subsequently find references to the plantar fascia, Achilles tendon, Iliotibial band, thoracolumbar aponeurosis, nuchal ligament, etc. Myers notes, 'These are just convenient labels for areas within the singular fascial web. They might qualify as ZIP codes, but they are not separate. You can talk about the Atlantic, the Pacific and the Mediterranean oceans, but there is really only one interconnected ocean in the world. Fascia is the same. We talk about individual nerves, but we know the nervous system reacts as a whole.' Your infinitely communicative integrated fascial webbing matrix functions as a unified system. This is also the nature of the Triple-Energizer Metasystem.

37.5 Like Yuan Qi, Our Fascial Web Will Still Be the Same Single Net We Started with

Regarding the longevity of our fascial web, Thomas Myers (28) advises:

> You can tear this net in injury, cut it with a surgeon's scalpel, feed and hydrate it well or clog it with high-fructose corn syrup. No matter how you treat it, it will eventually lose its elasticity. In your eye's lens, for instance, the net stiffens in a very regular way, requiring you to use reading glasses at about age 50. In your skin, the net frays to cause wrinkles. Key elements like hip cartilage may fail you before you die, and need replacement, but when you finally breathe your last breath your fascial web will still be the same single net you started with.

37.6 Allopathic Medical Science Has Not Recognized that the Fascial Connective-Tissue System Is an Organ of Extreme Complexity, Integrity, and Value

In an article in the *IDEA Fitness Journal* published in 2011, Thomas Myers (28) suggests that similar to the lymphatic system, nervous system, and the circulatory system that were created with complex communication, regulatory, and homeostatic mechanisms, it is truly amazing that modern medical science has not recognized that the fascial connective-tissue system is actually an organ of equal complexity, integrity, and value. It remains an unrecognized organ rather like an inconvenient plastic wrap that can be dissected out and discarded in the bin as a worthless waste. During a delicate surgical procedure, would a conscientious surgeon willingly exsanguinate 2 l of blood or discard a piece of healthy vital thyroid, kidney, or heart tissue in the bin?

Further along in the article, Myers (28) suggests, 'Connective tissue includes the blood and blood cells, and other elements not part of the structural net we are examining. Perhaps the closest term would be extra-cellular matrix (ECM), which includes everything in your body that isn't cellular.' Myers notes that the ECM is composed of three main components. The first component is strong pliable 'fibers' composed mainly of collagen, of which there are 12 varieties. Related types of fiber in connective tissue include elastin and reticulin. These fibers separate individual compartments while simultaneously binding them together. The second component of ECM is 'glue', which is a colloidal gel, composed of hyaluronic acid, heparin, and fibronectin. This glue allows for structural changes and makes up the substrate for nerve tissue and epithelial tissues, for example. The third component of ECM is water. The water borders and diffuses into all cells and acts as a medium for nutrient exchange. When mixed with the glue, various materials with differing properties are generated as required, and the water also moisturizes and softens the fibers to keep them malleable.

37.7 Definition of the Term *Fascia* according to Myers

In an article in the *IDEA Fitness Journal* published in 2011, Thomas Myers (28) advises that it must be noted that the term *fascia* also includes fibroblasts and mast cells. These cells form the necessary fibers and glue in the body, which are remodeled as required due to the ongoing requirements of injury, habit, and training. Collagen, elastin, and reticulin fibers are the principal structural elements in the extracellular matrix (ECM). Myers notes, 'Collagen is by far the most common of these, and by far the strongest. This is the white, sinewy stuff in meat. The collagen fiber is a triple helix; if it was half-inch thick, it would be about a yard long and look like an old three-strand rope.' Myers notes that collagen fibers can be organized in systematic, dense, regular directional rows—for example, in tendons or ligaments—or the collagen fibers can be organized in a dense or loose random crisscross fashion, like felt.

37.8 Glycoaminoglycans Are Mucopolysaccharides—Both Are Long Words for Snot

Regarding the properties and hydrophilic nature of mucopolysaccharides, Myers (28) explains that collagen fibers cannot truly stick to each other but need to be glued together with proteins called glycoaminoglycans (GAGs), which are mucopolysaccharides. He jokingly commented that both of these long words are synonymous with snot. Myers further noted that we are held together by the colloidal substance mucus, which possesses a surprising number of properties. The hydrophilic fernlike molecules of mucus are able to open and absorb water, or in the absence of water, they can close and bind themselves together. Depending on the specific chemical makeup, mucus can be thick and sticky to bind layers together, or mucus can be fluid and lubricating to allow layers to slide over one another.

37.9 Fascial Connective Tissue System Is Far More Innervated Than Muscle Tissue

The 2011 article by Myers (28) informs that because the fascial connective-tissue system is far more innervated than muscle, proprioception and kinesthesia are primarily fascial, not muscular. Research confirms that there are 10 times more sensory receptors in your fascial tissues than there are in your muscles. In spite of this known fact, surgeons still continue to regard fascia as inert wrapping fit for the bin. This ignorance reminds me of the quote by Serge Gracovetsky, who concluded, 'Medicine is perhaps the only human activity in which an attractive idea will survive experimental annihilation.' Myers (28) further reports that muscles have spindles that are able to measure the change in muscle length and also the rate of length change occurring in the muscles. These spindles are essentially fascial receptors, and each spindle has about 10 receptors located within the surrounding fascia. These protective receptors can be located in various tissue types, including the surface epimysium, the tendon and the attachment fascia, the ligaments nearby, and the superficial layers.

These protective receptors include the Golgi tendon organs (GTOs), which measure the load on the muscle by gauging the stretch in the fibers. Other protective receptors include the paciniform endings, which measure pressure, and Ruffini endings, which update the central nervous system (CNS) of shear forces in the soft tissues. Also included are the ubiquitous small interstitial nerve endings, which can report on all these receptors and also most probably pain messages too. Myers concludes, 'Think of your body as not having 600+ individual muscles, but one individual muscle divided up into 600+ different fascial pockets.'

37.10 Hydrophilic GAGs Contain Large Amounts of Water and Become Electronegative

In the 2012 book titled *Goldman's Cecil Medicine*, 24th Edition (133), in Chapter 267, entitled 'Connective Tissue Proteins and Macromolecules' under heading 'Proteoglycans', it's stated:

> These matrix macromolecules consist of a protein core to which short oligosaccharides and longer chains of glycosaminoglycans are covalently attached. The large variety of proteoglycans is based on the types and lengths of glycosaminoglycans as well as the sequence and length of the protein core. Chondroitin sulfate, keratan sulfate, heparan sulfate, and dermatan sulfate are four different forms of glycosaminoglycans [GAG's] that consist of repeating disaccharide units. *The sulfates present on the glycosaminoglycans create a highly negatively charged environment that is hydrophilic. Thus, connective tissues that contain large amounts of proteoglycans also contain relatively large amounts of water bound to the proteoglycans.* In addition to water, proteoglycans bind cationic proteins, which, in some cases, include growth factors resulting in a mechanism by which tissues can store growth factors in the matrix.

Over the last few years, several traditional Chinese medicine (TCM) researchers have been investigating the remarkable overlap between the fascial networks along the body and the TCM meridians. It appears that the acupuncture meridian system originates from within the fascial system. The fascial system is uninterrupted from the top of the head to the little toe. The layout of the fascial system explains why pain and dysfunction in one location of the body can be initiated by apparently unrelated fascial tightening in another location of the body.

I have discussed elsewhere that, from Professor Pollack's (27) research, he proved that Hydrophilic Materials + Water + Light = EZ battery energy production. The reference just quoted above shows that the four different forms of glycosaminoglycans are strongly hydrophilic and readily absorb water. With the addition of light in the form of biophotons, fourth phase EZ water should readily be created in situ. Thus, Pollack's mechanism allows the production of energy (Qi) to occur. Thereafter, the Qi could easily flow throughout the acupuncture meridian system. As stated above, it appears that the acupuncture meridian system originates from within the piezoelectric, fiber-optic, membranous fascial system, which would make the fascial system (Connective-Tissue Metasystem) the Mother of all the meridians. This makes sense as the Triple Burner (Triple-Energizer Metasystem) is also referred to as the Mother of the entire acupuncture meridian system. It thus appears that the Connective-Tissue Metasystem is intimately connected to the Triple-Energizer Metasystem.

37.11 Integrins Are Receptors Responsible for Communication via the Connective Tissues

In the 2010 article titled 'Notes on Anatomy and Physiology Function of the Thoracolumbar Fascia Part 2', Dr Bruce McFarlane (134) notes there are two ways of understanding the arrangement of cells. The current medical understanding proposes that the cellular elements are independent of one another and that no relationship exists between the cell and the surrounding connective tissue. The other option is that the *nucleus, cytoskeleton, and cell wall are all physically connected* and that integrins bridge the cell wall and connect the cell with the neighboring connective tissue called the extracellular matrix. Modern science has determined that integrins are receptors responsible for communicating information between a cell and the tissues that surround it, including other cells or the extracellular matrix (ECM).

37.12 Integrins Communicate Biochemical Intracellular and Extracellular Information

In the 2010 article cited above, Dr Bruce McFarlane (134) notes the integrins communicate information regarding the chemical composition of the ECM into the cell, and as such, they are involved in complex message signaling and regulating biological parameters, including cyclical stages, shape, and motility of cells. Typically, these integrin receptors inform a cell of the various molecules present in its environment, and subsequently, the cell responds accordingly. The integrins have a dual-communication functionality. While they communicate vital information about what is happening outside the cell to inside the cell, they also communicate information in the other direction by conversing cellular information to the ECM. Thus, they transduce information from the ECM to the cell as well as reveal the status of the cell to the outside, allowing rapid and flexible responses to the numerous continuous changes in the environment. An example of this swift communication is where platelets initiate blood coagulation when there is a rupture or tear in tissues anywhere in the body.

37.13 The Omnipresence of Fascia throughout the Body

Regarding the omnipresence of connective tissues throughout the entire body, in the web article titled 'More about Connective Tissue and Myofascial Release', Glenda Poletti (135) states:

> Fascia is one of the many types of connective tissue in the body. It is the thin yet very tough tissue that surrounds every muscle, bone, nerve and organ. It is also three dimensional, meaning it infuses every muscle, bone and organ, surrounding each cell in the body. There is just one fascial structure—it is contiguous from head to toe, from just below the skin, where it is called epimysium, to the deepest layer surrounding the spinal cord and brain, called the dura. Where it surrounds bone it is called periosteum. Where it surrounds each individual muscle cell it is called endomysium. *All of these structures, for which medical*

science has different names, are one, uninterrupted structure—the fascia. (Emphasis is mine)

The Triple-Energizer Metasystem (Triple Burner) was described by Ting Chin when he stated, 'The Triple Burner is a large bag supporting the organism from outside and holding it inside.' In the quote above, Poletti (135) succinctly states regarding the Connective-Tissue Metasystem, 'It is also three dimensional, meaning it infuses every muscle, bone and organ, surrounding each cell in the body.'

37.14 The Connection between 'Ground Substance' and Yuan Qi

Regarding the nature and properties of 'ground substance,' which permeates the entire body, in the article titled 'More about Connective Tissue and Myofascial Release', Glenda Poletti (135) explains that a viscous, transparent liquid resembling raw egg whites is present in all connective tissue. This ground substance is manufactured by *fibroblasts, which happen to be amongst the original cells to form in the embryo*. This mildly viscous liquid *ground substance surrounds every cell in the body* and allows nutrients to flow into the cells and toxic wastes to flow out. Further to this, the ground substance allows the transfer of antibodies and white blood cells, hormones, and gases.

Poletti raises several very important issues in her quote above. Firstly, she says, 'Ground substance is found in all connective tissue.' Ground substance is primarily composed of water. Just as the Connective-Tissue Metasystem is intimately composed of and involved with water distribution throughout the entire body, the same can be said for the Triple-Energizer Metasystem, where the *Su-wen* treatise 'Ling lan mi tien lun' states: 'The Triple Burner is the official responsible for the maintenance of the ditches. The waterways emerge from there.'

A second major point made by Poletti is that ground substance 'is produced by fibroblasts, which are among the earliest cells to develop in the embryo'. So the basic building block of connective tissue, ground substance, is produced by the same primordial fibroblasts which are 'among the earliest cells to develop in the embryo'. So here is a major component of the Connective-Tissue Metasystem (i.e. ground substance) present at conception and subsequent embryological development. In TCM, it is stated that Yuan Qi 'is the basic origin of the Triple Burner'. The Triple Burner then distributes Yuan Qi to the 12 conduits and all the organs throughout the body. So these facts indicate that the origin of the Triple-Energizer Metasystem is linked to the origin of the Connective-Tissue Metasystem. I do not believe that is a coincidence.

37.15 The Abundance and Structural Properties of Collagen

Concerning the abundance and structural properties of collagen, which is a primal building block of connective tissue, in the article titled 'More about Connective Tissue and Myofascial Release', Glenda Poletti (135) explains that while the fibroblasts produce the liquid ground substance, they also fabricate the solid protein collagen, which is abundant in all configurations of connective tissue. Collagen endows connective tissue with its structural integrity, pliability, and incredible strength, which can be up to 2,000 psi in some locations. Another remarkable property of collagen fibers involves their highly diversified morphology. Depending on the intrinsic properties required for the specific anatomical location, collagen fibers may be organized into rows, sheets, or blocks. They may be produced into a loose or dense formation and may be synthesized randomly or in well-ordered constructions.

Depending on the combination of a particular arrangement of collagen and elastin pooled with the fluid ground substance, many different structural components can be fabricated as these substances are the main building block ingredients in connective tissue. For example, very flexible elastin is the predominant substance that our earlobes are made of. Several other ingredients can be added to the mix—for example, hyaline for the formation of cartilage or different mineral salts, including calcium, to form much harder bone. For its weight, fascia is extremely strong. In some locations, the fascia can be thinner than nylon stockings, but in other locations depending on the function, it is much thicker. The Iliotibial band along the lateral thigh is quite thick and strong.

37.16 When Piezoelectric Connective Tissue Is Obstructed Disharmony and Illness Occur

Concerning the fact that connective tissue and especially collagen is piezoelectric, the article titled 'More about Connective Tissue and Myofascial Release', Glenda Poletti (135) notes that the energy or life force that flows through our bodies involves piezoelectricity, and she believes that the connective-tissue infrastructure is the substrate of acupuncture meridians. She explains that wherever connective tissue is injured or traumatized, it becomes more dense, thicker, or drier, and the piezoelectric flow is hampered and blocked. She suggests that the strong damaged fascia compresses nerves, blood vessels, organs, and muscles so that healthy cellular activity is interrupted, resulting in dysfunction, which leads to loss of function and pain.

37.17 Memories of Emotional Trauma May Be Entrapped within the Fascial Matrix

Concerning the fact that emotional trauma may be entrapped within the fascial matrix, the article by Glenda Poletti (135) explains that cellular communication throughout the body occurs within the fluid medium of the fascia. She believes that when fascia is injured and compresses, it appears to entrain the emotional response and memory related to the traumatic event. Subsequently, when stiffened fascia is released during a myofascial releasing treatment, the entrapped emotions and memories are liberated, a process she terms myofascial unwinding.

37.18 The Medical Community Does Not Appreciate the Importance of Fascia

In the article titled 'What Is Fascia?', the author, John Traino (136), advises that collagen is a strong substance with a lot of tensile strength and that, subsequently, ligaments and tendons are composed of it. He points out that elastin is reduced and we develop more collagen as we age. Subsequently, our bodies tend to stiffen and weaken, and we are more prone to injuries as we age. Typically, the medical community does not appreciate that fascia is an integral and vital component of the body and treats fascia as nothing more than packing material.

37.19 The Fascial System Is the 'Great Superhighway of the Body'

In the article titled 'What Is Fascia?', the author, John Traino (136), states that fascia acts as a soft skeleton and thus provides support throughout the entire body. Fascia thus determines our bodily shape and the freedom of movement that we are able to express. Traino likens our omnipresent fascial system to a 'Great Superhighway of the Body' because it makes up the omnipresent infrastructure that supports veins, arteries, nerves, and lymph vessels, which all function to supply nutrients to muscles and organs, to flush out metabolic wastes, and to provide communication instantaneously between the brain and every part of the body. He advises that a fascial system that is operating efficiently insures that all physiological systems throughout the body perform optimally, and subsequently, the digestive system, respiratory system, hormonal system, etc. are all dependent on the optimal functionality of the fascial system. This homeostatic function of the fascial system certainly sounds exactly like what the San Jiao functions are according to TCM.

37.20 What Causes Fascia throughout the Body to Become Unhealthy?

In the article titled 'What Is Fascia?', the author, John Traino (136), points out that to understand the pathology of the fascial system, we must realize that fascia is hydrophilic and that it associates with and reacts intimately with water. Traino notes that this invokes another major property of fascia. Fascia is thixotropic. When fascia throughout the body is healthy, it exists in a fluid state known as a *sol*, which is short for the *solvent* state. Traino continues, 'However, fascia can harden and stiffen, developing an unhealthy condition, known as a "gel" state. This property of fascia, its ability to go from a fluid to firm state and back again, is the key component of any thixotropic substance.' Traino likens the fascial system of the body to the Great Superhighway of the Body and notes that when pathology occurs within the fascial system such that a gel state occurs, the nervous system and the circulatory system are affected negatively and the body region affected stiffens and the range of motion is reduced. This causes pain to develop, and further injury becomes more likely to result. Traino advises that stress, trauma, lack of exercise, overexercise, and simple dehydration can cause the fascial system to go from a healthy sol state to an unhealthy gel state. He then assures readers that 'because of fascial thixotropic quality, a healthy fascial system can be restored and maintained'.

Summary of Chapter 37

Since the early 1960s, scientists have confirmed that numerous biological materials possess piezoelectric properties. Piezoelectric biological materials include keratin, elastin of the skin, ligaments, collagen in the tendon, actine and myosin in the skeletal muscles, hyaluronic acid, and even DNA molecules. Even some individual amino acids also exhibit piezoelectric properties. Scientists concluded that piezoelectricity is an innate property of most, if not all, tissues in the plant and animal kingdoms and that this piezoelectric property of fascia changes mechanical forces applied to the fascia into electric energy. The major component of fascia is collagen. Collagen possesses semiconductive, piezoelectric, and photoconductive properties. Research confirms that there are 10 times more sensory receptors in your fascial tissues than there are in your muscles. It is truly amazing that modern medical science has not recognized that the fascial connective-tissue system is actually an organ of equal complexity, integrity, and value as the lymphatic system, nervous system, and the circulatory system. The fascial system is uninterrupted from the top of the head to the little toe. It appears that the acupuncture meridian system originates from within the PVS fascial system. In spite of these known facts, surgeons still continue to regard fascia as inert wrapping material fit for the bin.

Fascia is ubiquitous and omnipresent throughout the body. Likening the fascial system to the established road system, Myers says every cell in the body is connected to every other cell in the body via the Connective-Tissue Metasystem. Like our original quota of Yuan Qi, our fascial web will still be the same single net we started with from birth. Communication is a primary function of the Connective-Tissue Metasystem. Integrins are receptors responsible for communication via the connective tissues. These integrin receptors inform a cell of the various molecules present in its environment, and subsequently, the cell responds accordingly. The integrins have dual communication functionality. While they communicate vital information about what is happening outside the cell to inside the cell, they also communicate information in the other direction by conversing cellular information to the ECM.

Regarding the Connective-Tissue Metasystem, Myers (28) reports that fascia is 'so variable among individuals that its actual architecture is hard to delineate'. Myers is stating that because of the extreme variability of the connective-tissue fascial system within individuals, it essentially has no form. So the Connective-Tissue Metasystem definitely exists, and 'it has a name but no form'. Regarding the same issue, in the commentaries on the 25th Difficult Issue on page 312 of Unschuld's (1) translation of the *Nan Ching*, Hsü Ta-ch'un says 'The text states that the Triple Burner has no form.'

A major component of the Connective-Tissue Metasystem is ground substance. It is present at conception and subsequent embryological development. In TCM, it is stated that Yuan Qi 'is the basic origin of the Triple Burner'. The Triple Burner then distributes Yuan Qi to the 12 conduits and all the organs throughout the body. So these facts indicate that the origin of the Triple-Energizer Metasystem is linked to the origin of the Connective-Tissue Metasystem. I do not believe that is a coincidence.

CHAPTER 38

Ming Men, Yuan Qi and the Origin and Inauguration of the Triple Energizer

Introduction
In the commentaries of the eighth Difficult Issue in Unschuld's (1) translation of the *Nan-ching*, on pages 136–137, regarding the origin of the Triple Burner, Katō Bankei says:

> When the *Nei-ching* states that drinks and food enter the stomach where their essence is transformed into subtle influences, [the movement of which] becomes apparent in the influence-opening, that is correct. However, while all the preceding paragraphs have focused their discussions on (these) influences [that are produced] by the stomach, this difficult issue emphasizes a very different question. *At the very beginning of the fetal [development], the true influences of heaven take their residence in the gate of life mansion between the kidneys. This is called the 'origin of the vital influences.' The mystery on which all beginnings depend, on which all life depends, works from here.* This applies even to [living beings] where one would not expect it—not only to humans! It is the same for all species! As soon as the young come to life they are fed with liquid or solid nourishment which leads to the formation of the depots and palaces, the conduits

> and network [vessels], the four extremities and hundred bones; all this depends on the foundation provided by those [original] influences. *Hence, they are called 'the gate of exhalation and inhalation' or 'the origin of the Triple Burner.'* (Emphasis is mine)

Here, Katō Bankei is saying that 'at the very beginning of the fetal [development]', the origin of the vital influences 'take their residence in the gate of life mansion between the kidneys' and constitutes 'the mystery on which all beginnings depend, on which all life depends'. And what is the origin of the vital influences? Katō Bankei clearly states they are called the gate of exhalation and inhalation or the origin of the Triple Burner. Katō Bankei is here clearly stating that the opening of the symbolic gate of exhalation and inhalation is synchronous with 'the origin of the Triple Burner'. Note that at this early stage of embryological development, the literal Lungs have not even differentiated, and the continuing life force is dependent on the maternal blood supply replete with all the necessary nutrients and oxygen to sustain life. For exhalation and inhalation to occur in a creature, that creature must be alive. When breathing stops, death very quickly follows. Thus, the term *the gate of exhalation and inhalation* is synonymous with the gate of continued breathing, or continued existence as it were, the so-called gate of life. Once the gate of life is opened, the Connective-Tissue Metasystem forms shortly thereafter and orchestrates the cascade of biochemical and morphological changes as cellular differentiation proceeds to produce all the necessary tissues, structures and organs that result in a healthy well-constructed living human life form. This makes sense as the TCM classics confirm that the kidney is responsible for ongoing inhalation. Note, however, that should the delivered newborn not take the first breath (in spite of the Triple Energizer being complete and intact in every way), the child would die, and the Triple Burner would never have been 'born' per se and continue to operate as a functional stand-alone entity free from the prior support and nourishment of the mother. I will now discuss in detail Ming Men, Yuan Qi, and the origin and inauguration of the Triple Energizer.

38.1 Embryologic Plasticity of Connective Tissue Is Maintained throughout Life

In the article entitled 'Bioelectric Responsiveness of Fascia: A Model for Understanding the Effects of Manipulation', the author, Judith A. O'Connell (111), stated:

> *Connective tissue is a major derivative of the mesodermal germ cell layer in the embryo.* As the name implies, it connects with mesodermal, endodermal, and ectodermal derivatives as the differentiation of organs and systems occurs. This special property of connective tissue allows it to act as an intermediary both embryologically and in the fully developed individual. There are four groups of connective tissue: connective tissue proper, cartilage, bone, and blood. Together they form the support structure for homeostasis that allows for normal function in the body. *A special property that connective tissue maintains throughout life is embryologic plasticity.* (Emphasis is mine)

O'Connell explains that 'connective tissue is a major derivative of the mesodermal germ cell layer in the embryo' and that it allows and supports all the required 'differentiation of organs and systems' for successful embryogenesis to occur. The 'embryologic plasticity' required for successful embryonic growth and development continues with us throughout our lifetime.

38.2 Fascia Is Ever Present from Our Embryological Beginnings to the Last breath We Take

In total agreement with what O'Connell stated above, on the web article entitled 'Fascia & Tensegrity', Myers (110) explains that the numerous elements of fascia constitute the biological infrastructure that engenders our very form. He rightly notes that humans are composed of 70 trillion individual cells harmoniously joined by a 3-D network of fascia composed of fibrous, gluey, and wet proteins that ensure that every single cell is maintained in the correct position. He stresses that this remarkable biomechanical regulatory feat of the fascial system is 'highly complex and under-studied', and he states, 'Understanding fascia is essential to the dance between stability and movement—crucial in high performance, central in recovery from injury and disability, and *ever-present in our daily life from our embryological beginnings to the last breath we take*' (emphasis is mine).

38.3 The Fascial System Starts about Two Weeks into Embryological Development

Regarding the origin of the embryological development of the fascial system, in the 2014 article entitled 'Fascia & Tensegrity', Myers (110) states:

> *Our single fascial system starts about 2 weeks into development as a fibrous gel that pervades and surrounds all the cells in the developing embryo. It is progressively folded by gastrulation and the rest of the motions of development into the complex layers of fascia we see in the adult.* In fact there is no discontinuity in the layers of fascia, so 'layers' is a useful but deceptive concept (like 'muscles'). . . . Genetics determine only what proteins are able to be manufactured; the local environment of how you use your body determines how they are arrayed from day to day. (Emphasis is mine)

Note that O'Connell and Myers place fascia at our embryological beginning, at our very origin of life. It is now known that the Primo Vascular System (PVS) is the very first system to develop after fertilization occurs. While the PVS is composed of connective tissue and is probably a specialized subset of the connective-tissue system, the fascial system proper 'starts about 2 weeks into development as a fibrous gel that pervades and surrounds all the cells in the developing embryo'. Note that the fascial system, or what I call the Connective-Tissue Metasystem, 'pervades and surrounds all the cells in the developing embryo' just as the Triple-Energizer Metasystem 'pervades and surrounds all the' organs and systems in the adult body.

38.4 Fascia Functions by Secreting and Excreting fluids—Vital and Destructive

Andrew Taylor Still, MD (1828–1917), an American physician, was the founder of osteopathic medicine. I believe the man was a sage. Regarding the embryologic plasticity of fascia, I will cite verbatim what Still (48) (page 4) said 115 years ago:

> Fascia functions by 'secreting and excreting fluid vital and destructive. By its action we live, and by its failure we shrink, swell, and die. . . . This connecting substance must be free at all parts to receive and discharge all fluids, if healthy to appropriate and use in sustaining animal life, and eject all impurities that health may not be impaired by the dead and poisoning fluids.'

Regarding the omnipresence of fascia, Still (48) (page 5) further stated:

> [Fascia] is almost a network of nerves, cells and *tubes*, running to and from it; it is crossed and filled with, no doubt, millions of nerve centers and fibers. . . . Its nerves are so abundant that no atom of flesh fails to get nerve and fluid supply therefrom. . . . *The cord throws out and supplies millions of nerves by which all organs and parts are supplied with the elements of motion, all to go and terminate in that great system, the fascia.* (Emphasis is mine)

In the first citation above, Still advises that the fascia control fluids in the body and that if the fascia fails, then we shrink, swell, or die. This indeed sounds exactly like a description of the Triple Energizer. In the second citation above, Still describes fascia as a 'network of nerves, cells and tubes, running to and from it' such that 'no atom of flesh fails to get nerve and fluid supply therefrom'. Once again, these properties of fascia (communication via the associated nerve connections and control over fluids) are identical to those of the Triple Energizer. Note that Still also stated that the fascia is a network of tubes. I believe that it is most likely that these 'tubes' also include the Primo Vascular System network of microtubules, a microsystem that was not known about in Still's time.

38.5 The Triple Heater Is the Fu of the Fire in the Middle of the Water

Regarding the properties of the Triple Heater, in the book *Heart Master Triple Heater*, Rochat de la Vallée (11), on pages 58–59, quotes Zhang Jiebin, who was a renowned doctor from the early seventeenth century. Zhang Jiebin wrote that while the Triple Heater is the *fu* of all the drainage and irrigation associated with the middle, it also 'gathers together and protects all the yang'. Regarding the Triple Heater, she then states, 'It does not only have the *regulation of fluids* under its command, but also the *regulation of the yang*, and it has as its title the *fire minister*. It is also the *fu of the fire in the middle of the water*, which is a way of indicating that the *fire is within the kidneys*, the fire of ming men between the two kidneys.' She explains that according to Zhang Jiebin, during its descending movements, the Triple Heater possesses a yin power which is reliant on the Bladder due to its relationship with the water of the Kidneys. This accords the Lower Heater the functionality of an irrigation canal. In the upper reaches of the body, the Triple Heater is yang in nature, and it unites with the enclosing network of the Heart (*bao luo*) and thus ensures free communication with the fire of the heart. The text concludes: 'The triple heater, its upper limit and its extreme point below, make it similar to the *six reunions or junctions, liu he, of the universe*, and there is nothing that it does not envelop or surround' (emphasis is mine).

This description of the Triple Burner, that 'there is nothing that it does not envelop or surround', describes the exact nature of fascia in that fascia surrounds every single cell and tissue in the body at the microscopic and macroscopic level. That the Triple Burner is 'also the fu of the fire in the middle of the water', I suggest, pertains to the fact that bulk water in touch with the omnipresent hydrophilic collagen of connective tissue generates the energized living liquid crystalline EZ water that generates the fire or energy-yielding battery that is EZ water. I believe it is this spark-of-life energy coming from the EZ water that accompanies the origin of life and sustains the subsequent life form. I discuss this topic in great detail elsewhere in this book.

Regarding 'the six reunions or junctions, liu he, of the universe', in the above paragraph, Rochat de la Vallée advises that the character for reunion, *he*, has the mouth radical below to suggest the blowing of breath. The six junctions are above and below the four directions. They are the manner in which the influences of heaven and earth intermingle their qualities and join their qi together to form the space in which life can develop. Regarding the formation of Adam in Genesis, the holy scriptures at Genesis 2:7 state that God 'breathed into his nostrils the breath of life and man *became* a living soul'. Note that Adam would have been completely formed prior to that life-initiating inhalation.

38.6 WHO Definition of *Yuan Qi*

The *WHO International Standard Terminologies on Traditional Medicine in the Western Pacific Region* (3) defines Yuan Qi as 'the original substance responsible for construction of the body'. Depending on which university you attended and/or the instructor that taught you, in TCM, Yuan Qi is also known as Ancestral Qi, Congenital Qi, Innate Qi, Prenatal Qi, Original Qi, Source Qi, Primordial Qi, Inborn Qi, or Genuine Qi. In Japanese, it is known as *genki*. It is the most important and fundamental of all qi forms. I personally still prefer to call it Yuan Qi.

38.7 The Kidneys Act as Dual Batteries—One Nonrechargeable and the Other Rechargeable

In the 2012 article by Dr Harreson Caldwell (137) titled 'Did You Know Your Kidneys Work like Batteries in Your Body?', he likens the Kidneys to a dual battery, where one battery is nonrechargeable and the other battery is rechargeable. He explains that the kidneys store energy just as a regular battery does and that the kidneys have two divisions for the storage of energy. He advises that our Congenital Qi is derived from our parents and that it is deposited in one section of our kidneys and is likened to the nonrechargeable component in the battery. This Qi form must be cherished and preserved as it is not rechargeable. He further advises that our Acquired Qi is another Qi form that is housed in our kidneys and that it is acquired from outside our body constantly via our life processes—for example, from our diet, from the air that we breathe, and from exposure to sunlight. This section of our Kidney energy store is the 'rechargeable battery'. The kidneys act as an energy storehouse for all our Qi, both Congenital Qi and Acquired Qi, and every single organ depends on this vital energy to perform their numerous life-sustaining functions. Caldwell gives the example that the Kidneys supply the essential energy for the Heart to be able to pump blood throughout the entire body. I thought it very noteworthy that the author states that 'the energy we acquire from the external' includes sunlight. I agree with him implicitly.

38.8 The Triple Burner Is the Activator and Circulator of the Yuan Qi

In traditional Chinese medicine contexts and in Chinese culture, yuan qi (元氣) is a description of innate or prenatal qi to distinguish it from acquired qi that a person develops over their lifetime. Thanks to the authoritative research by an international body of experts, the *WHO International Standard Terminologies on Traditional Medicine in the Western Pacific Region* (3), Yuan Qi is defined as 'the original substance responsible for construction of the body and generation of offspring, often referring to the reproductive essence, also called prenatal essence'.

The Small Intestine and Large Intestine organs house contents therein that are certainly 'burned and scorched' by the action of digestive processes and enzymes acting on the assorted contents of edible components that have been swallowed by the individual. All this biological activity is certainly exothermic and hence also warrants the term *Burner*.

The Yuan Qi is the vibrant life force that activates the diverse necessary biochemical activity of all the internal organs and is the genesis of activity and vitality. Yuan Qi circulates throughout the body to all cells, relying on the transporting system of the San Jiao (Triple Burner). It dwells between the two Kidneys, at the Gate of Vitality (Ming Men). Yuan Qi expedites the metamorphosis of Qi. Yuan Qi is the conductor of transformation, transmuting Zong Qi into Zhen Qi. Yuan Qi is integral in the manufacture of blood by enabling the transformation of *Gu Qi* into Blood.

38.9 The Triple Burner Is a Large Body-Wide Organ Complex
In his 2011 blog at Blogspot entitled 'The Triple Burner' Giovanni Maciocia (24) states:

> Chapter 38 of the *Nan Jing* reiterates the relationship between the Triple Burner and the Yuan Qi. It says: 'How come there are 5 Zang but 6 Fu? There are 6 Fu *because of the Triple Burner which stems from the Yuan Qi*. The Triple Burner governs all Qi in the body, it has a 'name but no form', it belongs to Hand Shao Yang, it is an 'external Fu' [or 'extra Fu']. That is why there are 5 Zang but 6 Fu.' (Emphasis is mine)

He goes on to relate that this Chapter essentially describes four important facets of the Triple Burner. He notes first that the Triple Burner is a Fu organ. He continues, 'Secondly, it stems from the Yuan Qi; thirdly, it governs all Qi of the body; fourthly, it has a "name but no form", i.e. it is a function rather than an organ (which actually contradicts the first point).'

However, I do not think there is a contradiction. I believe the Triple Energizer is an organ complex but not as we normally expect an organ to appear with defined morphology and a specific fixed anatomical location in the body. Take, for example, the skin. It is a very important organ. How would you define its location? What about the immune system of the body? What is its form? They are both extremely important organs which are not delegated to a fixed local abode, as is the left kidney, the stomach, or the spleen, for example. Are the skin and immune system any less organs because of their diversity, dispersed actual location, and large surface area in the case of the skin? No!

38.10 Which Anatomical Structure Constitutes the Root of the Triple Burner in/near the Kidney?

According to a quote from a brilliant website (138) where the author is anonymous, Tang Zonghai wrote in his somewhat palindromic work *A Refined Interpretation of the Medical Classics (Yijing Jingyi)*, 'The *root of the triple burner is in the kidney*, more precisely right between the two anatomical kidneys. Right there is a greasy membrane that is connected with the spine. It is called mingmen, and constitutes the *source of the three burners*' (emphasis is mine).

I suggest that this structure of 'greasy membrane' between the kidneys is an already known anatomical structure possessing an as-yet-unrecognized dynamic piezoelectric property—e.g. thoracolumbar fascia or perinephric fat(?)—that connects to the neural pathways where it is 'connected with the spine' and subsequently the brain. The significance of this structure will become obvious when I discuss the understated significance of the membranous neurofascial system that covers and interconnects with and communicates with every single cell in the body. As I believe that the Triple Burner is predominantly the Connective-Tissue Metasystem and as 'the root of the triple burner is in the kidney' (more precisely, medial to the right Kidney), it would make sense that this Kidney-engendering Triple Burner 'root', or origin, would be a greasy membranous connective-tissue structure medial to the right Kidney, proximal to the level of L2–L3. Time and a vigilant anatomist or physiobiologist will tell.

38.11 The Ming Men (Gate of Life) Has Many Other Names Due to Its Functions

In the article titled 'Ming-Men: An Acupressure Point with Power-Full Implications', the author (139) states, 'The Ming-men refers to an energy center located in the lower torso that is so power-full that it is associated with three acupuncture points (Governing Vessel 4 (Gv4), Conception Vessel 4 (Cv4) and Conception Vessel 5 (Cv5).' The author continues, 'The Chinese admiration for the Ming-men is captured in the disagreement on how to describe it. It has been called the Gate of Power, Proclamation Gate, Gate of Destiny, and Gate of Life. No matter how you define it, the healing ability and energy properties of the Mingmen abound.'

38.12 The 'Gate of Life' Is Actually in between the Two Kidneys

In the commentaries on the 39th Difficult Issue on page 402 of Unschuld's (1) translation of the *Nan Ching*, Ting Chin states, 'Obviously, Yüeh-jen linked the designation gate of life to the right kidney. *But the location of the gate of life is, in fact, in-between the two kidneys.* If that were not so, how could one say that it stores the essence and holds the womb? How could one say that its influences are identical with [those of] the kidney? I suspect the designation "gate of life" has been confused with the depot of the hand-heart-master, [i.e., the heart-] enclosing network. Hence, the text further down says: "The palace of the Triple Burner does not belong to the five depots." That makes it quite clear that it is associated with the depot of the [heart-] enclosing network. "Its influences are identical with [those of] the kidney" points out that the influences of the gate of life and of the right kidney are identical. All this becomes obvious by itself if one but carefully reads [the text]' (emphasis is mine).

38.13 The Connection of Ming men (Gate of Life) and Embryology

In answer to the question 'Can you say more about ming men?', on page 38 of the book *Heart Master Triple Heater*, Elisabeth Rochat de la Vallée (11) explains that ming men is the original, frontal, embryonic aspect, which cannot be expressed by itself. She explains that if ming men exists, then life exists, and subsequently, this 'fire of life' engenders the integral substance for the power of the Heart. Regarding how the transition is made in children, on page 39, Claude Larre (11) suggests that 'ming men is some sort of first condition of the embryo' and the subsequent further differentiation of all the constituents that engender a human being.

So these two learned scholars believe that the ming men or gate of life is the original primitive life force that initiates the beginning of life, or as Larre stated, the ming men is 'some sort of first condition of the embryo'. Regarding additional development of the embryo, Larre further explained that a separation occurs between the mother's heart and the child's heart. In the uterus, the baby is totally dependent on the heart of the mother to pump and supply life-sustaining oxygenated blood and nutrients to maintain life until birth occurs. At separation from the mother at birth, the neonate must initially sustain its own life by inhaling the first breath of life. That first breath sets in place the breathing cycle that will allow life to continue for that individual.

38.14 The 'Gate of Life' Is the Root of the Triple Burner

In the 36th Difficult Issue in the commentaries on page 385 of Unschuld's (1) translation of the *Nan Ching*, Yeh Lin states, 'The kidney has two lobes; one on the left and one on the right side. One masters the water; one masters the fire. They correspond to the mechanics of rise and fall. *The gate of life is the root of the Triple Burner* and the sea of the original influences of the 12 conduits; *it is the utensil that stores and transforms the essence*, and it is the place to which the womb, which conceives the embryo, is tied. Thus, it is the origin of man's life. Hence, it is called gate of life' (emphasis is mine).

The Gate of Life is not the ova in women and not the sperm in men. In women, every single month, ova are released, and no gate of life is opened such that an embryo forms. Subsequently, the unfertilized ova are broken down along with the endometrium that did not have to support a fertilized ovum. Similarly for men, sperm are regularly released and fail to fertilize an egg to generate a new life form. Note too that the gate of life is not the womb, but the gate of life is obviously tied to the womb when successful fertilization results in pregnancy.

When a successful fertilization does occur, the gate of life is opened, and the Prenatal Jing from the parents is passed to the embryo and nourishes the embryo and fetus during the pregnancy. There is also supplemental energy that is derived from the Kidneys of the mother during the pregnancy. The Prenatal Jing defines the basic constitution, strength, and vitality of the fetus. Prenatal Jing is fixed in quantity and is already determined at birth. Prenatal Jing cannot be supplemented or increased but can be conserved and should not be squandered or wasted. It is stored in the Kidneys (140).

I have discussed in great detail elsewhere in this book what I believe happens when the gate of life is opened. Briefly, the Primo Vascular System initiates the primordial differentiation process within hours of fertilization. By two weeks postfertilization, the Connective-Tissue Metasystem proper allows the tissues to differentiate into organs and systems. Regarding the nature and properties of 'ground substance', which permeates the entire body, in the article titled 'More about Connective Tissue and Myofascial Release', Glenda Poletti (135) advises that ground substance is found in all connective tissue and that it is produced by fibroblasts, which are among the earliest cells to develop in the embryo and that these fluid ground substances are the immediate environment of every cell in the body.

So the basic building block of connective tissue, ground substance, is produced by the same primordial fibroblasts which are 'among the earliest cells to develop in the embryo'. So here is a major component of the Connective-Tissue Metasystem (i.e. ground substance) present at conception and subsequent embryological development. Thus, the primordial fibroblasts generate the 3-D matrix in which all the organs and systems are suspended, surrounded by the Connective-Tissue Metasystem. I propose that the Connective-Tissue Metasystem, in amalgamation with the PVS, is one and the same as the Triple-Energizer Metasystem. Thus, simply stated, at conception, when the gate of life is opened, the primordial fibroblasts generate the Connective-Tissue Metasystem, which I believe is closely associated with the Triple Burner. Thus, it can be stated that the gate of life is the root of the Triple Burner.

38.15 The Triple Heater Has a Very Close Relationship with Ming Men

With reference to the relationship between the Triple Heater and ming men, in the book *Heart Master Triple Heater*, Rochat de la Vallée (11), on page 50, states, 'We will see, particularly in the *Ling shu*, that *the triple heater is rooted at the deepest level of being*. The *Nan jing* tells us that the triple heater has a very close relationship with ming men, that it develops throughout the whole body and that its influence extends right up to the outer layers of the skin' (emphasis is mine). She further states, 'We will also see that it represents the *mixing of water and fire*, allowing the *development of all life*' (emphasis is mine).

Thus, 'the triple heater is rooted at the deepest level of being' because the Triple Heater was formed at the beginning of life. In the presence of all-pervading, hydrophilic connective tissue, bulk water is converted into the fiery energy–supplying battery that is life-sustaining liquid crystalline EZ water, which flows within and throughout the Connective-Tissue Metasystem. This phenomenon accounts for 'the mixing of water and fire, allowing the development of all life'.

38.16 The Right Kidney Is Ming Men, Which Is the Power of Fire at the Level of the Kidneys

Regarding the relationship of the Triple Burner and ming men further, in the book *Heart Master Triple Heater*, Rochat de la Vallée (11), on page 57, states, 'So the triple heater as a fu is connected to the kidneys because they have a double storage, a double power which is like a couple. Commentators are sometimes more precise about this. They say that the left kidney is joined to the function of the bladder and the right kidney is linked to the triple heater.' She continues, 'We know from *Nan jing* 36 that the left kidney represents the kidneys as the power of water and the function of the bladder, which is in charge of all the liquids in the body, and the right kidney is ming men, which is the power of fire at the level of the kidneys. And the link between the triple heater and the fire of ming men is repeated in many texts.' Regarding the two former citations, I have discussed this relationship in depth under the previous heading.

38.17 The Triple Burner Engenders the Blood, and Ming Men Represents the Fundamental Unity of a New Living Being

Regarding the property of the Triple Burner in connection with blood, in the book *Heart Master Triple Heater*, Rochat de la Vallée (11), on page 54, states, 'And if the triple heater is really a very primitive intermediary between this being that does not yet have a form and the being which comes about and develops little by little, then it must have a certain influence on the quantity and quality of the blood.' This prompted the question 'In the embryo which comes first, *the organs or the heater*?' (emphasis is mine). She responds with the following answer:

> [i]t is really quite impossible to say, but we can speculate on what is one and two and three and four and five! One may be ming men, the representation of the fundamental unity of a new living being. Two could be two kidneys, yin yang, fire and water, heaven and earth. We know that ming men is between the two kidneys, between the power of the fire and the water and between yin and yang, two could be the primitive expression in an human being of this duality. Three could be the triple heater, this exchange between the two poles, between yin and yang, fire and water and so on, this exchange for communication and transformation. . . . We can continue like this, four limbs which are the extension of the form of the body, but there are also the four seas inside the body, or the first of the extraordinary meridians, du mai, ren mai, chong mai and dai mai. There are five zang and six fu, and six extraordinary Fu, and seven orifices. But all this is just speculation through numerology. The problem is that we have no Chinese texts which can answer occidental questions on embryology.

Rochat de la Vallée's answer is very poetic from a numerology point of view, but it is not plausible. How could the kidneys or any other organ possibly have come first if it did not have a supporting anchoring infrastructure to secure it in position and thus allow other organs and systems to be appropriately juxtaposed? Before windows and doors can be installed in a room, a solid foundation and appropriate infrastructure must be in place so that the door and window can be correctly fabricated in the right place and stay put. Likewise, a solid 3-D foundation and appropriate infrastructure obviously would have to be in place so that, at a later time, the organs and systems and structures could be fabricated by cellular differentiation. As I have discussed in several Chapters throughout this book, I believe that this 3-D foundation and appropriate infrastructure is produced by primordial fibroblasts and the PVS which engender the Connective-Tissue Metasystem, which is intimately associated with the Triple Heater, aka the Triple-Energizer Metasystem.

Pertinently, Rochat de la Vallée stated, 'If the triple heater is really a very primitive intermediary between this being that does not yet have a form and the being which comes about and develops little by little, then it must have a certain influence on the quantity and quality of the blood.' Note that all blood cells are made in the bone marrow which is the 'soft, flexible *connective tissue* within bone cavities'. So once again, specialized connective tissue is the exact location of the blood-manufacturing function attributed to the Triple Energizer.

38.18 The Fire of Ming Men Begets the Triple Burner to Engender Breath and Life

In the book *Heart Master Triple Heater*, regarding the properties of the Triple Burner, Claude Larre (11), on page 59, states, 'Heaven penetrates earth, it has to be outside but inside too. *When something is a living being then there is breath*, and the qi of earth is caused by the pressure of heaven upon earth.' (Emphasis is mine). In the chapter 2 of the same book (pages 60–61), regarding the interpretation of Ling shu by Zhang Jiebin, Rochat de la Vallée (11) states, 'Minister fire is linked with ming men and with the triple heater, and with all the network of communication around the heart. It is the way in which fire appears and circulates everywhere in the being. This fire is necessary for the proper functioning of each organ. It is said that without the fire of ming men no organ could function, or even exist.' She continues, 'So we can understand the deep relationship between the fire of ming men and the triple heater as this diffusion and blossoming of an efficient and active fire throughout all parts of the individual' (emphasis is mine).

As I have discussed above, when the ming men or gate of life is successfully opened and an embryological life form is initiated, the primordial fibroblasts and the Primo Vascular System initiate the development of the 3-D structural matrix required by all other systems, structures, and organs for them to be supported and positioned. Fire, in the form of highly activated liquid crystalline EZ, then diffuses within the connective-tissue component of the Triple Heater throughout the entire body. This allows all the other necessary systems, structures, and organs that develop later on to be correctly juxtaposed so that all the embryo's parts are in the right place and, at the same time, are supplied with water and fire from the same supportive structure. Obviously, yang fire is required to manufacture this complicated array of biological structures. Once parturition occurs and the baby is born, it must take its first breath to survive. As Larre states, 'When something is a living being then there is breath'.

38.19 The Triple Heaters Engender 'the Origin of the Life of a Being'

In the book *Heart Master Triple Heater*, Rochat de la Vallée (11), on page 52, states further, 'We will see in the *Ling shu* that the triple heater is connected to the kidneys. As you know, the kidneys are also the power of the origin and contain the origin of the life of a being. They have this double presence of water and fire which are called the authentic water and fire.'

From the above passages, it is seen that there is an intimacy between the Triple Heater and ming men (gate of life) and the kidneys, especially the right kidney. Embryogenesis explains why. After successful fertilization of the ova occurs, after about five days, blastocyst formation begins. The gestational sac is a structure that surrounds the embryo (which is a baby in the very early stages of development) and the yolk sac (umbilical vesicle), which supplies the first blood and circulation for the embryo. The gestational sac is filled with amniotic fluid, which helps to nourish and protect the developing baby throughout the pregnancy.

From the earliest embryological development, *membranous bags filled with fluid* are engendered. The description 'membranous bags filled with fluid' sounds exactly like a description of the Triple Heater, does it not? The fascial membranes support the tissues that undergo differentiation into the myriad of diverse systems, structures, and organs. I believe that the ming men, aka the gate of life, is the blastocyst that engenders life, and that the omnipresent membranous structures of the Triple Energizer constitute the organizational 3-D connective-tissue matrix that orchestrates the intricate tissue differentiation that allows life to begin, exist, and be perpetuated. Thus, I believe that the Primo-Vascular-System/Connective-Tissue-Metasystem complex constitutes the Triple Heater aka the Triple-Energizer Metasystem, and that it is the first TCM organ to form initially during embryogenesis.

38.20 Regarding Embryology, Chinese Ideographs Imply the Notion of the Origin of Life

In the *Chinese Medicine* article entitled 'Pi Wei Xiang Biao Li and the Trajectory of Zuyangming', under the heading 'Membranes. Yuan: Source, Origin', Electra Peluffo (125) states:

> Chinese people paid attention to human embryology because they were interested in concepts such as 'beginning, root, origin' . . . many Chinese ideographs imply the notion of origin or root or basic part: yuan source, basic idea in the source theory in *Ling Shu* 1 and in *Nan Jing* 66. They wanted to know where things came from, the source or origin of life, and embryology gave them an essential information, because *fetal organic development and its connective tissues, fascias, membranes allowed them to understand many of the classic ideas upon life, health and illness.* This evolution of the embryo was already announced in *Huai Nan Zi* which describes the human being development based on a previous text, *Guan Zi* from IV century BC where membranes and mesenteric fascias were not discussed as shown in *Huai Nan Zi*. (Emphasis is mine)

Peluffo continued, 'These mentions of the Classics show that, from ancient times, the anatomical presence of the interorganic membranes linking spleen and stomach together with their internal trajectories were well known.' It is important to note that in the text above regarding ancient Chinese embryology, Peluffo states, 'Because fetal organic development and its connective tissues, fascias, membranes allowed them to understand many of the classic ideas upon life, health and illness.' Peluffo shows in the text above that the ancient Chinese connected the earliest processes of embryogenesis with connective tissues, fascias, and membranes. This agrees with my theory that when the gate of life is opened, the primordial organ that forms is the Primo Vascular System, closely followed by the Connective-Tissue Metasystem. Throughout this book, I prove that the amalgamation of these two complexes is actually one and the same as the Triple-Energizer Metasystem (aka Triple Heater, Triple Burner, or San Jiao).

It would nicely fit my theory that the Kidney is the second organ to be differentiated (after the San Jiao) so that toxic biological by-products do not accumulate in the embryo and poison it. However, it is believed that the Heart begins to beat in humans by day 22–23; thus, *the Heart is generally considered to be the first **functioning** embryonic organ formed.* This is detailed in the article 'The Heart: Our First Organ' by author Stefan Jovinge (141), where he states:

> The heart is the first organ to form during development of the body. When an embryo is made up of only a very few cells, each cell can get the nutrients it needs directly from its surroundings. But as the cells divide and multiply to form a growing ball, it soon becomes impossible for nutrients to reach all the cells efficiently without help. The cells also produce waste that they need to get rid of. So the blood and circulatory system, powered by the heart, together form the first organ system to develop.

However, it is critical to note that in the 2013 article 'Kidney Development', the authors (142) explain that the term *nephrogenesis* pertains to the embryological development of the kidneys. Kidney development proceeds via a series of three successive developmental phases, whereby, through advancing phases, a more complex kidney develops. The pronephros is the primal kidney and develops in the cervical region of the embryo. Subsequent phases are the mesonephros and then finally the metanephros, which both develop lower in the embryo. The third and final metanephros form of the kidney is the most developed and remains as the adult kidney. The authors (142) state:

> *During approximately day 22 of human gestation, the paired pronephri appear towards the cranial end of the intermediate mesoderm.* In this region, epithelial cells arrange themselves in a series of tubules called nephrotomes and join laterally with the pronephric duct. This duct is fully contained within the embryo and thus *cannot excrete filtered material outside the embryo; therefore the pronephros is considered* **nonfunctional** *in mammals.* (Emphasis is mine)

Interestingly, by 'approximately day 22 of human gestation, the paired pronephri appear . . . [and] arrange themselves in a series of tubules called nephrotomes'. So by day 22, the paired primordial Kidneys (pronephri) have already formed, *but* it is believed that they 'cannot excrete filtered material outside the embryo; therefore the pronephros is considered non-functional in mammals'. So because they are supposedly not *functional*, according to embryologists, they are not considered to be the first organ to be formed. The Heart 'begins to beat in humans by day 22–23, thus the Heart is generally considered to be the first functioning embryonic organ formed'. Going out on an embryological limb (pun intended), I suggest that the primordial kidney, as the pronephros, actually is functioning as a simple pump, thanks to the novel mechanism described in detail by Professor Pollack (27) at KL 666–667 in his book. He stated, 'Water suspensions of tiny spheres pumped through (hydrophilic) gel tunnels mimicked the blood flowing through vessels.' His other book dedicated to the topic was titled *Cells, Gels, and the Engines of Life*. This novel research has not yet reached the embryology mainstream for their consideration.

I thus propose that immediately after the gate of life is opened, the first organ to be engendered is the Primo Vascular System and, shortly thereafter, the Connective-Tissue Metasystem. I propose that the amalgamation of these two organ complexes constitute the Triple-Energizer Metasystem or San Jiao and believe that this is the first TCM organ to be produced. I propose that the Second organ to be formed by the PVS is the Kidney in pronephros form. This would explain the intimate relationship that the gate of life has with the San Jiao and the Kidneys and Yuan Qi. I propose that, very shortly after the pronephros Kidneys are formed at day 22, the Third functioning organ, the Heart, is formed at day 22–23.

38.21 Embryonic Development and Tissue Differentiation

The article titled 'The Human Umbilical Vesicle ('Yolk Sac') and Pronephros—Are They Vestigial?' (143) discussed embryonic development. Once the sperm and ova have united, differentiation occurs, and body organs develop due to interactions between cells and tissues. Most often, one group of cells causes another set of cells or tissues to change their fate, a process called *induction*. One cell type or tissue is the *inducer* that produces the signal, and one cell type is the *responder* to that signal. As highly complex biochemical processes take place, accurate cell-to-cell signaling and information transfer is absolutely essential for induction, for conference of correct information to take place.

38.22 The Mammalian Embryonic Heart Is at First a Simple Tube

An interesting point that the article cited directly above revealed pertains to the embryonic heart. In the article, Douglas Dewar (143) advises that the reason *the mammalian embryonic heart is at first a simple tube* is that the embryo must have a simple functioning heart or pumping organ at a very early stage to keep the evolving being alive and that the head region requires a copious blood supply. Further heart developments occur such *that two tiny tubes are formed, which run parallel*. Those two tubes coalesce to form a single tube, and dedicated tissue differentiates to form valves. As further mammalian development proceeds, the design of the heart becomes a more complicated four-chambered heart to meet the increasing demands as rapid growth continues.

Note that the initial Heart pump organ was derived from a simple tube and then later two simple tubes. Note well that in Professor Pollack's (27) research, such simple hydrophilic tubes acted as fluid pumps and generate EZ living water.

38.23 The Three Anatomical Stages of the Embryonic 'Kidney' and the Formation of 'Living Water'

The article (143) states, 'Formation of the pronephric kidney (i.e., pronephros) lays the foundation for the induction of the mesonephric kidney (i.e., mesonephros), and it in turn lays the foundation for the induction of the metanephric kidney (i.e., metanephros). Hence, formation of a pronephric kidney is really the start of a developmental cascade leading to the formation of the definitive kidney.' The pronephros or head kidney is similar to simpler animals, but in humans, it degenerates by week 6. This simple kidney is necessary to remove metabolic wastes from the developing embryo.

This head kidney 'consists of a row of two or three nephridia on each side of the body. These nephridia are tubes, one end of which opens into the body-cavity and the other end into a common duct leading to the exterior. . . . As the embryo increases in size new nephridia are formed behind the first ones. These are of more complicated structure and are described as a second kidney, the mesonephros or middle-kidney.'

As the middle kidney matures, the pronephros gradually undergoes atrophy. Then as the far more complicated and final hind kidney develops, 'the mesonephros become absorbed, but their duct persists, being used to carry the male genital products'.

Note that remnants of the hind kidney are transformed into structures that carry male genital products, which agrees with the TCM concept that kidney controls such structures. I find it interesting that the primary pronephros kidney pump is derived from a simple tube. In Professor Pollack's research, such simple hydrophilic tubes acted as fluid pumps and generate EZ living water. I believe that this living water, in the form of $[H_3O_2]^-$, continues to be engendered by the Kidneys and circulated throughout the entire body by the San Jiao during our lives. This would explain why the Kidneys and the Triple Energizer has such a close relationship associated with the gate of life and why they are both associated with the constant circulation of life energy and Yuan Qi and with the circulation of fire in the water.

38.24 Prenatal Jing Is Ready to be Passed on to the Next Generation by as Early as the Third Week

In an article by creationist biologist Dr Gary Parker (143), he points out, 'The so-called "yolk sac" is the source of the human embryo's first blood cells, and death would result without it!' The so-called yolk sac is vitally important as it functions as the developmental circulatory system of the human embryo before an internal circulation system begins. The former term *yolk sac* is outdated, and most embryologists now call it the umbilical vesicle. The article (143) states that the presence of the umbilical vesicle is essential for several reasons:

- It has a role in the *transfer of nutrients to the embryo* during the second and third weeks when the uteroplacental circulation is being established.
- *Blood development first occurs in the well-vascularized extraembryonic mesoderm* covering the wall of the umbilical vesicle beginning in the third week and continues to form there until hemopoietic activity begins in the liver during the sixth week.
- During the fourth week, the endoderm of the umbilical vesicle is incorporated into the *embryo as the primordial gut. Its endoderm, derived from epiblast, gives rise to the epithelium of the trachea, bronchi, lungs, and digestive tract.*
- Primordial germ cells appear in the *endodermal lining of the wall of the umbilical vesicle* in the *third week* and subsequently *migrate to the developing gonads*. They differentiate into spermatogonia in males and oogonia in females. (Emphasis is mine)

Thus, the umbilical vesicle (yolk sac) remains a major structure associated with and supporting the developing embryo through the fourth week and performs many vitally important early functions. The mesoderm forming the outer layer of the umbilical vesicle, which is outside the embryo, is a major site of hematopoiesis (blood formation). So after the gate of life is opened, here at one place in the umbilical vesicle, there is a lot of life-giving creative activity, including nutrient transfer at week 2, blood manufacture at week 3, development of germ cells at week 3 (which produce gametes that further differentiate into eggs or sperm to define gender), and development of the primordial gut at week 4. These functions certainly are highly associated with TCM functions of the Triple Energizer. All this incredible subdivision and differentiation is happening within four weeks when the embryo is less than 6 mm long.

Primordial germ cells can already be found in the primary ectoderm (epiblast) of human embryos in as early as the second week. Remarkably, at such an early stage in embryological development, primordial germ cells differentiate into spermatozoa in males and oocytes in females and subsequently migrate to the developing gonads. Thus, the life force of future generations is already spawned by the third week after fertilization. Thus, when the gate of life is opened by the successful fertilization of the ova by a sperm, by the third week of differentiation, a small group of the primordial germ cells are 'put aside' to later form spermatozoa in males and oocytes in females. This essentially means that, after the gate of life is opened for one human life form, by the third week of differentiation, the gate of life's components (spermatozoa in males and oocytes in females) are already set in place for the next generation. This is certainly a representation of Yuan Qi in action. In TCM, it is believed that, at conception, the Prenatal Jing is passed to the fetus from the parents. This is certainly the case here in that the Prenatal Jing from the parents engenders the developing embryo to ensure the Prenatal Jing is ready to be passed on to the next generation by as early as the third week after fertilization. I find that fact truly amazing!

38.25 Embryology Confirms that the Kidneys Engender Female and Male Reproductive Organs

It is interesting to note that two uteri usually form initially in a female and a male fetus. Regarding the embryogenesis of the uterus as part of the internal female reproductive tract, the article titled 'Embryology Uterus Development' (144) states:

> Two paramesonephric ducts form from coelomic epithelium extending beside the mesonephric ducts. In the absence of Mullerian Inhibitory Factor these ducts proliferate and grow extending from the vaginal plate on the wall of the urogenital sinus to lie beside the developing ovary. The paired ducts begin to fuse from the vaginal plate end, forming the primordial body of the uterus and the unfused lateral arms form the uterine tubes. Recent research points to the paramesonephric ducts also being the entire embryonic origin of the vagina.

Regarding the two paramesonephric ducts, the article titled 'Paramesonephric Duct' (145) states, 'Paramesonephric ducts (or Müllerian ducts) are paired ducts of the embryo that run down the lateral sides of the urogenital ridge and terminate at the sinus tubercle in the primitive urogenital sinus. In the female, they will develop to form the uterine tubes, uterus, cervix, and the upper one-third of the vagina; in the male, they are lost'. Regarding the paramesonephric ducts, the article further states, 'Only in females do they develop into reproductive organs. They degenerate in males of certain species, but the adjoining mesonephric ducts develop into male reproductive organs.'

The mesonephric ducts were previously called Wolffian ducts. The mesonephros (Greek, 'middle kidney') is one of three excretory organs that develop in vertebrates, and it serves as a temporary kidney during human embryogenesis. So note that the female and male reproductive organs are derived from the kidneys while still differentiating through their early embryological stages. This is very important because TCM teaches that the main physiological functions and indicators of the kidneys are storing essence and controlling human reproduction.

38.26 The Gate of Life Is the Utensil that Stores and Transforms the Essence

In the 36th Difficult Issue in the commentaries on page 385 of Unschuld's (1) translation of the *Nan Ching*, Yeh Lin states, '*The gate of life is the root of the Triple Burner and the sea of the original influences of the 12 conduits*; it is the utensil that stores and transforms the essence, and it is the place to which the womb, which conceives the embryo, is tied. Thus, it is the origin of man's life. Hence, it is called gate of life' (emphasis is mine).

Yeh Lin states above that 'the gate of life . . . is the utensil that stores and transforms the essence, and it is the place to which the womb, which conceives the embryo, is tied'. I just mentioned above that the female reproductive organs, including the uterus, are derived from the Kidneys while the Kidneys are still differentiating through their early embryological stages. So the gate of life cannot be the uterus as some claim because as Yeh Lin states, the gate of life *is tied* to 'the womb, which conceives the embryo'. Yeh Lin states that 'the gate of life is . . . the utensil that stores and transforms the essence'. The 'utensil' is said to store the essence and transform the essence. The utensil cannot be an ovum, for every month, ova are released and do not open the gate of life unless they are fertilized. Likewise, the utensil cannot be the millions of sperm that are released at each ejaculation as they only open the gate of life when a successful fertilization occurs.

I suggest that the 'utensil' forms at conception, when the sperm successfully fertilizes the ovum, such that the DNA from both the successful sperm and the ovum commence the miracle of life. Only then is the gate of life opened, allowing the DNA that stores and transforms the essence in both the ovum and the sperm to initiate the marvellous cellular differentiation that produces all the necessary structures according to their allotted time thanks to the DNA code. As stated elsewhere in another Chapter of this book, I have shown that the Primo Vascular System and the fibroblasts are the first structures to develop and start the construction of what will become a human being. As the fibroblasts and the PVS are connective-tissue components, the first major TCM organ is the PVS/Connective-Tissue

Metasystem complex, which is intimately associated with the Triple-Energizer Metasystem (Triple Burner). Thus, the gate of life is the root of the Triple Burner. As the PVS/Connective-Tissue Metasystem amalgamation engenders all the structures within the body, the Connective-Tissue Metasystem (Triple Burner) also fabricates 'the sea of the original influences of the 12 conduits'. Remember that the Triple Burner is the very first TCM organ formed and is even responsible for the first membrane 'bag' or sac and the first accumulation of water in the new life form, namely the gestational sac and the embryonic fluid therein respectively. I propose that this new 'utensil' is called the blastocyst. What did Yeh Lin say about the blastocyst? He said, 'It is the utensil that stores and transforms the essence, and it is the place to which the womb, which conceives the embryo, is tied. Thus, it is the origin of man's life. Hence, it is called gate of life.'

According to the web article (138) where the author is anonymous, Tang Zonghai wrote in his somewhat-palindromic work *A Refined Interpretation of the Medical Classics* (*Yijing Jingyi*), 'The root of the triple burner is in the kidney, more precisely right between the two anatomical kidneys. Right there is a greasy membrane that is connected with the spine. It is called mingmen, and constitutes the source of the three burners' (emphasis is mine).

Subsequently, when the ancient classics define the location of ming men (the gate of life) as the description above, it would appear that it is located at the anatomical location of the original blastocyst after further tissue differentiation occurred up until parturition. Remember that the zygote developed into the morula, which then further developed into the blastocyst at five days after fertilization. For me, college was a long time ago. Maybe it was for you too. So I will briefly discuss the initial stages of human embryogenesis for clarity of terminology.

38.27 Description of the Origin and the Nature of the Two-Celled Human Zygote

Regarding the formation of the zygote after fertilization, the article titled 'Human Embryogenesis' (146) states:

Fertilization takes place when the spermatozoon has successfully entered the ovum and the two sets of genetic material carried by the gametes, fuse together, resulting in the zygote, (a single diploid cell). This usually takes place in the ampulla of one of the fallopian tubes. . . .

The zygote contains the combined genetic material carried by both the male and female gametes which consists of the 23 chromosomes from the nucleus of the ovum and the 23 chromosomes from the nucleus of the sperm. The 46 chromosomes undergo changes prior to the mitotic division which leads to the formation of the embryo having two cells.

This tiny two-celled mass called a zygote is the beginning of a human life form. The article continues, 'The zygote, which is defined as an embryo because it contains a full complement of genetic material, begins to divide, in a process called cleavage. A blastocyst is then formed and implanted in the uterus.' Thus, it can be clearly seen that the blastocyst 'is the utensil that stores and transforms the essence, and it is the place to which the womb, which conceives the embryo, is tied. Thus, it is the origin of man's life. Hence, it is called gate of life.'

38.28 Description of the Origin and the Nature of the Human Morula and Blastocyst

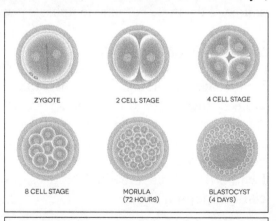

Figure 11. Development of fertilized egg. By megija. © Depositphotos.com.

After the formation of the two-celled zygote, it develops into the morula. Regarding the origin and the nature of the human sixteen-celled morula and its development into a blastocyst at five days, the article titled

'Blastocyst' (147) explains that the morula is the early stage of the embryo and is composed of sixteen undifferentiated cells which develop in the Fallopian tube. The morula then leaves the Fallopian tube and enters the uterus, where through cellular differentiation and cavitation, the morula differentiates into two sections, with the inner cell mass (ICM) developing on the interior of the blastocoel and the trophoblast cells growing on the exterior. This developmental phase is called the blastocyst. The side of the blastocyst where the inner cell mass resides is called the animal pole, and the opposite side is called the vegetal pole. Then a fluid cavity forms inside the embryo via a process called cavitation. The external trophoblast cells then pump sodium ions into the center of the developing embryo, which instigates the inflow of water via osmosis. This internal fluid-filled cavity is called the blastocoel. This characteristic internal blastocoel cavity surrounded by a cellular mass yield the hallmark morphology of the blastocyst. The article (147) further states:

> The blastocyst is a structure formed in the early development of mammals. It possesses an inner cell mass (ICM) which subsequently forms the embryo. The outer layer of the blastocyst consists of cells collectively called the trophoblast. This layer surrounds the inner cell mass and a fluid-filled cavity known as the blastocoele. The trophoblast gives rise to the placenta.
>
> In humans, blastocyst formation begins about 5 days after fertilization, when a fluid-filled cavity opens up in the morula, a ball consisting of a few dozen cells. The blastocyst has a diameter of about 0.1–0.2 mm and comprises 200–300 cells following rapid cleavage (cell division). After about 1 day, the blastocyst embeds itself into the endometrium of the uterine wall where it will undergo later developmental processes, including gastrulation.
>
> The use of blastocysts in in-vitro fertilization (IVF) involves culturing a fertilized egg for five days before implanting it into the uterus. It can be a more viable

method of fertility treatment than traditional IVF. The inner cell mass of blastocysts is also a source of embryonic stem cells.

38.29 The Gate of Life Is the Root of the Triple Burner and the Sea of the Original Influences of the Twelve Conduits

After the gate of life is opened, a cascade of molecular and morphological events controlling transformation and regression occurs. This cascade involves the highly precise cellular division and organ differentiation occurring within the connective-tissue matrix that is the developing embryo. In the article 'Fascia & Tensegrity' by Thomas Myers (110), he states, 'Our single fascial system starts about 2 weeks into development as a fibrous gel that pervades and surrounds all the cells in the developing embryo. It is progressively folded by gastrulation and the rest of the motions of development into the complex layers of fascia we see in the adult.' The connective-tissue matrix would continue to grow and develop along with the embryological development, including organ production. Thus, every substrate that was not an organ would be connective-tissue matrix that I propose is intimately associated with the Triple Energizer. Subsequently, the ming men would engender the Triple Energizer as stated in the TCM classics. Expressed another way, as stated above, 'the gate of life is the root of the Triple Burner and the sea of the original influences of the 12 conduits'. Why 'the sea

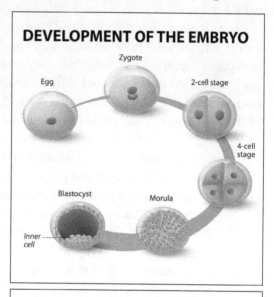

Figure 12. Early human embryonic development. Vector by edesignua. © Depositphotos.com.

of the original influences of the 12 conduits'? Because I believe that the acupuncture meridians are also contained within the connective-tissue/PVS matrix of the body in the form of collagen fibers that contain the numerous types of meridians and especially the twelve conduits or main meridians.

This gate of life is *not* the right kidney as many believe, but it is located between the two kidneys. Why is it the gate of life? If something goes wrong in the cellular differentiation process and the zygote or morula or blastocyst dies and then aborts, then the gate of life fails to open. So I propose that if the gate of life opens and stays open, the blastocyst will implant in the uterus and will continue to grow in utero and be successfully born at full term as a living individual. However, that is not the end of the matter. It is just the beginning of a new phase. The newborn has to accept independence from the mother and take that one first highly critical life-perpetuating breath that will trigger the sustained and ongoing independence of the new being. It is only when the individual takes their first breath upon the earth under the heavens that the Triple Energizer has completed its primary task of constructing the bodily infrastructure upon which every tissue, structure, and organ is suspended. After the first breath, the Triple Energizer must continue to support the individual by ensuring all the drinks and grains consumed are correctly processed into the necessary qi and fluids and circulated efficiently throughout the entire body to sustain life.

38.30 TCM Description of the Supposed Physical Anatomical Location of Ming Men

Based on the Chinese classics, the ming men (the Gate of Life) is where the original life essence of the individual is based. Authors have varying beliefs regarding ming men. These include the belief that Ming men lies between the kidneys, closer to the right Kidney and level with or just below the umbilicus (some say 2 cun), on the middle of the lower back, and it is composed of yellow fatty membrane and is attached to vertebra. It has been described as the 'tie between the kidneys'. The acupuncture point Governing Vessel 4 is called Ming Men and is located between the Shu points of the

Kidneys, Bladder 23. Some authorities believe that the Ming Men and the Dan Tien come into existence at birth.

So hypothetically, let's consider some of the possible literal anatomical structures that match that description in adults. As membranes feature dominantly when describing the Triple Energizer, if the ming men is a literal residual anatomical structure, I would predict the ming men involves some fascial membranous structure in the proximity of the Right kidney that is attached to vertebra. I would also think that it would most likely be intimately associated with the blastocyst, the origin of our life, and is a remnant of it. I suspect that the original zygote and the morula were too small and not yet implanted in the uterus and that insufficient cellular differentiation had occurred to make those two structures eligible.

38.31 Could Vestigial Organs of the Embryo Constitute 'the Tie between the Kidneys'?

In the book *The Human Embryo*, under the Chapter titled 'Development, Differentiation and Derivatives of the Wolffian and Müllerian Ducts', the authors (148), on pages 143–163, state, 'It has been experimentally shown that in vertebrates the Wolffian ducts are required for the induction of Müllerian duct formation and in their absence no Müllerian duct can develop. *Genes that are required for Müllerian duct formation are also found in kidney development.*' The authors continue to explain, '*The Müllerian ducts do not completely disappear.* The most cranial parts are supposed to *persist as appendices testis and the caudal parts as prostatic utricles.* In the female, where AMH [Anti-Müllerian Hormone] is lacking, the uniform Müllerian ducts differentiate in very specific segment to give rise to the uterine tubes (oviducts), the uterus, cervix and the vagina' (emphasis is mine). Regarding the vestigial organs, the authors (148) further advise that it was Morgagni who first discovered hydatids of genital organs. They reported that hydatids

> *are remnants of the cranial part of the Müllerian ducts and Wolffian ducts.* In males, the frequent appendices of the testes develop from the funnel region of the Müllerian

ducts. Likewise, in females, hydatids are found that are often pedunculated. They may occur at the fimbriae of the Fallopian tubes deriving from the Müllerian ducts, or as paratubular appendices vesiculosae or hydatids of Morgagni deriving their origin from the Wolffian ducts or mesonephric tubules. The clinical relevance of these structures is torsion of pedicle with acute syndrome within scrotum or abdomen. Tumors deriving from these vestigial structures are also described. The Wolffian ducts are the first appearing structures of the urogenital system and their migration and inductive properties are critical for the development of the permanent kidneys and the genital ducts in males and females.

Shortly after the onset of somite differentiation, the Wolffian duct anlagen separate from the intermediate mesoderm. During caudal migration they induce the pro- and mesonephroi within the ventral part of the intermediate mesoderm. Near the caudal entrance into the cloaca (sinus urogenitalis) an ureter bud sprouts out from each Wolffian duct and grows dorso-cranially to join the metanephric blastema. Each ureter bud divides in a special dichotomy manner and forms ureter, pelvis, calyces and collecting ducts of the permanent kidney. The Wolffian ducts need androgens for further differentiation.

In males, each duct forms a coiled ductus epididymidis and the straight vas deferens. Sprouting of the seminal vesicles occurs near the urogenital sinus. The ductus epididymidis together with some persisting tubules of the mesonephros differentiates into the epididymis, which via the rete testis is in close connection with the testis enabling transport and maturation of spermatozoa. Shortly before birth, the epididymis descends into the scrotum together with the testis. In females, the Wolffian ducts do not differentiate further, *but persists as rudimentary Gartner's ducts in the broad ligament lateral to the uterus.* The Gartners's ducts are found generally lateral to the uterus, but might also reach down to the wall of the vagina and can give rise to cysts.

In males, AMH induces apoptosis and epitheliomesenchymal transformation of the Müllerian ducts. *Only the most cranial and the most caudal parts frequently persist as appendix testis and utriculus prostaticus, demonstrating special properties of these regions. Early developing structures leave their trace in the adults as a vestigial organ which might be of clinical interest.*
(Emphasis is mine)

Note that the male and female reproductive organs are derived from the same tissue that the Kidneys are derived from. This explains their close association in TCM. Note too that the authors (148) point out that these vestigial organs (prostatic utricle and Gartners's ducts in the male and female respectively) 'demonstrate special properties of these regions' and 'might be of clinical interest'. Interestingly, the female Gartner's ducts are located 'in the broad ligament lateral to the uterus'. Could this be the 'greasy membrane' between the kidneys? It is of further note that the prostatic utricle is a small dimple located in the prostatic urethra near the ejaculatory ducts, the remains of the fused caudal ends of the paramesonephric ducts. It is the male homologue of the female uterus and vagina. It is believed that during coitus, the prostatic utricle contracts and allows the semen to be ejaculated. For a supposed vestigial organ, it sure has lots of synonyms, including *utriculus prostaticus, masculine uterus, Morgagni sinus, sinus pocularis, vagina masculina, vesica prostatica,* and the *Weber organ*. During embryological development, these vestigial organs are in close proximity and medial to the Kidneys. However, in an adult, the Kidneys occupy an anatomical location well away from and superior to these vestigial organs.

Perhaps masterful control over this supposedly vestigial prostatic utricle in males is associated with how the sages and devotees perform practices, like tantric sex or sexual qi gong, whereby climax is reached but ejaculation is avoided. Once mastered, this practice can prolong the sex act for many hours and is believed to preserve precious jing.

38.32 Anatomical Membranous Structures between the Kidneys and below the Umbilicus

Regarding the possible physical anatomical location of ming men after parturition, the posterior pararenal space is usually fat-filled, while the anterior pararenal space is often empty. The posterior pararenal space is developed by bluntly dissecting between the posterior renal fascia (Zuckerkandl's fascia, posterior perirenal fascia, retrorenal fascia) and the transversalis fascia that lines the posterior abdominal wall. The many organs and structures located between the kidneys include the vertebral column, including the neural canal, inferior vena cava, abdominal aorta (the half of the descending aorta below the diaphragm), anterior renal fascia, posterior renal fascia, two ureters, and the thoracolumbar fascia. Note that the ligamentum flavum is yellow and that some of it is between the kidneys. This ligament connects the laminae of adjacent vertebrae, all the way from the second vertebra, the axis, to the first segment of the sacrum. If an undiscovered section of this system between the kidneys is further differentiated more than the rest, it certainly matches the criteria to be the possible location of ming men.

38.33 Does Ming Men Involve the Thoracolumbar Fascia?

In his two articles on the thoracolumbar fascia (134, 149) Dr Bruce McFarlane explains, 'The largest fascia in the body is the thoracolumbar fascia, which is a diamond-shaped *yellowish-clear membrane sheet that forms part of the deep fascia*. It is most developed in the lumbar region, and is composed of *multiple layers of crosshatched collagen fibers* that cover the various back muscles in the lower thoracic and lumbar region before sliding through these muscles where it *attaches to the sacral bone*' (emphasis is mine).

McFarlane explains that due to its unique centrality and its large array of interconnection with body components in all directions, the thoracolumbar fascia acts as a cardinal point. It communicates with the entire body, the head and neck, shoulders, arms and hands, abdomen, pelvis, hamstrings, and legs and feet. At the very core, the spine is located lying as a central hub.

Remarkably, western medicine only discovered the functional significance of the thoracolumbar fascia in the last 40 years. The thoracolumbar fascia links the hands to the feet, the back to the front, and the outside to the inside. The thoracolumbar fascia acts as a scaffolding device and adds both supportive structure and power to the vertebral column. McFarlane explains with tensegrity in mind, 'We can begin to appreciate how the thoracolumbar fascia creates structure through tension, whereby it connects—the bones of all four limbs, spine, pelvis and skull. *The thoracolumbar fascia also acts as a source of power that energizes the vertebral column.* Because the fascia is elastic in nature, once the fascial fibres have been stretched, elastic-recoil energy is stored in the fibres, and they want to shorten and return to its original length and uncharged state' (emphasis is mine).

One of the roles served by the thoracolumbar fascia is that of a *sensing device*. McFarlane explains that the thoracolumbar fascia and its extensions into the head and neck, upper and lower limbs, abdomen, low back, and pelvis are equipped with sensors that fire with stretching. The second function of the thoracolumbar fascia is as a *regulator of biological processes* (emphasis is mine). McFarlane explains, 'The interior of every cell of the body has the same tent-like structure enjoyed by the body as a whole. Each cell's cytoskeleton, or internal form, is furnished with tension and compression elements that create tensegrity within the cell.' He continues, 'Incoming signals, we now know, produce fundamental biochemical and genetic changes within the cell. Hence, simple moving about causes the thoracolumbar fascia and other connective tissue to tug on the interior of all the cells with which they come in contact, altering cell behaviour in significant ways. Once again, we see that there are no independent structures in the body'.

We can see from the information above that the thoracolumbar fascia is a yellow membrane that inserts between the kidneys to the vertebra and is involved with strength and vitality and is part of a major tensegrity communication device that extends throughout the body. However, I can find no reference to a major component of the thoracolumbar fascia being present proximal to the right kidney. Hypothetically, there may be an as-yet-unrecognized structure within the thoracolumbar fascia that constitutes ming men in an adult.

38.34 The Triple Heater Does Exist before the First Breath Is Taken

While discussing the 62nd Difficulty of *Nan Jing*, in the book *Heart Master Triple Heater*, Elisabeth Rochat de la Vallée (11), on page 117, comments on her personal belief regarding the origin of the Triple Heater by saying, 'My personal feeling is that the triple heater does not exist until after the first breath is taken, although this is never spoken about in the texts.'

I do not agree with this comment by Rochat de la Vallée, stating, 'The triple heater does not exist until after the first breath is taken.' I have shown elsewhere in this book that the Triple-Energizer Metasystem is the first embryological TCM organ complex to develop and that the subsequent differentiation of all the other organs and structures occurs in close connection with the Primo Vascular System. The Triple-Energizer Metasystem then manages all the necessary qi and fluid movements to ensure that embryogenesis proceeds successfully, including controlling the amniotic sac, which is just another 'bag of water' for the Triple Heater to control until full term. I believe that there is a swap-over at birth, whereby the Kidneys initiate the first inhalation 'in a single breath' and then continue to initiate involuntary inhalations for the rest of our life. While in the fluid-filled amniotic sac in utero, the baby is completely dependent on the mother to supply every single nutrient to sustain life. However, the inauguration of the new life form is when the first single breath was taken, and all the organs then have to operate independent of the mother. Because the Triple Heater was the original first-formed organ, Yang could rightfully state, '"Origin" is an honorable designation for the Triple Burner.' This primordial organ should rightfully be responsible for governing and distributing primordial qi (Yuan Qi or Original Qi).

Summary of Chapter 38

In the commentaries of the eighth Difficult Issue in Unschuld's (1) translation of the *Nan-ching*, on pages 136–137, regarding the origin of the Triple Burner, Katō Bankei says, 'At the very beginning of the fetal [development], the true influences of heaven take their residence in the gate of life mansion between the kidneys. This is called the "origin of the vital influences." The mystery on which all beginnings depend, on which all life depends, works from here.' What is the origin of the vital influences? Katō Bankei clearly states they are called the gate of exhalation and inhalation or the origin of the Triple Burner. Katō Bankei is here, clearly stating that the opening of the symbolic 'gate of exhalation and inhalation' is synchronous with the *origin* of the Triple Burner. Note that at this early stage of embryological development, the literal Lungs have not even differentiated and the continuing life force is dependent on the maternal blood supply replete with all the necessary nutrients and oxygen to sustain life. For exhalation and inhalation to occur in a creature, that creature must be alive. When breathing stops, death very quickly follows. Thus, the term *the gate of exhalation and inhalation* is synonymous with the *gate* of continued breathing or continued existence as it were, the so-called gate of life.

In the article entitled 'Bioelectric Responsiveness of Fascia: A Model for Understanding the Effects of Manipulation', the author, Judith A. O'Connell (111), stated, 'Connective tissue is a major derivative of the mesodermal germ cell layer in the embryo. . . . A special property that connective tissue maintains throughout life is embryologic plasticity.' The embryologic plasticity required for successful embryonic growth and development continues with us throughout our lifetime. In total agreement with what O'Connell states, Myers (110) states, 'Understanding fascia is essential to the dance between stability and movement—crucial in high performance, central in recovery from injury and disability, and ever-present in our daily life from our embryological beginnings to the last breath we take.'

Regarding the origin of the embryological development of the fascial system, in the 2014 article entitled 'Fascia & Tensegrity', Myers (110) states, 'Our single fascial system starts about 2 weeks into development as a fibrous gel that pervades and surrounds all the cells in the developing embryo. It is progressively folded by gastrulation and the rest of the motions of development into the complex layers of fascia we see in the adult.' Note that O'Connell and Myers place fascia at our embryological beginning, at our very origin of life. It is now known that the Primo Vascular System (PVS) is the very first system to develop after fertilization occurs. While the PVS is composed of connective tissue and is probably a specialized subset of the connective-tissue system, the fascial system proper 'starts about 2 weeks into development as a fibrous gel that pervades and surrounds all the cells in the developing embryo'. Note that the fascial system, or what I call the Connective-Tissue Metasystem, 'pervades and surrounds all the cells in the developing embryo' just as the Triple-Energizer Metasystem pervades and surrounds all the organs and systems in the adult body.

Andrew Taylor Still, MD (1828–1917), an American physician and the founder of osteopathic medicine, taught 115 years ago that the fascia control fluids in the body and that if the fascia fails, then we shrink, swell, or die. This indeed sounds exactly like a description of the Triple Energizer. Still described fascia as a 'network of nerves, cells and tubes, running to and from it' such that 'no atom of flesh fails to get nerve and fluid supply therefrom'. Once again, these properties of fascia (communication via the associated nerve connections and control over fluids) are identical to those of the Triple Energizer. Note that Still also stated that the fascia is a network of tubes. I propose that it is most likely that these 'tubes' also include a part of the Primo Vascular System's network of microtubules, a microsystem that was not known about in Still's time.

That the Triple Burner is 'also the fu of the fire in the middle of the water', I suggest, pertains to the fact that bulk water, in touch with the omnipresent and hydrophilic collagen of connective tissue, generates the energized, living liquid crystalline EZ water that generates the fire or energy-yielding battery that is EZ water. I suggest it is this spark-of-life energy coming from the EZ water that accompanies the origin of life and sustains the subsequent life form.

The Yuan Qi is the vibrant life force that activates the diverse necessary biochemical activity of all the internal organs and is the genesis of activity and vitality. Yuan Qi circulates throughout the body to all cells, relying on the transporting system of the San Jiao (Triple Burner). It dwells between the two Kidneys, at the Gate of Vitality (Ming Men). Yuan Qi expedites the metamorphosis of Qi.

Yuan Qi is the conductor of transformation, transmuting Zong Qi into Zhen Qi. Yuan Qi is integral in the manufacture of blood by enabling the transformation of Gu Qi into Blood.

Tang Zonghai wrote in his work *A Refined Interpretation of the Medical Classics (Yijing Jingyi)*, 'The root of the triple burner is in the kidney, more precisely right between the two anatomical kidneys. Right there is a greasy membrane that is connected with the spine. It is called mingmen, and constitutes the source of the three burners.' Which anatomical structure constitutes the root of the Triple Burner in/near the Kidney? I suggest that this structure of greasy membrane between the kidneys is an already known anatomical structure possessing an as-yet-unrecognized dynamic piezoelectric property—e.g. thoracolumbar fascia or perinephric fat(?)—that connects to the neural pathways, where it is connected with the spine and subsequently the brain. The significance of this structure will become obvious when I discuss the understated significance of the membranous neurofascial system that covers and interconnects and communicates with every single cell in the body. As I believe that the Triple Burner is predominantly the Connective-Tissue Metasystem and as 'the root of the triple burner is in the kidney' (more precisely, medial to the right Kidney), it would make sense that this Kidney-engendering Triple Burner root or origin would be a greasy, membranous connective-tissue structure medial to the right Kidney, proximal to the level of L2–L3. Time and a vigilant anatomist or physiobiologist will tell.

In the article titled 'Ming-Men: An Acupressure Point with Power-Full Implications', the author (139) states, 'The Chinese admiration for the Ming-men is captured in the disagreement on how to describe it. It has been called the Gate of Power, Proclamation Gate, Gate of Destiny, and Gate of Life. No matter how you define it, the healing ability and energy properties of the Mingmen abound.'

On page 38 of the book *Heart Master Triple Heater,* Larre and Rochat de la Vallée (11) both believe that the ming men or gate of life is the original primitive life force that initiates the beginning of life and that the ming men is 'some sort of first condition of the embryo'.

I suggest that the Gate of Life is not the ova in women and not the sperm in men. In women, every single month, ova are released, and no gate of life is opened such that an embryo forms. Subsequently, the unfertilized ova are broken down along with the endometrium that did not have to support a fertilized ovum. Similarly for men, sperm are regularly released and fail to fertilize an egg to generate a new life form. Note too that the gate of life is not the womb but that the gate of life is obviously tied to the womb when successful fertilization results in pregnancy.

I have discussed in great detail elsewhere in this book what I believe happens when the gate of life is opened. Briefly, the Primo Vascular System initiates the primordial differentiation process within hours of fertilization. By two weeks postfertilization, the Connective-Tissue Metasystem proper allows the tissues to differentiate into organs and systems. Regarding the nature and properties of 'ground substance,' which permeates the entire body, in the article titled 'More about Connective Tissue and Myofascial Release', Glenda Poletti (135) advises that ground substance is found in all connective tissue and that it is produced by fibroblasts, which are among the earliest cells to develop in the embryo, and that these fluid ground substances are the immediate environment of every cell in the body.

I propose that the Connective-Tissue Metasystem, in amalgamation with the PVS, is one and the same as the Triple-Energizer Metasystem. Thus, simply stated, at conception, when the gate of life is opened, the primordial fibroblasts generate the Connective-Tissue Metasystem, which I believe is closely associated with the Triple Burner. Thus, it can be stated that 'the gate of life is the root of the Triple Burner'. Thus, 'the triple heater is rooted at the deepest level of being' because the Triple Heater was formed at the beginning of life. In the presence of all-pervading, hydrophilic connective tissue, bulk water is converted into the fiery energy–supplying battery that is life-sustaining liquid crystalline EZ water that flows within and throughout the Connective-Tissue Metasystem. This phenomenon accounts for 'the mixing of water and fire, allowing the development of all life.'

As I have discussed above, when the ming men or gate of life is successfully opened and an embryological life form is initiated, the primordial fibroblasts and the Primo Vascular System initiate the development of the 3-D structural matrix required by all other systems, structures, and organs for them to be supported and positioned. Fire, in the form of highly activated liquid crystalline EZ, then diffuses within the connective tissue component of the Triple Heater throughout the entire body. This allows all the other necessary systems, structures, and organs that develop later on to be correctly juxtaposed so that all of the embryo's parts are in the right place and, at the same time, are supplied with water and fire from the same supportive structure. Obviously, yang fire is required to manufacture this complicated array of biological structures. Once parturition occurs and the baby is born, it must take its first breath to survive. As Larre states, 'When something is a living being then there is breath.'

From the earliest embryological development, membranous bags filled with fluid are engendered. The description 'membranous bags filled with fluid' sounds exactly like a description of the Triple Heater, does it not? The fascial membranes support the tissues that undergo differentiation into the myriad of diverse systems, structures, and organs. I suggest that the ming men, aka the gate of life, is the life force (blastocyst) that engenders life and that the omnipresent membranous structures of the Triple Energizer constitute the organizational 3-D connective-tissue matrix that orchestrates the intricate tissue differentiation that allows life to begin, exist, and be perpetuated. Thus, I believe that the Primo Vascular System/Connective-Tissue Metasystem is the first TCM organ (Triple Energizer/Triple-Energizer Metasystem) to form initially during embryogenesis.

I thus propose that immediately after the gate of life is opened, the first organ to be engendered is the Primo Vascular System and, shortly thereafter, the Connective-Tissue Metasystem. I propose that the amalgamation of these two organ complexes constitute the Triple-Energizer Metasystem or San Jiao and believe that this is the first TCM organ to be produced. I propose that the Second organ to be formed by the PVS is the Kidneys in the pronephros form.

This would explain the intimate relationship that the gate of life has with the San Jiao and the Kidneys and Yuan Qi. I propose that very shortly after the pronephros Kidneys are formed at day 22, the Third functioning organ, the Heart, is formed at day 22–23.

Douglas Dewar (143) advises that the reason the mammalian embryonic heart is at first a simple tube is because the embryo must have a simple functioning heart or pumping organ at a very early stage to keep the evolving being alive and because the head region requires a copious blood supply. Further heart developments occur such that two tiny tubes are formed, which run parallel. Those two tubes coalesce to form a single tube, and dedicated tissue differentiates to form valves. As further mammalian development proceeds, the design of the heart becomes a more complicated four-chambered heart to meet the increasing demands as rapid growth continues.

During initial embryological development of the Kidney, the head kidney 'consists of a row of two or three nephridia on each side of the body. These nephridia are tubes, one end of which opens into the body-cavity and the other end into a common duct leading to the exterior. . . . As the embryo increases in size new nephridia are formed behind the first ones. These are of more complicated structure and are described as a second kidney, the mesonephros or middle-kidney'. As the middle kidney matures, the pronephros gradually undergoes atrophy. Then as the far more complicated and final hind kidney develops, 'the mesonephros become absorbed, but their duct persists, being used to carry the male genital products'.

When the gate of life is opened by the successful fertilization of the ova by a sperm, by the third week of differentiation, a small group of the primordial germ cells are put aside to later form spermatozoa in males and oocytes in females. This essentially means that, after the gate of life is opened for one human life form, by the third week of differentiation, the gate of life components (spermatozoa in males and oocytes in females) are already set in place for the next generation. This is certainly a representation of Yuan Qi in action. In TCM, it is believed that at conception, the Prenatal Jing is passed to the fetus from the parents. This is certainly the case here in that the Prenatal Jing from the parents engenders the developing embryo to ensure the Prenatal Jing is ready to be passed on to the next generation by as early as the third week after fertilization. I find that fact truly amazing!

It is interesting that the female and male reproductive organs are derived from the kidneys while they are differentiating through their early embryological stages. This is very important because TCM teaches that the main physiological functions and indicators of the kidneys are storing essence and controlling human reproduction.

Remember that the Triple Burner is the very first TCM organ formed and is even responsible for the first membrane 'bag' or sac and the first accumulation of water in the new life form, namely the gestational sac and the embryonic fluid therein respectively. I propose that this new utensil is called the blastocyst. What did Yeh Lin say about the blastocyst? He said, 'It is the utensil that stores and transforms the essence, and it is the place to which the womb, which conceives the embryo, is tied. Thus, it is the origin of man's life. Hence, it is called gate of life.'

The gate of life is *not* the right kidney as many believe, but it is located between the two kidneys. Why is it the gate of life? If something goes wrong in the cellular differentiation process and the zygote or morula or blastocyst dies and then aborts, then the gate of life fails to open. So I propose that if the gate of life opens and stays open, the blastocyst will implant in the uterus and will continue to grow in utero and be successfully born at full term as a living individual. However, that is not the end of the matter. It is just the beginning of a new phase. The newborn has to accept independence from the mother and take that one first highly critical life-perpetuating breath that will trigger the sustained and ongoing independence of the new being. It is only when the individual takes their first breath upon the earth under the heavens that the Triple Energizer has completed their primary task of constructing the bodily infrastructure upon which every tissue, structure, and organ is suspended. After the first breath, the Triple Energizer must continue to support the individual by ensuring all the drinks and grains consumed are correctly processed into the necessary qi and fluids and circulated efficiently throughout the entire body to sustain life.

Regarding the possible physical anatomical location of ming men after parturition, the posterior pararenal space is usually fat-filled, while the anterior pararenal space is often empty. The posterior pararenal space is developed by bluntly dissecting between the posterior renal fascia (Zuckerkandl's fascia, posterior perirenal fascia, retrorenal fascia) and the transversalis fascia that lines the posterior abdominal wall. The many organs and structures located between the kidneys include the vertebral column, including the neural canal, inferior vena cava, abdominal aorta (the half of the descending aorta below the diaphragm), anterior renal fascia, posterior renal fascia, two ureters, and the thoracolumbar fascia. Note that the ligamentum flavum is yellow and that some of it is between the kidneys. This ligament connects the laminae of adjacent vertebrae, all the way from the second vertebra, the axis, to the first segment of the sacrum. If a section of this system or any of the other anatomical structures located between the kidneys is further differentiated and more specialized than the rest, it certainly matches the criteria for a possible hypothetical anatomical location of ming men after parturition. Note too that the authors (148) point out that the vestigial organs (prostatic utricle and Gartners's ducts in the male and female respectively) 'demonstrate special properties of these regions' and 'might be of clinical interest'. Interestingly, the female Gartner's ducts are located 'in the broad ligament lateral to the uterus'. Could this be the 'greasy membrane' between the kidneys? It is of further note that the prostatic utricle is a small dimple located in the prostatic urethra near the ejaculatory ducts, the remains of the fused caudal ends of the paramesonephric ducts. It is the male homologue of the female uterus and vagina. For a supposed vestigial organ, it sure has lots of synonyms, including utriculus prostaticus, masculine uterus, Morgagni sinus, sinus pocularis, vagina masculina, vesica prostatica, and the Weber organ. During embryological development, these vestigial organs are in close proximity and medial to the Kidneys. However, in an adult, the Kidneys occupy an anatomical location well away from and superior to these vestigial organs.

Remember, many new organs have been recently uncovered (gut microbiome, PVS) and many new functions of existing organs have been elucidated (adipose tissue has an endocrine function). It would be appropriate that the residual structure derived from the original blastocystic ming men would occupy a location between the umbilicus anteriorly and Governing Vessel 4 (Ming Men) posteriorly, and that via a conduit of PVS/Connective-Tissue Metasystem/Acupuncture Meridian Metasystem the residual ming men component communicates with Governing Vessel 4 (Ming Men) and other systems and structures throughout the body.

Rochat de la Vallée believes that 'the triple heater does not exist until after the first breath is taken'. I do not agree with that comment, and I have shown elsewhere in this book that the Triple-Energizer Metasystem is the first embryological TCM organ complex to develop and that the subsequent differentiation of all the other organs and structures occurs in close connection with the Primo Vascular System. The Triple-Energizer Metasystem then manages all the necessary qi and fluid movements to ensure that embryogenesis proceeds successfully, including controlling the amniotic sac, which is just another 'bag of water' for the Triple Heater to control until term. I believe that there is a swap-over at birth, whereby the Kidneys initiate the first inhalation 'in a single breath' and then continue to initiate involuntary inhalations for the rest of our life. While in the fluid-filled amniotic sac in utero, the baby is completely dependent on the mother to supply every single nutrient to sustain life. However, the inauguration of the new life form is when the first single breath was taken, and all the organs then have to operate independent of the mother. Because the Triple Heater was the original first-formed organ, Yang could rightfully state, '"Origin" is an honorable designation for the Triple Burner.' This primordial organ should rightfully be responsible for governing and distributing primordial qi (Yuan Qi or Original Qi).

CHAPTER 39

Relationship Between the Triple-Energizer Metasystem, the Connective-Tissue Metasystem, the Primo Vascular System, the Acupuncture Meridian System, and the Kidneys

Introduction
I propose that the Primo Vascular System (PVS), the Connective-Tissue Metasystem (CTM), the Triple Burner (Triple-Energizer Metasystem), and the Acupuncture Meridian System are all intimately related and interconnected. It will take much more research to accurately delineate each of these entities and their exact composition and 3-D locations throughout the body. In the meantime, the following can be said about each of these entities. Please note that many sections in this Chapter have been directly copied from throughout the book in an effort to summarize each of the five different components under discussion.

39.1 Summary of the Primo Vascular System
Primo Vascular System development and proliferation precedes the formation of *all* other structures during embryogenesis. Regarding

the critical initial formation of the PVS during embryogenesis, Soh et al. (7), on page 13 of their book, state:

> *Proliferation of the [PVS] meridians takes place ahead of proliferation of any other organs, such as the blood vessels and the nervous system.* Embryo development follows the following steps: the step for the formation of the primo vessel blast cell occurs 7–8 h after fertilization; the step for primordial primo vessel occurs 10 h after fertilization; the step for the formation of primitive primo lumens occurs 15 h after fertilization; and the final step for the completion of the primo lumens occurs 20–28 h after fertilization. The fact that the proliferation of the PVS precedes the formation of other structures suggests the PVS plays an important role during development of an organism. (Emphasis is mine)

The PVS is shown here to be intimately and inextricably involved with embryogenesis from as early as 7–8 hours postfertilization. I suggest that this is the initiation of the gate of life and also of the primordial formation of the Triple-Energizer Metasystem. This PVS structural conception is obviously tied to 'the original influences which man has received from father and mother', as discussed in the eighth Difficult Issue of the *Nan Ching*. As this omnipresent PVS structure is ubiquitous throughout the adult body, the embryological vestige is truly dispersed throughout the entire body via the omnipresent Connective-Tissue Metasystem that I propose is the domicile of the Triple-Energizer Metasystem, which is intimately entwined with the PVS. Avijgan and Avijgan (106) state, 'The PVS has been suggested to be responsible for embryonic development. . . . Accordingly, this report tries to present the PVS as the ruler of embryonic cell division and development, which regulates all complicated events during that period of life.' They further state, 'The PVS rules the *differentiation of epiblast cells* into the three (next) layers *ectoderm, mesoderm and endoderm* of an embryo' (emphasis is mine).

Stefanov et al. (84) also believe that the PVS is 'the oldest morphological functional system'. I consider that the Triple Burner is the primordial, original TCM organ to be formed during embryogenesis and subsequently deserves the honorable designation of 'origin'. The PVS has been shown to be the primordial organ formed very shortly after fertilization. However, due to the minuteness of the hollow strands of the omnipresent Primo Vascular System, I consider that the PVS structure requires support and protection from a larger more concrete structure, the likewise omnipresent Connective-Tissue Metasystem. Interestingly, the PVS is itself composed of connective tissue. Thus, I propose that the PVS and the various components of the associated supportive Connective-Tissue Metasystem constitute the Triple Burner or Triple-Energizer Metasystem.

In the commentaries on the 66th Difficult Issue on page 569 of Unschuld's (1) translation of the *Nan Ching*, while discussing the yuan points (origin or source points) on the 12 meridians, scholar Hua Shou states, '"Origin" represents an honorable designation for the Triple Burner.' If Avijgan and Avijgan (106) are correct, then the Primo Vascular System would be conceived from the very first day of embryonic life. That certainly is at the origin of life. As I believe that the PVS and the Triple Burner are intimately connected, this absolutely would explain why the term *origin* does represent 'an honorable designation for the Triple Burner'.

Regarding the anatomical location of the Primo Vascular System, Kwang-Sup Soh et al. (34) reported the IV primo vascular system (PVS) has been *detected floating inside* the caudal vena cava of rabbits, rats, and mice and that, 'more importantly, the PVS in the atrium of a bovine heart was *found to form a floating network*' (emphasis is mine). It thus appears that the Bonghan channels have a lot in common with the Triple-Energizer Metasystem in that they are composed of membranes which 'steam' and lubricate numerous body structures, including hair, joints, eyes, skin, and even heart valves.

Concerning the various tissues where the PVS has been found, Kwang-Sup Soh (7), on page 25, reports that the primo vascular system (PVS) has been observed in the 'nerve system, cardio-vascular system, lymphatic system, fascia in the abdominal cavity, adipose tissue, generative system (testis), skin and abdominal wall, primo fluid and microcells, egg vitelline membrane, and cancer'. It thus appears that the PVS is omnipresent within animals and humans. Generalizing the aspects of Bong-Han Theory, Jong-Su Lee (7), on page 23, states, 'All cells are connected to Bong-Han system, Bong-Han canal penetrate to nuclei of cells. If we have problems of hematological or endocrine systems, Bong-Han systems solve these problems.'

With respect to the Physiological function of the Primo Vascular System, I suggest that the Bonghan ducts are an integral component of the Triple-Energizer Metasystem—for example, in their tissue-lubricating function, their healing capacity, and their communication facility. I believe that due to their presence on the major organs of the body, the Bonghan ducts act as the mediators between the organs and the acupuncture meridian system. The fact that Primo microcells (P-microcells) flowing in primo vessels are hypothesized to function like very small totipotent embryonic-like adult stem cells could account for the healing component of Triple-Energizer Metasystem.

Regarding the properties of the PVS, in the 2013 *Journal of Acupuncture and Meridian Studies* article titled 'The Primo Vascular System as a New Anatomical System', the authors (84) note that recent (2013) research confirms that the PVS is composed of ubiquitous connective tissue and are predominantly collagen and as such is intimately related to the omnipresent Connective-Tissue Metasystem that pervades the body. Note that the authors suggest that the light propagation function of the PVS accounts for the instantaneous effect after needling acupoints, suggesting that the acupuncture meridians are indeed made up from or involved with the PVS. The PVS is *not* the same as the Triple-Energizer Metasystem but does contain collagen too.

Referencing the extreme complexity of the convoluted structure of the Primo Vascular System, in the article '50 Years of Bong-Han Theory and 10 Years of Primo Vascular System', the authors (34) reported that there are five subclass PVS networks and that the nature of the communication among those five subclass PVS networks is extremely complicated. The authors stated, 'All PVS observed and listed previously were *floating in the body fluid*, such as blood, lymph, abdominal fluid, and CSF' (emphasis is mine). Where else has the PVS been observed? The authors state, 'The presence of the *PVS entering the adipose tissues around the rat small intestine* was first noticed via optical imaging using Alcian blue, which was injected intravenously at the femoral vein. The Alcian blue entering a *PVS floating inside a blood vessel reached to a PN in an adipose tissue . . .*' (emphasis is mine). Note that often the PVS is located floating within body fluids. Yet many researchers insist that the Connective-Tissue Metasystem houses most of the interstitial fluids of the body.

Regarding the embryological development of the connective-tissue fascial system, in the 2014 article entitled 'Fascia & Tensegrity', Myers (110) states, *'Our single fascial system starts about 2 weeks into development as a fibrous gel that pervades and surrounds all the cells in the developing embryo. It is progressively folded by gastrulation and the rest of the motions of development into the complex layers of fascia we see in the adult.* In fact there is no discontinuity in the layers of fascia, so "layers" is a useful but deceptive concept (like "muscles")' (emphasis is mine).

Likely, initially, after fertilization, the miniscule Primo Vascular System conduits are able to support themselves inside the miniscule fertilized egg. A human oocyte is only about 100 μm in diameter. As further cellular differentiation and then gastrulation occurs, a more stable and resilient supporting and protective infrastructure would be required. Enter the supportive, omnipresent connective-tissue structure to secure and protect the delicate Primo Vascular System.

Regarding the structures for exterior primo vessels, the book (7), on page 11, states, 'These are composed of ducts and corpuscles along the blood vessels and nervous system. They are covered by a thick membrane of connective tissues.' The PVS is intimately associated with the connective tissues. Kwang-Sup Soh reported that the extravascular primo vessels running along the outside of the blood vessels were located just outside the connective tissues of the blood vessel. The book (7), on page 31, states there is a 'close relation between the PVS and the fascia'. I consider that the PVS throughout the entire body will be found to be intimately associated with and directly or indirectly connected to the omnipresent Connective-Tissue Metasystem.

Regarding the possible mechanism that explains how numerous physical activities produce healing throughout the body, in the article titled 'Can the Primo Vascular System (Bong Han Duct System) be a Basic Concept for Qi Production?', the authors (122) state:

> DNA included in the PMC [primo micro cell, sanal or granule] could be the source and storing place of biophotons which could be suggested to be called Qi. Based on this concept, any Qi (Biophotons) producing stimulators, for example, light (Laser or Infra Red ray), movement (massage), heat (Moxabustion) and needling, and so on, could stimulate the production of Qi. By having a wide web network, such as PVS, any point of body, which contains the PMC and DNA in it, could produce Qi (Biophotons), and this Qi could be distributed. If we were to have a map of PVS, this distribution could be controlled and Qi could be directed to target organs, which is the main purpose of acupuncture.

Note that the sanals have only been located in the hollow PVS ducts and *not* in the Connective-Tissue Metasystem fluid channels. So Yuan Qi circulation throughout the body pertains to the PVS conduits and not to the Connective-Tissue Metasystem conduits specifically. Thus, it appears that the piezoelectricity and fiber optic properties of the abundant connective-tissue collagen in the Connective-Tissue Metasystem are present to support the PVS and concurrently assist the PVS with information transfer throughout the body.

If the PVS actually is a major component of the original San Jiao, as I postulate, this agrees with TCM theory about San Jiao's function of maintaining harmony throughout all body structures. Avijgan and Avijgan (106) believe that the PVS 'controls *all developmental processes of the products of conception,* including trophoblast, inner cell mass differentiation, zona hatching and differentiation' (emphasis is mine). The San Jiao is believed to be the Fu organ that is responsible for numerous functions of homeostasis throughout the adult body, including energy production, digestion, fluid metabolism and synthesis, immune protection, hormonal balance, temperature regulation, detoxification through urine and feces formation, nutrient synthesis, etc. It would make sense that the PVS/San Jiao's function of maintaining harmony throughout all body structures, systems, and organs would continue from embryogenesis throughout the rest of the creature's life.

There is a strong water connection between the Primo Vascular System and the San Jiao. Primo Vascular System vessels are surrounded by membrane with high concentrations of fatty hyaluronic acid, which acts as a cushioning and lubrication agent. Hyaluronic acid molecules have the unique ability to attract and hold more than 1,000 times their weight in water, which is more than any other known biological substance. It is remarkable that here is a substance that readily holds such a vast amount of water present in the biological system of the PVS, which I consider is a major component of the San Jiao, the ruler of the waterways throughout the body according to TCM.

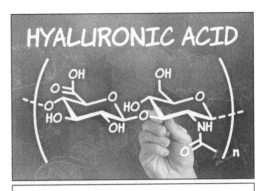

Figure 13. The chemical formula of hyaluronic acid. Photo by Zerbor. © Depositphotos.com.

39.2 Evidence of PVS Structures Uniquely Responsible for the Effect of Acupuncture Is Lacking

In the 2013 article titled 'The Primo Vascular Structures Alongside Nervous System: Its Discovery and Functional Limitation', the authors (150) state:

> Unfortunately, there was no evidence of the role of primo vascular structures in the effect of acupuncture. Recently,

Wang et al. (2012) examined the effects of stimulation of the primo vessels on gastric motility as well as the roles of the primo vessels in mediating the effects of acupuncture on gastric motility. In their results, direct electrical stimulation of the primo vascular structures on the stomach did not affect the gastric motility, and the inhibitory or stimulatory effects induced by acupuncture at CV12 or ST36 were unchanged after the primo vascular structures had been cut. They suggested that the primo vascular structures are not involved in the acupuncture modulation of gastric motility. The possibility that the primo vascular system might utilize other biological processes to achieve the therapeutic effects of acupuncture on gastric disorders cannot be excluded because Wang et al.'s study used only one measurement of gastric motility and the primo vascular system was implicated in a range of biological processes, such as the immune responses, hormone, and regeneration. Therefore, the functional roles of the primo vascular system in the effects of acupuncture on gastric motility are unclear.

As of 2013, no researchers have positively confirmed that the PVS is the actual substrate of the acupuncture meridian system. There appears little doubt that the PVS is intimately associated with acupoints and the acupuncture meridian system in some way. However, the specific mechanism is unclear. Many researchers believe that the PVS is a fiber-optic communication system of the body that regulates all cellular functionality throughout the entire body. The connective-tissue system has not been ruled out as the substrate for acupoints and the acupuncture meridian system. In fact, numerous studies of the low-resistance properties of acupuncture points have been performed. In 2008, 'Ahn et al. took ultrasonic images of acupuncture points and proposed, anatomically, that a collagenous band in connective tissues was related to the low-impedance property of acupuncture points' (151).

I personally do not believe that the Bonghan vessels (primo vessels) of the PVS constitute the acupuncture meridian system in its entirety.

For a start, the ducts are only 50–100 μm thick; that is only 0.05–0.1 mm wide. Acupuncture meridians are much wider and more forgiving than that. I believe that the acupuncture meridians directly communicate with and are intimately involved with the Bonghan system, which is immersed and floating within a subsystem of the Connective-Tissue Metasystem.

I trust it is only a matter of time before it is confirmed that the many different functions of the different types of acupuncture channels are somehow due to the circuitry of the PVS ducts. The PVS has been confirmed to be present in numerous deep internal organs (7). The PVS has been confirmed to be present in the superficial hypodermis (7). Remember too that research in 2009 (34) revealed that 'the flow of the primo fluid in a certain path was demonstrated in the study using Alcian blue, from the rat acupoint BL-23 in the dorsal skin to the PVS on the surface of internal organs'. The PVS engenders defined individual meridians connecting internal organs to the external skin surface. This is in keeping with the TCM theory of Main meridians and other meridians.

Several authors believe that the acupuncture meridian system is affiliated with the PVS. In the 2013 *Journal of Acupuncture and Meridian Studies* article titled 'The Primo Vascular System as a New Anatomical System', Stefanov et al. (84) state, 'Morphological science is greatly challenged to offer a new biomedical theory that explains the possible existence of new bodily systems such as the primo vascular system (PVS). The PVS is a previously unknown system that integrates the features of the cardiovascular, nervous, immune, and hormonal systems. It also provides a physical substrate for the acupuncture points and meridians.' The exact affiliation that connects the PVS to the functionality of the Acupuncture Meridian System still has to be elucidated.

39.3 Summary of the Connective-Tissue Metasystem
The origin of the Connective-Tissue Metasystem is very thought-provoking. The connective-tissue fascial system (Connective-Tissue Metasystem (CTM)) develops about two weeks into embryological

development. Regarding the embryological development of the connective tissue fascial system, on the 2014 web article entitled 'Fascia & Tensegrity', Myers (110) states, 'Our single fascial system starts about 2 weeks into development as a fibrous gel that pervades and surrounds all the cells in the developing embryo. It is progressively folded by gastrulation and the rest of the motions of development into the complex layers of fascia we see in the adult.' Emphasizing that fascia is continuous throughout the entire body, Myers stresses, 'In fact there is no discontinuity in the layers of fascia, so "layers" is a useful but deceptive concept (like "muscles")'. Likely, initially, after fertilization, the miniscule Primo Vascular System conduits are able to support themselves inside the miniscule fertilized egg. A human oocyte is only about 100 μm in diameter. As further cellular differentiation and then gastrulation occurs, a more stable and durable supporting and protective infrastructure would be required. Subsequently, the supportive omnipresent connective tissue structure enters the embryological scene to secure and protect the delicate hollow Primo Vascular System thread-like conduits.

Regarding the embryological derivation of the Connective-Tissue Metasystem, Avijgan and Avijgan (106) state, 'The PVS rules the differentiation of epiblast cells into the three (next) layers ectoderm, mesoderm and endoderm of an embryo.' Connective tissue is a major derivative of the mesodermal germ cell layer in the embryo. Thus, the PVS rules the differentiation of epiblast cells into mesoderm cells, which then engender connective tissues. The mesoderm connects with 'endodermal and ectodermal derivatives as the differentiation of organs and systems occurs'. By means of the mesoderm generating connective tissues, an intricate omnipresent supporting infrastructure is generated that enmeshes every cell, system, and organ within the body. Remember too that the mesoderm 'has a great responsibility by inducing some other internal organs and structures like the heart, kidney, spleen and connective tissue' (106).

Resembling connective tissue, the San Jiao envelops all cells, organs, systems, and structures throughout the entire body. Regarding the nature of the Ying Qi of TCM, the author of the book *Gua Sha: A Traditional Technique for Modern Practice*, Arya Nielsen, (20) on page 34, states, 'Like connective tissue, it is said there is nothing the San Jiao does not envelope, including the vessels that hold the Blood and conduct the Ying Qi.'

In the 2014 *Journal of Chinese Medicine* article, the authors, Steven Finando and Donna Finando (103), quoted Findley et al. when they stated, 'Fascia is defined as the soft tissue component of the connective tissue system that permeates the human body . . . [It includes] aponeuroses, ligaments, tendons, retinaculae, joint capsules, organ and vessel tunics, the epineurium, the meninges, the periosteal and all the endomysial and intramuscular fibers of the myofasciae.' They continue, 'The fascia of the human body is a continuous sheath of tissue that moves, senses and connects every organ, blood vessel, nerve, lymph vessel, muscle and bone. It is a continuous, three-dimensional, whole-body matrix, a dynamic metasystem that interpenetrates and connects every structure of the human body, an interconnected network of fibrous collagenous tissues that are part of a whole-body tensional force transmission system.'

Every single tissue in the body is either connective tissue or other than connective tissue. Connective tissue exists throughout the entire body and enmeshes and covers every tissue type and every organ that is *not* connective tissue. Interestingly, Avijgan and Avijgan (2014) (106) suggest that all human tissues and cells are linked to the PVS. Thus, both connective tissue and the PVS appear to directly connect to every human cell in the body simultaneously. I ask myself, is the miniscule delicate omnipresent PVS being supported, housed, and protected by the larger, sturdier, omnipresent Connective-Tissue Metasystem? As noted above, wherever the Connective-Tissue Metasystem exists, so does the Triple-Energizer Metasystem. Thus, the three systems—the PVS, the Connective-Tissue Metasystem, and the Triple-Energizer Metasystem—always seem to be at the same place at the same time, like the Three Amigos.

Regarding the omnipresence of connective tissue/fascia and the Primo Vascular System, in the 2013 *Journal of Acupuncture and Meridian Studies* article titled 'The Primo Vascular System as a New Anatomical System', the authors (84) confirm that the PVS is composed of ubiquitous connective tissue and that the PVS is intimately associated with and is abundant in loose connective tissue, serous membranes, and fascia. As it is likely 'distributed as a web among all body systems, including the tissues of organs,' this puts the PVS/connective-tissue network in the same location that the Triple Energizer occupy. This greatly supports my theory that a combination of the elements of the Connective-Tissue Metasystem, coupled with elements of the Primo Vascular System, constitutes the Triple-Energizer Metasystem or San Jiao.

While some researchers (84) believe that the PVS is the major substrate for the acupuncture meridian system, there appears to be no doubt that the Connective-Tissue Metasystem is also involved in the relationship. Perhaps because the components of the PVS are so very small in themselves, they require the supportive aspect of the Connective-Tissue Metasystem to perform their task. One could also ask, if the connective-tissue matrix is only present for the support of the bioelectrical PVS communication system, as some researchers assert, why should it be necessary for the predominant collagen fibers to possess such incredible bioelectrical characteristics, including fiber-optic and piezoelectric properties?

I believe the major point regarding my personal hypothesis is what Margulis (102) stated when she quotes James Oschman, who said, 'Connective tissue reaches every cell in the body.' This is by no means an original thought. Ho and Knight (82) state that 'collagens and bound water form a global network' throughout the entire body and that 'the "ground substance" of the entire body may provide a much better intercommunication system than the nervous system'. In complete harmony with Pollack (27), they also state that 'the hydrogen-bonded water network of the connective tissues is actually linked to ordered water dipoles in the ion-channels of the cell membrane' and beyond due to the 'connective tissue–intracellular matrix continuum.' Finando and Finando (103) state, 'The fascia of the human body is a continuous sheath of tissue that moves, senses and connects every organ, blood vessel, nerve, lymph vessel, muscle and bone. It is *a continuous, three-dimensional, whole-body matrix, a dynamic metasystem that interpenetrates and connects every structure of the human body*, an interconnected network of fibrous collagenous tissues that are part of a whole-body tensional force transmission system.' The authors (103) further state, *'Research has demonstrated that fascia should actually be considered as an organ that provides a unified environment contributing to the functioning of all body systems.* All of these comments show that there is a continuum between all the omnipresent connective tissue "membranes" that permeate the entire body right down to the intracellular matrix continuum' (emphasis is mine).

The book The Primo Vascular System: Its Role in Cancer and Regeneration (7), on page 31, states that there is a 'close relation between the PVS and the fascia'. I believe that the PVS throughout the entire body will be found to be intimately associated with and directly or indirectly connected to the omnipresent Connective-Tissue Metasystem, and I theorize that this amalgamation is synonymous with the Triple-Energizer Metasystem. I further consider that the PVS will be found to be responsible for body tissue types and zangfu relationships as per the example given by Qu Lifang and Mary Garvey (43), when they stated, 'The flesh and muscles belong to the Spleen, whilst the skin and soft body hair, at the most external layer of the body, belong to the Lung.'

Resembling the San Jiao, Loose Connective Tissue regulates all the body fluids. Findley and Shalwala (48) reported a remarkable property of fascia (pages 4–5). They stated, 'Loose connective tissue harbors the vast majority of the 15 liters of interstitial fluid.' I ponder over whether the fluid supposedly inside the Connective-Tissue Metasystem is actually inside the PVS but so far misidentified as the researchers were not been aware of or did not consider the existence of the fluid-filled PVS.

In the commentaries on the 25th Difficult Issue on page 313 of Unschuld's (1) translation of the *Nan Ching*, Ting Chin declares, 'In the Ling-shu and in the Su-wen, the treatise "Pen-shu" states: "The Triple Burner is a palace acting as central ditch; the passage-ways of water emerge from it." It is associated with the bladder and it constitutes the palace of uniqueness.' From detailed information discussed by Findley and Shalwala (48), it becomes blatantly obvious that references pertaining to the proven properties of fascia, including 'loose connective tissue harbors the vast majority of the 15 liters of interstitial fluid', 'interstitial flow regulates nutrient transport', 'towards lymphatic capillaries', 'blood flow to skeletal muscle', 'inelastic fascia can promote lymphatic flow', 'fluid volume is regulated', 'resulting in edema formation', and 'fluid flow can increase almost 100 fold within minutes' all pertain to the *passageways of water*. From the above research findings, it has been shown that the Connective-Tissue Metasystem exercises control over interstitial fluid, blood, and lymph, the three most voluminous liquids in the body. So here we can see that the recent scientific research findings associated with the connective tissues of the body (Connective-Tissue Metasystem) acting as a central ditch and regulating the *passageways of water* are all functions attributed to the Three Heaters' control of all the various fluid distribution throughout the body.

Andrew Taylor Still, MD (1828–1917), an American physician, was the founder of osteopathic medicine. I believe the man was a sage. Regarding the embryologic plasticity of fascia, 115 years ago, Still (48) (page 4) said, 'Fascia functions by secreting and excreting fluid vital and destructive. By its action we live, and by its failure we shrink, swell, and die. . . . This connecting substance must be free at all parts to receive and discharge all fluids, if healthy to appropriate and use in sustaining animal life, and eject all impurities that health may not be impaired by the dead and poisoning fluids.' Regarding the omnipresence of fascia, Still (48) (page 5) further states, '[Fascia] is almost a network of nerves, cells and tubes, running to and from it; it is crossed and filled with, no doubt, millions of nerve centers and fibers. . . . Its nerves are so abundant that no atom of flesh fails to get nerve and fluid supply therefrom. . . . The cord throws out and supplies millions of nerves by which all organs and parts are supplied with the elements of motion, all to go and terminate in that great system, the fascia.'

In the first citation above, Still advises that the fascia control fluids in the body and that if the fascia fails, then we shrink, swell, or die. This indeed sounds exactly like a description of the Triple Energizer.

In the second citation above, Still describes fascia as a 'network of nerves, cells and tubes, running to and from it' such that 'no atom of flesh fails to get nerve and fluid supply therefrom'. Once again, these properties of fascia (communication via the associated nerve connections and control over fluids) are identical to those of the Triple Energizer. Note that Still also stated that the fascia is a network of tubes. I believe that it is most likely that these 'tubes' also include a part of the Primo Vascular System network of microtubules, a microsystem that was not known about in Still's time.

Glenda Poletti (135) advises that ground substance is found in all connective tissue and that it is produced by fibroblasts, which are among the earliest cells to develop in the embryo and that these fluid ground substances are the immediate environment of every cell in the body. So the basic building block of connective tissue, ground substance, is produced by the same primordial fibroblasts, which are 'among the earliest cells to develop in the embryo'. So here is a major component of the Connective-Tissue Metasystem (i.e. ground substance), present at conception and subsequent embryological development. It is paramount to note that via control from the PVS, the mesoderm induces or creates connective tissue initially, and then later the mesoderm induces or creates the Kidney organ and other major organs.

When a successful fertilization does occur, the gate of life is opened, and the Prenatal Jing from the parents is passed to the embryo and nourishes the embryo and fetus during the pregnancy. There is also supplemental energy that is derived from the Kidneys of the mother during the pregnancy. The Prenatal Jing defines the basic constitution, strength, vitality of the fetus, and later, the adult. Prenatal Jing is fixed in quantity and is already determined at birth. Prenatal Jing cannot be supplemented or increased but can be conserved and should not be squandered or wasted. It is stored in the Kidneys (140). I have discussed in great detail elsewhere in this book what I believe happens when the gate of life is opened. Briefly, the Primo Vascular System initiates the primordial differentiation process within hours of fertilization. By two weeks postfertilization, the Connective-Tissue Metasystem proper allows the tissues to differentiate into organs and systems.

So the basic building block of connective tissue, ground substance, is produced by the same primordial fibroblasts, which are 'among the earliest cells to develop in the embryo'. So here is a major component of the Connective-Tissue Metasystem (i.e. ground substance) present at conception and subsequent embryological development. Thus, the primordial fibroblasts generate the 3-D matrix in which all the organs and systems are suspended, surrounded by the Connective-Tissue Metasystem. For this reason, I believe that a combination of the elements of the Connective-Tissue Metasystem, coupled with elements of the Primo Vascular System, constitutes the Triple-Energizer Metasystem or San Jiao. Thus, simply stated, at conception, when the gate of life is opened, the primordial fibroblasts generate the Connective-Tissue Metasystem, which I believe is intimately related to the Triple Burner. Thus, it can be stated that 'the gate of life is the root of the Triple Burner'.

I found it fascinating that in the Chapter 'Biophotons and the Highly Ordered Nature of Cellular Communication', the different authors presented analogous evidence and concluded that the acupuncture channels allow communication internally and with the external environment (44), communication throughout the entire body and beyond is mediated by the Triple Heater (11) (83), and that the PVS allows communication between living organisms and the environment (84). From these analogous comments, it appears that acupuncture channels, the Triple Heater, and the Primo Vascular System are all responsible for communication within the body and with the external environment. I believe that these three systems are intimately related and interconnected in that the primordial Primo Vascular System actually initially engendered the Connective-Tissue Metasystem as a support structure, and thereafter, the Connective-Tissue Metasystem, in association with the Primo Vascular System, came to constitute the Triple-Energizer Metasystem. I believe further that the Acupuncture Meridian System is a combined derivative of the collagenous Connective-Tissue Metasystem in intimate association with subbranches of the Primo Vascular System. I propose that due to the hydrophilicity of the omnipresent Connective-Tissue Metasystem, the omnipresent water in the body is transformed into liquid crystalline EZ, and this derived ubiquitous material permits biophotonic information transfer for both the Triple-Energizer Metasystem and the Acupuncture Meridian Metasystem to perform their delineated roles.

39.4 Summary of the Triple-Energizer Metasystem (San Jiao)

TCM teaches that the primordial Yuan Qi circulates via the Triple Energizer to every main acupuncture meridian and that Yuan Qi is dominant and surfaces at the Yuan (Source) points and that needling the Yuan point of an ailing hollow or solid organ will benefit that specific organ. Remarkably, sanals (p-microcells or stem cells) circulate through the PVS conduits and are able to transform into any cell type that has been damaged and cause tissue rejuvenation. I suggest that the p-microcells constitute what the ancient Chinese sages called the Yuan Qi and that the PVS conduits represent to some degree the Triple Energizer.

In the commentaries on the 31st Difficult Issue on page 354 of Unschuld's (1) translation of the *Nan Ching*, regarding the root and the origin of the Triple Burner, scholar Yeh Lin states, 'It is a *fatty membrane* emerging from the *tie between the kidneys*' (emphasis is mine). Also in the commentaries for the 31st Difficult Issue on page 355, Huang Wei-san said, 'The Triple Burner encloses all the depots and palaces externally. *It is a fatty membrane covering the entire physical body from the inside*' (emphasis is mine). In the commentaries on the eighth Difficult Issue on page 132 of Unschuld's (1) translation of the *Nan Ching*, regarding the origin of the vital influences or the transmutation of bulk water into different biological fluids and liquids, scholar Yeh Lin relates:

> Together with the fire of the heart, these [yang influences] *cause the water of the bladder to rise as steam*. [Hence, the water] is transformed into [volatile] influences. These ascend via the through-way and 'employer [vessels].' They pass the diaphragm and enter the lung whence, in turn, they leave [the body] through mouth and nose. [Some of the] influences which ascend in order to leave [the body] are transformed into liquids in the mouth and by

the tongue and in the depots and palaces. These [liquids] leave [the body] through the skin [and its] hair by way of the 'influence paths' (ch'i-chieh). They serve to steam the skin and to soften the flesh. They constitute the sweat. [All of] this occurs [in accordance with] the principle that fire is transformed into [volatile] influences when brought into water. (Emphasis is mine)

Regarding the root and the origin of the Triple Burner, note that scholar Yeh Lin states, 'It is a fatty membrane emerging from the tie between the kidneys.' Yeh Lin makes two important comments about the Triple Burner. Firstly, he says the Triple Burner is a fatty membrane, wherein I conclude that the Triple Burner is derived of fatty connective-tissue membrane. That is not too big a stretch of the imagination. Then he states that it emerges 'from the tie between the kidneys'. He does not say that it originates *from* the right Kidney. The Triple Heater is also said to originate from the Gate of Life, which is also considered to reside in close proximity to the right Kidney. I believe that this description pertains to the location of Triple Burner in an adult. During embryogenesis, the fatty connective-tissue membranes originate prior to the derivation and formation of the Kidney, which is formed only a few days later. I suggest that the Bonghan ducts are an integral component of the Triple-Energizer Metasystem—for example, in their tissue-lubricating function, their healing capacity, and their communication facility.

Regarding the issue of the Triple Burner transmitting the original influences (Yuan Qi) through the 12 conduits and 'the influences on which man's life depends', in the commentaries on the eighth Difficult Issue on pages 135–136 of Unschuld's (1) translation of the *Nan Ching*, scholar Tamba Genkan relates, 'The Triple Burner is a special envoy [transmitting] the original influences.' Here, Tamba Genkan clearly shows that 'the Triple Burner is a special envoy transmitting the original influences' (Yuan Qi) throughout the 12 conduits, originating from the 'kuan-yüan section', which is the acupuncture point CV 4 (Origin Pass). Amazingly, Soh (109) showed that the PVS is 'a conduit for a fluid, which, according to proteomic analyzes contains remarkably high levels of carbohydrate metabolic derivatives that are usually associated with stem cells.' The article poignantly continued that 'the PVS has the potential to transport growth and communication factors' throughout the body. These exact functions have been attributed to the Triple-Energizer Metasystem for thousands of years.

In the commentaries on the 66th Difficult Issue on page 561 of Unschuld's (1) translation of the *Nan Ching*, it is again stated, 'The Triple Burner is the special envoy that transmits the original influences.' Treating the source points of a particular meridian will remediate the injured organ of that meridian. For example, the yuan point (source point) of the Liver, Liver 3, will benefit the Liver organ when there is illness there. So thanks to the Triple-Energizer Metasystem, the healing Yuan Qi arrives at Liver 3 and then the Yuan Qi (healing Primo Vascular System P-microcells (Sanals)) is directed to the ailing Liver. Remember, these pluripotent P-microcells (stem cells) are the main agent for wound healing and regeneration.

Many Chinese scholars referred to the Triple Heater as the origin. For example, in the commentaries on the 66th Difficult Issue on page 569 of Unschuld's (1) translation of the *Nan Ching*, while discussing the yuan points (origin or source points) on the 12 meridians, scholar Hua Shou states, '"Origin" represents an honorable designation for the Triple Burner.' The authors (84) quoted above believe that the PVS is 'the oldest morphological functional system'. I believe that the Triple Burner is the primordial, original TCM organ to be formed during embryogenesis and subsequently deserves the honorable designation of 'origin'. Thus, I propose that a portion of the PVS in association with the Connective-Tissue Metasystem constitute the Triple Burner or Triple-Energizer Metasystem.

Chapter 38 of the *Nan Jing* discusses the relationship between the Triple Burner and the Yuan Qi. It says: 'There are 6 Fu because of the Triple Burner which stems from the Yuan Qi. The Triple Burner governs all Qi in the body.' Recent research confirms that proliferation of the Primo Vascular System is the very first phase of embryological development and differentiation, commencing as early as seven to eight hours after fertilization. Soon thereafter, the connective-tissue fascial system originates its formation and differentiation as early as two weeks after fertilization. Thus, the PVS and the Connective-Tissue Metasystem are present at the very beginning of life, actually when the gate of life is opened. As the PVS is composed of connective tissues anyway, it is essentially the Primo Connective Tissue System. So it appears that the comingled life forces of the Sperm + Ova engender the Primo Vascular System. Later, the process called gastrulation occurs, whereby the blastocyst (once implanted into the endometrium) develops the three germ layers—ectoderm, endoderm, and the mesoderm—which later give rise to all the different organs, systems, and structures of the human body. I believe that this is aka the Yuan Qi engendering the Triple Burner, as stated in Chapter 38 of the *Nan Jing*, where it says, 'The Triple Burner stems from the Yuan Qi [gate of life].' Should the embryo continue to grow and prosper to term, it is via the PVS and the Connective-Tissue Metasystem that the life force is allowed to be molded or woven into the numerous differentiations that become all the individual tissues, organs, and structures that engender a human being. It is for this reason that the Connective-Tissue Metasystem surrounds every single organ and structure like an outer wall or like a wrapping bag. Subsequently, I suggest that the omnipresent Connective-Tissue Metasystem is intimately associated with the omnipresent Triple-Energizer Metasystem.

In the commentaries on the 38th Difficult Issue on page 396 of Unschuld's (1) translation of the *Nan Ching*, Li Chiung states, '*The Triple Burner represents nothing but membranes attached to the upper, central, and lower opening of the stomach*. The Triple Burner has a name, but no real form' (emphasis is mine). In this text, Li Chiung states very clearly that 'the Triple Burner represents *nothing but membranes*' (emphasis is mine). This is what I believe to be the case where the Triple-Energizer Metasystem is a ubiquitous maze of omnipresent membranes or connective tissue that permeates the entire body by wrapping all the numerous organs and structures with connective-tissue fascia like a bag, thereby allowing communication between every cell in the body via the fascia and the lymphatic system, which is supported and held in place by the structural framework and scaffolding of the omnipresent connective-tissue fascia.

This is also the view of Peluffo (125). She states, 'The connective tissue is a versatile organic system, quite widespread and omnipresent which, being both well-known and well analyzed by modern medicine, interconnects all parts of human body in each level, from the macro to microscopic one and can be found even in the simplest of the organelles, in each cell and in each texture or body framework. Of greater importance for Eastern medicine—both theoretical and practical—are the now known properties these tissues show: to generate energy and to conduct it.'

The Triple Burner 'externally completely encloses the stomach'. This is also true for all the other organs too. Because the Triple Burner surrounds the Stomach like an outer wall or like a large wrapping bag, there is an intimately close relationship that allows the Stomach to perform all its attributed functions thanks to the proximity and mutual assistance of the Triple Burner. However, due to the macerated mixture of drinks and grains, the Stomach empowers the Triple Burner. I believe the influences sent out by the Stomach that empower the Triple Burner are predominantly the bulk water consumed for hydration, along with the 80% water present in most foods (grains), which are transmuted into energized, living liquid crystalline EZ water, along with the biophotons encapsulated in the foods.

Regarding the flow of Qi inside the spaces (cou) within the Sanjiao network, in the *Journal of Chinese Medicine* article titled 'The Location and Function of the Sanjiao', the authors, Qu Lifang and Mary Garvey (43), cite *Jinkui Yaolue Fanglun* (clauses 1–2), which says, 'The spaces (cou) are a place of the Sanjiao where yuan qi and zhen qi circulate and converge, the spaces are suffused with blood and qi.' They then cite *Neijing Suwen* (chapter 58), which states, 'Big convergence of muscle is called valley; small convergence of muscle is called brook, circulation of ying wei and convergence of big qi take place in between the sheaf of muscle and joint of valley and brook.' The third citation they make to clarify their point is from *Nan Jing*, Difficulties 31 and 36, which says, 'The Sanjiao is a place of qi starting and ending. The Sanjiao is in charge of passing through the three qi that go through the five zang and six fu.' The authors then explain, 'Here, "big qi" and "three qi" refer to yuan qi, zhen qi, and wei qi. Whilst wei qi is concerned with defense against pathogenic influences, the yuan qi and zhen qi transmit the functional aspect of jing-essence from ming men to the zangfu and body tissues. Yuan, then, and wei qi circulate and converge throughout the body, and the Sanjiao supplies a site and thoroughfare for this.'

Here again it makes perfect sense that the three big qi—that is, yuan qi, zhen qi, and wei qi—all flow through the channels of the Connective-Tissue Metasystem and the Primo Vascular System's hollow, highly-organized conduits (cou), which pervade the body as an organized network (li) from the superficial skin (where the cou are the pores) to the couli network of interstitial spaces that pervade the body. Also included is the couli network of synovial joints of bones throughout the body to the deepest organs, allowing the different forms of fluid and qi to be directed to the required locations so that they can perform their respective functions. All these are under the control of the ubiquitous, omnipresent Triple-Energizer Metasystem, which occupies the same space as and is intimately affiliated with the omnipresent Connective-Tissue Metasystem.

What is the energetic mechanism that makes the Triple Heater work? This is the pivotal concept that I believe embodies the Triple Energizer. In the presence of omnipresent hydrophilic connective tissue, water throughout the body is converted into its fourth phase, EZ water, by free electrons from the earth and biophotons from the sun. This confirms our absolute dependence on the sun and the earth to maintain optimal health and well-being. Note that the free electrons from the earth (along with light from the sun) generate EZ water, which holds the energy derived from the earth (or sun), much like a battery, so that the energized water is circulated throughout the Triple Energizer via the ubiquitous, omnipresent, hydrophilic PVS/Connective-Tissue Metasystem, which pervades our being and covers and communicates with every single cell in the body. All food and water that we consume has been irradiated and energized by the sun so that the biophotons therein are what are absorbed from the Stomach and circulated throughout the entire body via the Triple Heater cycle. These biophotons further empower the formation of

EZ water, which nourishes and empowers every cell in the body and, along with the sanals present in the PVS, optimize rejuvenation and self-healing.

39.5 Summary of the Acupuncture Meridian System

There appears to be no doubt that PVS conduits are somehow involved with acupoints and the acupuncture meridian system. Researchers (7) (34) have traced injected dyes flowing from the skin surface at defined acupoints to the inner organs deep inside the body. As the dye has been shown to flow through the defined PVS conduits, there is no doubt that the PVS conduits are somehow involved with the acupoints and the acupuncture meridian system. As PVS ducts are only 50–100 µm thick (only 0.05–0.1 mm wide), this is too minute to constitute an acupuncture meridian in its entirety. Acupuncture meridians are much wider and more forgiving than that. However, concerning the confirmation of the Primo Vascular System being intimately associated with the conception vessel acupuncture meridian in some way, on page 29 of their book, Soh et al. (7) report:

> Primo nodes at CV 12, 10, 8 were observed, and basic histological study with H&E and Mason's trichrome revealed that they were different from lymph nodes. By injecting FNP [fluorescent nanoparticles] into the primo nodes, we traced the flow of nanoparticles along the CV line to the ligament wrapping the bladder in the primo vessels. Thus, we established the presence of extravascular primo vessels along the blood vessels just outside the connective tissues of the blood vessel. In this experiment, the PVS ran along the CV fat line to the bladder.

Due to the minuteness of the PVS conduits, they could not be supported and protected without help from the existing omnipresent framework, namely the Connective-Tissue Metasystem. As acupuncture is holistic, a structure that allows instant communication throughout the entire body is essential. It has been proposed by researchers that the Primo fluid allows communication throughout the entire body via biophotons stored within the sanals. It is noteworthy that the predominant connective-tissue component, collagen, is piezoelectric and has fiber optic properties. These bioelectrical properties of collagen most likely complement and synergize the bioelectrical properties of the miniscule PVS thread-like conduits in their communication role. Regarding the fact that successful communication within the body must involve every single cell, Robertson (44) poignantly states:

> Where exactly are the channels located? Expanding upon the assertion in the Inner Classic that the channels are 'subsumed in the spaces around the flesh — deep and not visible', Professor Wáng Jūyì proposes that we should conceive of the channels as being found in the spaces surrounding all of the structures of the body (i.e. not just in the areas where acupuncture points are found but even within and around the organs). When reviewing modern research into possible anatomical correspondences to the channels, we see promising leads by those considering the physiological role of fascia and connective tissue. Nevertheless, *strictly speaking we might say that it is not the fascia to which the Inner Classic is referring above, but instead to the quality of the fluids moving within and around these connective tissues.* The Inner Classic abounds in metaphors likening qi circulation to water systems. Thus as acupuncturists looking to find the channels, our attention should focus more on the nature of fluids within the connective tissues and less on the surrounding solid structures. (Emphasis is mine)

Based on Professor Wáng Jūyì's theory, note that Robertson suggests that the acupuncture meridians occupy 'the spaces surrounding all of the structures of the body'. More specifically, he singles out 'the fluids moving within and around these connective tissues' and confirms that 'the Inner Classic abounds in metaphors likening qi circulation to water systems'. This is very plausible, remembering that Professor Pollack (27) has proven that water in contact with hydrophilic materials (which includes the collagen of connective tissues) is transformed into the fourth-phase liquid crystalline water (EZ) in situ. Thus, Pollack's explanation allows the production of energy (Qi) to occur. Thereafter, the Qi could easily flow throughout the continuous acupuncture meridian system.

Note too the title of an article written by Dr Ho (61) in 2012, namely 'Super-Conducting Liquid Crystalline Water Aligned with Collagen Fibres in the Fascia as Acupuncture Meridians of Traditional Chinese Medicine'. While Dr Ho believes that collagen fibers in the fascia aligned with EZ water constitute the acupuncture meridians, I believe one step further that the omnipresent connective tissues, collagen fibers, fascia, and membranes make up the Triple-Energizer organ complex. Dr Ho also said, 'It is water that fuels the dynamo of life; water is the basis of the energy metabolism that powers all living processes, the chemistry and the electricity of life.' Again that sounds like the major function of the Triple Energizer organ complex, which is 'the official in charge of irrigation and it controls the water passages'.

Proposing a scientific hypothesis for the description of Qi, in the 2013 *Journal of Acupuncture and Meridian Studies* article titled 'The Primo Vascular System as a New Anatomical System', the authors (84) state, 'We offer a new point of view concerning the type of vital energy Qi is. Because the PVS may be an optical channel for photon emission, an electromagnetic field that travels throughout the PVS and throughout the DNA in the PVS may be the mysterious vital energy Qi that can be distributed throughout the entire body.' Stefanov et al. (84) further suggest that 'the ancient vital energy Qi probably is an electromagnetic wave that is transported through the PVS, and the information obtained from that electromagnetic wave may be stored in the DNA of the PVS microcells'. They suggest that 'the PVS is furthermore the physical substrate for the acupuncture points and meridians and is involved in the development and the functioning of living organisms'.

Could the acupuncture meridian system be composed of the collagen of connective tissues in affiliation with water that has been transmuted into EZ water? James L. Oschman (152) noted that, in collaboration with Peter Gascoyne and Ron Pethig, Albert Szent-Györgyi (Nobel Prize winner in physiology or medicine in 1937) demonstrated that the protein collagen is a semiconductor. As a biophysicist and cell biologist, Oschman reported that he has always found connective tissue fascinating and noted that the collagen that makes up connective tissue is the most abundant protein on the planet.

He continued, '*Fascia forms the largest system in the body as it is the system that touches all of the other systems.* It can be described as the construction fabric of the animal body' (emphasis is mine). Oschman noted, 'The role of water in relation to the living matrix has been explored by Mae-Wan Ho in collaboration with David Knight. They have suggested that the water associated with collagen constitutes a major part of the acupuncture meridian system and is involved in memory functions.' The living matrix cited above is defined as 'the continuous molecular fabric of the organism, consisting of fascia, the other connective tissues, extracellular matrices, integrins, cytoskeletons, nuclear matrices and DNA'.

A remarkable feature of fascia is that it is highly organized and tightly bound. In fact, fascia is so regular and repeating in its organization that fascia meets the definition of being *liquid crystalline* in nature. Fascia actually even performs as a liquid crystal. Fascia within our body can generate and transmit piezoelectricity. Physical stress and pressure applied to a piezoelectric material generates electricity. Subsequently, movement within our bodies generates an electrical signal that is transmitted to literally every cell in our bodies.

Pollack (27) (KL 1593) believes that 'the EZ battery could be a *versatile supplier* of much of *nature's energy*' (emphasis is mine). Is Professor Pollack discussing the functions of the Triple-Energizer Metasystem, or is he touting his theory on the nature of Exclusion Zone (EZ) Water in biological systems? I believe they are one and the same. I propose that the Triple Energizer is predominantly composed of hydrophilic connective tissue and membranes, which are omnipresent throughout the body. Pollack has proved beyond doubt that bulk water (present in drink and foods) in contact with hydrophilic materials is converted to EZ water. What Professor Pollack just stated could easily describe the machinations of how the Triple Energizer distribute water and fluids throughout the entire body and simultaneously supply energy throughout the entire body as the various differentiated forms of Qi. This mechanism of Pollack's allows for both features of the Triple Energizer to emerge simultaneously via the same medium, namely the energized fourth phase of water, life-giving liquid crystalline EZ water. Eureka! I believe I now know where the Triple-Energizer Metasystem has been hiding all this time.

It is likely that acupuncture meridians pervade the entire body, but only the most influential meridians have been defined and their pathways and their associated acupoints specified as they have the most influence on health. Note that there are hundreds of extrameridian acupoints that have been reported, and they include hundreds of auricular, nose, scalp, hand and foot, and scalp points, for example. Some authorities believe that there are specific defined acupuncture meridians that flow within each ear. Iridology and hand-and-foot reflexology also rely on information from the entire body being relayed to the eyes, hands, and feet respectively. Is it the PVS or the Connective-Tissue Metasystem that communicate the status of all the organs to all the reflex locations? Or is it *simply* the nervous system, which as it turns out may be controlled by the PVS anyway? (84)

Regarding the many types of acupuncture meridians or channels within the body, Margulis (102) advises that there are many different channels. Some channels are very superficial, and some are very deep. The differing forms of channels include cutaneous channels, minute collateral channels, sinew channels, luo-connecting channels, primary meridian channels, divergent channels, and the eight extraordinary channels. There also exist the deep pathways of the primary channels and the divergent channels. Margulis (102) concluded the article by saying, 'Two months ago you heard James Oschman describe how connective tissue reaches every cell in the body. *I believe that at least some acupuncture channels occur along connective tissue pathways*' (emphasis is mine).

While there are 361 traditionally defined acupuncture points on the main acupuncture channels, there are hundreds of secondary acupuncture points known as extrameridian points. This established fact alone shows that the acupuncture meridian system is more all-encompassing than just the defined 12 main meridians of TCM. This is in keeping with the fact that the Primo Vascular System is composed of an omnipresent maze of thin threads that is also all-encompassing and pervades the entire body and, as many researchers believe, has an intimate connection with every tissue and cell within the body.

It has been proposed that acupuncture meridians are a specialized information network within the 'living matrix'. Regarding the research of Ho and Knight cited above, in his manuscript, James L. Oschman (152) further states:

> They also proposed that there is a body consciousness, possessing all of the hallmarks of consciousness—sentience, intercommunication, and memory, existing alongside 'brain consciousness.' They propose that 'brain consciousness' which we usually consider to be the only consciousness, is embedded within this 'matrix consciousness' and is coupled to it. In the context of such a system the acupuncture meridians are regarded as a specialized information network, based on liquid crystalline resonant pathways, that link and coordinate the various structures and functions within the organism, separate from or along with neural communications.

While it is incredible to think that our bodies contain more 'empty space' than anything else, this fact is actually true. In his manuscript, James L. Oschman (152) states, 'The largest generator of electricity in the body is the heart. It is obvious that the matrix is one system that conducts the "music of the heart" to all parts of our bodies. Research from the Institute of HeartMath has documented the emotional aspects of the harmonics produced by the heart and these harmonics have important roles in healing.' He continues, 'The new research on epigenetics is teaching us that the way we think about ourselves and even the words used by the people around us can cause changes at the level of our DNA molecules. It is said that the DNA in every cell in your body is listening to every word you say.' For these reasons,

Oschman believes that 'one can see that the vibratory living matrix probably plays a key role in delivering the vibrations of our words and thoughts to every DNA molecule in our bodies'. Regarding the very nature of our body composition, Oschman stated, 'Indeed, our bodies contain more "empty space" than anything else, and quantum physics tells us that empty space is not, in fact, empty. Space is alive with energy and information and connects us to the deeper levels of consciousness and healing at a distance.'

A recent (2015) research finding is that acupuncture points and the PVS both have a higher density of mast cells (MCs). It is known that mast cells (MCs) play a major role in allergic reactions. Research has confirmed that acupuncture points have a higher density of MCs compared with nonacupoints in the skin. This finding is consistent with the fact that acupuncture treatment amplifies the immune function. Researchers (153) hypothesized that the primo vascular system (PVS), which was proposed as the anatomical structure of the acupuncture points and meridians, should have a high density of MCs. To test that hypothesis, the team 'investigated the primo nodes isolated from the surfaces of internal organs, such as the liver, the small and the large intestines, and the bladder'. During the study, harvested primo nodes were stained with toluidine blue, and the MCs were easily recognized by their red-purple stains and their characteristic granules. The authors reported, 'The results showed a high density of MCs in the primo nodes and confirmed the hypothesis. The MCs were uniformly distributed in the nodes. The relative concentration of the MCs with respect to other cells was ~15%.' The authors (153) concluded that 'the current work suggests that the PVS may participate in the immune response to allergic inflammation, which closely involves MCs'. This research lent support to the hypothesis that the PVS is involved with the anatomical structure of the acupuncture points and meridians as both entities have a higher density of MCs compared with nonacupoints in the skin.

Within the water that is present within the body, it can be stated there is Fire in the Water and Yang within the Yin. Let me explain. The Water element Kidneys are the organ of choice to process and control water and fluids in the body. But the Triple Energizer is also assigned that role, which is an obvious choice due to the water content of the ubiquitous, omnipresent, hydrophilic (water-loving) maze of connective tissues that pervade the body and occur only where something other than connective tissue wishes to reside. Also obvious is the further connection with the bladder, which holds the wastewater of the body. Interestingly, all these organs—the yang right Kidney, the Triple-Energizer Metasystem, and the Bladder—are resplendent with fire and yang. The highly energized and longest meridian is the Bladder meridian with its 67 points. As these 3 organs have a major role in water and fluid circulation, they are in constant contact with highly charged EZ Water [H_3O_2]⁻, which, by its very nature, is highly energized. It performs as a battery that is contained within the Triple-Energizer Metasystem network, which permeates the entire body and is only micrometers from every cell in the body that requires an energy top-up. The sagacious Yu Tuan was most enlightened when he considered that the Triple Heater was the fire that circulated throughout the body to develop all life. I believe that modern science, through Professor Pollack (27), has confirmed that this fire is contained within the very water that the Triple Energizer circulate and manage in the form of EZ Water, also known as the hydrated hydroxide ion or hydroxide hydrate anion [H_3O_2]⁻.

So what is the proposed composition of the acupuncture meridian system? I do not believe that the Bonghan vessels (primo vessels) constitute the acupuncture meridian system in its entirety. For a start, the ducts are only 50–100 μm thick—that is, only 0.05–0.1 mm wide. Acupuncture meridians are much wider and more forgiving than that. I propose that the acupuncture meridians directly communicate with and are intimately involved with the Bonghan system, which is immersed and floating within a subsystem of the Connective-Tissue Metasystem, which is itself the Triple-Energizer Metasystem. I suggest that the Bonghan ducts are an integral component of the Triple-Energizer Metasystem—for example, in their tissue-lubricating function, their healing capacity, and their communication facility. I believe that due to their presence on the major organs of the body, the Bonghan ducts act as the mediators between the organs and the acupuncture meridian system. The fact that Primo microcells (P-microcells) flowing in primo vessels are hypothesized to function like very small totipotent, embryonic-like adult stem cells could account for the healing component of Triple-Energizer Metasystem. I propose that the P-microcells within the microtubules of the Primo Vascular System constitutes what the ancients called Yuan Qi circulation constituting 'the original influences which man has received from father and mother', as Li Chiung put it.

I further believe that all the acupuncture meridians constitute a part of the Connective-Tissue Metasystem conduits, which are somehow associated with the PVS conduits, and that the EZ water flows alongside then and through them and that biophotons and piezoelectricity are transmitted along them. Sanals within the PVS are also involved in the healing aspect of the meridian complex.

39.6 Summary of the Origin of the Kidneys

Embryologically, it is paramount to note that the Primo Vascular System develops first and then the Connective-Tissue Metasystem is formed, prior to the formation of the Kidney organ. Importantly, it is via control from the PVS that the mesoderm induces or creates the connective-tissue matrix initially, and shortly after that, the mesoderm induces the formation (creation) of the primal paired pronephri (Kidneys) close to the neck.

The original influences (Yuan Qi) ascend from the Kidneys through the Chong Mai. In the commentaries on the eighth Difficult Issue on page 132 of Unschuld's (1) translation of the *Nan Ching*, regarding the 'moving influences between the two kidneys', scholar Li Chiung relates, *'The moving influences between the two kidneys are the original influences which man has received from father and mother.* Furthermore, the [movement of the] influences in the through-way vessel emerges from between the kidneys. The body's five depots and six palaces have twelve separate conduits. Their root and foundation are, indeed, the kidneys' (emphasis is mine). Li Chiung continues, 'The [influences of the] kidneys ascend, on both sides of the employer vessel, toward the throat. [The throat is responsible for] passing the breath; it serves as the gate of exhalation and inhalation. The

Triple Burner of man is patterned after the three original influences of heaven and earth; *the kidneys are the basic origin [of the Triple Burner]* (emphasis is mine). I understand this quotation to mean that 'the kidneys are the basic origin of the Triple Burner'; otherwise stated, the original formation of the Triple Burner is the Kidneys. This is in keeping with embryological chronological organ formation. Regarding this matter, note that there is a subtle disagreement between the commentaries on the 8th Difficult Issue and the 31st Difficult Issue. In the commentaries on the 31st Difficult Issue on page 354 of Unschuld's (1) translation of the *Nan Ching*, regarding the root and the origin of the Triple Burner, scholar Yeh Lin relates, '*It is a fatty membrane* emerging from the tie between the kidneys.'

So what could this '*fatty membrane* [emphasis is mine] emerging from the tie between the kidneys' be? So note here that the origin of the Triple Burner 'is a fatty membrane emerging from the tie between the kidneys'.

In the commentaries on the 66th Difficult Issue on page 569 of Unschuld's (1) translation of the *Nan Ching*, while discussing the yuan points (origin or source points) on the 12 meridians, scholar Hua Shou states, '"Origin" represents an honorable designation for the Triple Burner.' If Avijgan and Avijgan (106) are correct in this matter, then the Primo Vascular System would be conceived from the very first day of embryonic life. That certainly is at the origin of life. As I believe that the PVS and the Triple Burner are intimately connected, this absolutely would explain why the term *origin* does represent 'an honorable designation for the Triple Burner'. I propose that the PVS and the acupuncture meridian system and the Triple Burner are all intimately connected.

Which anatomical structure constitutes the root of the Triple Burner between the Kidneys? According to a quote (138) where the author is anonymous, Tang Zonghai wrote in his somewhat-palindromic work *A Refined Interpretation of the Medical Classics (Yijing Jingyi)*, '*The root of the triple burner is in the kidney, more precisely right between the two anatomical kidneys. Right there is a greasy membrane that is connected with the spine. It is called mingmen,* and constitutes the source of the three burners' (emphasis is mine). I suggest that this structure of greasy membrane between the kidneys is an already known anatomical structure possessing an as-yet-unrecognized dynamic piezoelectric property—e.g. thoracolumbar fascia or membrane associated with the perinephric fat(?)—that connects to the neural pathways where it is connected with the spine and subsequently the brain. It is a fact that the membranous neurofascial system surrounds and communicates with every single cell in the body. As I believe that the Triple Burner is actually the Connective-Tissue Metasystem and as 'the root of the triple burner is in the kidney' (more precisely, medial to the right Kidney), it would make sense that this root or origin of the Triple Burner would be a greasy membranous connective-tissue structure medial to the right Kidney, proximal to the level of L2–L3. Time and a vigilant anatomist interested in electrophysiology will tell.

Note that there is a gross disconnect between the embryological derivation of organ formation and location to that present in the adult. The original kidneys, the pronephri, actually originate in the cervical region of the embryo and then migrate downwards to where they finally reside in the lower back region of the adult.

After the Connective-Tissue Metasystem (San Jiao) initially forms, what is the second organ formed? It would nicely fit my theory that the Kidney is the second organ to be differentiated (after the Connective-Tissue Metasystem or San Jiao) so that toxic biological by-products do not accumulate in the embryo and poison it. However, it is often stated that the Heart begins to beat in humans by day 22–23. Thus, the Heart is generally considered to be the first functioning embryonic organ formed. For example, this is detailed in the article 'The Heart: Our First Organ' by author Stefan Jovinge (141), where he states, 'The heart is the first organ to form during development of the body. When an embryo is made up of only a very few cells, each cell can get the nutrients it needs directly from its surroundings. But as the cells divide and multiply to form a growing ball, it soon becomes impossible for nutrients to reach all the cells efficiently without help. The cells also produce waste that they need to get rid of. So the blood and circulatory system, powered by the heart, together form the first organ system to develop.'

However, it is critical to note that in the 2013 article 'Kidney Development', the authors (142) explain that the term *nephrogenesis* pertains to the embryological development of the kidneys. Kidney development proceeds via a series of three successive developmental phases, whereby, through advancing phases, a more complex kidney develops. The pronephros is the primal kidney and develops in the cervical region of the embryo. Subsequent phases are the mesonephros and then finally the metanephros, which both develop lower in the embryo. The third and final metanephros form of the kidney is the most developed and remains as the adult kidney. The authors (142) state:

> *During approximately day 22 of human gestation, the paired pronephri appear towards the cranial end of the intermediate mesoderm.* In this region, epithelial cells arrange themselves in a series of tubules called nephrotomes and join laterally with the pronephric duct. This duct is fully contained within the embryo and thus *cannot excrete filtered material outside the embryo; therefore the pronephros is considered nonfunctional in mammals.* (Emphasis is mine)

Interestingly, by 'approximately day 22 of human gestation, the paired pronephri appear . . . [and] arrange themselves in a series of tubules called nephrotomes'. So by day 22, the paired primordial Kidneys (pronephri) have already formed, *but* it is believed that they 'cannot excrete filtered material outside the embryo; therefore the pronephros is considered nonfunctional in mammals'. So because they are not *functional*, according to embryologists, they are not considered to be the first organ to be formed. The Heart 'begins to beat in humans by day 22–23, thus the Heart is generally considered to be the first functioning embryonic organ formed.' Going out on an embryological limb (pun intended), I suggest that the primordial kidney, as the pronephros, actually is functioning as a simple pump, thanks to the novel mechanism described by Professor Pollack (27)

(KL 2050–2059), as discussed in an earlier chapter of this book. This information has not yet reached the embryology mainstream.

I thus propose that immediately after the gate of life is opened, the first organ to be engendered is the Primo Vascular System and, shortly thereafter, the Connective-Tissue Metasystem. I propose that the amalgamation of these two organ complexes constitute the Triple-Energizer Metasystem or San Jiao and believe that this is thus the first TCM organ to be produced. I propose that the Second organ system to be formed by the PVS is the Kidneys in the pronephros form. This would explain the intimate relationship that the gate of life has with the San Jiao and the Kidneys and Yuan Qi. I propose that very shortly after the pronephros Kidneys are formed at day 22, the Third functioning organ, the Heart, is formed at day 22–23.

Note that remnants of the hind kidney are transformed into structures that carry male genital products, which agrees with the TCM concept that kidney controls such structures. I find it interesting that the primary pronephros kidney pump is derived from a simple tube. In Professor Pollack's research, such simple hydrophilic tubes acted as fluid pumps and generate living EZ water. I believe that this living water in the form of $[H_3O_2]^-$ continues to be engendered by the Kidneys and circulated throughout the entire body by the San Jiao during our lives. This would explain why the Kidneys and the Triple Energizer have such a close relationship associated with the gate of life and why they are both associated with the constant circulation of life energy and Yuan Qi.

CHAPTER 40

Taste Receptors in the Stomach Determine Necessary Nutrient Absorption

Introduction
In the commentaries on the 25th Difficult Issue on page 314 of Unschuld's (1) translation of the *Nan Ching*, Ting Chin states:

> It was said further that 'the constructive [influences] emerge from the central burner; the protective [influences] emerge from the lower burner.' *The constructive [influences] become the blood because they are [generated from] the essence of the taste [influences] of the grains.* The protective [influences] are [volatile] influences [because they are] generated from the [volatile] influences of the grains. All these [transformations occur] because of the [activities of the] stomach. But how could the stomach be stimulated to perform these transformations if it were not for the fact that the Triple Burner externally completely encloses [the stomach] and manages the movement of the influences? (Emphasis is mine)

Note well exactly what Ting Chin states above regarding the 25th Difficult Issue, namely 'the constructive influences become the blood because they are generated from the essence of the taste-influences of the grains . . . because of the activities of the stomach'.

The taste influences of the grains are determined by the taste receptors, whereby the constructive influences occur because the stomach determines what elements are required to build the blood or the Ying qi, to reinforce the bones, or to nourish the flesh and muscles. But can the stomach actually taste? Let's see what modern science has revealed. Be prepared to be stunned!

40.1 Research Reveals Taste Receptors in the Gut Can Taste Artificial and Natural Sweet Flavour

In the 2007 article in *Science Daily* entitled 'Your Gut Has Taste Receptors' (154), the Summary stated that researchers in the Department of Neuroscience at Mount Sinai School of Medicine have identified taste receptors present in human intestines. Scientists have formerly confirmed that the absorption of dietary sugars inside the intestine is mediated by a protein—a sugar transporter—that varies in response to the sugar content present in foods. The intestine uses a glucose-sensing system to monitor these sugar variations, but until now, the nature of this biological system was unknown. The taste receptor T1R3 and the taste G protein gustducin are critical to perceive sweet taste in the tongue. This new research confirms that these two sweet-sensing proteins are also expressed in specialized taste cells of the gut, where they sense the presence of glucose within the intestine.

These new findings were published online in the *Proceedings of the National Academy of Sciences* and could have implications for the treatment of diabetes and obesity. The lead author of the research, Robert F. Margolskee, MD, PhD (a professor of neuroscience at Mount Sinai School of Medicine), reported that it is now known that sweet receptors that detect sugar and artificial sweeteners are not only located on the tongue as previously believed. He noted that their research findings were very important and had major implications in the new field of gastrointestinal chemosensation, whereby cells in the gut are able to perceive and react to the sweet taste and other nutrients. Margolskee further announced that the same mechanism used by taste cells to taste glucose on the tongue is used by taste cells in the gut. As the research shows that insulin secretion and appetite-regulating hormones are governed by taste receptors in the gut, Margolskee believes that their research reveals how sugar uptake from our diets is controlled and how our blood sugar level is regulated.

Dr Margolskee continued, 'This work may explain why current artificial sweeteners may not help with weight loss, and may lead to the production of new non-caloric sweeteners to better control weight. . . . Sensing glucose in the gastrointestinal tract is the first step in regulating blood sugar levels. Having discovered the identity of the gut's sweet receptors may open the way for new treatment options for obesity and diabetes.' The article advised that prior to this research in 2007, the intestinal sugar sensors were unknown.

40.2 Umami, the Fifth Flavour Other Than Sweet, Sour, Bitter, and Salty

The 2014 article titled 'The Surprising Food Flavor that Can Help You Shed Pounds', Dr Mercola (155) advises, 'More than 100 years ago, a Japanese chemist named Kikunae Ikeda discovered the secret that made dashi, a classic seaweed soup, so delicious. It was glutamic acid, which, in your body is often found as glutamate.' Mercola advises that 'Ikeda called this new flavor "umami," which means "delicious" in Japanese, but it wasn't until 2002 that modern-day scientists confirmed umami to be a fifth taste, along with sweet, sour, bitter, and salty'. While most foods do contain some glutamate, some foods contain more than others. Rich sources of glutamate include protein-rich meat, eggs, poultry, milk, cheese, and fish, along with sea vegetables, ripe tomatoes, and mushrooms. Umami has the property of enhancing the flavour of foods and making them taste better. Mercola notes that 'when an umami-rich food like seaweed is added to soup stock, for instance, it makes the broth heartier, more "meaty" and more satisfying'. However, umami has a bit of a dark side because it is chemically similar to the synthetic food additive monosodium glutamate (MSG), which is often added to foods to impart more of that sought-after umami flavour. Mercola reports, 'Umami in its natural form glutamate or glutamic acid, may boost post-meal satiety, helping you to eat less, and possibly lose weight, over time'. Five healthy umami-rich foods that may help you eat less include mushrooms, truffles, green tea, seaweed, and ripe tomatoes.

40.3 Taste Cells and Receptors Have Been Found in the Small Intestine and the Pancreas

In the 2011 article 'Can the Stomach Taste?' (156), it was noted that because the taste of food is a major cause of food ingestion, it follows that taste is a pertinent factor in the study of obesity. Due to this fact, for the last twenty years, the tongue has been the principal focus of organoleptic research. However, at the Obesity 2011 symposium, the focus was on nonoral sensory receptors, specifically sweet and bitter taste receptors that have recently been found much further down the gastrointestinal tract than the taste receptors known to be present on the tongue. For example, the molecular biologist from Monell Chemical Senses Center in Philadelphia, Dr Robert Margolskee, presented an overview of known taste receptors present in the tongue. He reported that the average human tongue and oral cavity houses 2,000 to 8,000 taste buds. Each of these taste buds is composed of 50 to 100 taste receptor cells. As is well known, these taste cells on the tongue are the initial chemosensors of all food types entering the alimentary tract. Different taste receptors in the cells on the tongue discern the five different taste qualities. Dr Margolskee advised that the five diverse tastes and their associated receptors are salty (ENaC/Deg receptors), sour (several receptors), umami or savory (T1R1 + T1R3), sweet (T1R2 + T1R3), and bitter (T2R).

Dr Margolskee proceeded to review basic science research studies which confirm that there are also taste cells and receptors in the gut, specifically in the small intestine and the pancreas. He advised that there are also multiple taste-signalling proteins present in the gut, but these receptors don't provide a conscious sense of taste. He believes that they may integrate physiologic responses to digestion. It was Dr Margolskee who discovered gustducin, which is a G protein important in tasting bitter tastes on the tongue. He described how *gustducin is found in the tongue and also in other cells located throughout the foregut, including alpha cells in the pancreas.*

40.4 Taste Signaling Receptors Present in the Tongue Are Also Found in the Gut Enteroendocrine Cells

The 2011 article (156) further discussed that every taste-signaling receptor that is present in the tongue is also found in the gut enteroendocrine cells and that they are performing a functional role, including regulating endocrine function. Interestingly, the artificial sweetener sucralose and other artificial sweeteners also initiate insulin secretion from pancreatic islets in the same way that natural glucose does, but this is in a dose-dependent fashion. So even though artificial sweeteners might have zero calories, they are metabolically active and may have long-term negative clinical effects that are as yet undetermined.

40.5 Taste Receptors Are Intertwined with Metabolic Processes in Very Interesting Ways

Dr Timothy Osborne from the Sanford-Burnham Institute also gave a presentation at the Obesity 2011 symposium which focused on bitter taste receptors. Dr Osborne emphasized that taste receptors are intertwined with metabolic processes in very interesting ways that are only now beginning to be understood. The mechanism of Bitter perception appears to be much more complex than the perception of Sweet, and at least 35 genes affect bitter perception in mice and at least 25 genes in humans. Scientists studying bitter taste use the very bitter substance phenylthiocarbamide (PTC) in their experiments. Phenylthiocarbamide is found in green vegetables, for example, broccoli. Interestingly, only 70% of people can taste PTC, which may affect their vegetable consumption. In the 2011 article (156), Dr Osborne reported that some studies have shown that tastant-induced signaling for bitter taste occurs in the mammalian gut, and he has recently confirmed the expression of these bitter receptor proteins in the murine small intestine. He has further confirmed that bitter substances stimulate the secretion of gut peptides, including glucagon-like peptide 1 (GLP-1) and cholecystokinin (CCK). It is proposed that the bitter-tasting receptors present in the gut may be a protective mechanism to prevent the absorption of toxic bitter molecules if they do happen to get past the tongue and are swallowed. The presence of bitter substances in the gut stimulates the release of CCK and GLP-1. These substances induce vomiting, delay the emptying of gastric contents, and decrease the feeling of hunger. These effects would cause less bitter (and potentially toxic) substance to be consumed. The article continued:

> In addition to taste cells existing lower down the gastrointestinal tract than previously thought, Margolskee also presented data showing that there are also endocrine cells in the tongue, which secrete a variety of gut hormones including glucagon, GLP-1 and PYY. He is now starting to study these gut taste receptors in bariatric surgery models in mice.

40.6 Taste Receptors Can Trigger Memories, Impact Respiration and Affect Sperm

Eating a food that we enjoy can trigger powerful memories of pleasure, lust, and even love. However, all it takes is one bad oyster in a meal and a night over the toilet bowl to make you avoid eating oysters for the rest of your life. Neuroscientists who specialize in the study of taste are only now just beginning to comprehend how and why the interaction of a few chemical molecules that stimulated taste buds on your tongue can trigger instinctive behaviors or intense memories.

In the commentaries on the 23rd difficult issue on page 293 of Unschuld's (1) translation of the *Nan Ching*, Yü Shu states, 'When [the text states: "blood and influences"] start from the central burner, that refers to [a location] directly between the two breasts, called the tan-chung hole. . . . The *Su-wen* states: "The tan-chung is the emissary among the officials." That is to say, *the stomach transforms the taste [of food] into influences which are transmitted from here upward to the lung*' (emphasis is mine).

According to the 2012 *Nature* journal article entitled 'Neuroscience: Hardwired for Taste' (157), researchers have recently developed a 'gustotopic map' based on the idea that there are regions of the brain that are similarly dedicated to taste perception of the taste buds on our tongue, which can differentiate the basic tastes, including sweet, salty, bitter, sour, umami, and arguably, a few others. A recent research revelation in organoleptic research is that the taste receptors that detect bitter, sweet and umami are not only restricted to the tongue. Amazingly, they are distributed throughout the stomach, intestine, and pancreas, where they assist various digestive processes—for example, by influencing the appetite and regulating insulin production. These taste receptors are also present in the airways, where they have an impact on respiration. Note what Yü Shu stated above, namely, 'The stomach transforms the taste of food into influences which are transmitted from here upward to the lung.' Compare that with what recent research has elucidated: 'These taste receptors are also *present in the airways*, where they have an impact on respiration' (emphasis is mine).

Even more surprising, the *taste receptors have been detected on the sperm*, where they affect maturation. The article stated, 'A better understanding of what they [i.e. the taste receptors throughout the body] do and how they work could have implications for treating conditions ranging from diabetes to infertility.'

40.7 Sperm Can Carry the Memory of a Father's Environment and Lifestyle Patterns to an Embryo

It is very well known that an expectant mother's diet can affect her unborn offspring. A 2013 Canadian study led by Dr Sarah Kimmins at McGill University found significant findings regarding the impact of the father's nutritional status on his unborn child. The research, involving mice, revealed that dad's folate levels (vitamin B_9) may likely be just as significant as the mom's to the future health and development of their offspring. The article (158) reporting Kimmins' findings stated:

> Sperm carry a 'memory' of the father's environment and possibly even of his diet and lifestyle choices to the embryo. Researchers were surprised to witness an almost 30 percent increase in birth defects in the offspring sired by fathers whose levels of folates were low, including

severe skeletal abnormalities that included cranio-facial and spinal deformities.

It is well known that folate is critical for brain and overall neurological health, the development of memory, learning, and other cognitive processes. Dr Kimmins has very sound advice for all males wishing to become a father. She says, 'Our research suggests that fathers need to think about what they put in their mouths, what they smoke, and what they drink—and remember they are caretakers of generations to come.'

40.8 Taste Receptors in the Intestine Detect Glucose and Trigger the Release of Hormones

It is not at all surprising that some of the better-understood types of taste receptors are present in the digestive system. In the article titled 'Neuroscience: Hardwired for Taste', the author, Bijal Trivedi (157), explains that the sweet receptor (T1R2/T1R3) present on K-type and L-type enteroendocrine cells located in the intestine secrete hormones called incretins, which subsequently stimulate the production of insulin. This recent finding enlightens physiologists who have been mystified for more than 50 years regarding the fact that swallowing glucose triggers a significantly higher insulin release compared to injecting glucose directly into the bloodstream. This phenomenon is called the incretin effect. In 2007, neuroscientist Robert Margolskee of the Monell Chemical Senses Center in Philadelphia, Pennsylvania, confirmed his hypothesis that sweet receptors on L cells in the human duodenum actually detect glucose and trigger the release of gastrointestinal incretin hormone GLP-1, which stimulates the production of insulin and sends a satiety signal to the brain. If these sweet receptors are blocked, the amount of insulin release is reduced.

40.9 The Stomach Contains T1R3 Receptors, Which Detect Sugars and Amino Acids and Taste Sweet a Second Time

While experimenting with mice alongside a group from the University of Liverpool, UK, Margolskee showed that the sweet receptors that detect glucose in the intestines also react to artificial sweeteners and trigger a spike in insulin. The article (157) continues:

> 'Sweet receptors, traditionally associated with just the mouth, were in the gut and essentially "tasting" the sugar a second time,' says Anthony Sclafani, a behavioural neuroscientist at the City University of New York. This 'second tasting' triggers glucose transport into the cells and bloodstream, and the faster this happens, the more insulin will be released. 'It's an incredibly important finding for the control of blood sugar,' he says, adding that it was surprising that artificial sweeteners, which were thought to influence only the tongue, also trigger changes in the gut.
>
> In the stomach, cells carrying the T1R3 receptor, which aids detection of both sugar and amino acids, secrete the hunger hormone ghrelin when they encounter carbohydrates and protein, encouraging eating when important nutrients are available.
>
> For bitter tastes, however, the T2R receptors in the digestive system seem to have contradictory functions. In 2011, Belgian researchers showed that bitter-tasting compounds that reach the stomach of mice initially trigger the release of ghrelin, stimulating eating as usual. But after 30 minutes, food intake decreased, as did gastric emptying, keeping the food in the stomach. This curbs the appetite by prolonging the sense of fullness and satiety—perhaps to prevent the ingestion of toxic food.

It is a known fact that the Romans drank wine infused with bitter herbs to prime the appetite and prevent overeating. The authors speculate, 'Stimulating bitter receptors in the gut could potentially be used to treat certain eating disorders.'

40.10 Nutrient Sensors in the Stomach that Are Separate from Taste Buds Can Detect Nutrients

An article in *The Scientist* journal in 2013 titled 'Sensing Calories Without Taste' (159) reported that Scientists engineered fruit flies that can't taste sugar in their taste buds. Initially, the fruit flies showed no preference between consuming sugar water and plain water. However, after 15 hours without any food, thanks to nutrient sensors in the stomach, the fruit flies began to show a preference for the sugar water, seemingly sensing the fact that the sugar water contained life-sustaining calories even though they couldn't taste anything. 'They sense calories in a way that is somehow independent of tasting the calorie itself,' said Ivan de Araujo, a neurobiologist at Yale University. The research described above was published in the *Proceedings of the National Academy of Sciences* in 2011. There have been several other recent studies that have proven that mice, rats, and fruit flies all have sensors that are separate from taste buds and are capable of detecting nutrients. It isn't simply that the sensation of satiety is reached, according to de Araujo. There is something more that is occurring as the animals are not reacting to the stomach stretching or to the release of hormones, including insulin, or to any other negative satiety indicators that signal the animals that they are full and should stop eating. The article states, 'Rather, the signals that de Araujo and others in his field are studying seem to be positive, rewarding messages that help the animals learn to repeatedly choose nutritive options.' The presence of the nutrient sensors in the stomach may help animals detect what's nutritionally beneficial to eat even if the taste buds don't quite get the message. Tony Sclafani, who directs the Feeding Behavior and Nutrition Laboratory at Brooklyn College said, 'For example, the sensors could work for certain starches that don't have strong flavors, but still offer useful calories for the body.' But he cautions that 'we're still trying to figure out the mechanisms, let alone evolutionary meaning.'

40.11 Taste Buds in Your Stomach Don't Send a Signal to Your Brain about 'Taste'

Dr Terry Simpson is a Phoenix and Scottsdale based weight loss surgeon. In the 2013 article entitled 'Taste Buds: In Your Tongue & Gut—Cats Can't Taste Sweets', Simpson (160) wrote, 'Bitter is the most sensitive of the tastes. The classic bitter taste is quinine (think tonic water).' In the article, he continued:

> Bitter tastes are thought to have evolved to help us avoid toxins. Eat some bad shellfish, and while their bitter taste may not 'taste' bad to your tongue, the taste buds in the stomach will sense them. *The taste buds in your stomach don't send a signal to your brain about 'taste' but instead are used for other functions. The stomach taste bud sends a signal that stop the muscles of the stomach from pushing the bad food past your stomach and then trigger a center in your brain for nausea* (emphasis is mine). Once you regurgitate the bad food, and the stomach no longer senses that bitter taste, the signal to the brain stops, and you no longer have nausea. Somewhere in your brain, as you recover from the bad shellfish, a memory is linked and it is unlikely you will eat shellfish in that form again (be it a bad clam, or restaurant, or city).

40.12 Taste Buds in the Stomach Ensure Constructive Influences Occur Directly in the Gut

Dr Simpson (160) proceeds to explain that the taste buds in the stomach act to ensure that beneficial and constructive influences occur directly in the neighborhood of the gut without input from the cranial brain. The gut *is* acting as the local brain without relying on the cranial brain to make decisions. Regarding this, he explains further:

> The taste buds present in the stomach and the first part of the intestine (duodenum), don't provide a 'taste' sensation to the brain, *but help prepare the body for the glucose that is coming.* Glucose is the most important fuel of the body- every cell uses it, every cell is dependent upon it. Because of that, there is an evolutionary *advantage to absorbing glucose as rapidly as possible.* The *intestinal taste buds facilitate the transport of glucose from the intestine into the blood stream*. There are cells that actively transport glucose into the bloodstream. When the intestinal taste

bud senses the sweetness, it also signals those cells to turn on so they are prepared to actively transport glucose. People who have undergone gastric bypass surgery for obesity, or a Roux En Y gastric bypass for ulcer disease, have bypassed these taste buds in the duodenum, and when tested, these individuals have less ability to respond to glucose challenges. This is thought to be one of the mechanisms for weight loss from the gastric bypass operation. (Emphasis is mine)

40.13 Artificial Sweeteners Cause Metabolic Confusion because No Calories Arrive

Artificial sweeteners do not help you lose weight because your body is not fooled by sweet taste without the accompanying calories. The 2014 article titled 'How Artificial Sweeteners Confuse Your Body into Storing Fat and Inducing Diabetes' (161) states:

> When you eat something sweet, your brain releases dopamine, which activates your brain's reward center. The appetite-regulating hormone leptin is also released, which eventually informs your brain that you are 'full' once a certain amount of calories have been ingested. However, when you consume something that tastes sweet but doesn't contain any calories, your brain's pleasure pathway still gets activated by the sweet taste, but there's nothing to deactivate it, since the calories never arrive.
>
> Artificial sweeteners basically trick your body into thinking that it's going to receive sugar (calories), but when the sugar doesn't come, your body continues to signal that it needs more, which results in carb cravings.
>
> Besides worsening insulin sensitivity and promoting weight gain, aspartame and other artificial sweeteners also promote other health problems associated with excessive sugar consumption, including: cardiovascular disease and stroke, and Alzheimer's disease.

Summary of Chapter 40
With reference to the 25th Difficult Issue of the *Nan Ching*, Ting Chin states, 'It was said further that "the constructive [influences] emerge from the central burner".' Yü Shu states, 'The stomach transforms the taste of food into influences which are transmitted from here upward to the lung.' Research performed in 2007 by Robert Margolskee, professor of neuroscience, showed that the five known taste sensors have been found throughout the body away from the tongue. Amazingly, every taste-signalling receptor that is present in the tongue is also found in the gut enteroendocrine cells. The body cannot be tricked by industrial chemists, and it is now known that the taste receptors that sense sugar and artificial sweeteners are not limited to the tongue and that artificial sweeteners may not help with weight loss at all as they initiate the same biochemical responses as sugar does in the gut.

So what has modern science elucidated about 'the constructive influences [that] emerge from the central burner'? The non-oral sensory receptors have been found in the Stomach and the Spleen (in the central burner). Interestingly, these taste receptors don't provide a conscious sense of taste to the brain. Dr Osborne emphasized that taste receptors are intertwined with metabolic processes in very interesting ways that are only now beginning to be understood. Dr Osborne has shown that bitter substances stimulate secretion of gut peptides, such as cholecystokinin (CCK) and glucagon like peptide 1 (GLP-1). The bitter-stimulated release of CCK and GLP-1 induces vomiting, delays gastric emptying, and decreases hunger. This mechanism may prevent the ingestion of toxic food. These taste-signaling receptors are certainly performing a multipronged functional role, including integrating physiologic responses to digestion, regulating endocrine function, and regulating the secretion of insulin and hormones that regulate appetite. These taste receptors help prepare the body for the glucose that is coming. Glucose is the most important fuel of the body as every cell uses it and every cell is dependent upon it. In the Stomach, dedicated cells carrying the T1R3 receptor aids detection of both sugars and amino acids and secrete the hunger hormone ghrelin when they encounter carbohydrates and protein, encouraging eating when important nutrients are available.

Strangely, these taste receptors are also present in the airways, where they have an impact on respiration. The taste-influences of the grains certainly are determined by the nonoral sensory receptors, whereby the constructive influences occur. Subsequently, the stomach determines what proteins, fats, and carbohydrates are required to manufacture the blood or the Ying qi, reinforce the bones, or nourish the flesh and muscles, etc.

In the commentaries on the 14th Difficult Issue on page 190 of Unschuld's (1) translation of the *Nan Ching*, Yü Shu states, 'The spleen takes in the five tastes. It transforms them to produce the five influences [for the] depots and palaces, and to make flesh and skin grow. Here now because of the injury, the tastes are not transformed and, hence, the flesh becomes emaciated.' Thus, it can be seen that the Spleen processes the five tastes and transforms them into biological influences to allow the organs of the body (depots and palaces) to manage their portfolio as Yü Shu shows here the Spleen's portfolio is 'to make flesh and skin grow'. Dr Simpson (160) mentioned above explained that 'people who have undergone gastric bypass surgery for obesity ... have bypassed these taste buds in the duodenum' and lose large amounts of weight so that, in place of their former obesity, 'the flesh becomes emaciated'.

This new scientific research is an important advance for the newly discovered and rapidly developing field of gastrointestinal chemosensation, a process whereby the cells of the gut detect and respond to sugars and numerous other nutrients derived from the drinks and the grains. Many unanswered questions are finally being answered since 2007 now that scientists and doctors have come up to speed with what TCM was aware of thousands of years ago. Now that biomedical researchers are aware that the 'the spleen takes in the five tastes' and 'transforms them to produce the five influences for the depots and palaces', what other 'constructive influences that emerge from the central burner' will be elucidated? In the next Chapter, I will discuss this matter further.

CHAPTER 41

Taste Receptors in Numerous Organs Necessary for Biological Control Throughout the Body by the Zang and Fu

Introduction
In the 2013 article titled 'Mysterious Taste Sensors Are Found All Over The Body', Jennifer Welsh (162) opened her *Business Insider Australia* article with the stunning pronouncement indicating that taste receptors that perceive salty, sweet, and bitter foods aren't only present on our tongues. She said, 'Recently researchers are finding them present all over the body, from the mouth to the anus. Literally.' She advised that taste receptors have been found in several locations in the body apart from the traditional tongue and that the locations included the stomach, intestines, pancreas, lungs, and brain, and intriguingly, researchers really don't know what they are there for. So let's consider some of the recent scientific findings in light of what was stated by ancient sages skilled in the art and science of Traditional Chinese Medicine.

41.1 The Biological Function of These Extra-Orally Located Taste Receptors Is Generally Unknown
Study researcher Bedrich Mosinger of the Monell Chemical Senses centre advised *Business Insider Australia* that while taste receptors

and signaling proteins away from the tongue are able to sense the presence of sugars or amino acids, their actual function is still uncertain. (162)

41.2 Taste Proteins for Sweet and Umami Present in Testes Are Essential for Fertility

The article (162) further advised that new research published on 1 July 2013 in the journal *Proceedings of the National Academy of Sciences* found that taste proteins for the detection of sweet and umami (the amino acid taste of soy sauce) also exist in murine testes and that they play an important role in mouse fertility. The researchers were originally aiming to develop mice that didn't have these taste receptors for use in oral taste-related studies. However, the researchers found that if these taste receptors were removed from the mouse testes or if their taste receptor function was blocked, the mice became infertile and were unable to reproduce. Regarding this remarkable finding, Mosinger said, 'The males are sterile, their sperm count is low, and spermatozoa are not developed properly.'

A very concerning finding of this research is that the drug that the researchers used in the experiments to block the taste receptors in the testes is of a class of drugs that are used to treat high-blood cholesterol in humans. The dire consequence is that these anticholesterolemic drugs could be interfering with human fertility, the researchers said. Knowing how important this interaction is, this could make way for new treatments for infertility or even lead to male birth control. Study researcher Robert Margolskee of the Monell Chemical Senses centre reported that their current research findings pose more questions than answers and that further research is required to determine the pathways and mechanisms present in testes that employ these taste genes so that a greater understanding can be gained regarding why their loss results in infertility. (162)

41.3 The Stomach Can 'Taste' Umami, Glucose, Carbohydrates, Proteins, and Fats

A 2014 article entitled 'Stomach as Nutrition Sensor' (33) states:

> The stomach can 'taste' sodium glutamate using glutamate receptors, and this information is passed to the lateral hypothalamus and limbic system in the brain as a palatability signal through the vagus nerve. *The stomach can also sense, independently to tongue and oral taste receptors, glucose, carbohydrates, proteins and fats. This allows the brain to link nutritional value of foods to their tastes.* (Emphasis is mine)

This is an incredible finding. That 'the stomach can also sense, independently to tongue and oral taste receptors, glucose, carbohydrates, proteins and fats' categorically confirms that the ancient sages were eons ahead of their time regarding the biochemical and physiological characteristics of the digestive system.

41.4 Major Nutrient Absorption at the Large Intestine/Small Intestine 'Screen Gate'

In the commentaries on the 44th Difficult Issue on page 428 of Unschuld's (1) translation of the *Nan Ching*, regarding the location and function of the seven through gates of the body, Ting Te-yung says, '[The *Nan-ching* states further:] "Where the large and small intestines meet is the screen-gate." . . . The place where the large and the small intestines join is [the location] where water and grains are separated [to be transformed into] essence and blood; each has its destination to move toward. Hence [the *Nan-ching*] speaks of a "screen-gate."'

This anatomical location is more in keeping with what we know to be the case when it comes to nutrient absorption from the gut to be utilized in the manufacture of biological fluids and blood.

41.5 The Stomach Truly Does Transform the Five Tastes into Influences Transmitted to the Organs

In the commentaries on the first Difficult Issue on page 73 of Unschuld's (1) translation of the *Nan Ching*, regarding the location and function of the seven through gates of the body, Yü Shu states:

> [Food items carrying] the five tastes enter the stomach. There they are transformed to generate the five influences. The five tastes are sweet, salty, bitter, sour, and acrid; the five influences are rank, frowzy, aromatic, burned, and foul.' These are the influences and tastes [associated with] the Five Phases. After the tastes have been transformed into influences, [the latter] are transmitted [from the stomach] upward into the hand-great-yin [conduit]. The great-yin [conduit] is responsible for the influences. It receives the five influences in order to pour them into the five depots. If the stomach loses its harmony, it cannot transform [taste into] influences. As a consequence, there is nothing for the hand-great-yin [conduit] to receive.

Note here in the very first Difficult Issue that Yü Shu directly states, 'Food items carrying the five tastes enter the stomach. There they are transformed to generate the five influences.' These five influences are transmitted into or poured into the five depots or solid organs. Incredibly, then notice what Yü Shu states, namely, 'If the stomach loses its harmony, it cannot transform [taste into] influences. As a consequence, there is nothing for the hand-great-yin (Lung) [conduit] to receive.' It is an established fact that when individuals have had gastric bypass surgery, they are subject to malabsorption of many essential nutrients and minerals thereafter. For example, in the article 'Nutritional Deficiencies after Gastric Bypass Surgery', the authors (163) state in the Abstract:

> Although perioperative complications associated with gastric bypass surgery are generally low (<1%), the postoperative complications can be quite high. For example, because bariatric surgery often involves gut manipulation that alters the natural absorption of nutrients, nutritional

deficiencies can develop. The most common deficiencies are vitamin B12, folate, zinc, iron, copper, calcium, and vitamin D and can lead to secondary problems, such as osteoporosis, Wernicke encephalopathy, anemia, and peripheral neuropathy.

To avoid such complications, dietary supplementation often begins shortly after surgery, while the patient is still in the hospital. . . . One study found that 3 years after gastric bypass surgery, even with multivitamin supplementation, as many as 50% of patients had iron deficiency, while nearly 30% had cobalamin deficiency.

I would be very confident that after traumatic gastric bypass surgery, 'the stomach loses its harmony' and would subsequently be unable to 'transform [taste into] influences' mainly because the taste receptors have been removed or negated due to the surgical procedure. Remember that the 'influences transmitted from the stomach' include the ability to absorb necessary nutrients to sustain biological functioning, including the manufacture and synthesis of all the essential fluids of life (blood, lymph, cerebrospinal fluid, interstitial fluid, saliva, etc.).

41.6 The Stomach Delegates the Transformation Role to the Spleen

In the commentaries on the 14th Difficult Issue on page 190 of Unschuld's (1) translation of the *Nan Ching*, regarding injury to the spleen, Yü Shu states, 'The spleen takes in the five tastes. It transforms them to produce the five influences [for the] depots and palaces, and to make flesh and skin grow.' In the case of injury to the Spleen, he states, 'Here now, because of the injury, the tastes are not transformed and, hence, the flesh becomes emaciated.' Note in this regard that the other solid organs also manifest pathological changes if they are injured.

41.7 Balanced Intake of the Five Tastes Ensures Strong Libido and High Fertility

In the commentaries on the 49th Difficult Issue on page 462 of Unschuld's (1) translation of the *Nan Ching*, regarding injury to the spleen, Yü Shu states:

> The spleen is the official responsible for the granaries. The [preferences for any of the] five tastes originate from there. That is to say, [the spleen] receives the five tastes and transforms them, generating the five influences in order to nourish the human body. Here, drinking and eating [without restraint], as well as weariness and exhaustion, have led to harm that is caused by oneself. It is for this reason that the sages paid great attention to a balanced [intake of the] five tastes in order to keep their bones straight and their muscles tender. They paid great attention to conducting their lives in accordance with the rules. Their [existence on earth through] heavenly mandate was long; how could they ever have done harm to themselves? And how could anybody not be careful on his way of nourishing his life?

Hua Shou concurred with Yü Shu when he states: 'The spleen rules food and drink and the four limbs. Hence, drinking and eating [without restraint], as well as weariness and exhaustion, harm the spleen.'

In the commentaries on the 56th Difficult Issue on page 508 of Unschuld's (1) translation of the *Nan Ching*, regarding the question 'The spleen is responsible for the tastes ... What does that mean?', Wang Ping commented on this passage:

> [w]hen the [large] intestine and the stomach develop an illness, it will be taken over by heart and spleen. When the heart takes it over, the blood will cease flowing. *When the spleen takes it over, the [food carrying the] tastes will not be transformed. When the blood does not flow, the females do not have their monthly period; when the [food carrying the] tastes is not transformed, the males*

> *have little essence [i.e., semen].* Hence the most private
> and concealed matters cannot occur. (Emphasis is mine)

In this case of spleen illness, the taste influences in the food are not transformed to initiate the necessary influences, and females experience amenorrhea and low libido along with infertility, while under the same conditions, males develop low sperm count, most likely along with low libido and infertility also. It is remarkable that Wang Ping answers the question 'What does that mean?' regarding 'the spleen is responsible for the tastes'. He immediately connects *tastes* with blood manufacture and amenorrhoea in females. Profoundly, Wang Ping states, 'When the [food carrying the] tastes is not transformed, the males have little essence [i.e., semen]. Hence the most private and concealed matters cannot occur.' Wang Ping is here directly connecting tastes with poor sperm parameters and infertility and impotence. Can you comprehend that? Does that make sense to you? Read on *Macbeth*.

As stated elsewhere in this book, new research (162) published on 1 July 2013 in the journal *Proceedings of the National Academy of Sciences* found that taste proteins for the detection of sweet and umami (the amino acid taste of soy sauce) also exist in murine testes and that they play an important role in mouse fertility. The researchers found that if these taste receptors were removed from the mouse testes or if their taste receptor function was blocked, the mice became infertile and were unable to reproduce. Regarding this remarkable finding, Mosinger says, 'The males are sterile, their sperm count is low, and spermatozoa are not developed properly.' How on earth did Wang Ping know that correct perception of the five tastes, along with a balanced diet involving all the tastes, was critical for a strong libido and high fertility?

41.8 Scent Receptors Occur throughout the Entire Body and Transform Smells into Influences

Since 2005, Jennifer Pluznick, a researcher at Yale School of Medicine in New Haven, Connecticut, has investigated why taste and scent receptors occur throughout the entire body and not only on the tongue and olfactory organs as previously thought. In her article titled 'The Startling Sense of Smell Found All Over Your Body', science writer Veronique Greenwood (164) noted that hundreds of different scent receptors are present in our noses and that these sensitive scent receptors allow us to discriminate between a considerable number of odours. This continuous sensing of the ambient environment powerfully connects us to the physical world around us. Similar to the lock and key anology, specific molecules emitted from different substances fit specific receptors within our nose, and specialized neurons then send messages to the brain so we'll recognize what we have just smelled.

Over the last decade, research biologists have found all over our body these scent receptors as well as taste receptors previously thought to be only found in the olfactory organ of our nose and on taste buds of our tongue respectively. Incredibly, in 2003, bitter taste receptors were found in sperm. Greenwood (164) reported further that research biologists at the University of California, San Diego, identified the presence of sour receptors in the spine, while Pluznick isolated scent receptors in the kidney. Subsequently, researchers found taste receptors in the bladder and the gut that detected sweet taste and taste receptors in the sinuses, airways, pancreas, and brain that detected bitter taste. Scent receptors were even found in muscle tissue. Scientists do not know what taste receptors and scent receptors are doing in these diverse tissue types so far away from the tongue and nose.

Research by Pluznick and others confirms that in at least some tissues, these receptors are not passive. It appears that numerous types of tissues spread throughout our inner body are actually 'smelling' and 'tasting' the local environment deep inside of us and that this sensory mechanism is crucial for good health. Greenwood further explained that Professor Yehuda Ben-Shahar from Washington University in St. Louis isolated cells in the human airway that are equipped with bitter receptors. These airway cells are surrounded by microscopic hair-like protrusions that are called cilia. When dangerous chemicals are inhaled, the cilia flap in an effort to flush the potential danger out of the system. Regarding Ben-Shahar, Greenwood continued,

'He and his collaborators found that when receptors in these cells were exposed to certain noxious molecules, it triggered a cascade of events that culminated in the flapping of the cilia.' Greenwood (164) further reported:

> Work by Noam Cohen, an ear, nose and throat doctor at University of Pennsylvania Medical School, suggests an intriguing role for bitter receptors recently discovered in the sinuses. He found one particular type can intercept the chemical signals that bacteria send to each other when they are coming together to form biofilms, a manoeuvre that greatly strengthens their defences against immune-system attacks.... [w]hen the bitter receptors in the sinuses pick up these signals, they set in motion an attack against the bacteria, causing cells to release toxic gas and cilia to flap.

Greenwood (164) continued, 'Pluznick's research has revealed the existence of Olfr 78 [scent] receptors beyond the kidney, in blood vessels in the skin, heart and muscle. Scent receptors may be even more widely distributed, Firestein thinks, in blood vessels right across in the body.' It appears that these olfactory receptors act as a general-purpose chemical sensor throughout the entire body. Pluznick acknowledges there is still a lot to discover about the broader functions of what we still currently call scent and taste receptors. She pointed out, 'We just happened to come across the receptors in the nose and the mouth first.'

In the commentaries on the 34th Difficult Issue on page 368 of Unschuld's (1) translation of the *Nan Ching*, regarding the determination of odours and tastes in the body, Chang Shih-hsien states, 'The heart masters the odors ... The spleen masters the tastes.' Note what Greenwood stated above regarding scent receptors. Firestein believes that scent receptors are distributed throughout the entire body inside the blood vessels. As the Heart controls the blood vessels in TCM, it makes perfect sense that 'the Heart masters the odors', exactly as Chang Shih-hsien states.

Summary of Chapter 41

So how important are tastes and smells to the maintenance of good health throughout the entire body? Massively so! Sometimes, as practitioners, we may become a bit blasé about our TCM knowledge and relegate it to interesting information without much merit, but as can be seen from modern research, although our patients may have looked at us strangely when we said that salty flavour feeds the kidneys, I believe it is only a matter of time when researchers confirm this. Perhaps the removal of sweet and umami from mouse testes affecting the kidney-controlled testes is related to the effect of the mother and grandmother not feeding and controlling water? Time will tell. What this certainly does mean is that for infertile patients, diverse and wholesome foods are essential to increase their fertility. This research also confirms that aromatherapy certainly has scientific veracity for its proposed power to influence every part of our body, including our emotional state. These recent scientific findings suggest that numerous organs and structures within our bodies are 'smelling' and 'tasting' things deep inside of us all the time and that these abilities are crucial to our health. How did the ancient TCM scholars and practitioners know this fact?

CHAPTER 42

How the Gut Microbiome Complements the Triple Energizer

Introduction
With regard to the true natural healing practices of Hippocrates versus the practices of allopathetic medicine, it is truly ironic that Allopathic medicine claims Hippocrates of Cos as the 'father of medicine' when the practices of the allopathic school of medicine go totally counter to many of his most prized paradigms. For example, he stated, 'The natural healing force within each of us is the greatest force in getting well.' Allopathic Medicine reduces this powerful concept to 'It's *just* the placebo effect.' If we could all potentiate our own placebo effect, we would all be more vital and bounding with wellness and completely free from the necessity of consuming pharmaceutical medications by the handful as modern doctors are trained to dispense. Millions of individuals have died by simply taking their prescribed pharmaceuticals exactly as directed by their doctors. My mum was one of them. While 70% of modern medications are derived from plants and herbs, drug manufacturers foolishly and naively believe that there is only one active ingredient. That millions of medical patients die by following medical instructions doesn't prick the conscience of doctors or the legalized labs that supply the toxic cocktails.

It's ironic that their 'Father' Hippocrates stated 'Primum non nocerum', which means 'First, do no harm.' Michael Jackson, my mum, and millions of other individuals would be alive today if those three words 'Primum non nocerum' were applied. Rather than prescribing unnatural synthetic xenobiotic substances called drugs, Hippocrates favored the use of foods and natural herbs and substances. The sage stated, 'Let food be thy medicine and medicine be thy food.' I can't recall when my doctor advised me to drink green tea and eat more purple carrots, blueberries and have quinoa for breakfast. I am constantly having the need for statins thrust at me by my GP at my annual checkup. If Dr Hippocrates was still in practice, he would realize that high cholesterol is a marketing farce, but if the cholesterol was inordinately high, I am sure he would suggest red yeast rice with its natural statin known as lovastatin, the active ingredient in the prescription drug Mevacor. Always ahead of his time, Hippocrates stated, 'Walking is man's best medicine.' I believe that if Hippocrates was still alive today, he would be a naturopath or an acupuncturist. Having a scientific background, I appreciate that antibiotics, statins, antihistamines, steroids, etc. are all modern and powerful medications and have their place, but I do not believe for a second that they should be consumed like jelly beans on a daily basis. Every xenobiotic medication has the potential to behave as antibiotics and wreak havoc on our microbiome—some more than others. A massive amount of research is being performed on our microbiome by numerous diverse sectors of the medical arena. Many researchers conclude that they have discovered a new organ, namely our gut microbiome. In this Chapter, I discuss in detail many of the remarkable findings being made on a daily basis by researchers investigating the gut microbiome.

42.1 The Vast Majority of Our Human Physiology Comes from Our Gut Microbiome

The Human Genome Project revealed that the *vast majority of our human physiology* remarkably *came from somewhere other than our 25,000 genes*. This revolutionary finding lead to an understanding of epigenetics, which is the body's capability of manifesting variable genetic expression within our epigenome without changes to the primary nucleotide sequences of our DNA itself. The greatest interface with our environment involves our *gut microbiome*, which is composed primarily of *bacteria that amplify our genomic library*

150 fold. In the article 'How to Build a Healthy Microbiome, before, during, and after Birth', Mercola (165) states, 'The gut is also responsible for metabolizing food that we couldn't otherwise, for producing nutrients that we couldn't otherwise, and even for detoxing chemical exposures that we couldn't otherwise.'

42.2 The Remarkable Fact That Microbial Genes Are Found in Human DNA

Regarding the fact that bacteria, fungi, and viruses may be part of the missing link in the progress of humans, in the article titled 'Fiber Provides Food to Your Gut Microbes That They Ferment to Shape Your DNA', Dr Joseph Mercola (166) advises that many of our genes 'slipped into our DNA from microbes living in our bodies'. This mechanism is called horizontal gene transfer. Mercola continues, 'Bacteria slip genes to each other, and it helps them evolve. And scientists have seen insects pick up bacterial genes that allow them to digest certain foods . . . Humans may have as many as hundreds of so-called foreign genes they picked up from microbes.'

Mercola explains that the human genome consists of about 23,000 genes. He then emphasizes that the combined genetic material of the human gut microbiome is somewhere between 2 million and 20 million. He states, 'According to the researchers, these extra genes may have played a role in helping to diversify our own DNA.' He further reports, 'Researchers at the University of Cambridge identified 128 "foreign" genes in the human genome, including the gene that determines your blood type (A, B, or O).' This important gene that determines your blood type has apparently been transferred into your human gene pool from microbes that dwell in your gut.

42.3 The Role of Bacteria in Early Embryological Development

There is no doubt that the status of the gut is largely responsible for human health. But how is our gut composition initiated? What are the contributing factors to gut health? When does our microbiome start? Mercola (165) answers these questions by stating, 'Adding to a long list of "oops!" in the history of medicine, it was long-held that the womb was a sterile environment. We now have broadened our appreciation of the ubiquitous nature of microbes to encompass their special place in the placenta, umbilical cord, and fetal membranes.' Mercola continues, 'In this way, the pregnant, mother-to-be's long life of antibiotics, birth control, gluten, GMOs, chronic stress, and vaccination all conspire to set the stage for the antepartum, intrapartum, and postpartum passage of microbial information for gene expression.'

42.4 Psychological Health Arises in the Womb!
In March 2011, a study published in *Neurogastroenterology & Motility* (167) found that having sufficient amounts of probiotic gut bacteria from birth may be essential for future psychological health. Germ-free mice that lacked gut bacteria demonstrated neurochemical changes in their brains and subsequently behaved differently from normal mice, engaging in what would be referred to as 'high-risk behavior'. According to the authors, 'Acquisition of intestinal microbiota in the immediate postnatal period has a defining impact on the development and function of the gastrointestinal, immune, neuroendocrine and metabolic systems. For example, the presence of gut microbiota regulates the set point for hypothalamic-pituitary-adrenal (HPA) axis activity.' This healthy microbial loading in the gut is very important as many women of reproductive age are deficient in probiotics, a deficiency that transfers to their offspring and may set the stage for a vast array of medical and psychological problems throughout their life. Some authors (22) believe that the Triple Heater is linked to the hypothalamus.

A human study (168) involving seven healthy volunteers found that consumption of a dairy drink containing three different strains of probiotic microorganisms caused changes in the activity of hundreds of genes six hours after consumption of the drinks. Genetic changes were analyzed from biopsies taken from the upper part of the small intestine (duodenum). The changes resembled the therapeutic effects of pharmaceutical medications in the human body, including medications that positively influence the immune system and medications for reducing high blood pressure. The researchers used dairy drinks containing the commercial probiotics *Lactobacillus acidophilus, L. casei,* and *L. rhamnosus* compared to a placebo drink. Professor Michiel Kleerebezem of NIZO Food Research said that 'probiotics cause a local reaction in the mucosa of the small intestine' and that 'these effects are similar to the effects of components that the pharmaceutical industry applies to medicines, but less strong'.

42.5 Presence of Gut Bacteria during Infancy Permanently Alters Gene Expression

A 2011 animal study entitled 'Normal Gut Microbiota Modulates Brain Development and Behaviour' (169) found that gut bacteria may influence mammalian early brain development and behavior and that the absence or presence of gut microorganisms during infancy *permanently alters gene expression.* Hence, their results suggest that the microbial colonization process initiates signaling mechanisms that affect neuronal circuits involved in motor control and anxiety behavior.

In a Press Release in 2013 entitled 'Bacteria in the Gut May Influence Brain Development', Dr Rochellys Diaz Heijtz (170) (lead author of the study) stated, 'Normal gut microbiota modulates brain development and behaviour.' Through gene profiling, the researchers were able to discern that absence of gut bacteria altered genes and signaling pathways involved in learning, memory, and motor control. This suggests that gut bacteria are closely tied to early brain development and subsequent behavior, which is what Dr Campbell-McBride (59) has found as well. These behavioural changes could be reversed as long as the mice were exposed to normal microorganisms *early in life*. But once the germ-free mice had reached adulthood, colonizing them with bacteria did not influence their behavior. Dr Heijtz (170) said, 'The data suggests that there is a critical period early in life when gut microorganisms affect the brain and change the behavior in later life.' Professor Sven Pettersson, coordinator of the study, further noted in the article that colonizing bacteria appear to regulate neurotransmitters, including serotonin and dopamine. And importantly, the colonizing bacteria also appear to regulate nerve synapse function as well. It was further noted that this information currently applies only to mice and, at this stage, can't be extrapolated to include the effect that human gut bacteria have on the human brain.

42.6 Correct Gut Microbiome during Infancy Essential for Future Psychological Health, Brain Development, Learning, and Behaviour

It has been estimated for some time that 80–85% of our immune system is located in our gut, so reseeding our gut on a daily basis with healthy probiotic organisms is extremely important for our overall physical health. However, note what researchers have recently discovered from the references above regarding psychological health. Amongst the statements are 'sufficient amounts of probiotic gut bacteria from birth may be essential for future psychological health' (167); 'the presence of gut microbiota regulates the set point for hypothalamic-pituitary-adrenal (HPA) axis activity' (167); 'probiotic microorganisms caused changes in the activity of hundreds of genes six hours after consumption of the drinks' (168); 'gut bacteria may influence mammalian early brain development and behaviour' (169); 'presence of gut microorganisms during infancy permanently alters gene expression' (169); 'gut bacteria altered genes and signaling pathways involved in learning, memory, and motor control' (170); and 'there is a critical period early in life when gut microorganisms affect the brain and change the behavior in later life' (170).

A robust immune system is our number-one defense against *all* diseases—from the common cold or influenza virus affecting our Lungs in our Upper Heater to pathogenic organisms (*Helicobacter pylori*) causing ulcers in our Stomach (Middle Heater) and through to food-borne pathogenic bacteria or viruses affecting our Small Intestine in our Lower Heater. This immunological protection via the Three Heaters embraces the entire body to protect against the formation of chronic diseases, autoimmune diseases, and cancer anywhere in the body. In light of this, I recommend optimizing our gut microflora, especially our gut bacteria, by consuming traditionally made, unpasteurized versions of fermented foods on a daily basis to all my patients.

42.7 Definition of Yüan Qi

In traditional Chinese medicine and Chinese culture, yuán qì [元氣] is a description of one form of qi. It is usually described as innate or prenatal qi to distinguish it from acquired qi that a person may develop over their lifetime.

42.8 The Connection between 'Yüan Qi' and the Triple Energizer

In the commentaries on the 31st Difficult Issue of Unschuld's (1) translation of the *Nan Ching* on page 348, Yang states, 'Chiao ["burner"] stands for yüan ["origin"]. *Heaven has the influences of the three originals; they serve to generate and form the ten thousand things*. Man reflects heaven and earth. Hence, he too has the influences of the three originals *to nourish the form of the human body*. All three [sections of the Triple] Burner occupy a definite position and still they do not represent a proper depot' (emphasis is mine).

In this reference, Yang shows the connection between 'yüan' and the Triple Energizer. I have pointed out in this book that, after the sperm fertilized the ovum and life began as a zygote, the Primo Vascular System formed first, and shortly thereafter, the Connective-Tissue Metasystem differentiated. I propose that these two metasystems as an amalgamation embody the Triple-Energizer Metasystem and is the first TCM organ to be created. That explains why the Triple Burner is referred to as origin and why the original qi (Yuan Qi) continues to be controlled and circulated throughout the body by the Triple-Energizer Metasystem. Between the actions of bacteria and other microbiome members and their enzyme secretions in the intestines in the Lower Burner, 'ten thousand things' are produced in the form of a diverse array of biochemicals, including nutrients to produce blood and nourish the body and neurotransmitters to message body requirements throughout the entire body. I discuss this plethora of biochemicals later. Thus, 'the influences of the three originals' and their connection with the Triple Energizer certainly do 'nourish the form of the human body'.

42.9 The Genetic Code of Our Gut Microflora Overrides Human Genetic Code

The article entitled 'Microbiome' (171) advises that we essentially use the genetic code of the microflora in our gut to do numerous things that we can't do ourselves—for example, making different vitamins and digesting food components that our own human enzymes cannot digest. Even our moods and cravings can be dictated by our gut microbiome. For example, when pathogenic yeasts causing thrush in our gut or vagina require more sugar to sustain their own growth, they secrete molecules that make us crave sugar and refined carbohydrates so they can sustain their own selfish needs. Imagine that intimidating fact—our human cravings and human feelings and moods are often being controlled and manipulated by our gut's nonhuman microflora.

Your gut bacteria provide a significant source of many essential nutrients, including amino acids, vitamins, lactase, coenzyme Q10, and three other major substances, including the energy source substance (adenosine triphosphate or ATP), tryptophan, and serotonin.

42.10 Your Human DNA Is Largely Controlled by Your Gut Flora
Mercola (83) further points out, 'beneficial bacteria also play an enormous role in your genetic expression—continuously helping flip genes off and on as you need them. Your genetic expression is 50 to 80 percent controlled by how you eat, think and move, and your genes change daily—if not hourly.' He poignantly continues, 'You may be surprised to learn that 140 times more genetic influence comes from your microbes as from the DNA in your own cells. Your genes control protein coding, which determines your hormones, weight, fertility, mood and others.'

42.11 Is the Microbiome System of the Body a Component of the Triple-Energizer Metasystem?
It is interesting that we believe that our body is our own, but in reality on a cellular basis, only one tenth of our cells are human (172). The other 90% of the cells in our body are all hitchhikers, mainly made up of harmless bacteria and other not-so-harmless organisms that have made our body their residence. There are 100,000 billion of these organisms in our guts. That means, on a cell-to-cell basis, we are 90% microbial and 10% human! Professor Jeffrey Gordon (172) from the St. Louis School of Medicine at the Washington University points out that as there are 10 times more microbial cells in our

bodies than there are human cells, there are also 100 times more functional microbial genes in our bodies than there are genes in our entire human genome.

Many of these microorganisms play a crucial role in our health and well-being by helping to metabolize our food and convert it into energy. They also protect us by excluding or at least reducing and controlling bacteria and other microorganisms that are harmful to us. Some of these commensal organisms even produce substances, including serotonin, which makes us happier individuals. In this symbiotic relationship, we provide these multitudinous hitchhikers with shelter, and we even share our food with them. These hitchhikers are most likely responsible for the differences in our ability to digest food and how we react to and resist disease.

While the ancients were aware that Xie Qi (Evil Qi, Incorrect Qi) existed, they perceived its presence as Fong Qi in its numerous forms (Cold, Hot, Damp, Dry). They were not aware as we are now that there are thousands of different individual microscopic life forms that can hitchhike on us for months, years, or a lifetime. They were not aware these organisms could poison us with their potent enterotoxins and neurotoxins to produce symptoms and signs specific to that organism. These potential 'Lingering Pathogens' include viruses, bacteriophages, prions, virions, bacteria, yeasts, fungi, moulds, dinoflagellates, spirochetes, mites, algae, paramecium, nematodes, etc.

Likewise, the ancients were not aware that there are hundreds of varieties of beneficial microflora living in our gut, lungs, mouth, nasal passages, ear canal, sinuses, fallopian tubes, vagina, stomach, and on our skin that are absolutely essential for us to maintain optimal health and emotional well-being. Our body is teeming with an incomprehensibly large number of beneficial microorganisms, and these microorganisms control a vast number of biological functions that scientists and biologists have previously attributed to being under the control of established human organ systems. Of course, this is obvious in the Lungs (UH), Stomach (MH), and Large Intestine (LH). I believe that many of these numerous biological functions managed by the pylogenetically diverse microbial communities are the same functions and properties attributed to the Triple Energizer and described by the ancient sages.

Scientist Martin Blaser (173) wrote an article entitled 'Why Antibiotics Are Making Us All Ill'. In the article, he stated:

> We need to look closely at the micro-organisms that make a living in and on our bodies—massive assemblages of competing and co-operating microbes known collectively as the microbiome. In ecology, a 'biome' refers to the sets of plants and animals in a community, such as a jungle, forest, or coral reef. An enormous diversity of species, large and small, interact to form complex webs of mutual support. When a 'keystone' species disappears or goes extinct, the ecology suffers. It can even collapse.
>
> Each of us hosts a similarly diverse ecology of microbes that, over eons, co-evolved with our species. They thrive in the mouth, gut, nasal passages, ear canal and on the skin. In women, they coat the vagina. The microbes that constitute your microbiome are generally acquired early in life; surprisingly, by the age of three, the populations within resemble those of adults. Together, they play a critical role in your immunity and ability to combat disease. In short your microbiome keeps you healthy.
>
> And parts of it are disappearing. The reasons are all around us, including overuse of antibiotics in humans and animals, caesarean sections, and the widespread use of sanitisers and antiseptics, to name just a few. Mothers give their microbes to their babies when they pass through the birth canal, but babies born by C-section miss that.

While many individuals know about the importance of taking probiotics daily for optimal gut health, few are aware that the inner surfaces of our mouth, sinuses, stomach, bowels, lungs, etc. are covered by a vast array of different microorganisms. For example, the article titled 'Lung Microbiome' (174) reports that microorganisms—including bacteria, yeasts, viruses, and bacteriophages—make up the complex human lung microbiome present in the lower respiratory tract, especially on the epithelial surfaces and within the mucus layer. The bacterial members of the microbiome have been studied more deeply, and nine core bacterial genera predominate. They include *Prevotella, Sphingomonas, Pseudomonas, Acinetobacter, Fusobacterium, Megasphaera, Veillonella, Staphylococcus,* and *Streptococcus*. This human lung microbiome is greatly variable in individuals and is composed of about 140 distinct families, whereby, for instance, the bronchial tree encompasses a mean of 2,000 bacterial genomes per square centimeter of surface area. The most significant harmful or potentially harmful bacteria that can cause respiratory disorders in humans include *Moraxella catarrhalis, Haemophilus influenzae,* and *Streptococcus pneumoniae*. Exactly how these organisms continue to survive in the lower airways in healthy individuals is unknown. In situations when the human immune system is compromised, these bacteria can proliferate and become pathogenic.

The article (174) further states:

> *The commensal bacteria are nonpathogenic and defend our airways against the pathogens.* There are several possible mechanisms. Commensals are the native competitors of pathogenic bacteria, because they tend to occupy the same ecological niche inside the human body. Secondly, they are *able to produce antibacterial substances called bacteriocins which inhibit the growth of pathogens*. Genera *Bacillus, Lactobacillus, Lactococcus, Staphylococcus, Streptococcus,* and *Streptomyces* are the main producers of bacteriocins in the respiratory tract. Moreover, commensals are known to induce Th1 response and anti-inflammatory interleukin (IL)-10, antimicrobial peptides, FOXP3, secretory immunoglobulin A (sIgA) production. (Emphasis is mine)

Note the incredible significance of that. These friendly commensal bacteria happily live in our airways and lungs, producing natural antibiotics (bacteriocins—Wei Qi) to fight pathogens (Fong Qi). Their function is 'to induce Th1 response and anti-inflammatory interleukin (IL)-10, antimicrobial peptides, FOXP3, secretory immunoglobulin A (sIgA) production'. Could this western scientific jargon be a portion of what the ancients called Wei Qi? Now let's consider health aspects associated with the microflora that inhabits our mouth.

42.12 The Complex Microbial Ecosystem That Inhabits Our Mouth

Regarding the microbial ecosystem that exists in our mouth, in the article titled 'Oil Pulling the Key to Oral Health', author Dr Valerie Malka (175) explains that the human mouth is a unique ecosystem composed of billions of microorganisms, including bacteria, fungi, viruses, and parasites. While many oral bacteria are benign and even beneficial, many predispose to causing gum disease and may even initiate other serious chronic medical conditions. The one liter of saliva that we produce daily helps regulate pathogenic bacteria. However, we produce very little protective saliva during the nighttime. The oral microbiome is unique to each person due to many factors, including lifestyle and dietary habits, gender, hormonal levels, weight, general medical conditions, stress levels, and level of hydration. Poor diet high in sugar and refined carbohydrates leads to substantial proliferation of damaging acid-producing bacteria responsible for dental caries and gum disease.

42.13 Ninety-five Percent of People in Modern Society Have Some Form of Tooth Decay or Gum Disease

In the article titled 'Oil Pulling the Key to Oral Health', author Dr Valerie Malka (175) states:

> More than 95 per cent of people in modern society have some form of tooth decay or gum disease. Medical research has shown a clear and direct link between oral health and a multitude of chronic diseases, something ancient healing traditions have known for centuries. Chinese, Tibetan and

Indian medicine pay particular attention to examination of the tongue, teeth, gums and oral cavity to diagnose a wide variety of disease states. The good news is that an ancient daily oral practice known as 'oil pulling' can promote not only mouth health but general health as well.

Oil pulling, the simple daily routine of swishing vegetable or coconut oil in your mouth, spitting it out and rinsing, has been reported to also prevent, improve or treat chronic diseases such as asthma, diabetes, headaches, chronic fatigue and colitis.

42.14 What Other Medical Researchers Have to Say about the Practice of Oil Pulling

Naturopathic physician Dr Bruce Fife is the author of the book *The Coconut Oil Miracle* and, more recently, the bestseller *Oil Pulling Therapy*. In the article by Dr Malka (175), she explains that Dr Fife has spent many years researching the powerful healing advantages of oil pulling. He believes that the practice is one of the most powerful and effective detoxification and healing procedures available. Recent research is confirming that our oral microbiome has a profound influence on our general health and that oil pulling absorbs toxins and eliminates pathogenic bacteria present on our teeth and gums and outperforms any mouthwash or toothbrush. Dr Fife notes that brushing the teeth accounts for cleaning only 10–15% of our oral cavity. The pathogenic organisms in the other 85–90% of the oral cavity are not affected by brushing. These bacteria can find their way to elsewhere in the body and overcome the immune system and produce serious illness.

42.15 Pathogenic Organisms in Our Mouth Cause Disease States throughout the Entire Body

Regarding the mechanism by which pathogenic organisms in our mouth can cause disease states throughout the entire body, in the article titled 'Oil Pulling the Key to Oral Health', author Dr Valerie Malka (175) explains that modern medicine is realizing that poor oral and dental health can lead to serious chronic and degenerative medical conditions as well as produce adverse outcomes during pregnancy. Just as small ulcers or cuts can permit disease-causing microorganisms entry into the bloodstream, so too can abscesses, gum disease, and even tooth decay. If these conditions exist, brushing and flossing the teeth may lead to bleeding, and pathogenic organisms present can enter the bloodstream and be deposited and cause infection in any organ or tissue throughout the body, including respiratory illnesses in the lungs, heart valve infections, liver abscesses, and bone infections. Malka further notes that these circulating microorganisms can also produce medical conditions and diseases that don't even appear to be related to infectious organisms. These pathologies include arthritis, skin conditions, kidney disease, asthma, stroke, dementia, osteoporosis, adrenal fatigue, diabetes, coronary artery disease, inflammatory bowel disease, and cancer.

42.16 The Underlying Mechanism of How Oil Pulling Works

Concerning the mechanism by which the powerful yet simple inexpensive process of oil-pulling works, author Valerie Malka (175) explains that oil-pulling engenders balance and harmony of the oral microbiome. She adds that a 'healthy diet and lifestyle, good oral hygiene, the correction of nutritional deficiencies and proper hydration' are all absolutely essential to ensure that optimal health is the outcome. She further explains that oil pulling also works 'by removing the load of disease-causing microorganisms and toxins so that the immune system is no longer burdened, overwhelmed and overworked, and inflammation is reduced'. She explains that oil pulling works 'by attracting the fatty membranes of these organisms, much like a magnet, pulling them from their hiding places and keeping them in solution'. Malka cautions to never swallow the oil as, at the end of the process, it is filled with toxins and pathogenic microorganisms.

42.17 Description of the Oil Pulling Technique

The oil-pulling technique is very simple. In the article titled 'Oil Pulling the Key to Oral Health', author Dr Valerie Malka (175) advises that because oral bacterial counts are greatest first thing in the morning, ideally, oil pulling should be performed in the morning, before breakfast. She suggests having a glass of water prior to oil pulling to stimulate more saliva, which is essential for the cleansing process. Malka advises using 1 tbsp of vegetable oil or coconut oil. A few drops of essential oil (e.g. oregano oil) may be added to the oil if preferred. The oil should be moved slowly and methodically inside your mouth, making sure that all surfaces of the oral cavity are treated. It is essential to pull the oil through the teeth and allow it to contact all surfaces of the oral cavity for 15–20 minutes. Continuing the oil-pulling process for the full twenty minutes ensures the best results. Unprocessed and cold-pressed sesame oil and coconut oil have been shown to be the best oils to use for oil pulling. Ensure that the oil is not gargled or swallowed. The oil becomes thinner and whiter during the procedure. After the 15–20 minutes, spit out the oil, and rinse out your mouth with freshwater several times to remove all liberated toxins and microorganisms. As with all cleansing and detoxifying procedures, a 'healing crisis' may occur. Symptoms may include unwellness, nausea, headaches, diarrhea, or fatigue, and some existing health complaints may initially intensify for a few days. And if your toxin load is high, it may take a few weeks. After the detox process, you will feel great.

42.18 Some More Medical Conditions That Oil Pulling Has Helped

Regarding the medical conditions that have been benefited by oil pulling, Malka (175) explains that because over 95% of people have suboptimal oral health, nearly everyone who practices oil pulling regularly attests to amazing results. Recent acute conditions generally resolve more quickly, and long-term chronic complaints generally resolve with more time. Added benefits include less inflammation and infection inside the oral cavity, fresher breath, less plaque, and whiter teeth. Continuing the practice daily results in numerous benefits, including improved energy levels, lessening of many diverse long-standing medical conditions (allergies, skin conditions, digestive complaints, and headaches), stabilization of blood-sugar levels, and improvement of sinusitis, asthma and bronchitis. Others have experienced relief from debilitating arthritis, inflammatory bowel disease, cardiovascular disease, chronic fatigue syndrome, and fibromyalgia. Malka cautions that oil pulling is by no means a quick cure-all and notes that it may take a long while to see improvement if the medical condition is very longstanding. She emphasizes that if the medical condition did not originate from an oral imbalance initially, then a positive outcome may not be realized.

42.19 Microorganisms Can Travel throughout the Body by Various Means

In his 2014 article entitled 'Gut Bacteria and Fat Cells May Interact to Produce "Perfect Storm" of Inflammation That Promotes Diabetes and Other Chronic Disease', Dr Joseph Mercola (176) states, 'Here, bacteria are again playing a preeminent role, as periodontal disease is the result of the colonization of certain bacteria in your mouth. This bacterial profile, by the way, is again linked to an imbalance of beneficial and pathogenic bacteria in your gut.' Thus, the bacteria present in the gut are able to find their way to the periodontal region of the mouth. What is the pathway? I believe that in addition to the circulatory system, an intricate omnipresent maze of conduits associated with both the Primo Vascular System and the Connective-Tissue Metasystem are able to carry fluids and nutrients along with microorganisms, including bacteria, to every cell throughout the body—in this case, to the periodontal region of the mouth—in the same way that the gut microbiome dispenses beneficial bacteria from the gut of new mothers to the breast milk to inoculate the neonate. There simply is no other known defined biological conduit network that performs these roles. But how is communication by the gut microbiome maintained throughout the entire body, and how do the good bacteria in the gut know when they need to rally at a certain location to fight off the bad bacteria or other microorganisms?

42.20 How Microorganisms Communicate Using Chemical 'Words'

As mentioned previously (172), humans have 30,000 genes, while the numerous diversified bacteria that hitchhike on us have 100 times more genes, expressing themselves throughout our entire body at all times. Bacteria are definitely not passive hitchhikers. They cover our numerous internal and external surfaces in armor that keeps environmental insults out so that we can stay healthy. Bacteria educate our immune system, digest our food, synthesize vitamins and neurotransmitters for us, and fight off invaders for us. But they are far too small to have an impact, beneficial or adverse, if they act as individuals.

The microorganism *Vibrio fischeri* is a bioluminescent bacterium from the sea that makes light. When the colony numbers are low, no light is made at all. But when a certain population size is reached, they then simultaneously all emit light. How could these organisms 'know' when the others are about to emit light? Researcher Bonnie Bassler (177) has determined that the *Vibrio* bacteria 'talk' to each another in their own language. Each bacterium secretes chemical substances, which could be likened to hormones. The extracellular concentration of the substances increases in concentration as the individual cells increase in number. When the molecule reaches a certain threshold that tells the bacteria how many neighbors there are, all the bacteria turn on the light in synchrony. So the bioluminescence occurs because the bacteria talk with one another using these chemical molecules as words.

The mechanism the bacteria use is a process called Bacterial Quorum Sensing. A protein enzyme inside the bacterium makes the hormone molecule that is secreted into its surroundings. The bacterium has a sensor for that molecule. The molecule fits the sensor specifically like the correct key fits the lock it was designed for. When sufficient molecules abound, the lock is opened in all the bacteria, and all the bacteria get the message and thus simultaneously emit light. All bacterial niches have dynamic mechanisms like this. So all bacteria can 'talk' to one another using different chemical 'words', and then they are able to turn on group behaviors when the threshold for the specific chemical word is reached. But the process is only successful when all the cells participate in unison. It is like every cell votes, the votes get counted, and then all the cells respond to the vote. Bassler (177) believes there are hundreds of different actions and functions that bacteria carry out in this collective behaviour of communication.

The major adverse impact of this scenario for humans is virulence. Because we are so large relative to microscopic organisms, the toxins secreted by a few pathogenic organisms would have no effect on us. What happens is that pathogenic organisms invade us and then reproduce, and when they are in sufficient numbers, they secrete their chemical 'words' until the 'sentence' is complete, whereby they simultaneously launch their virulent attack in a unified front. That way, they will be successful at overcoming an enormous host. From Bassler's work (177), she believes it is likely that bacteria control their pathogenicity with quorum sensing, using specific molecules.

While bacterial intraspecies communication has been shown to be mediated using a chemical that is specific for that species, it appears that all bacteria use molecules that are identical at one end of their chemical structure. For example, bacterial species—including *Serratia* spp., *Vibrio* spp., *Agrobacterium* spp. and *Pseudomonas* spp.—all 'talk' using a molecule that is the same at the left-hand end of the molecule. However, the right-hand end of the molecule is different in every species. This endows the molecules with exquisite species specificity to these languages. So the bacteria can have intraspecies communications that are private secret conversations only with other members of their species, and no other bacterial species can 'hear' the conversation.

So each bacterium is able to count its own siblings such that members of the same species have social behaviors. However, in nature, bacteria live amongst hundreds of other species at any given time. Bassler (177) determined that all bacteria are actually multilingual and liaise with all their neighbors. They all have a species-specific system that says 'me' as discussed above. But running in parallel to that is a second system. All bacteria have a generic interspecies system. They have a second enzyme that makes a second signal, and they all have a receptor for that molecule. That molecule is the universal language used by all different bacteria used for interspecies communication. So bacteria are essentially able to count 'how many of me' and 'how many of you' are present in a specific biosystem at any given time.

From that information, they can determine what tasks to carry out depending on who is in the majority and who is in the minority of any given population. Bassler (177) determined that a very small five-carbon molecule is produced by every type of bacteria. They all possess the same enzyme that produces the exactly same molecule. So all the bacteria in a mixed population use this molecule for interspecies communication, similar to a bacterial Esperanto. Using their incredibly complicated chemical lexicon, they know how many of their own siblings are around and how many nonsiblings are around at any given time. So they know not to mount an attack when they don't have the numbers to win the fight.

42.21 Is Defensive Wei Qi Produced by the Gut Microbiome in the Lower Heater?

On the other side, when their specific molecule advises them that they have sufficient number to successfully mount an attack, they simultaneously release their toxin into the system and express their virulence. Bacteria have collective behaviors and can carry out tasks collectively that they could not accomplish if they acted as individuals of small groups. This is the same as what happens in our body. The different cells in the body recognize other similar cells and know what their tasks are and perform defined desirable functions. If we beef up the conversations between the good bacteria that live as mutualists with us, their beneficial protective functions will be optimized, and we will be made more healthy. I suggest that it is via this interspecies communication within our gut microbiome component present in the Lower Heater of the Triple Energizer that appropriate timely communication to occur to ensure all the necessary neurotransmitters and life-sustaining biochemicals are manufactured by the microorganisms.

But as I mentioned in another Chapter of this book, research is suggesting that other faster communication systems using biophotons exist within our intricate and complicated body. It is known that very weak biophoton emission is a general property of all plant and animal life. In the article titled 'The Science of Biophotons', Dr Rüdiger Paschotta (178) discussed the research findings of Professor Popp, where Popp proposes that the biophotons present in food convey order to the consuming organism. Popp goes as far as saying that high-quality food is not primarily a chemical supply of energy but rather a means of transporting oscillations in the form of biophotons, which decrease the entropy of the consuming organism. Low entropy is associated with higher order in our body cells. So the obvious conclusion is that the healthier the food we eat, the healthier our body will be. Further, the more natural the food we consume, the larger amount of biophotons we will have available in our digestive system to energize necessary biochemical and biological functions and assist in the necessary biofeedback communication between the commensal gut microbiome organ component of the Triple Energizer and our human organs.

It is interesting to note that both the Small Intestine and the Three Heaters are Fire elements and yang in nature. I believe that a major contributing component of the Triple-Energizer Metasystem is the microbiome that resides predominantly in the Lower Heater inside the Small Intestine and Large Intestine as there is an enormous number of microorganisms existing and flourishing within the seething mass of transmuted food that we eat; it is vigorously digested and processed into necessary nutrients and biochemical molecules by an immense number of diverse microflora. It is known that 80–85% of our defensive immune system is due to the gut microbiome in the Lower Heater. In harmony with this scientific fact, in TCM it is stated that the defensive Wei Qi is produced in the Lower Heater. I discuss this in greater detail elsewhere in this book.

42.22 The Gut Microbiome Produces Essential Substances and Nutrients for Our Benefit

Prior to elimination, the fecal contents are composed of a vast number of numerous microorganisms that are unique to us alone. The microflora in the microbiome are performing an enormous job generating energy directly, supplying nutrients (e.g. CoQ10) for our mitochondria to further produce energy, providing massive immune support (80–85%), and synthesizing a vast array of diverse life-sustaining biochemical substances. These include neurotransmitters to communicate between our gut-brain and the head-brain via the vagus nerve, lactase to break down lactose in dairy products, B-group vitamins, CoQ10, etc.

By-products of the bacterial fermentation process in the gut by our microbiome include lactic acid and acetic acid. These are profoundly powerful nutrients for the bacteria. They convert lactic and acetic acid into butyric acid and butyrate derivatives, which enterocytes absolutely require to thrive. Your enterocytes, the cells that line your gut wall, have a rapid turnover and only live for a few days. The butyric acid and butyrate derivatives increase the turnover of cells in the wall of the large intestine. These substances also neutralize the activity of carcinogenic nitrosamines, which are produced by bacteria in individuals consuming a high-protein diet (179). Further to this is the fact that lactic acid itself acts as a bacteriocin or a naturally occurring antibiotic that stops harmful bacteria and other pathogenic organisms from growing to become the dominant species, which would result in disease. Beneficial commensal microbes all produce metabolites that act as bacteriocins. Lactic acid is one of these natural antibiotics, which are produced to crowd out the more pathogenic bacteria in its own strain. Specific bacteriocins are produced by microorganisms in a group to wipe out and control other members of that group of organisms that are more hostile and pathogenic in nature. It is a self-regulating, natural feedback mechanism which disallows nasty pathogenic organisms from becoming dominant in the body and launching an attack.

Note from the above information that the bacteria produce the nutrients that allow the enterocytes to thrive and prosper and that it is the function of the enterocytes to allow beneficial predigested nutrient-rich fluid broths into the body from the lumen of the intestines. I suggest that these nutrient-rich fluids are used to manufacture the various components of the TCM or San Jiao fluid distribution system, including blood, lymph, interstitial fluid, saliva, digestive juices, etc.

The colon is not highly innervated. It doesn't have as large a blood supply as the small intestine does. The large intestine needs that lactic acid, butyric acid, and acetic acid to nourish itself effectively.

Because it doesn't receive a lot of nourishment from the blood supply, it is dependent upon having large colonies of bacteria to directly supply these nutrients for continued survival and functioning.

The neurotransmitter serotonin is involved in the regulation of mood control, depression, and aggression. Thus, it is very interesting that the greatest concentration of serotonin in the body is actually present in your intestines and not in your cranial brain as psychiatrists previously thought! So when it comes to improving your mood, this recently established fact lends further support to the prudence of nourishing your gut flora rather than reaching for a prescription drug. That's fermented food for thought.

42.23 The Diverse Biological Functions of Microbiome Microorganisms in Animals and Humans

Following is a list of the presently known functions of the gut microflora that I have collated. Scientists and researchers are being stunned on a daily basis as to exactly what the gut microbiome does for us to optimize our weight, wellness, fertility, immune system, and emotions on an hourly basis. The numerous and diversified microorganisms that make up our gut microbiome are responsible for the following biological functions in the maintenance of health of animal models and humans alike.

Microorganisms in the microbiome defend the body from disease.
- maintain the correct balance between harmful and beneficial bacteria in the gut, thus maintaining a healthy digestive system
- prevent diarrhea when overseas or after consuming contaminated foods
- stimulate the immune system, thereby reducing infections
- prevent colon cancer
- prevent the formation of substances that cause cancer (e.g. nitrosamines)
- control the growth of harmful bacteria throughout the body
- reduce skin infections, such as athlete's foot and candidiasis
- control the growth of pathogenic bacteria, fungi, and viruses

- neutralize metabolic and environmental toxins
- assist with the production of antibodies and anticarcinogens
- produce many B-group vitamins, such as folic acid, niacin, biotin, B_5, B_6, and B_{12}, which are important for energy production
- reduce infections in the respiratory system, intestines, and vaginal and urinary tracts
- slow down parasitic infections
- act as natural antibiotics, helping to fight off bad bacteria in the gut
- boost our immune system (85% of our immune system is located in our digestive tract)
- engage in a complex and vital interplay with the immune system
- maintain proper development and function of the immune system
- protect against overgrowth of other microorganisms that could cause disease
- develop the immune system
- educate our immune system.

Microorganisms in the microbiome optimize digestion and optimize nutrition.
- improve overall food digestion and absorption
- minimize gas formation and bloating
- improve the digestion of lactose and overcome lactose intolerance
- improve the digestion of fats, carbohydrates, and proteins
- improve mineral and nutrient absorption from foods
- improve digestion by stimulating peristalsis, the rhythmic contractions of the large intestine, which helps to propel food along the colon
- are responsible for the manufacturing of Vitamin K, necessary for blood clotting
- aid in the elimination of digestive complaints, including gassiness, constipation, diarrhea, and IBS
- synthesize vitamins that their hosts' bodies can't produce

- ensure adequate digestion and absorption of certain carbohydrates
- ensure the adequate elimination of environmental and metabolic toxins
- help detoxify compounds we eat and are exposed to
- perform countless metabolic tasks
- influence villous length in the intestines
- alter colonic barrier function by modifying colonic wall permeability
- play important roles in harvesting energy from the diet
- stimulate the proliferation of the intestinal epithelium
- regulate fat storage in the body.

Microorganisms in the microbiome control and regulate body functions.
- regulate blood cholesterol levels
- lower blood pressure in the human body
- perform a role in balancing sex hormones, thus improving fertility
- reduce the risk for developing preeclampsia during pregnancy
- have a natural pain-relieving action on the gut
- produce heat from metabolic processes
- manufacture energy for cellular use
- turn food into energy
- prevent allergies
- stimulate the growth of new blood vessels
- allow proper tissue development
- protect the integrity of the skin
- regulate secretion and absorption
- regulate cell composition
- influence mitotic activity
- promote vaginal health in women
- may permanently alter gene expression.

While the above list pertains to the scientifically confirmed properties and functions of the gut microbiome, from my practical understanding of TCM, these properties and functions are in part or mostly attributed to functions of the Triple Burner organ. Thus,

I propose that the gut microbiome greatly assists the Lower Heater portion of the Triple-Energizer Metasystem.

It has been established that diverse commensal bacteria, predominantly *Bacteroides*, occupy our Large and Small Intestines at birth, thanks to the bacterial loading inherited from our father and primarily our mother during the birthing process. The dominant *Bacteroides* microbiome we inherit at birth has recently been shown to have a large effect on the expression of our DNA and also molds our emotional makeup and is with us until the day we die. It has been established that there is a connection between the neurotransmitters synthesized by the gut microbiome and the way these neurotransmitters allow communication between the gut brain and the head brain neural systems via the vagus nerve. It is amazing that the gut microbiome we inherited from our parents has a large impact on many physiological and emotional aspects of our future life. I believe that the gut microbiome we inherited in some way impacts our own human DNA inheritance in a much greater manner than scientists give credit for.

42.24 The Gut Microbiome Is Considered a Newly Discovered 'Human Organ'

Our gut microbiome is composed of a diverse ecological community of commensal, symbiotic, and pathogenic microorganisms that literally share our body space. The human body contains over ten times more microbial cells than human cells, although the entire microbiome weighs between 200 g and 1,400 g (171). Some researchers (58) consider the gut microbiome to be a newly discovered organ since its existence was not generally recognized until the late 1990s and it is only now being understood to have potentially an overwhelming impact on human health.

The author of *The Economist* article (58) entitled 'The Human Microbiome: Me, Myself, Us' succinctly stated what I have believed for the last three years. He/she stated:

One way to think of the microbiome is as an additional human organ, albeit a rather peculiar one. It weighs as much as many organs (about a kilogram, or a bit more than two pounds). And although it is not a distinct structure in the way that a heart or a liver is distinct, an organ does not have to have form and shape to be real. The immune system, for example, consists of cells scattered all around the body but it has the salient feature of an organ, namely that it is an organised system of cells. The microbiome, too, is organised.

The *PubMed* abstract (180) stated:

The human organism is a complex structure composed of cells belonging to all three domains of life on Earth, Eukarya, Bacteria and Archaea, as well as their viruses. Bacterial cells of more than a thousand taxonomic units are condensed in a particular functional collective domain, the intestinal microbiome. The microbiome constitutes the last human organ under active research. Like other organs, and despite its intrinsic complexity, the microbiome is readily inherited, in a process probably involving 'small world' power law dynamics of construction in newborns. Like any other organ, the microbiome has physiology and pathology, and the individual (and collective?) health might be damaged when its collective population structure is altered. The diagnostic of microbiomic diseases involves metagenomic studies. The therapeutics of microbiome-induced pathology include microbiota transplantation, a technique increasingly available. Perhaps a new medical specialty, microbiomology, is being born.

One 2015 article (181) cited five different microorganisms, including bacteria, viruses, and yeasts that are known to be linked to cancers when their numbers and aggressivity are not controlled by commensal microflora. They include *Helicobacter pylori* in gastric cancer. While the International Agency for Research on Cancer defines this microbe as a carcinogen, *H. pylori* has also been linked to a reduced risk of esophageal adenocarcinoma, demonstrating the complexity involved and the organ-specific effects that microbes can have when it comes to their impact on cancer formation. Chronic *Salmonella enterica* infection has been linked to gallbladder cancer, while Hepatitis C virus (HCV) is strongly associated with hepatocellular carcinoma (HCC). *Haemophilus influenzae* and *Candida albicans* are implicated in lower–respiratory tract tumors.

The science article entitled 'Microbiome' (171) states, 'The human microbiome may be implicated in auto-immune diseases like diabetes, rheumatoid arthritis, muscular dystrophy, multiple sclerosis, fibromyalgia, and perhaps some cancers. Common obesity might also be aggravated by a poor mix of microbes in the gut.' The article continued, 'Since some of the microbes in our body can modify the production of neurotransmitters known to be found in the brain, we may also find some relief for schizophrenia, depression, bipolar disorder and other neuro-chemical imbalances.' The article (171) proceeded to relate:

> Human skin represents the most extensive organ of the human body, whose functions include protecting the body from pathogens, preventing loss of moisture, and participating in the regulation of body temperature. Considered as an ecosystem, the skin supports a range of microbial communities that live in distinct niches. Hair-covered scalp lies but a few inches from exposed neck, which in turn lies inches away from moist hairy underarms, but these niches are, at a microbial level, as distinct as a temperate forest would be compared with savannah and tropical rain forest. Studies characterizing the microbiota that inhabit these different niches are providing insights into the balance between skin health and disease.

The article (171) continued:

> The human microbiome consists of about 100 trillion microbial cells, outnumbering human cells 10 to 1. Thus it can significantly affect human physiology. For example,

in healthy individuals the microbiota provide a wide range of metabolic functions that humans lack. In diseased individuals altered microbiota are associated with diseases such as neonatal necrotizing enterocolitis, inflammatory bowel disease and vaginosis. Thus studying the human microbiome is an important task that has been undertaken by initiatives such as the Human Microbiome Project.

So essentially our body uses the genetic code of the microflora in our gut to do things that we can't do ourselves—for example, making different vitamins. Remember too that our feelings and 'emotions' are essentially due to chemical molecules in our bloodstream, the very same molecules that lowly bacteria and yeasts also produce in our gut. But we have trillions of organisms discharging similar chemical messages into our intestines, the very organs that specialize in absorbing, and some of these chemical messages are identical to our very own. For example, when pathogenic yeasts causing thrush in our gut or vagina require more sugar to sustain their own growth, they secrete molecules that make us crave sugar and refined carbohydrates so they can sustain their own selfish needs. Imagine that—our cravings and our moods are being controlled and manipulated by our gut microflora! It certainly appears that we have a very powerful and diverse organ operating inside of us, indeed a new organ, which has only recently been discovered.

42.25 A Faulty Microbiome Can Poison Us, Causing 'Foggy Brain' and Psychiatric Conditions

Myhill (182) points out that some patients have bacteria, yeasts, and possibly other parasites present in their upper gut, so rather than foods being digested, they are being fermented into ethyl alcohol, propyl alcohol, butyl alcohol, and possibly methyl alcohol. These would subsequently be metabolized by the liver into acetaldehyde, propylaldehyde, butyraldehyde, and possibly formaldehyde respectively. Alcohol and aldehydes result in 'foggy brain', 'toxic brain', feeling 'poisoned', and so on. Alcohol also disturbs blood sugar levels, which makes the sufferer crave sugar and refined carbohydrates, which are the very foods the contaminating microorganisms in the upper gut require to ensure their own survival. Myhill concludes, 'This is arguably a clever evolutionary ploy by bugs to ensure their own survival!'

It is also possible that our gut microflora may cause psychiatric conditions. Japanese researcher Katsunari Nishihara (183) proposes that some psychiatric conditions are initiated by gut microbes fermenting neurotransmitters to create amphetamine and LSD-like substances. Nishihara proposes:

> If neurons are contaminated with large numbers of bacteria (several thousands) or viruses (several hundreds of thousands)—that is, comparable to the population of mitochondria—the monoamine metabolic path deteriorates and the amounts of dopa and dopamine become unstable. Neurotransmitters noradrenaline or adrenaline become oxidized or hydroxidized into amphetamine or methamphetamine, which is a stimulant.

He further states, 'The metabolism of the neurotransmitter serotonin is disturbed by contaminating enteromicrobes, one consequence of which is that serotonin changes into hallucinogenic LSD.'

42.26 Our Gut Microbiome Digests Food Components That Human Enzymes Cannot

An article written in 2012 in *The Economist* titled 'The Human Microbiome: Me, Myself, Us' (58) discussed how our gut microbiome is also better at extracting nutrition from mother's milk because various bacteria produce an enzyme known as glycoside hydrolase, which converts carbohydrates called glycans, of which milk has many, into usable sugars. Glycans cannot be digested by any of the enzymes that are encoded in the 23,000 genes that humans possess. Only bacterial enzymes can convert glycans into usable nutrients in the human gut. A lot of other carbohydrates would also be indigestible and unavailable to us if we only had access to human-derived enzymes. The massively more diverse genome of our gut microbiome has correspondingly greater digestive capabilities, and even complex carbohydrates succumb to its power. Carbohydrates of all sorts are ceaselessly munched and their reduced units churned out as small fatty-acid molecules—particularly formic acid, acetic acid, and butyric acid—which can pass through the gut wall into the bloodstream, where biochemical pathways release energy from them. It is estimated that 10–15% of the energy used by an average adult is generated via this process, thanks to our gut microflora.

42.27 Microbiome Organ Transplants are Powerful against Deadly Infection

The article in *The Economist* (58) stated:

> According to America's Centres for Disease Control and Prevention, *C. difficile* kills 14,000 people a year in America alone. The reason is that many strains are resistant to common antibiotics. That requires wheeling out the heavy artillery of the field, drugs such as vancomycin and metronidazole. These also kill most of the patient's gut microbiome. If they do this while not killing off the *C. difficile*, it can return with a vengeance.

Dr Mark Mellow of the Baptist Medical Centre in Oklahoma City uses fecal transplants to fight infections of *Clostridium difficile*, a bacterium that causes severe diarrhea, particularly among patients that are already in hospital. Dr Mellow treats sick patients with an enema composed of feces from a healthy individual. The transplanted commensal bacteria multiply rapidly in the lower intestine and dominate the *C. difficile*. Dr Mellow and his colleagues have performed this procedure on 77 patients in five hospitals with an initial success rate of 91%. When the seven patients who did not initially respond were given a second fecal transplant, six were cured.

42.28 Links between a Faulty Microbiome and Human Health in Autistic Individuals

According to the article (58), a striking claim regarding links between the microbiome and human health involves the brain. It is an established fact that autistic individuals generally have intestinal problems as well, which are generally associated with abnormal microbiomes that are generally rich in various species of *Clostridia*. There is always conflict and competition in every ecosystem, even stable, productive ones. *Clostridia* spp. kill competing bacteria residing in their niches, using toxic chemicals called phenols. Unfortunately, phenols are also very poisonous to human cells too and have to be neutralized. The body accomplishes this detox procedure by adding sulfate molecules to them. Having large colonies of *Clostridia* present in the gut, producing large amounts of toxic phenols, will deplete the reserves of sulfur in the body. Sulfur has many other beneficial functions in the body, including brain development and management, and where there is a sulfur deficiency, abnormal brain development may result. It is known that many autistic people have a genetic sulfur-metabolism defect. It is possible that the interference by the large toxic *Clostridia* load in their guts is the tipping point that greatly exacerbates their problem, causing abnormal brain development.

42.29 Bacterial Fermentation in the Gut Produces 500 kcals of Energy per Day

In the 2014 article titled 'Fermentation in the Gut and CFS', Dr Sarah Myhill (182) notes that the stomach, duodenum, and small intestine are generally free from microorganisms, including bacteria, yeasts, and parasites. She also points out that having an acidic stomach to digest protein efficiently also kills acid-sensitive microbes, that an alkali duodenum to digest carbohydrates also kills the alkali-sensitive microbes with bicarbonate, and that bile salts are also toxic to microbes. The small intestine does more digesting and also absorbs the amino acids, fatty acids, glycerol, and simple sugars that result. In the article, Myhill (182) states:

> Anaerobic bacteria, largely Bacteroides, flourish in the large bowel, where foods that cannot be digested upstream are then fermented to produce many substances highly beneficial to the body. Bacteroides ferment soluble fibre to produce short chain fatty acids—over 500 kcals of energy a day can be generated. This also creates heat to help keep

us warm. The human body is made up of 10 trillion cells, (while) in our gut we have 100 trillion microbes or more, i.e. ten times as many! Bacteria make up 60% of dry stool weight, there are over 500 different species, but 99% of bugs are from 30–40 species.

42.30 Even the Gut Has Its Own Predator-Prey Balance!

In her 2014 article titled 'Faecal Bacteriotherapy', Dr Sarah Myhill (184) states, 'Up to 15% of microbes residing in our bodies may be viruses—not the more familiar human pathogens which are responsible for viral illness and infections, but viruses which predate on bacteria. These are called bacteriophages, or phages for short. Phages have been greatly studied and used in Russia and Eastern Europe as natural antibiotics.' Myhill remarked that there is a predator–prey balance occurring even in the human gut!

42.31 Bacteroides Are the Most Abundant and Most Important Microbiome Organisms

Microbiology had identified about 100 different groups of bacteria, which are known as phyla. They all have a different range of biochemical proficiencies. Interestingly, human microbiomes are dominated by only four of these different phyla: the Bacteroidetes, Actinobacteria, Firmicutes, and Proteobacteria. Myhill (182) advises that the most important and most abundant microorganisms in the gut are the obligate anaerobes *Bacteroides*, which make up 90% of the gut flora, along with *E. coli*, *Lactobacilli*, and *Bifidobacteria* predominantly making up the rest. *Bacteroides* allow us to digest soluble fiber to produce short-chain fatty acids, which is the main source of food for the colonocytes, the specialized cells lining the bowel. Atrophy of the colon results when short-chain fatty acids are deficient. Short-chain fatty acids also protect us from becoming hypoglycemic.

As *Bacteroides* are vital for recycling bile acids, low levels of bile acids are indicative of low levels of *Bacteroides*. These organisms reside on the surface of the gut lining and prevent pathogenic species (including *Salmonella*, *Shigella*, and *Clostridia* spp.) from adhering to the gut lining and causing infection. As *Bacteroides* are anaerobic organisms, they cannot exist outside the human gut for more than a few minutes as oxygen kills them quickly. To ensure we have sufficient *Bacteroides* it is essential to feed the gut with sufficient prebiotics, such as those found in pulses, nuts, seeds, and vegetables, including onions and leeks on a daily basis. Given the right substrate, bacteria can double their numbers every 20 minutes.

Myhill (182) states, 'Probiotics have not lived up to their therapeutic potential. I suspect this is because the single most important probiotic is *Bacteroides* and this cannot exist outside the human gut. Oxygen kills it within minutes. There is no probiotic on the market which contains *Bacteroides*.'

A critical point to note is that we acquire our *Bacteroides* colonies at birth and retain those bacteria for the rest of our life (182). As mentioned previously (172), humans have 30,000 genes, while the numerous diversified bacteria that hitchhike on us have between them 3,000,000 genes (which is 100 times more genes) expressing themselves throughout our entire body at all times. Could they be associated with and contributing to our Yuan Qi? Further, *Bacteroides* are a literal powerhouses of energy production as *Bacteroides* are estimated to generate more than 500 Kcal of energy per day, which makes them a very significant source of energy (182)! Could this be one of the ways that the Triple Burner produces qi, which is circulated throughout the entire body?

42.32 Other Important Gut Microflora

While some *E. coli* are very pathogenic, commensal *E. coli* are very essential for our continued good health. One gram of stool should contain between 7 million and 90 million *E. coli* organisms (182). These maligned bacteria produce lactase (to ferment lactose in dairy products), folic acid, vitamin K_2 (which protects against osteoporosis), Co-enzyme Q10 (which is essential for energy production by mitochondria), together with three amino acids, namely tyrosine and phenylalanine (these are precursors of dopamine, lack of which results in low mood) and tryptophan, a precursor of serotonin, which is responsible for gut motility and mood elevation. Myhill (182) states, 'If there are low counts of *E-coli*, one can expect problems in all the above areas, i.e. osteoporosis and bone problems, (poor) mitochondrial function, low mood and poor gut motility. Dr Butt told me about a study done in Germany where *E-coli* probiotics were given in the treatment of constipation and there was a dramatic improvement from 1.6 motions a week to 6, illustrating the effects on motility! *E-coli* is contained in the probiotic Mutaflor™ produced commercially in Germany.'

Lactobacilli spp. ferment sugars to lactic acid, which provides an acid environment in the large bowel to protect against infection. *Lactobacilli* spp. are abundant in Kefir and are highly protective against bowel cancer.

Bifidobacteria spp. assists digestion and protects against the development of allergies and cancer. When the human gut is nourished with regular intake of *Lactobacilli* spp. and *Bifidobacteria* spp., then good health and wellness will prevail. When these species are deficient, other bacteria will flourish and become pathogenic and cause disease. These organisms are plentiful in many naturally fermented foods. Details of how to prepare fermented foods can be found in the book *Gut and Psychology Syndrome*. (59)

Streptococcus spp. ferment suitable substrate to produce large amounts of lactic acid, which may lead to a predisposition to acidosis. Lactic acid is metabolized in the liver by lactate dehydrogenase, so high levels of this may indicate bowel overgrowth with *Streptococcus* spp. Fermentation produces two lactic acid isomers. The D-lactate is the problem as the body cannot metabolize this, and it accumulates in mitochondria and inhibits energy production.

Prevotella spp. are *Bacteroides* growing in the upper gut. It is believed that their fermentation products include the poisonous rotten-egg gas, hydrogen sulfide, which inhibits mitochondrial function directly. So 'a positive hydrogen sulphide urine test shows there is a severe gut dysbiosis probably due to overgrowth of *Prevotella*' (182).

42.33 Our Gut Microbiome Produces Numerous Biochemicals Necessary for Life

Chronic colitis was associated with anxiety-like behaviour in mice. In the article titled 'The Anxiolytic Effect of *Bifidobacterium longum* NCC3001 Involves Vagal Pathways for Gut-Brain Communication', the authors, Bercik et al. (185), reported that the probiotic *Bifidobacterium longum* NCC3001 normalizes anxiety-like behavior and hippocampal brain-derived neurotrophic factor (BDNF) in mice with infectious colitis. The authors concluded that in this colitis model, anxiety-like behavior is mediated via the vagus nerve. The anxiolytic effect of *B. longum* requires a functioning vagus nerve. As *B. longum* decreases excitability of enteric neurons, it may signal to the central nervous system by activating vagus nerve pathways at the level of the enteric nervous system.

In the 2014 article titled 'The Amazing Healing Properties of Fermented Foods', the author (186) notes that beneficial bacteria that make up the microbiome in our gut break down antinutrients and secrete enzymes that we can't produce ourselves. Further to the benefits derived from the probiotic microorganisms themselves, consuming fermented food generates an entirely novel assembly of nutrients and medically important phytocompounds. Fortunately, there is also the group of about 3,500 food metabolites with low molecular weight that are produced by our microbiome when we consume foods. For example, our gut microbiome converts the lignans present in flaxseed into at least two important phytoestrogens— namely, enterolactone and enterodial. These possess tumour-regressive properties against breast and prostate cancers. Due to the extreme importance of the gut microbiome within the human alimentary canal and the emmensity of the microbial numbers present, some scholars have proposed that we reclassify humans as a 'metaorganism'. The author further stated:

> Ultimately, a return to fermented foods is a return to our own ground of being, and well-being. There are profound challenges that stand in our way, of course. The modern world nukes its food, yes, with literal nuclear waste.

We microwave, we cook, we fry, we dehydrate, we spray our food into certain death. And now new research shows that even the very food starter bacteria normally found within healthy soil are being decimated by Monsanto's ROUNDUP herbicide glyphosate, which is destroying its microbial biodiversity and hence fertility.

42.34 TV Advertisements Brainwash Us into Having Our Skin Washed Far Too Excessively

In the 2015 article titled 'Antiperspirants Can Make You Smell Worse by Altering Armpit Bacteria', the author, Dr Joseph Mercola (187), states, 'Sometimes "the cure" leads to a worsening of the very problem you're trying to solve. Such may be the case when it comes to antiperspirants. As reported by Real Clear Science, antiperspirants affect the bacterial balance in your armpits, leading to an even more foul-smelling sweat problem.' The active ingredient in most antiperspirants is aluminum compounds, which kills off the bacteria that cause less odor, thus allowing bacteria that produce more pungent odours to thrive instead. The article advised that individuals who used antiperspirants produced a definitive increase in *Actinobacteria*, which are bacteria that are hugely responsible for foul-smelling armpit odor. Research showed that study participants who abstained from using antiperspirants for a month developed armpits where the population of odor-causing *Actinobacteria* had dwindled into virtual nonexistence.

In 2014, a test group investigated, using a living bacterial skin tonic created by AOBiome. The tonic looked, felt, and tasted like water, but each spray bottle contained billions of cultivated *Nitrosomonas eutropha*, an ammonia-oxidizing bacteria (AOB) that is most commonly found in dirt and untreated water. The article stated, 'AOBiome scientists hypothesize that it [i.e. *Nitrosomonas eutropha*] once lived happily on us too—before we started washing it away with soap and shampoo—acting as a built-in cleanser, deodorant, anti-inflammatory and immune booster by feeding on the ammonia in our sweat and converting it into nitrite and nitric oxide.' For the test, one study subject 'agreed to mist her face, scalp, and body with the live bacteria twice a day for a month'. The article continued, 'The theory that adding rather than eradicating bacteria from your body might produce better results seems rather logical, considering what we now know about the gut microbiome, and how the bacterial balance in your armpits affects your sweat odor.' Long-term results of using the product look very promising. The article reported on several individuals that incorporated *N. eutropha* into their hygiene routine several years ago and related that the most extreme case was David Whitlock, the MIT-trained chemical engineer who invented AO+. Incredibly, Whitlock had not showered for the past twelve years but takes an occasionally sponge bath. He believes that the *Nitrosomonas eutropha* looks after his hygiene sufficiently. The proof appears to be in the pudding, so to speak, as the author noted that in spite of such radical bathing habits, none of the men appeared 'unclean' either visually or in the olfactory sense.

I discussed in another Chapter in this book that toxic chemicals are eliminated through our skin in the sweat. Sodium lauryl sulfate is an inexpensive and very effective foaming agent found in many personal care products, including soaps, shampoos, toothpaste, etc. It may be the deadliest to *N. eutropha*, but nearly all common liquid cleansers remove at least some of the bacteria. Many food additives and most pharmaceutical drugs kill off beneficial bacteria in the gut, and when they are eliminated through the skin, they kill off sensitive microorganisms, like *N. eutropha*, and more potentially harmful microbes become the dominant species on the skin. In this regard, the article stated, 'There's a strong correlation between eczema flare-ups and an increased number of *Staphylococcus aureus* bacteria on the skin, which has scientists pondering the possibilities for treating the skin disorder with the appropriate skin bacteria.'

Summary of Chapter 42
The Human Genome Project revealed that the vast majority of our human physiology remarkably came from somewhere other than our 25,000 genes. This revolutionary finding led to an understanding of epigenetics, which is the body's capability of manifesting variable genetic expression within our epigenome without changes to the

primary nucleotide sequences of our DNA itself. The greatest interface with our environment involves our gut microbiome, which is composed primarily of bacteria that amplify our genomic library 150-fold. Many of our genes 'slipped into our DNA from microbes living in our bodies' (166). This mechanism is called horizontal gene transfer. Mercola states, 'Bacteria slip genes to each other, and it helps them evolve. And scientists have seen insects pick up bacterial genes that allow them to digest certain foods. Humans may have as many as hundreds of so-called foreign genes they picked up from microbes.' Mercola explains that the human genome consists of about 23,000 genes. He then emphasizes that the combined genetic material of the human gut microbiome is somewhere between 2 million and 20 million. It is incredible that even the gene that determines your blood type (A, B, or O) is a foreign gene. A human study (168) involving seven healthy volunteers found that consumption of a dairy drink containing three different strains of probiotic microorganisms caused changes in the activity of hundreds of genes six hours after consumption of the drinks.

There is no doubt that the status of the gut is largely responsible for human health. But how is our gut composition initiated? What are the contributing factors to gut health? When does our microbiome start? Mercola (165) answers these questions by stating, 'Adding to a long list of "oops!" in the history of medicine, it was long-held that the womb was a sterile environment. We now have broadened our appreciation of the ubiquitous nature of microbes to encompass their special place in the placenta, umbilical cord, and fetal membranes.'

It has been estimated for some time that 80–85% of our immune system is located in our gut, so reseeding our gut on a daily basis with healthy probiotic organisms is extremely important for our overall physical health. However, note what researchers have recently discovered regarding psychological health. Amongst the statements are 'sufficient amounts of probiotic gut bacteria from birth may be essential for future psychological health' (167); 'the presence of gut microbiota regulates the set point for hypothalamic-pituitary-adrenal (HPA) axis activity' (167); 'probiotic microorganisms caused changes in the activity of hundreds of genes six hours after consumption of the drinks' (168); 'gut bacteria may influence mammalian early brain development and behaviour' (169); 'presence of gut microorganisms during infancy permanently alters gene expression' (169); 'gut bacteria altered genes and signaling pathways involved in learning, memory, and motor control' (170); and 'there is a critical period early in life when gut microorganisms affect the brain and change the behavior in later life' (170). According to the authors of one study (167), 'Acquisition of intestinal microbiota in the immediate postnatal period has a defining impact on the development and function of the gastrointestinal, immune, neuroendocrine and metabolic systems.'

It is astounding that the genetic code of our gut microflora overrides human genetic code. The article entitled 'Microbiome' (171) advises that we essentially use the genetic code of the microflora in our gut to do numerous things that we can't do ourselves—for example, making different vitamins and digesting food components that our own human enzymes cannot digest. Even our moods and cravings can be dictated by our gut microbiome. Your gut bacteria provide a significant source of many essential nutrients, including amino acids, vitamins, lactase, coenzyme Q10, and three other major substances, including the energy source substance (adenosine triphosphate or ATP), tryptophan, and serotonin.

Your human DNA is largely controlled by your gut flora. Mercola (83) points out, 'Beneficial bacteria also play an enormous role in your genetic expression—continuously helping flip genes off and on as you need them. Your genetic expression is 50 to 80 percent controlled by how you eat, think and move, and your genes change daily—if not hourly.' He poignantly continued, 'You may be surprised to learn that 140 times more genetic influence comes from your microbes as from the DNA in your own cells. Your genes control protein coding, which determines your hormones, weight, fertility, mood and others.'

It is interesting that we believe that our body is our own, but in reality, on a cellular basis, only one tenth of our cells are human (172). The other 90% of the cells in our body are all hitchhikers, mainly made up of harmless bacteria and other not-so-harmless organisms that have made our body their residence. There are 100,000 billion of these organisms in our guts. That means, on a cell-to-cell basis, we are 90% microbial and 10% human! Professor Jeffrey Gordon (172) from the St. Louis School of Medicine at the Washington University, points out that as there are 10 times more microbial cells in our bodies than there are human cells. There are also 100 times more functional microbial genes in our bodies than there are genes in our entire human genome.

While the ancients were aware that Xie Qi (Evil Qi, Incorrect Qi) existed, they perceived its presence as Fong Qi in its numerous forms (Cold, Hot, Damp, Dry). They were not aware as we are now that there are thousands of different individual microscopic life forms that can hitchhike on us for months, years, or a lifetime. They were not aware these organisms could poison us with their potent enterotoxins and neurotoxins to produce symptoms and signs specific to that organism. These potential 'Lingering Pathogens' include viruses, bacteriophages, prions, virions, bacteria, yeasts, fungi, moulds, dinoflagellates, spirochetes, mites, algae, paramecium, nematodes, etc. Likewise, the ancients were not aware that there are hundreds of varieties of beneficial microflora living in our gut and in our lungs, mouth, nasal passages, ear canal, sinuses, fallopian tubes, vagina, stomach, skin that are absolutely essential for us to maintain optimal health and well-being. Note the incredible significance of that. These friendly commensal bacteria happily live in our airways and lungs, producing natural antibiotics (bacteriocins—Wei Qi) to fight pathogens (Fong Qi). Their function is 'to induce Th1 response and anti-inflammatory interleukin (IL)-10, antimicrobial peptides, FOXP3, secretory immunoglobulin A (sIgA) production'. Could this western scientific jargon be a portion of what the ancients called Wei Qi? Now let's consider health aspects associated with the microflora that inhabits our mouth.

In the article titled 'Oil Pulling the Key to Oral Health', the author, Dr Valerie Malka (175), states, 'More than 95 percent of people in modern society have some form of tooth decay or gum disease. Medical research has shown a clear and direct link between oral health and a multitude of chronic diseases, something ancient healing traditions have known for centuries. Oil pulling, the simple daily routine of swishing vegetable or coconut oil in your mouth, spitting it out and rinsing, has been reported to also prevent, improve or treat chronic diseases such as asthma, diabetes, headaches, chronic fatigue and colitis.'

Bacteria educate our immune system, digest our food, synthesize vitamins and neurotransmitters, and fight off invaders for us. But they are far too small to have an impact, beneficial or adverse, if they act as individuals. The mechanism the bacteria use is a process called Bacterial Quorum Sensing. A protein enzyme inside the bacterium makes the hormone molecule that is secreted into its surroundings. The bacterium has a sensor for that molecule. The molecule fits the sensor specifically like the correct key fits the lock it was designed for. When sufficient molecules abound, the lock is opened in all the bacteria, and all the bacteria get the message and thus simultaneously act. All bacterial niches have dynamic mechanisms like this. So all bacteria can 'talk' to one another, using different chemical 'words', and then they are able to turn on group behaviors when the threshold for the specific chemical word is reached. But the process is only successful when all the cells participate in unison. It is like every cell votes, the votes get counted, and then all the cells respond to the vote. Bassler (177) believes there are hundreds of different actions and functions that bacteria carry out in this collective behaviour of communication.

It has been established that diverse commensal bacteria, predominantly *Bacteroides*, occupy our Large and Small Intestines at birth, thanks to the bacterial loading inherited from our father and primarily our mother during the birthing process. The dominantly *Bacteroides* microbiome we inherit at birth has recently been shown to have a large effect on the expression of our DNA and also moulds our emotional makeup and is with us until the day we die. I believe that the gut microbiome we inherited in some way impacts our own human DNA inheritance in a much greater manner than scientists give credit for.

The author of *The Economist* article (58) entitled 'The Human Microbiome: Me, Myself, Us' succinctly stated what I have believed for the last three years. He/she stated:

> One way to think of the microbiome is as an additional human organ, albeit a rather peculiar one. It weighs as much as many organs (about a kilogram, or a bit more than two pounds). And although it is not a distinct structure in the way that a heart or a liver is distinct, an organ does not have to have form and shape to be real. The immune system, for example, consists of cells scattered all around the body but it has the salient feature of an organ, namely that it is an organised system of cells. The microbiome, too, is organised.

A faulty microbiome can poison us, causing 'foggy brain' and psychiatric conditions. Myhill (182) points out that some patients have bacteria, yeasts, and possibly other parasites present in their upper gut, so rather than foods being digested, they are being fermented into ethyl alcohol, propyl alcohol, butyl alcohol, and possibly methyl alcohol. These would subsequently be metabolized by the liver into acetaldehyde, propylaldehyde, butyraldehyde, and possibly formaldehyde respectively. Alcohol and aldehydes result in 'foggy brain', 'toxic brain', feeling 'poisoned', and so on. Alcohol also disturbs blood sugar levels, which makes the sufferer crave sugar and refined carbohydrates, which are the very foods the contaminating microorganisms in the upper gut require to ensure their own survival. Myhill concludes, 'This is arguably a clever evolutionary ploy by bugs to ensure their own survival!'

An article written in 2012 in *The Economist* titled 'The Human Microbiome: Me, Myself, Us' (58) discussed how a lot of carbohydrates would also be indigestible and unavailable to us if we only had access to human-derived enzymes. The massively more diverse genome of our gut microbiome has correspondingly greater digestive capabilities, and even complex carbohydrates succumb to its power. Carbohydrates of all sorts are ceaselessly munched and their reduced units churned out as small fatty-acid molecules—particularly formic acid, acetic acid, and butyric acid—which can pass through the gut wall into the bloodstream, where biochemical pathways release energy from them. It is estimated that 10–15% of the energy used by an average adult is generated via this process, thanks to our gut microflora.

A critical point to note is that we acquire our *Bacteroides* colonies at birth and retain those bacteria for the rest of our life (182). As mentioned previously (172), humans have 30,000 genes, while the numerous diversified bacteria that hitchhike on us have between them 3,000,000 genes (which is 100 times more genes), expressing themselves throughout our entire body at all times. Could they be associated with and contributing to our Yuan Qi? Further, *Bacteroides* are literal powerhouses of energy production as *Bacteroides* are estimated to generate more than 500 Kcal of energy per day, which makes them very significant source of energy! (182). Could this be one of the ways that the Triple Burner produces qi, which is circulated throughout the entire body?

In 2014, a test group investigated, using a living bacterial skin tonic created by AOBiome. The tonic looked, felt, and tasted like water, but each spray bottle contained billions of cultivated *Nitrosomonas eutropha*, an ammonia-oxidizing bacteria (AOB) that is most commonly found in dirt and untreated water. The article stated, 'AOBiome scientists hypothesize that it [i.e. *Nitrosomonas eutropha*] once lived happily on us too—before we started washing it away with soap and shampoo—acting as a built-in cleanser, deodorant, anti-inflammatory and immune booster by feeding on the ammonia in our sweat and converting it into nitrite and nitric oxide.' One subject had not showered for twelve long years. Did he stink? No, the bacterial spray he had engineered kept him smelling fresh. There is no doubt that gut and skin microorganisms are absolutely essential for continued good health and vitality and for smelling good into the bargain.

CHAPTER 43

How the Enteric Nervous System Complements the Triple Energizer

Introduction
An article in *Scientific American* by Adam Hadhazy (188) shows that there is an often-overlooked massive network of neurons lining our guts that is so extensive some scientists have nicknamed it our second brain. This second brain is able to control gut behavior independently of the brain. The article states, 'Technically known as the enteric nervous system, the second brain consists of sheaths of neurons embedded in the walls of the long tube of our gut, or alimentary canal, which measures about nine meters end to end from the oesophagus to the anus. The second brain contains some 100 million neurons; this is more than what's in either the spinal cord or the peripheral nervous system.' The article further states, 'This multitude of neurons in the enteric nervous system enables us to "feel" the inner world of our gut and its contents.' A major role of the enteric nervous system is involved in managing the processes of digestion. The digestion of complex foods, absorption of fluids and nutrients, and expelling of metabolic waste products requires a vast amount of processing. It seems to me that these features involve characteristics associated with the ancient description of the Triple Energizer.

43.1 Ninety Percent of the Vagus Nerve Fibers Carry Information from the Gut to the Brain

'The system is way too complicated to have evolved only to make sure things move out of your colon,' says Emeran Mayer, professor of physiology, psychiatry, and biobehavioral sciences at the David Geffen School of Medicine at the University of California, Los Angeles (UCLA). Scientists were recently shocked to learn that about 90% of the fibers in the primary visceral nerve, the vagus nerve, carry information from the gut to the brain and not the other way around. '*A big part of our emotions are probably influenced by the nerves in our gut,*' Mayer says.

Scientists are making incredible findings about how *the gut controls so many biological parameters within the body*. The *enteric nervous system uses more than thirty neurotransmitters*, just like the brain, and in fact, 95% of the body's serotonin is found in the bowels. '*It was totally unexpected that the gut would regulate bone mass to the extent that one could use this regulation to cure—at least in rodents—osteoporosis,*' says Gerard Karsenty, lead author of the study and chair of the Department of Genetics and Development at Columbia University Medical Center (emphasis is mine).

43.2 The Same Genes for Neuron Synapse Formation in the Brain Are Involved in the Gut

Michael Gershon, chairman of the Department of Anatomy and Cell Biology at New York–Presbyterian Hospital/Columbia University Medical Center, has discovered that the same genes involved in synapse formation between neurons in the brain are involved in the alimentary synapse formation. 'If these genes are affected in autism,' he says, 'it could explain why so many kids with autism have GI motor abnormalities in addition to elevated levels of gut-produced serotonin in their blood.'

Cutting-edge research is currently investigating how the second brain mediates the body's immune response; after all, at least 70% of our immune system is aimed at the gut to expel and kill foreign invaders. UCLA's Mayer is doing work on how the trillions of bacteria in the gut communicate with enteric nervous system cells (which they greatly outnumber). His work with the gut's nervous system has led him to think that, in coming years, psychiatry will need to expand to treat the second brain in addition to the one atop the shoulders.

Note that the bacteria of the gut microbiome communicate with enteric nervous system cells. The microflora in the gut is responsible for synthesizing many, if not all, of the neurotransmitters that the second brain uses to communicate with the cranial brain.

Summary of Chapter 43
An article in *Scientific American* by Adam Hadhazy (188) shows that there is an often-overlooked massive network of neurons lining our guts that is so extensive some scientists have nicknamed it our second brain. This second brain is able to control gut behavior independently of the brain. This second brain contains some 100 million neurons; this is more than what's present in either the spinal cord or the peripheral nervous system.

Scientists were recently shocked to learn that about 90% of the fibers in the primary visceral nerve, the vagus nerve, carry information from the gut to the brain and not the other way around. 'A big part of our emotions are probably influenced by the nerves in our gut,' Mayer (188) says. This recent scientific finding is in keeping with ancient TCM philosophy.

Scientists are making incredible findings about how the gut controls so many biological parameters within the body. The enteric nervous system uses more than thirty neurotransmitters, just like the brain, and in fact, 95% of the body's serotonin is found in the bowels.

Note that the bacteria of the gut microbiome communicate with enteric nervous system cells. The microflora in the gut is responsible for synthesizing many, if not all, of the neurotransmitters that the second brain uses to communicate with the cranial brain. Many of the functions of the enteric nervous system are common functions of the San Jiao.

CHAPTER 44

How the Lymphatic System Complements the Triple Energizer

Introduction
The Britannica.com website, under the heading 'Lymphoid Tissue' (189), advises that lymphoid tissue (cells and organs that make up the lymphatic system) includes white blood cells (leukocytes), bone marrow, thymus, spleen, and lymph nodes, which are all part of the immune system. Confirming that lymphatic tissue is a specialized and highly organized connective tissue, the article reports that the lymphoid tissue present in the spleen is composed of a cylinder of loose organized cells surrounding small arteries. While the thymus and lymph nodes contain the most highly organized lymphoid tissues, in the bone marrow, lymphoid tissues are mixed with the blood-forming cells, and there is no apparent organization. The article states, 'The most diffuse lymphoid tissue is found in the loose connective-tissue spaces beneath most wet epithelial membranes, such as those that line the gastrointestinal tract and the respiratory system.' Many of the lymphatic system cells in these spaces wander freely to defend against microorganisms and foreign material. They form localized cell production centers called nodules in response to such microbial invasions. These cell production centers (nodules) should not to be confused with nodes, which constitute an entirely different structure. Some nodule formations become relatively permanent tissue structures and include the tonsils, appendix, and

Peyer's patches, which occupy the lining of the small intestine. Most nodule structures develop and then vanish in response to specific local needs. The lymphoid system is made up of several different types of cells and include reticular cells and white blood cells (macrophages and lymphocytes).

44.1 Hippocrates Was the First Western Physician to Describe the Lymphatic System

In the article titled 'The New Era of the Lymphatic System: No Longer Secondary to the Blood Vascular System', regarding the first person to describe the lymphatic vessels, the authors (190) state:

> Hippocrates (460–377 B.C.) first described the lymphatic vessel as 'white blood' and coined the term 'chyle' (from the Greek *chylos*, meaning juice). Chyle is a milky tissue fluid consisting of emulsified fats and free fatty acids, collectively called lymph, which is formed in the digestive system and taken up by the specialized lymph vessels known as lacteals.

44.2 Reticular Cells Provide Structural Support for the Defensive Lymphatic System

Regarding the structural aspect of the lymphatic system, the Britannica.com website, under the heading 'Lymphoid Tissue' (189), advises that because reticular cells produce and maintain a framework of thin fibers, they provide structural support for most lymphoid organs throughout the body. Macrophages engulf foreign materials via a process called phagocytosis and then initiate the immune response. These cells may reside at one location—for example, in lymph nodes—or conversely, they may roam around in the loose connective-tissue spaces. Lymphocytes are the most common lymphoid-tissue type. Both macrophages and lymphocytes are derived from stem cells present in the bone marrow. They leave the bone marrow and are then disseminated in the blood to the lymphoid tissue. B lymphocytes actually mature within the bone marrow and then proceed directly to the lymphoid organs. However,

T lymphocytes undergo maturity inside the thymus before continuing to the other lymphoid organs—for example, the spleen. In the presence of infectious microorganisms, they both play a major role in the immune response.

44.3 The Nature and the Function of Lymphatic Fluid in the Body

Regarding the nature and function of lymph, in the article titled 'Tips for Reducing Cellulite with a Power Plate', Dr Joseph Mercola (191) states:

> Lymph is a clear, colorless fluid that serves as the transport medium of your lymphatic system. Your lymphatic system is part of your circulatory system and has a number of functions, including the removal of interstitial fluid, the extracellular fluid that bathes most of your tissues. Lymph also transports white blood cells between your 600 to 700 lymph nodes and other areas of your body.
>
> Lymph moves between your cells, tissues, and organs along a lymphatic 'highway' of specialized capillaries. About three-quarters of these capillaries are superficial, located near the surface of your skin. This is why treating cellulite requires that you give some special care to your lymphatic system!
>
> Unlike your blood, which has a beating heart to push it along, your lymph requires actual movement from you in order to keep it flowing. So, if you don't move much, your lymph doesn't move much either . . . and cellular 'garbage' begins piling up. Your fat cells are particularly susceptible to the build-up of fluids and toxins. You have 500 to 1,000 times more toxins in your fat than in the rest of your tissues. If your lymph becomes sluggish, then your fat cells may swell—and one of the consequences is cellulite.

44.4 The Presence of Cellulite Is an Indicator Your Lymphatic System Is Backed Up

In the 2014 article 'Tips for Reducing Cellulite with a Power Plate', Dr Joseph Mercola (191) notes that cellulite occurs due to a buildup of toxic materials within your tissues and, more particularly, inside your fat cells. The presence of cellulite indicates that your detoxification system needs support. This especially applies to your liver and lymphatic system. In the article, Mercola states, 'Your lymphatic system is essentially your body's "sewage processing plant," responsible for removing waste, toxins, and other unwanted material out of your cells and tissues so they can be flushed out of your body.' The detoxification procedure (detox) involves the mobilization and elimination of metabolic wastes, toxic materials, pathogenic microorganisms, and other unwanted debris from your body. Subsequently, if you wish to reduce cellulite, it is important to improve the health of your liver and ensure that your lymph flows more efficiently. Mercola reports, 'Women with cellulite typically have lymphatic deficiencies. Another study found that 80 percent of women have sluggish lymphatic systems.'

44.5 Lymph Is Essentially Recycled Blood Plasma But Is Protein Poor

The article 'Lymphatic System' (13) notes, 'The lymphatic system is part of the circulatory system, comprising a network of lymphatic vessels that carry a clear fluid called lymph . . . directionally towards the heart.' While the cardiovascular system is a closed system, the lymphatic system is an open system. The article further states:

> The circulatory system processes an average of 20 litres of blood per day through capillary filtration which removes plasma while leaving the blood cells. Roughly 17 litres of the filtered plasma get reabsorbed directly into the blood vessels, while the remaining 3 litres are left behind in the interstitial fluid. The primary function of the lymph system is to provide an accessory route for these excess 3 litres per day to get returned to the blood. Lymph is essentially recycled blood plasma but is protein poor.

44.6 Lymph Collected from Interstitial Fluid around the Body Drains into the Subclavian Veins in the Upper Heater before Entering the Heart

The article titled 'Lymphatic System' (13) notes:

> The blood does not come into direct contact with the parenchymal cells and tissues in the body, but constituents of the blood first exit the microvascular exchange blood vessels to become interstitial fluid, which comes into contact with the parenchymal cells of the body. Lymph is the fluid that is formed when interstitial fluid enters the initial lymphatic vessels of the lymphatic system. The lymph is then moved along the lymphatic vessel network by either intrinsic contractions of the lymphatic passages or by extrinsic compression of the lymphatic vessels via external tissue forces (e.g. the contractions of skeletal muscles), or by lymph hearts in some animals. The organization of lymph nodes and drainage follows the organization of the body into external and internal regions; therefore, the lymphatic drainage of the head, limbs, and body cavity walls follows an external route, and the lymphatic drainage of the thorax, abdomen, and pelvic cavities follows an internal route. Eventually, the lymph vessels empty into the lymphatic ducts, which drain into one of the two subclavian veins (near the junctions of the subclavian veins with the internal jugular veins).

So it is seen that eventually the lymph collected from throughout the body enters a large collecting tube, the thoracic duct, located near the heart. From the thoracic duct, the lymph empties into the blood circulatory system itself at the left subclavian vein. This fluid entering the blood circulatory system comprises of lymph, a fluid of water, solutes taken from the lymphatic system, and also the chylomicrons or chyle that are created inside the intestines from lipids and dietary fat.

Thus, the body fluids meet in the Upper Heater, where they are returned to the heart. The filtered lymph from the two subclavian veins enters the heart, where it is mixed with venous blood also entering the heart. From here, the mixture is pumped to the Lungs to be oxygenated and then returned to the heart. At this point, the combined fluids have been purified, filtered, mixed, and oxygenated so that it is now rich red Blood that can be circulated through the body once again to bring new life to every cell in the body. The lymphatic system performs a second function also. Fats that have been absorbed in the small intestine enter lymph vessels in that organ. Those fats are then carried through the lymphatic system back into the blood circulatory system.

44.7 The Never-Ending Cycle of Lymphatic Fluid Movement throughout the Body

The Betterhealth website (192) of the government of Victoria, Australia, informs that blood vessels carry freshly oxygenated life-giving red blood away from the heart and allow the seepage of fluid in the blood into surrounding tissues. The lymphatic system, which is a subset of the circulatory system, drains off any extra fluid to stop the tissues from puffing up. The feet in particular are prone to puffiness. The human lymphatic system is composed of a complex network of tubes distributed throughout the body except for the central nervous system. The lymphatic system drains lymphatic fluid from all the tissues and discharges it back into the bloodstream. Similar to the valves in veins, some lymphatic vessels possess valves, which stop the lymph from running countercurrent, away from the direction of the heart. The major functions of the lymphatic system include balancing the fluid volume throughout the body, filtering out, and trapping unwelcome bacteria and other microorganisms in the lymph nodes, where they can be attacked and destroyed by white blood cells, which call the lymph nodes home.

44.8 The Lymphatic System Filtration Organs of the Body

The word *lymphatic* comes from the Latin word *lymphaticus*, meaning 'connected to water,' because lymph is clear. Lymphatic fluid is filtered through various dedicated organs, including the spleen, thymus, and lymph nodes before being discharged into the blood. The spleen, which is largest lymphatic organ, is located in the upper-left quadrant of the abdomen just above the kidney and under the diaphragm and protected under the lower left ribs. This is one of the purifying and filtering organs of the blood, ensuring that microbes are removed. The spleen is also responsible for destroying old or damaged red blood cells from circulation. The thymus lies under the rib cage, just behind the breastbone. The thymus is another blood-filtering organ. It houses many white blood cells called lymphocytes. The Victorian government's *Betterhealth* website (192) further states:

> Lymph nodes are found at various points around the body, including the throat, armpits, chest, abdomen and groin. All lie close to arteries. Bacteria picked up from the tissues by the lymph are trapped in the lymph node. White blood cells called lymphocytes can then attack and kill the bacteria. This is why your lymph nodes tend to swell if you have an infection. Viruses and cancer cells are also trapped by lymph nodes.

44.9 When the Lymphatic System Comes under Attack
When microorganisms are recognized in the lymph fluid, the defensive lymph nodes produce more infection-fighting white blood cells, which can cause the nodes to swell. Swollen lymph nodes can be felt in the neck, underarms, and groin when a fight is proceeding. There are several disorders of the lymphatic system. Humans can live without a spleen although people who have lost their spleen to disease or injury are more prone to infections. The Victorian government's *Betterhealth* website (192) advises that some common problems can occur when the lymphatic system malfunctions. Glandular fever is the common term used to describe a viral infection called infectious mononucleosis. The virus that causes glandular fever is known as Epstein-Barr virus. Symptoms include tender lymph nodes and possibly enlargement of the spleen. Hodgkin's disease is a type of cancer of the lymphatic system. When edema occurs, there is swelling in the tissues due to excessive fluid retention. Infection of the tonsils in the throat is called tonsillitis.

44.10 The Unidirectional Upward Flow of Lymphatic Fluid within the Body

On the livescience.com website, Kim Ann Zimmermann (193) succinctly explains the flow of lymphatic system through the body. She states:

> Unlike blood, which flows throughout the body in a continuous loop, lymph flows in only one direction—*upward toward the neck*—within its own system. It flows into the venous blood stream through the subclavian veins, which are located on either sides of the neck near the collarbones. Plasma leaves the cells once it has delivered its nutrients and removed debris. Most of this fluid returns to the venous circulation through the venules and continues as venous blood. The remainder becomes lymph. Lymph leaves the tissue and enters the lymphatic system through specialized lymphatic capillaries. About three-quarters of these capillaries are superficial capillaries that are located near the surface of the skin. There are also deep lymphatic capillaries that surround most of the body's organs. (Emphasis is mine)

Zimmermann noted that there are two different and very dissimilar drainage areas that constitute the lymphatic system. The right lymphatic drainage area processes only the right arm and chest. The left lymphatic drainage area performs the lion's share as it processes the remaining body areas, including both legs and feet, the lower trunk of the body, the upper left region of the chest, and the left arm and hand.

44.11 The Unidirectional Continuous Upward Flow of Lymph within the Body

The lymphatic system does not have a pumping organ like the heart. Subsequently, its upward movement toward the neck depends on the motions of the muscle and joint pumps. Regarding this unidirectional continuous upward flow of lymph, in the article titled 'The Lymphatic System', the authors (194) explain that while blood flows throughout the entire body in a uninterrupted loop, the lymphatic system allows lymph flow in only one direction, which is *only ever upwards toward the neck*, where it enters the venous bloodstream via the subclavian veins proximal to the collarbones.

Note that both of the authors above confirm that the lymphatic flow is upwards to the Upper Heater. As the Spleen is the major lymphatic organ from a western viewpoint, it is very interesting that it ensures that the Lymph is directed upwards. It is a common fact that in traditional Chinese medicine, it is believed that by ensuring smooth flow of Qi in all directions of the body, the Liver ensures that Spleen Qi flows smoothly upwards in its right direction.

44.12 Functions of the Lymphatic System in the Body

In the article titled 'The Lymphatic System', the authors (194) note that 'the lymphatic system aids the immune system in removing and destroying waste, debris, dead blood cells, pathogens, toxins, and cancer cells'. Further to this, the author advised that 'the lymphatic system absorbs fats and fat-soluble vitamins from the digestive system and delivers these nutrients to the cells of the body where they are used by the cells'. So the lymphatic system plays a part in supplying nutrients from the digestive system to cells in the body, a feature attributed to the Triple Energizer.

44.13 The Evolution and Transformation of Circulating Body Fluids

In the article titled 'The Lymphatic System', the authors (194) further state:

> Arterial blood carries oxygen, nutrients, and hormones for the cells. To reach these cells it leaves the small arteries and flows into the tissues. This fluid is now known as *interstitial fluid* and it delivers its nourishing products to the cells. Then it leaves the cell and removes waste products. After this task is complete, 90% of this fluid returns to the circulatory system as venous blood.

44.14 Another Explanation as to the Origin of Lymph

Regarding the origin of lymph, the authors (194) explain that after the arterial blood flows out of and away from the heart, its flow rate reduces as it passes through a capillary bed. Due to this slowing, some of the fluid portion of blood, known as plasma, leaves the small arteries (arterioles) and moves into the tissue spaces, where it becomes tissue fluid, or extracellular fluid. This extracellular fluid flows in between the cells but does not flow into cells. It delivers numerous nutrients, oxygen, and hormones to individual cells as their demands dictate. When leaving the cellular environs, the fluid carries away cellular waste materials and protein cells. About 90% of this extracellular fluid flows away as plasma in the small veins of the venous circulation of the circulatory system. The residual 10% of the remaining fluid is called lymph.

Thus, we can see that when the oxygenated arterial blood has left the arteries to deliver nutrients, hormones, and oxygen, it becomes extracellular fluid. Then when this extracellular fluid gives up nutrients and picks up metabolic waste materials, 90% of it returns to the circulatory system as venous blood. The lingering 10% of the fluid that remains in the tissues as a clear to yellowish somewhat-sticky fluid is known as lymph. For example, the liquid formed in a blister is lymph. So the one initial liquid has assumed four different liquid forms of slightly different composition, depending on what stage it is in and its relative location in its never-ending cycle. Those four fluids are sequentially arterial blood to extracellular fluid to venous blood in the circulatory system, with lymph being left in the interstitial space, ready for collection by the lymphatic system.

44.15 Location of the Lymphatic Capillaries throughout the Body

With respect to the location of the lymphatic capillaries throughout the body, in the article cited just previously, the authors (194) explain that lymph can leave the tissues only via the specialized lymphatic capillaries of the lymphatic system. The lymphatic capillaries originate as blind-ended tubes only a single cell-layer thick. These cells slightly overlap one another, similar to the shingles on a tiled roof, where each cell is attached to nearby tissues by a securing filament. About 70% of these lymphatic capillaries are located in close proximity to the skin. The other 30% make up the deep lymphatic capillary system and surround most of the organs of the body. Note that 70% of the defensive lymphatic capillaries are superficial and located near or just under the skin. These probably constitute part of or at least assist with the circulation of the Wei Qi defensive energy discussed in the TCM classics.

44.16 Unequal Lymphatic-System Drainage Areas of the Body
Regarding the very unequal lymphatic system drainage areas, in the article titled 'The Lymphatic System', the authors (194) point out that while the lymphatic drainage system is organized into two separate drainage area sections, those sections are very dissimilar. While the right lymphatic drainage area drains only the right arm and right chest, the left drainage area drains the very large remaining areas of the body, including both legs, the lower trunk, the upper left side of the chest, and the left arm.

44.17 The Lymphatic System Has Been Neglected by Both Scientific and Medical Communities
In the Concluding Remarks of the article titled 'The New Era of the Lymphatic System: No Longer Secondary to the Blood Vascular System', the authors (190) state:

> *Since its original description by Hippocrates, the lymphatic system has been neglected by both scientific and medical communities because of its vagueness in structure and function.* Even after its rediscovery 400 years ago, the lymphatic system was considered a secondary vascular system that supports the blood vascular system. However, a series of landmark discoveries in lymphatic research has significantly advanced our understanding of not only the organogenesis, function, and anatomic structure of the system, but also the cellular and molecular biology of LECs [lymphatic endothelial cells]. In particular, substantial attention has been given to the elucidation of

the molecular control of physiological and pathological lymphangiogenesis, re-evaluating its essential roles in human health and well-being. This paradigm shift simultaneously forced us to take a *brand-new look at the lymphatic system as the other, not the secondary, vascular system.* Considering the vital functions that the lymphatic system engages in and how little knowledge we have regarding the system, lymphatic research is truly a gold mine that invites ambitious young scientists and clinicians. (Emphasis is mine)

I found it remarkable that the authors stated that 'the lymphatic system has been neglected by both scientific and medical communities *because of its vagueness in structure and function'* (emphasis is mine). Interestingly, while many TCM practitioners consider that the Triple Energizer constitute a defined group of functions performed by the other tangible Zang and Fu organs, TCM practitioners that actually do believe that the Triple Energizer is also a real tangible organ are at a loss to describe its structure because of its vagueness when being described by the ancient medical classics. Not one knowledgeable person would even doubt the existence of the Lymphatic System, but its vagueness leaves them befuddled. I hope by this stage that you now grasp the pure simplicity of the structure of the Triple-Energizer Metasystem. It is the omnipresent predominantly connective-tissue maze that is 'nothing but membranes', and its ubiquitous entire-body 3-D infrastructure acts like a bag or an external wall that contains all the bodily systems, structures, and organs that are themselves not connective tissue. Thus, I propose that a combination of the elements of the Connective-Tissue Metasystem coupled with elements of the Primo Vascular System constitutes the Triple-Energizer Metasystem or San Jiao.

44.18 The Lymphatic System Is Finally Appreciated as a Stand-Alone Vascular System

In the article titled 'The New Era of the Lymphatic System: No Longer Secondary to the Blood Vascular System', regarding the 'elusive morphology and mysterious pathophysiology' of the lymphatic system the authors (190), state:

> Although the blood system has been studied extensively, the lymphatic system has received much less scientific and medical attention *because of its elusive morphology and mysterious pathophysiology*. However, a series of landmark discoveries made in the past decade has begun to change the previous misconception of the lymphatic system to be secondary to the more essential blood vascular system. In this article, we review the current understanding of the development and pathology of the lymphatic system. We hope to convince readers that the lymphatic system is no less essential than the blood circulatory system for human health and well-being. . . . Modern molecular, cellular, and genetic approaches as well as the state-of-the-art imaging technologies have allowed true appreciation of the value of the lymphatic system as the other vascular system, no longer secondary to the blood vascular system.
>
> Moreover, lymphatic vessels such as lacteals in the intestines absorb and transport large molecules, fats, and lipids in the digestive system mainly in the form of lipoprotein such as chylomicrons—large lipoprotein particles that are created by the enterocytes of the intestine and consist of triglycerides, phospholipids, cholesterol, and proteins. Notably, lymph fluid and chylomicrons can stimulate adipocyte differentiation. (Emphasis is mine)

44.19 The Lymphatic System Is Embryologically Derived from the Cardiovascular System

In the article titled 'The New Era of the Lymphatic System: No Longer Secondary to the Blood Vascular System', regarding the origin of the lymphatic system, the authors (190) state:

> The controversy regarding the origin of the lymphatic system continued for a hundred years until 1998, when

the first mutant mouse with failed lymphatic development supported Sabin's hypothesis of the blood vascular origin of the lymphatic system discovered that deletion of a homeodomain transcription factor, Prox1, resulted in arrest of lymphatic endothelial differentiation at an early stage and that Prox1 knockout mice fail to develop a lymphatic system.

The mesoderm germ layer initiates the origin of the Connective-Tissue Metasystem (CMT) of the body. I have shown elsewhere how I believe that the PVS/Connective-Tissue Metasystem is the primordial TCM organ (i.e. Triple-Energizer Metasystem) formed during embryogenesis. Development of the heart and cardiovascular system begins very early in embryogenesis from the lateral plate mesoderm, which is in the middle of the three primary germ layers. The intermediate mesoderm develops into kidneys and gonads. Thus, the Kidneys and Heart develop from the same germ tissue (mesoderm) that the connective-tissue metasystem develops from. This likely explains the close relationship between the Connective-Tissue Metasystem (aka Triple-Energizer Metasystem) and the Kidneys. From the above citation, it can be seen that it was recently (1998) confirmed that the lymphatic system is derived from differentiation of the blood vascular system. Thus, the mesoderm germ layer primarily formed the Connective-Tissue Metasystem, which then begat the cardiovascular system, which then begat the lymphatic system via the Triple Energizer.

44.20 Like Earth's Springs, Streams, Rivers, and Oceans the Lymphatic System Is Cleansing

As if he was reading from 62nd Difficult Issue in the *Nan Ching*, regarding how the Triple Burner channels Yuan Qi and fluids through the meridians associated with wells, rapids, brooks, and streams, in the article titled 'Heal Your Lymphatic Ocean', the author, David Yarrow (195), poetically states:

> Lymph keeps our inner world wet and clean. It actually washes and bathes the entire insides of your body. Lymph

> fluid comes from our bloodstream—tiny capillaries between arteries and veins let some of the water in blood leak out, while blood cells pass through from artery to vein. These leaking capillaries are the springs which feed the rivers of lymph inside. . . .
>
> The lymph system is a network of vessels, ducts and glands to collect and circulate this water. These springs, streams and rivers of lymph fluid flow into our lower abdomen. Thus, the intestines are your lymphatic ocean, and your navel is the center of this inner sea. Lymph nodes are the ponds and lakes where our inner lymph fluids pool for a moment on their way to the ocean.

While the Triple Energizer has the role of supplying nutrients, moisturizing, and cleaning the body right down to individual cells, the Lymphatic System certainly performs the lion's share of this highly important life-sustaining commission.

44.21 Accumulated Waste in the Lymphatic System Causes Abnormal Mucus Discharges

Regarding the formation and nature of mucus, in the article titled 'Heal Your Lymphatic Ocean', the author, David Yarrow (195), expressively states:

> Mucous is a common term for the wet, sticky, slippery ooze secreted from the lymph system. Everyone has it, but few know what it is, or why they have it. Not even doctors seem to know. Mucous is synthesized from the sugars, fats and proteins floating in our lymph fluids. A certain amount is needed to lubricate and moisten our body's inner surfaces. But often there is too much, or it's an inferior quality, so it becomes waste. Like carbon dioxide in our modern 'greenhouse effect' atmosphere, excess and poor quality mucous is a primary cause of 'environmental' disease and allergy. Sometimes our grime fighters are too weak to do an effective job. But more

> often we produce too much by eating improperly. Often digestion puts 'stuff' in our blood and lymph which our cells have no use for. 'Garbage in, garbage out' (GIGO) is as true for cell metabolism as for data-processing. . . . The result is an extremely inordinate amount of 'sticky stuff' trying to leave the body. Accumulated waste in the lymph system causes abnormal mucous discharges. Among the sicknesses this causes are colds, sore throats, ear infection, hayfever, rhinitis, flu, pneumonia, constipation, sinus headache, acne, arthritis, bronchitis, asthma, to name only a few. Mucous also accumulates in the joints, producing swelling, aches and arthritis. These problems afflict many people today, yet few understand what all this mucous is about.

While mucus is an often despised substance, it is a product of high importance throughout the body with the role of lubrication and moistening our body's inner sensitive surfaces. When metabolic wastes and introduced toxins accumulate in the body, the lymphatic system performs its assigned function, collects the rubbish material, and then naturally has to dispose of the abnormally large mucus discharges. As Yarrow mentions above, the 'sicknesses this causes are colds, sore throats, ear infection, hayfever, rhinitis, flu, pneumonia, constipation, sinus headache, acne, arthritis, bronchitis, asthma, to name only a few'. Thank goodness for the cleansing property of the Lymphatic System, the Triple-Energizer Metasystem, and mucus.

Summary of Chapter 44
The flow path of the superficial lymphatic system sums up the description of TCM Triple Energizer fluid movement after it has left the Stomach, whereby fluids from the Spleen are directed up to the Upper Heater to the Heart to be pumped to the Lungs for oxygenation and subsequently returned to the Heart, after which clean, pure, filtered, oxygenated, live-giving red Blood is pumped to the body to sustain life.

Lymphoid tissue (cells and organs that make up the lymphatic system) includes white blood cells (leukocytes), bone marrow, the thymus, spleen, and lymph nodes, which are all part of the immune system. The most diffuse lymphoid tissue is found in the loose connective-tissue spaces beneath most wet epithelial membranes, such as those that line the gastrointestinal tract and the respiratory system. Macrophages help eliminate invaders by engulfing foreign materials and initiating the immune response. These cells may be fixed in one place, such as lymph nodes, or they may wander in the loose connective-tissue spaces. The most common cell type in the lymphoid tissue is the lymphocyte.

Hippocrates (460–377 BC) first described the lymphatic vessel as containing 'white blood' and coined the term *chyle* (from the Greek *chylos*, meaning 'juice'). Lymph is a clear, colorless fluid that serves as the transport medium of your lymphatic system. Your lymphatic system is part of your circulatory system and has a number of functions, including the removal of interstitial fluid, the extracellular fluid that bathes most of your tissues. Lymph also transports white blood cells between your 600 to 700 lymph nodes and other areas of your body. There are 500 to 1,000 times more toxins in fat cells than in the rest of the tissues. If your lymph becomes sluggish, then your fat cells may swell—and one of the consequences is cellulite. The presence of cellulite indicates that your detoxification system needs support. This especially applies to your liver and lymphatic system.

The body fluids move upwards and meet in the Upper Heater, where they are returned to the heart. As the Spleen is the major lymphatic organ from a western viewpoint, it is very interesting that it ensures that the Lymph is directed upwards. It is a common fact that in traditional Chinese medicine, it is believed that by ensuring smooth flow of Qi in all directions of the body, the Liver ensures that Spleen Qi flows smoothly upwards in its assigned direction. The filtered lymph from the two subclavian veins enters the heart, where it is mixed with venous blood also entering the heart. From here, the mixture is pumped to the Lungs to be oxygenated and then returned to the heart. At this point, the combined body fluids have been purified, filtered, mixed, and oxygenated so that at this phase in the

Heart, the processed fluids have become rich red Blood that can be circulated through the body once again to bring new life to every cell in the body. This mirrors TCM theory, which states that blood is made red in the heart.

I found it remarkable that the authors (190) stated that 'the lymphatic system has been neglected by both scientific and medical communities *because of its vagueness in structure and function*' (emphasis is mine). Interestingly, while many TCM practitioners consider that the Triple Energizer constitute a defined group of functions performed by the other tangible Zang and Fu organs, TCM practitioners that actually do believe that the Triple Energizer is also a real tangible organ are at a loss to describe its structure because of its vagueness when being described by the ancient medical classics. Not one knowledgeable person would even doubt the existence of the Lymphatic System, but its vagueness leaves them befuddled. I hope by this stage that you now grasp the pure simplicity of the structure of the Triple-Energizer Metasystem. It is the omnipresent predominantly connective-tissue maze that is 'nothing but membranes' and its ubiquitous entire-body 3-D infrastructure acts like a bag or an external wall that contains all the bodily systems, structures, and organs that are themselves not connective tissue. Thus, I am convinced that a combination of the elements of the Connective-Tissue Metasystem coupled with elements of the Primo Vascular System constitutes the Triple-Energizer Metasystem or San Jiao.

Regarding the *elusive morphology and mysterious pathophysiology* of the lymphatic system, researchers (190) now believe that the lymphatic system is finally being appreciated as a stand-alone vascular system. They advised that 'a series of landmark discoveries made in the past decade has begun to change the previous misconception of the lymphatic system to be secondary to the more essential blood vascular system'. It is intriguing that both the San Jiao and the Lymphatic System are believed to have elusive morphology and mysterious pathophysiology and that both organ systems have 'been neglected by both scientific and medical communities because of its vagueness in structure and function'. It is more intriguing that both organ systems have a major role in managing water and fluids throughout the body.

As if he was reading from 62nd Difficult Issue in the *Nan Ching*, regarding how the Triple Burner channels Yuan Qi and fluids through the meridians associated with wells, rapids, brooks, and streams, in the article titled 'Heal Your Lymphatic Ocean', the author, David Yarrow (195), poetically states, 'Lymph keeps our inner world wet and clean. It actually washes and bathes the entire insides of your body.'

While mucus is an often despised substance, it is a product of high importance throughout the body with the role of lubrication and moistening our body's inner sensitive surfaces. When metabolic wastes and introduced toxins accumulate in the body, the lymphatic system performs its assigned function, collects the rubbish material, and then naturally has to dispose of the abnormally large mucus discharges. Yarrow (195) advises that this cleanup throughout the body results in numerous diverse sicknesses, including 'colds, sore throats, ear infection, hayfever, rhinitis, flu, pneumonia, constipation, sinus headache, acne, arthritis, bronchitis, asthma, to name only a few'. Thank goodness for the cleansing property of the Lymphatic System, the Triple-Energizer Metasystem, and mucus.

CHAPTER 45

How Brown Adipose Tissue (BAT) and White Adipose Tissue Complement the Triple Energizer

Introduction
At this stage of scientific endeavor, there are three different defined types of adipose tissue, which are named based on the color of the tissue. They are Brown Adipose Tissue, Beige Adipose Tissue, and White Adipose Tissue. White adipose tissue is by far the most voluminous of the three types of fatty tissue. When I went to college, fat was just fat and was quite unremarkable. Recent research has confirmed that there are actually three subcategories of fat and that the properties of the different forms of fat are truly remarkable. I will now discuss recent scientific findings about adipose tissue.

45.1 Brown Adipose Tissue (BAT) Is Mainly Found around the Neck Areas (Front and Back)
In the article 'What Is Brown Fat? What Is Brown Adipose Tissue?', the author, Christian Nordqvist (196), states that Brown Adipose Tissue (BAT) is 'mainly found around the neck areas (front and back)'. In the article, it says, 'Scientists are only now just beginning to understand what the functions of brown fat are.' The author went

on to say, 'Brown fat generates heat by burning calories. When it is cold, brown fat's lipid reserves are depleted, and its color gets darker.'

45.2 The Upper Heater Circulates Qi to Warm Up the Divisions of the Skin—the Fen Rou

In the book 'Heart Master Triple Heater', Elisabeth Rochat de la Vallée (11), on page 99, quotes Ling shu chapter 81, which says, 'The upper heater makes the qi come out in order to warm up the divisions of the flesh.' She then says that the word used for 'divisions of the flesh' is *fen rou* and explains:

> It is all the flesh, but considered as muscular masses which are separated or distinguished from one another, and which give the idea of ravines and valleys and the possibility for circulation. The most gross image is that of muscular masses, *but within each muscle there is that which allows the possibility of circulation in all the spaces*. In each little piece of flesh the qi can weave its way through. The flesh is not just a compact mass, but has the capacity to be infiltrated by qi. *The Chinese always think like this, they are obsessed with the idea of movement*. (Emphasis is mine)

Rochat de la Vallée summarizes by saying, 'So the upper heater makes the qi circulate in order to warm up these divisions of the skin that we talked about, the fen rou, to nourish the bony articulations, and be in free communication with the cou li.'

Here it is likely that the Brown Adipose Tissue (BAT), which predominates in the Upper Heater, is cooperating with and under the control of the Upper Heater and allows the heat generated in the BAT to be circulated just as stated above.

45.3 Scientists Find Body Fat Very Intriguing, Especially Brown Adipose Tissue (BAT)

The article by Kathleen Doheny (197) titled 'The Truth about Fat' notes that scientists find fat very intriguing and more so every day. 'Fat is one of the most fascinating organs out there,' says Aaron Cypess, MD, PhD, an instructor of medicine at Harvard Medical School, and a research associate at the Joslin Diabetes Center in Boston. He said, 'We are only now beginning to understand fat.' 'Fat has more functions in the body than we thought,' agrees Rachel Whitmer, PhD, research scientist at the Kaiser Permanente Division of Research in Oakland, California, who has studied the links between fat and brain health. The article advised that brown fat is now thought to behave more like muscle tissue than like white fat. When brown fat is activated, it actually burns white fat. 'A 150-pound person might have 20 or 30 pounds of fat,' Cypess says, but 'they are only going to have 2 or 3 ounces of brown fat'. However, that 2 ounces of BAT could burn off 300 to 500 calories a day if maximally stimulated, enough to lose up to a pound in a week.

Note that Cypess stated, 'Fat is one of the most fascinating organs out there,' thus showing that he feels that the adipose system present throughout the body warrants being considered an organ.

45.4 Metaboloregulatory Thermogenesis Due to BAT Is under Hypothalamic Control

Regarding the heat-producing property of BAT, a 2004 PubMed article (198) states:

> The function of brown adipose tissue is to transfer energy from food into heat; physiologically, both the heat produced and the resulting decrease in metabolic efficiency can be of significance. Both the acute activity of the tissue, i.e., the heat production, and the recruitment process in the tissue (that results in a higher thermogenic capacity) are under the control of norepinephrine released from sympathetic nerves.

The article further reported that brown adipose tissue is essential for classical nonshivering thermoregulatory thermogenesis. Heat production from brown adipose tissue is activated whenever the organism is in need of extra heat (e.g. postnatally) during entry into a febrile state and during arousal from hibernation, and the rate of thermogenesis is centrally controlled via a pathway initiated in the hypothalamus. Consuming food (in TCM, *grains*) 'also results in activation of brown adipose tissue'. The article states, 'This metaboloregulatory thermogenesis is also under hypothalamic control.' When brown adipose tissue is active, large amounts of lipids and glucose (grain) are combusted in the tissue. The article further states, 'The development of brown adipose tissue with its characteristic protein, uncoupling protein-1 (UCP1), was probably determinative for the evolutionary success of mammals, as its thermogenesis enhances neonatal survival and allows for active life even in cold surroundings.'

45.5 Thermogenic BAT Is Distributed on the Most Yang Region of the Most Yang Body Surface

The article titled 'Brown Adipose Tissue' (199) states, 'Brown fat cells and muscle cells both seem to be derived from the same stem cells in the embryo.' We can see from the classics the Triple Burner and its all-encompassing system of fatty membranes are responsible for energy production. How does modern science describe how energy (Qi) is manufactured in the body? The article states:

> The mitochondria in a eukaryotic cell utilize fuels to produce energy (in the form of ATP). This process involves storing energy as a proton gradient, also known as the proton motive force (PMF), across *the mitochondrial inner membrane*. This energy is used to synthesize ATP when the protons *flow across the membrane* (down their concentration gradient) through the ATP synthase enzyme; this is known as chemiosmosis. (Emphasis is mine)

Note that *mitochondrial inner membranes* are required to produce energy in the form of ATP. In warm-blooded animals, body heat is maintained by signaling the mitochondria to perform their function via the protein aptly named thermogenin (i.e. producer of heat). Brown adipose tissue is highly specialized for this thermogenesis as each BAT cell has a higher number of mitochondria compared to more typical cells. Secondly, these BAT 'mitochondria have a higher-than-normal concentration of thermogenin in the *inner membrane*' (emphasis is mine). In neonates (newborn infants), brown adipose tissue makes up about 5% of the body mass and is *located on the back, along the upper half of the spine and toward the shoulders.* This BAT 'is of great importance to avoid lethal cold (hypothermia is a major death risk for premature neonates)'. As I mentioned previously, Li Chiung stated, 'The Triple Burner represents nothing but membranes.' The article cited above states that '*mitochondrial inner membranes* are required to produce energy in the form of ATP' (emphasis is mine). I believe that the Triple-Energizer Metasystem is aka the PVS/Connective-Tissue Metasystem and that the role of the Triple Heater to produce energy and heat (thermogenesis) extends down microscopically to the 'mitochondrial inner membranes'.

45.6 Anatomical Distribution of BAT Is Likely to Confer Survival Value by Protecting Organs

In the Abstract of the article 'Anatomical Locations of Human Brown Adipose Tissue', regarding the anatomical distribution of BAT, the authors, Sacks and Symonds (200), state:

> *Its anatomical distribution is likely to confer survival value by protecting critical organs from hypothermia by adaptive thermogenesis.* Ultimately, the location and function will be important when considering therapeutic strategies for preventing and treating obesity and type 2 diabetes, in which case successful interventions will need to have a significant effect on BAT function in subjects living in a thermoneutral environment. In view of the diverse locations and potential differences in responsiveness between BAT depots, *it is likely that BAT will be shown to have much more subtle and thus previously overlooked functions and regulatory control mechanisms.* (Emphasis is mine)

45.7 Why Are the Main Brown Adipose Tissue Depots within the Supraclavicular Region?

In the article 'Anatomical Locations of Human Brown Adipose Tissue', regarding the anatomical distribution and functional relevance of brown adipose tissue (BAT), the authors, Sacks and Symonds (200), explain that up until about ten years ago, it was believed that BAT was biologically active only in neonates and young children, where it generated heat during cold exposure to preserve the optimal body temperature. It was further believed that BAT degenerated with aging and converted into white adipose tissue (WAT) such that BAT deposits present in adults was not deemed to be significant in energy production and metabolism. The authors, Sacks and Symonds, (200) state:

> *It was therefore demonstrated that the main BAT depot was within the supraclavicular region*, although as detailed below a number of perhaps less important depots were identified. Because more attention has been given to the physiology, pathophysiology, and clinical characteristics of human BAT rather than its anatomy, the purpose of this Perspective is to review information about and *to consider hypotheses why BAT is located where it is in humans as well as the functional relevance and therapeutic implications of its locations.* (Emphasis is mine)

45.8 Location and Distribution of Thermogenic Visceral BAT in Humans

In the article 'Anatomical Locations of Human Brown Adipose Tissue', regarding the three major different anatomical locations of BAT distribution (namely, around blood vessels, around hollow organs, and around solid organs), the authors, Sacks and Symonds (200), state that Visceral BAT includes the following:

> 1) Perivascular BAT around the aorta, common carotid artery, and brachiocephalic artery; in anterior mediastinum (paracardial) fat; and around epicardial coronary artery and cardiac veins as well as

medium-sized muscular arteries and veins including the internal mammary and the intercostal artery branches from the subclavian and aorta. The intercostal veins drain blood from the chest and abdominal walls into the azygous veins, the left joining the main right azygous vein in the latter's thoracic cephalad course closely adjacent to the inferior vena cava before emptying into the superior vena cava.

2) Viscus BAT, defined as BAT surrounding a hollow muscular organ other than blood vessels, situated in variable amounts in the epicardium around the heart and in the esophago-tracheal groove, as well as greater omentum and transverse mesocolon in the peritoneal cavity.

3) BAT around solid organs, namely, kidney, adrenal, pancreas, liver, and splenic hilum including paravertebral fat, which was not examined in Heaton's series but can be seen on CT scans of the thorax adjacent to periaortic fat. It lies next to the intercostal artery from which a spinal branch supplies the spinal cord. To our knowledge, BAT has not been described in the meninges covering the brain and spinal cord or in the subcutaneous tissue of the scalp.

45.9 Location and Distribution of Thermogenic Subcutaneous BAT in Humans as They Age

In the 2013 article 'Anatomical Locations of Human Brown Adipose Tissue', regarding further subcutaneous anatomical locations of BAT distribution, the authors, Sacks and Symonds (200), state that BAT includes the following:

> Subcutaneous BAT includes depots lying between the anterior neck muscles and in the supraclavicular fossa posterior to the brachial plexus; under the clavicles; in the axilla; in the anterior abdominal wall; and in the inguinal

area. Importantly, BAT is present in all the aforementioned sites in infants and children under 10 years old and its distribution decreases in variable amounts with increasing age so that by 80 years old, few individuals have BAT at any site. Brown adipocytes defined by histology disappear from the interscapular fat pad at infancy and at 10 years old from inguinal and anterior abdominal wall areas.

These findings aptly explain why humans feel the cold more as they age.

45.10 Speculations on the Functional Relevance of the Locations of Human BAT

In the article 'Anatomical Locations of Human Brown Adipose Tissue' (200), regarding blood vessel temperature regulation, the article advises that the diverse and specific distribution of BAT throughout the body of warm-blooded animals affords the basis for comprehending the raison d'être of the anatomic locations of human BAT. The authors quote Smith and Horwitz from their extensive 1969 review of brown fat and thermogenesis in mammals, where they said, 'In brief, the brown fat provides an internal heating jacket that overlies parts of the systemic vasculature and *on signal becomes an active metabolic heater applied directly to the flowing bloodstream as it passes to and from the cooler periphery*' (emphasis is mine).

So it appears that the BAT organ system is specifically designed to generate heat throughout the entire body, in the Upper Heater ('around epicardial coronary artery and cardiac veins as well as medium-sized muscular arteries and veins including the internal mammary and the intercostal artery branches from the subclavian and aorta'), in the Middle Heater ('as well as greater omentum and transverse mesocolon in the peritoneal cavity'), and in the Lower Heater ('in the inguinal area'). As the Triple Heater has the role of heating the body, it can be seen that the BAT organ system compliments the Triple Heater organ in accomplishing this biological function.

45.11 The Warming of Venous Blood from the Upper Extremities Returning to the Heart

Regarding colder blood returning from the upper extremities, in the article 'Anatomical Locations of Human Brown Adipose Tissue', the authors (200) advise, 'On the venous side of the human circulation, blood with lower-than-normal temperature in the subclavian and jugular veins resulting from cold-exposed upper extremities, head, and neck would be warmed by axillary, supra- and subclavicular, and superior mediastinal BAT.' As to why warming this venous blood is necessary, the authors state, 'The collection of BAT in the cervical-axillary region is strategically placed to offset the influx of cooled venous blood from the upper body into the right atrium, which confers survival value because the myocardium is susceptible to life-threatening arrhythmias resulting from a decrease in coronary blood temperature.'

So it appears that the BAT distribution in the Upper Heater is designed so that cooled blood returning from upper extremities does not quench Heart Fire when it returns to the Heart. Here again, it is shown that the Triple Heater's role of heating the body is reinforced and complimented by the BAT organ system.

45.12 Natural Increasing Temperature Gradient in the Large Veins as They Approach the Heart

Regarding the decrease of overall body mass of BAT as humans age, the article 'Anatomical Locations of Human Brown Adipose Tissue' (200) explains that it is likely that the age-related decline of overall BAT mass and in epicardial UCP1 (Thermogenin) explains in part why the elderly feel the cold more and are predisposed to hypothermia. The authors note that venous blood coming from the abdominal and chest walls is heated by BAT located around the intercostal veins and along the azygous venous system. Noteworthy is the fact that the azygous venous system also collects blood originating from the fat pad between the scapulae, which in rodents is the primary BAT heat source. The authors further suggest that BAT present subcutaneously in the inguinal fossa might promote warming of the blood returning from the lower limbs in young children via the femoral vein. Because inguinal BAT wanes in adults, it is believed that venous blood coming from the lower extremity is heated within the inferior vena cava due to the warmer ambient temperatures of the intra-abdominal and intrathoracic cavity. Importantly, there is an increasing temperature gradient of the blood flowing in the large veins of the body as the blood approaches the heart.

45.13 New (2014) Research Reveals that Brown Adipose Tissue Has Several Roles

Regarding the findings of new scientific research, in the 2014 article titled 'Researchers Find Healthy Brown Fat Regulates Your Blood Sugar', Dr Joseph Mercola (201) advises that generally, lower percentage of white body fat is associated with healthier life outcomes. However, he reports that the dynamics of thermogenic brown fat is being investigated with regard to maintaining healthy metabolism as a weight loss agent and much more. While it was generally believed that BAT was essential to help newborns to regulate their body temperature, newer scientific research reveals that BAT has previously unrealized physiological roles beyond body heat generation.

45.14 Brown Adipose Tissue (BAT) May Help Regulate Blood Sugar Levels in Humans

Regarding the findings of new scientific research, in the 2014 article titled 'Researchers Find Healthy Brown Fat Regulates Your Blood Sugar', Dr Joseph Mercola (201) advises that one recent study confirmed that individuals with higher BAT levels possess a faster metabolic rate, have higher insulin sensitivity, and regulate blood sugar levels more effectively. It appears, however, that BAT must be triggered to generate optimal results, and one trigger involves exposure to cold temperatures. The research confirmed that for males who have significant amounts of BAT depots, exposure to mildly cold temperatures for up to eight hours activated the BAT deposits, which resulted in increased resting metabolic rate, improved insulin sensitivity, along with improved glucose processing, and increased removal of glucose from the circulation.

45.15 Failing Brown Adipose Tissue May Cause Middle-Age Spread

In 2014, Dr Joseph Mercola (201) published the article titled 'Researchers Find Healthy Brown Fat Regulates Your Blood Sugar'. The article advises that as humans age, the thermogenic activity of BAT gets less and less. A major reason there's a tendency to increased weight gain with age is likely due to a decrease in the functionality of BAT. Mercola cited the editor-in-chief of *The FASEB Journal* and explained that because BAT function declines as we age, seniors report that they generally have to work twice as hard with their diets and exercise to get half of the rewards that younger people gain. Interestingly, Mercola noted that another newly discovered cell type of fat known as beige adipocytes or beige fat has been identified and knowledge about its beneficial properties and biochemistry in humans is in its infancy. It does have similar thermogenic properties of BAT. So here again is yet another previously undiscovered tissue type that has been launched into the scientific research arena.

45.16 Characteristics of People with Increased Amounts of Brown Adipose Tissue

In the 2014 article cited above, Dr Joseph Mercola (201) advises that it is believed that practically all adults have some BAT in their body. Those individuals with more activated BAT than others tend to have healthier metabolic measures. Mercola further advised that slender individuals have relatively more BAT than obese individuals, younger people have more BAT than elderly people, and individuals whose blood sugar level is normal have more BAT than individuals with high blood sugar. An interesting comment he made was that individuals who medicate their high blood pressure with beta-blocker drugs have less active BAT, and he suggested this outcome was because the hormones released as part of your body's natural fight-or-flight response (catecholamines) are blocked by the beta-blocker medication. As catecholamines are known to stimulate BAT, this would explain why people on beta-blocker drugs usually tend to put on weight more readily.

45.17 Natural Methods to Raise the Levels and Activity of Brown (and Beige) Fat

Further in the 2014 article, Mercola (201) states, 'So knowing that BAT is beneficial in the body how can you get more of it, or potentiate what you have got?' Mercola suggests several noninvasive methods that have been proven to encourage BAT production and its activation. He explained how exposure to cold has been repeatedly shown to activate brown fat in adults. He cites Swedish research published in 2009 that showed 'cold-induced glucose uptake was increased by a factor of 15'. Based on animal models, 'researchers estimated that just 50 g of brown fat (which is less than what most study volunteers have been found to have) could burn about 20 percent of your daily caloric intake—and more if "encouraged"'. Mercola cited four options to 'apply the heat' to your BAT by cooling down your body. He suggested (1) placing an ice pack over the upper back and upper chest for 30 minutes daily, (2) consuming about 500 ml of ice water each morning, (3) taking cold showers or taking ice water baths up to the waist for 10-minute intervals three times per week.

From the TCM point of view, I would strongly advise not to practice the second suggestion above as that would profoundly affect your Stomach energy negatively. Another potential method to elevate your BAT activity is exercise. Mercola reported that one study found that mice converted white fat into BAT simply by exercising but noted that it has not been confirmed if the same would be the case for humans. Mercola cited a BAT researcher who stated, 'Our results showed that exercise doesn't just have beneficial effects on muscle, it also affects fat. . . . It's clear that when fat gets trained, it becomes browner and more metabolically active. We think there are factors being released into the bloodstream from the healthier fat that are working on other tissues.'

Yet another method that Mercola suggests to elevate your BAT activity involves producing enough melatonin as it 'stimulates the appearance of "beige" fat, which, the researchers of one study believe, may explain why melatonin helps control body weight, along with its metabolic benefits'. Mercola quoted from another study published in *Science Daily* which found that long-term administration of the sleep hormone melatonin improves the thermogenic effect of cold exposure, increases the thermogenic effect of exercise and, subsequently, is very beneficial therapy against obesity. As opposed to white fat, the mitochondria in beige fat release UCP1 protein (formerly called Thermogenin), which is responsible for burning calories and generating heat. This would account for the fact that the lack of sleep is linked to obesity, and if you're getting insufficient sleep, there's a strong likelihood that your melatonin production is insufficient too.

45.18 The Many Functions of the White Adipose Tissue beneath the Skin

The fat layer just under your skin helps to insulate your body and protect you from extreme temperature conditions. The subdermal fat layer also helps attach the dermis layer of your skin to your muscles and bones and also acts as a shock absorber in the event of a fall. Further to those functions, fat is stockpiled energy and acts as an ever-ready energy store.

45.19 That Fat Cells (Adipocytes) Comprise a Key Endocrine Organ Is Unquestionable

A large amount of recent scientific research proves beyond doubt that much-maligned fat tissue has a double life and is actually also an endocrine organ. For example, the satiety hormone leptin is produced by fat cells and is secreted into our bloodstream and reduces a person's appetite. The recently discovered adipocyte-secreted energy-regulating hormone called adiponectin (also known as adipoQ or ACRP30) is an insulin-sensitizing and antiatherosclerotic hormone.

The 2015 article titled 'Adipose Tissue' (202) advises that a large array of different hormones is released from adipose tissue and that they are responsible for various functions within the body. The 2015 article advised that Adiponectin is released from adipose tissue. Adiponectin protects against developing type-2 diabetes as it increases the body's sensitivity to the effects of insulin. Apolipoprotein E and Lipoprotein lipase are released from adipose tissue. They play a part in fat storage and metabolism and the processing of fat stores to release energy. The article reported the following as further examples of these:

- Aromatase, which is involved in sex hormone metabolism.
- TNF alpha, IL-6 and leptin, which are collectively termed 'cytokines' and are involved in sending messages between cells.
- Plasminogen activator inhibitor-1, which is involved in the clotting of blood.
- Angiotensin, which is involved in blood pressure control.

45.20 Skin Fat Cells (Adipocytes) Might Protect You from Infections

Regarding the outcome of when skin damage occurs and infections may result, in the 2015 article titled 'Fat Beneath Skin May Ward Off Infections', the author, Dr Joseph Mercola (203), states that apart from specialized immune system cells called neutrophils and monocytes arriving at the scene to gobble up pathogens, other cell types already present in the area of the wound, including mast cells and leukocytes, provide a more immediate response against invaders. However, it turns out that these are not your body's only line of defense. Dermal fat cells, or adipocytes, which are just below your skin, may be the first defensive responders against potential invaders.

45.21 Fat Cells Produce Defensive Germ-Fighting Antimicrobial Peptides

Mercola (203) cites information from a new study presented in a 2015 article in *Science*, where Richard Gallo, MD, PhD, professor and chief of dermatology at UC San Diego School of Medicine (the study's lead researcher), said that it was incorrectly believed that if the defensive skin barrier was damaged, it became the solitary responsibility of circulating white blood cells (including neutrophils and macrophages) to protect the wound from becoming septic. However, it does take a while for neutrophils and macrophages to be recruited to the wound site. Gallo advised that a totally unexpected finding was that fat stem cells are also responsible for protecting us from infections. It was previously unknown that an adipocyte produces almost as much antimicrobials as a neutrophil.

In the 2015 article, Mercola (203) reported that the current study by Gallo revealed that fat layers in the skin of mice thickened after being exposed to pathogenic *Staphylococcus aureus* bacteria and mice that were incapable of forming new fat cells were more prone to infection. Gallo reported, 'The differentiating fat cells secreted a small-molecule peptide called cathelicidin, specifically in response to the infection.' Thus, the fat cells manufactured high levels of an antimicrobial peptide (AMP) called cathelicidin antimicrobial peptide, or CAMP. The article states, 'AMPs are used by your body's innate immune response to kill invading pathogens, including bacteria, viruses, fungi and other pathogens.' The article continues, 'Further tests showed that human fat cells also produced CAMP, which suggests the animal findings will hold true in humans as well—and your fat may serve as an important germ-fighting barrier against infection.' This is in keeping with the TCM concept that defensive Wei Qi circulates in the superficial layers of the body.

45.22 Too Much CAMP Is Pro-Inflammatory
In the 2015 article, Mercola (203) continued by noting that endogenous antimicrobial peptides (AMPs) are very useful for fighting infections as they are able to destroy the protective cell walls of viruses, bacteria, and fungi. Human skin also manufactures AMPs, so containing them in the fat layer would deliver a further protective mechanism to prevent infection due to microbial invasion. Mercola noted that cathelicidin has both antimicrobial and pro-inflammatory properties, so too much CAMP is associated with autoimmune and inflammatory medical conditions. The study determined that obese mice had higher CAMP levels in their blood than those mice of normal weight. According to Gallo, 'In humans it is becoming increasingly clear that the presence of AMPs can be a double-edged sword, particularly for CAMP. Too little CAMP and people experience frequent infections.' He continued, 'But too much CAMP is also bad. Evidence suggests excess CAMP can drive autoimmune and other inflammatory diseases like lupus, psoriasis and rosacea.'

Summary of Chapter 45

The article by Kathleen Doheny (197) titled 'The Truth about Fat' notes that scientists find fat very intriguing and more so every day. 'Fat is one of the most fascinating organs out there,' says Aaron Cypess, MD, PhD, an instructor of medicine at Harvard Medical School and a research associate at the Joslin Diabetes Center in Boston. He said, 'We are only now beginning to understand fat.' Note that Cypess stated, 'Fat is one of the most fascinating organs out there,' thus showing that he feels that the adipose system present throughout the body warrants being considered an organ.

Note that mitochondrial inner membranes are required to produce energy in the form of ATP. In warm-blooded animals, body heat is maintained by signaling the mitochondria to perform their function via the protein aptly named thermogenin (i.e. producer of heat). Brown adipose tissue is highly specialized for this thermogenesis as each BAT cell has a higher number of mitochondria compared to more typical cells. Secondly, these BAT 'mitochondria have a higher-than-normal concentration of thermogenin in the inner membrane'. In neonates (newborn infants), brown adipose tissue makes up about 5% of the body mass and is located on the back, along the upper half of the spine and toward the shoulders. This BAT is of great importance to avoid lethal cold as hypothermia is a major death risk for premature neonates. Li Chiung stated, 'The Triple Burner represents nothing but membranes.' The article cited above states that 'mitochondrial inner membranes are required to produce energy in the form of ATP.' I believe that the Triple-Energizer Metasystem is aka the PVS/Connective-Tissue Metasystem and that the role of the Triple Heater to produce energy and heat (thermogenesis) extends down microscopically to the 'mitochondrial inner membranes'. It is an interesting scientific finding that thermogenic BAT is distributed on the most yang region of the most yang body surface.

Regarding the distribution of BAT throughout the body, authors Sacks and Symonds (200) state, 'Its anatomical distribution is likely to confer survival value by protecting critical organs from hypothermia by adaptive thermogenesis. . . . In view of the diverse locations and potential differences in responsiveness between BAT depots, it is likely that BAT will be shown to have much more subtle and thus previously overlooked functions and regulatory control mechanisms.' According to the authors (200), the three major anatomical locations of BAT distribution are around blood vessels, around hollow organs, and around solid organs.

It appears that the BAT organ system is specifically designed to generate heat throughout the entire body, in the Upper Heater ('around epicardial coronary artery and cardiac veins as well as medium-sized muscular arteries and veins including the internal mammary and the intercostal artery branches from the subclavian and aorta'), in the Middle Heater ('as well as greater omentum and transverse mesocolon in the peritoneal cavity'), and in the Lower Heater ('in the inguinal area'). As the Triple Heater has the role of heating the body, it can be seen that the BAT organ system compliments the Triple Heater organ in accomplishing this biological function.

Regarding the findings of new scientific research, in the 2014 article titled *'Researchers Find Healthy Brown Fat Regulates Your Blood Sugar'*, Dr Joseph Mercola (201) advises that brown adipose tissue (BAT) may help regulate blood sugar levels in humans and that failing Brown Adipose Tissue as individuals age may be the cause of 'middle-age spread'.

In 2015, Mercola (203) reported that the fat layers in the skin of mice thickened after being exposed to pathogenic *Staphylococcus aureus* bacteria and mice that were incapable of forming new fat cells were more prone to infection. It thus appears that subcutaneous fat cells secrete a small-molecule peptide called cathelicidin, specifically in response to infection. The research suggests that human subcutaneous fat cells may serve as an important germ-fighting barrier against infection. This is in keeping with the TCM concept that defensive Wei Qi circulates in the superficial layers of the body.

That fat cells (adipocytes) comprise a key endocrine organ is unquestionable. A large amount of recent scientific research proves beyond doubt that much-maligned fat tissue has a double life and is actually also an endocrine organ. For example, the satiety hormone leptin is produced by fat cells and is secreted into our bloodstream and reduces a person's appetite. The recently discovered adipocyte-secreted energy-regulating hormone called adiponectin (also known as adipoQ or ACRP30) is an insulin-sensitizing and anti-atherosclerotic hormone.

The 2015 article titled 'Adipose Tissue' (202) advises that a large array of different hormones is released from adipose tissue and that they are responsible for various functions within the body. The 2015 article advised that Adiponectin is released from adipose tissue. Adiponectin protects against developing type-2 diabetes as it increases the body's sensitivity to the effects of insulin. Apolipoprotein E and Lipoprotein lipase are released from adipose tissue. They play a part in fat storage and metabolism and the processing of fat stores to release energy. The article reported the following as further examples of these:

- Aromatase, which is involved in sex hormone metabolism.
- TNF alpha, IL-6 and leptin, which are collectively termed 'cytokines' and are involved in sending messages between cells.
- Plasminogen activator inhibitor-1, which is involved in the clotting of blood.
- Angiotensin, which is involved in blood pressure control.

CHAPTER 46

How Earthing or Grounding Complements the Triple Energizer

Introduction
Grounding or Earthing™ is accomplished simply by standing on the ground barefoot. Whether you stand barefoot on dirt, bare soil, grass, sand, rocks, concrete, brick, or ceramic tiles (especially when humid or wet), you will be grounded. The earth is electronegative, storing an infinitely large reservoir of electrons. When you ground yourself to the electron-enriched earth, electron flow into your body, and an improved balance of the sympathetic and parasympathetic nervous system occurs immediately throughout every single cell and surface within your entire body.

The earth is a powerful source of electrons and subtle electrical fields, which are absolutely essential for proper functioning of our immune system, circulation, synchronization of biorhythms, and other physiological and biological processes. As free electrons from the earth neutralize damaging free radicals just as antioxidants do, grounding is likely the most natural and most abundant antioxidant available. Most chronic, degenerative medical conditions have been linked to low-grade chronic inflammation occurring throughout the body. Clint Ober, the discoverer of grounding, calls this phenomenon electron deficiency syndrome. But how does Earthing™ affect the Triple Energizer?

46.1 Grounding Is a Key Mechanism by Which Your Body Maintains Health

In the 2014 article by Dr Mercola (204) titled 'Grounding Is a Key Mechanism by Which Your Body Maintains Health', he states, 'Have you ever noticed how good it feels to walk barefoot on a sandy beach, or in a forest? There is a reason for that—it's called the grounding effect. The reason for this sense of well-being is due to the fact that you're receiving a surge of potent healing electrons from the ground.' This is because the earth is a reservoir of electrons and possesses a slightly negative charge. Standing barefoot on the earth essentially allows a transfusion of healing electrons into your body. In the article, Mercola continues, 'Modern science has thoroughly documented the connection between inflammation and all the chronic diseases, including the diseases of aging and the aging process itself.' Pointedly, Mercola states, 'It is important to understand that inflammation is a condition that can be reduced or prevented by grounding your body to the Earth, the way virtually all your ancestors have done for hundreds if not thousands of generations.'

46.2 Grounding Reduces Blood Viscosity and Powerfully Treats TCM 'Blood Stasis' Condition

Regarding the negative effects of free-radical stress and the subsequent chronic inflammation throughout the body, Mercola (204) explains that the blood thickens when your body experiences a lot of free-radical stress and positive electrical charges concentrate in your body. The outcome is chronic inflammation, which engenders the greater majority of chronic and degenerative diseases. Because grounding supplies electrons, it reduces inflammation and thins the blood by making the red blood cells electronegative, so they would repel one another and prevent 'stickiness', which, when left untreated, can result in blood clots that can produce a heart attack or stroke.

At no time in human history has the problem associated with chronic inflammation and the plague of chronic degenerative disease and autoimmune disorders been more prolific than now. Note that from the TCM point of view, the etiology of many medical conditions is due to what Chinese Medicine terms stuck blood, blood stagnation, or blood stasis. From the TCM perspective, such blood stasis leads to heat and inflammation and a panoply of medical conditions (which include abdominal masses or immobile masses under the skin; precisely localized sharp stabbing pain that is worse with applied pressure; dark complexion or dusky face; rough, dry, scaly skin; ulcers; abscesses; dry, course hair; purplish lips and fingernails). There is generally a dark purple tongue, and the pulse feels sluggish or choppy.

Regarding the health benefits of Grounding, modern research has demonstrated that it takes as little as 80 minutes for the free electrons from the earth to thin down unhealthy thick accumulated stuck blood making the blood flow more smoothly throughout your bloodstream. Subsequently, thinner blood means reduced blood pressure as the Heart does not need to pump so strongly.

46.3 Grounding Charges Every Single Cell in Your Body with Energy for Self-Healing

In the 2014 Grounding article, Dr Mercola (204) eloquently connects the concepts of Grounding and the effect that the free electrons have on the structure of water in the body when he explains that individuals that try grounding usually note various health benefits, including improved sleep, pain relief, and a heightened sense of well-being. Grounding has been shown to improve heart rate variability and calm the sympathetic nervous system. Mercola also noted that Gerald Pollack, PhD, and author of *The Fourth Phase of Water: Beyond Solid, Liquid, and Vapor*, has confirmed that the negatively charged free electrons supplied from grounding modifies the structure of water molecules throughout the entire body so that the negatively charged water molecules can effectively store energy, much like a battery, and then can deliver energy as required. Essentially, when you ground yourself, every single cell in your body is being charged with energy that your body can utilize for biological processes, including self-healing.

This is the pivotal concept that I believe embodies the Triple Energizer such that the water throughout the body is converted into its fourth form, EZ water, by free electrons from the earth and from light from the sun, showing our absolute dependence on the sun and the earth to maintain optimal health and well-being. Note that the free electrons from the earth (along with light from the sun) generate EZ water, which holds the energy derived from the earth (or sun) much like a battery, so that the energized water is circulated throughout the Triple Energizer via the ubiquitous, omnipresent, hydrophilic PVS/Connective-Tissue Metasystem that pervades our being and covers and embellishes every single cell in the body. All food and water that we consume has been irradiated and energized by the sun so that the biophotons therein are what are absorbed from the Stomach and circulated throughout the body via the Triple Heater cycle. These biophotons further empower the formation of EZ water, which nourishes and empowers every cell in the body and optimizes self-healing.

Note that grounding is not a treatment or a cure for any particular disease or disorder. Rather, grounding is an essential requirement of your body that maintains homeostasis and engenders optimal health. Humans have been in constant contact with the earth for millennia, and your body needs this continuous supply of free electrons in order to function properly. Walking or exercising barefoot outdoors daily is one of the most wonderful, inexpensive, and powerful ways of incorporating grounding into your life. You can also incorporate grounding products inside by obtaining any of the numerous grounding products supplied by Earthing Heaven™ in Australia or any of the Earthing products suppliers globally.

46.4 Exercise Physiologist Discusses the Many Benefits of Earthing™ or Grounding

Roy Stevenson has a master's degree in exercise physiology from Ohio University. He has taught exercise science and nutrition at Seattle University, University of Puget Sound, Highline Community College, and Lake Washington Technical College. Stevenson (205) wrote an article in *WellBeing* magazine in February 2012 called 'Barefoot Earthing'. He advises that barefoot earthing or Grounding, as it is also known, simply involves taking off your shoes and standing barefoot on green grass or sand along a beach so that you can effectively earth yourself. Currently, there is a lot of interest about inflammation and disease processes from the biomedical community. In the article, Stevenson says you can 'decrease your muscle inflammation, prevent free-radical damage, heal your injuries faster, reduce muscle stiffness and tension, help you sleep better and improve your health and energy levels? It almost sounds too good to be true, doesn't it?' He continues, 'If you were healthier, had more energy and your injuries healed faster, you would be a lot closer to achieving your genetic lifespan potential or at the very least have a better quality of life.'

Extraordinary claims have been made regarding barefoot Earthing™, and a large amount of research is being performed to validate these claims. What scientific principles could make such a simple and natural activity as standing barefoot on the surface of Mother Earth so beneficial to your health? The scientific theory behind barefoot earthing involves electrons present on the earth's surface and how they interact with your body. Stevenson (205) explains that the Global Electric Circuit (GEC) that bathes the earth is continuously renewed due to the 1,000–2,000 continually active lightning storms occurring at any one time around the globe. The positive charges generated from these electrical storms accumulate in the upper atmosphere, while the negative charges accumulate in the form of free electrons at the surface of the earth. These negatively charged free electrons bathe the earth's surface and flow freely into all conductive surfaces in contact with the earth, including mineral-based materials (concrete and metal structures) and even trees, plants animals, and humans if they are barefoot or wearing conductive footwear. Unfortunately, most human wear insulated footwear that denies the entry of these naturally occurring healing free electrons.

46.5 TCM's Earth Deficiency Syndrome Can Be Improved by Grounding or Earthing™

Electrophysiologists and biological researchers using a voltmeter have confirmed that when humans stand on the earth in bare feet, they become electrically grounded. Intimate physical contact with the earth returns our body to the same neutral electrical voltage of the earth. We have all noticed how more relaxed and refreshed we feel after walking barefoot on green grass or spending a day walking barefoot on the beach. If we are electrically balanced, this gives credence to the term of *being grounded.*

Your entire body is intermeshed with its own electrical circuitry. Due to thunderstorms, the entire surface of the Earth is charged with free electrons. Electrophysiologists believe that the free electrons present at the earth's surface ideally should enter our body and neutralize the positively charged free radicals that cause so much damage to our health. For millennia, we walked barefoot and happily slept on the earth. However, due to the practices of modern society, we rarely make intimate physical contact with the earth. By wearing synthetic-soled shoes and living in an environment that insulates us from the earth's subtle electron charge, such as artificial grass and bitumen on the ground and timber or carpeted floors in homes and high-rise buildings, our bodies are deprived of the natural, free electron flow from the earth. Because we are not protected by the earth's electric stabilizing field with its natural supply of free electrons, we literally become ungrounded, and the electrical stability in the body is disrupted. From a TCM viewpoint, we quite literally become 'Earth Deficient'.

46.6 Earthing™ Neutralizes the Free Radicals That Induce Inflammation in the Body

Scientific research claims that barefoot earthing reduces muscle soreness and inflammation from acute and even chronic injury. What is the possible scientific explanation for barefoot earthing being so beneficial for us? It is a fact that free radicals can induce inflammation in the body. It is also a fact that free radicals are neutralized by electrons provided from any source. Free radicals act as electrophiles that attract free or mobile electrons from any source, and it is theorized that contact with the earth through grounding can supply ample electrons to neutralize the positive free radicals. It is likely that this flow of free electrons into the body is what reduces inflammation by blocking the formation of free radicals and therefore speeding up the repair process of damaged inflamed tissues. Regarding the positive healing outcome of Earthing™ on various musculoskeletal disorders, as verified by thermal imaging photos of the patients before and after treatment, Stevenson (205) reported that the president of the International Academy of Clinical Thermography, Dr William Amalu, conducted a series of medical case studies on 20 patients with a diversity of inflammatory medical conditions. All the patients were grounded two to three times each week for as little as 30 minutes at a time. Amalu took thermal imaging photographs of all the patients prior to and after the grounding treatments and the outcome was 'nothing short of dramatic'. Analysis of the results showed that many of the patients in Amalu's study had pain reduction after only one treatment, while 80% of the patients experienced less pain within two to four weeks of grounding. Interestingly, most of these patients had attained poor results previously while receiving standard physical therapy and other standard medical treatments.

46.7 Scientific Trial Confirms Earthing™ Greatly Improved Sleep Parameters of Test Subjects

With respect to a blinded trial (n = 60) pertaining to the measurement of 3 sleep parameters and 3 pain and general health parameters over a 30-day period, where half of the subjects were unknowingly Earthed™, Stevenson (205) reports:

> Ober (2000) looked at 60 test subjects (22 men, 38 women) with sleep problems and a variety of joint and muscle pain. Half of the group slept on grounded pads, while the control group slept on ungrounded pads. The experiment lasted 30 days and the results were, again, extraordinary. Of the grounded volunteers, 85 per cent went to sleep more quickly, 93 per cent slept better during the night, 82 per cent had a significant reduction in muscle stiffness, 74 per cent experienced reduction or elimination of chronic back and joint pain, 100 per cent reported feeling more rested upon waking and 78 per cent felt that their general health had improved.

This trial conclusively demonstrates that sleeping Earthed™ has an extraordinary effect on at least six parameters associated with sleep quality, muscle stiffness, elimination of chronic back and joint pain, and the sense of well-being.

46.8 Double-Blinded Study Confirms Instantaneous Reductions in Muscle Stress and Tension

Concerning muscle stress and tension, Stevenson (205) discussed the results of a study performed by Chevalier et al. in 2006. Stevenson noted that the double-blind study produced statistically significant results for 58 healthy adults who received almost instantaneous reduction in overall muscle stress and tension in the body after they were earthed. Using electromyography, Chevalier et al. verified that while the subjects were earthed, hypertonic muscles returned to normal tension, and tensed muscles relaxed. Several of the studies reported that the subjects slept more soundly when they slept in a grounded state. As sleep is when our body mends from the daily assault of life, anything that allows us to sleep better is going to improve our overall health and wellness. To date, there are no studies that conflict with these findings.

46.9 Earthing™ Helps to Synchronize the Circadian Rhythm especially Associated with Jet Lag

As an interesting side note, Stevenson (205) noted that a practical application of earthing affects travelers crossing time zones and developing jet lag. The simple procedure to negate jet lag is to find a patch of moist green grass, remove your shoes, and walk around barefoot for 30 minutes. Scientist theorize that doing so helps the desynchronized circadian body clock and biorhythms to resynchronize to the electrical rhythm of the earth at the new location.

46.10 There Are Numerous Health Benefits Associated with Daily Earthing™

If you live by the beach, then walking along the beach, especially along the wet sand, for at least 30 minutes is the perfect way to remain grounded and earthed. For those individuals that do not live at the seaside, walking barefoot on green grass for at least 30 minutes is the easiest way to ensure that you get your electron fix for the day. Sitting on grass or sitting on a chair with your feet touching the grass works just as well. Intimately touching the earth in this manner several times a day is very beneficial and relaxing. If you live in a concrete jungle without access to a strip of green grass, there are products available, including earthing pads, sheets, mattresses, beds, and recovery bags. All of these grounding products are available within Australia from Earthing Heaven at http://thenaturalmedicalhealthwell.com/earthing-heaven/.

Recent scientific research suggests nocturnal Earthing™ during sleep is beneficial for individuals by:

- reducing inflammation and improving the symptoms of inflammatory disorders
- eliminating or reducing chronic pain
- improving sleep in many cases
- increasing energy levels
- lowering stress levels by calming down the nervous system
- normalizing the body's biological rhythms
- improving blood pressure by thinning the blood
- relieving headaches and muscle tension
- improving recovery and repair from intense athletic activities
- relieving the symptoms of premenstrual tension
- accelerating healing
- preventing bed sores
- stopping or reducing jet lag
- protecting against potentially harmful EMFs (Electromagnetic frequencies).

Stevenson (205) believes the time is coming when regular barefoot earthing sessions will be as common as working out at a gymnasium. I remember a time when joggers were a curious thing, and bystanders wondered what the individual was running from.

While it has not been confirmed at this stage, it is very likely that most of the Earthing™ effect of electrons being transferred to every single cell in the body is via the electroconductive connective tissues and fascial membranes that wrap every cell, structure, and organ in the body like a bag, just as the San Jiao is said to do.

46.11 Dr Joseph Mercola Interviews Clint Ober, the Father of Earthing™/Grounding

In an interview with Clint Ober in 2012, Dr Joseph Mercola (206) discussed the following information on recent grounding research. Earthing or grounding results in synchronizing your internal biological clocks, hormonal cycles, and physiological rhythms, and it suffuses your body with healing, negatively charged, free electrons abundantly present on the surface of the Earth. For 25 years, Clint Ober participated in the spectacular rise of the cable television industry. His own business was the leading cable installation company in the United States, and he was renowned for giving customers images that were superior to broadcast television. After his retirement, Mr. Ober began to consider how his experience with cable TV might apply to the human body. He realized that most people wear synthetic-soled shoes that insulate their bodies from the Earth's energy field that stabilizes not only cable TV but all electrical equipment—industrial or residential—throughout the world. This seemingly simple realization inspired a scientific adventure that has resulted in what is perhaps the greatest health discovery of our time.

46.12 Failing to Contact the Earth Leads to Electron Deficiency Syndrome, or Earth Deficiency

By isolating and insulating yourself from the natural grounding effect of contacting the earth, you have developed what could be called electron deficiency syndrome. The damaging effects of EMF irradiation can be greatly eliminated by the use of grounding pads. The earth is a source of an infinitely large number of free electrons and can donate the electrons to us if we allow the electrons to enter us by making direct intimate contact with the earth. All plants and wild animals making contact with the earth absorb electrons from the earth and equalize. Since the 1960s, with the mass production of inexpensive plastics and polymers, we have been getting more and more cut off from the natural earthing effect that Nature provides. As recently as 1954, 95% of the soles of shoes were salt-treated leather. Leather-soled shoes are relatively conductive and do allow some transfer of electrons when in contact with the ground. Salt-laden sweaty feet help to increase the conductivity as well. As I recall, shoes were only worn to go to school, church, or work. Otherwise, shoes were removed at the first opportunity, and going barefoot was the order of the day.

Today 95% of the soles of shoes are synthetic. Interestingly, the parabolic growth of sales of synthetic-soled shoes over that same period correlates with the increase of diabetes (inflammation). When you are grounded, you are electrically neutral, and it is impossible to have a positive charge accumulate in the body. The majority of individuals in western, 'civilized' countries rarely literally touch the earth. The major time this occurs for Mr or Mrs Citizen today is during their 10-to-15-minute shower, and grounding at that time only occurs if copper pipe is used in the plumbing for the building. With the high cost of copper in recent years, plumbing pipework is now generally nonconductive specialized plastic pipes that are flexible. I suggest that owners of dwellings with these plastic pipes will not feel as refreshed and energized after a shower as they used to when the shower water was delivered through highly conductive copper tubing, which allowed a vast quantity of free electrons to be carried to your shower so that you were bathed in free electrons from the earth. They call that progress. It is no wonder that electron deficiency syndrome, or Earth Deficiency, is rife and that chronic low-grade inflammation and diseases of autoimmunity are massively on the increase.

46.13 Natural Antioxidants from Foods Are Necessary to Neutralize Damaging Free Radicals

Free radicals are atoms or groups of atoms with an unpaired number of electrons and can form when oxygen interacts with certain molecules. Free radicals are especially produced via the immune system as the white blood cells oxidize pathogens and generate free radicals. Once free radicals form, they are highly reactive and can initiate destructive chain reactions, like a stack of dominoes falling over to destabilize the next domino. The major issue is the large amount of damage that they can cause when they react with essential cellular constituents, such as DNA or the cell membrane. Cells may be severely damaged and become dysfunctional or die if this assault proceeds unchecked. To overcome the constant free-radical damage that the body encounters during metabolic processes, an inbuilt defense mechanism has been set in place that involves the assistance of antioxidants. There are many natural antioxidants present in the body. They are molecules which are able to safely interact with free radicals and prevent the chain reaction from causing severe damage to critical molecules. There are several enzyme systems present in the body that scavenge free radicals. Major players of this protective system include vitamin E, beta-carotene, and vitamin C. The trace element selenium also performs a major role in this regard. These substances must be supplied in the diet as they can't be manufactured by the body.

46.14 Actually Touching the Earth Generates the 'Antidote' to Damaging Free Radicals

In 1960 only 5% of visits to the doctor were due to stress-related disorders. Today 95% of the population is affected by stress-related and inflammatory health disorders and medical conditions. Something is stressing the body so that health can't be maintained as it previously was. The oxidation of healthy tissues leads to free-radical formation, which leads to cellular stress and inflammation. Having access to a reservoir of free electrons in the body prevents the oxidation of healthy tissues and counteracts cellular stress and inflammation. Free radicals don't just remain stored in our body. They are generated within nanoseconds, and they then attach to tissue components within nanoseconds. The actual result is free-radical damage to tissues and cells. They don't travel in your bloodstream, for example. It is all down to electrical transfer at the cellular level. Healthy bodily functionality is a foundational, fundamental electrical state where the free electrons absorb the pathologic by-products of biological and metabolic processes and keep everything clean and balanced. The immune system is the primary producer of free radicals in its job of destroying pathogenic organisms and toxic substances in our body. The immune system is the major generator of these damaging free radicals, which cause the cellular stress and inflammation, which cause disease. But the immune system is, after all, only doing its job as the creator intended. The process, however, necessitated that while this protective war was going on, we would automatically be grounded so that the damage from the free radicals would not occur. However, this condition is now uncontrolled, and we have moved away from Nature, and as we are no longer grounded on a daily basis, the outcome has been an ever-increasing load on the bodily biosystem with chronic inflammation and cellular stress—the 'reward' from moving away from what Nature intended. As TCM says, we are not living in harmony with the Tao, the natural way and truth of the universe. Grounding restores the electrical stability of the body so that we can synchronize with the earth and subsequently live in harmony with the Tao, the reward of which is good health and well-being. So, grasshopper, take this book outside, take off your shoes, and touch the earth barefoot, and then continue reading.

46.15 Earthing Reduces the Zeta Potential of Red Blood Cells

When you are unearthed, the red blood cells tend to clump and thicken the blood as its viscosity increases. This feature has been likened to the difference between red ketchup and red wine. Earthing restores the normal zeta potential after two hours of earthing, and the red blood cells are less attracted to one another and thin out to individual cells separated from other red blood cells. In keeping with this finding, there is a 40% increase in stroke in people that live in multistory buildings as their blood gets thicker.

46.16 Precautions for Those on Medications When They Commence Daily Earthing™

1. A major effect of electron deficiency syndrome is thyroid deficiency. Any person already on thyroid medications may have to reduce the dosage of their thyroid medication once they start using regular grounding treatment; otherwise, they may start to experience heart palpitations. They should liaise with their doctor and may need to reduce their thyroid medication.

2. A second consideration involves individuals on blood-thinning medication, like Coumadin (Warfarin). Because grounding changes the zeta potential of the red blood cells and reduces its viscosity, thus thinning the blood, you could end up with the blood being too thin. So it is essential to have your doctor monitor your blood-thinning medication levels very carefully.

3. Thirdly, grounding affects blood glucose levels. So hypoglycemics need to be aware of this effect of grounding and monitor their blood glucose levels regularly. The lack of daily grounding is believed to be responsible for the groundswell of epidemic proportion of prediabetes and diabetes in modern societies. If you are taking oral hypoglycemic medications, consult with your doctor to ensure that your blood sugar levels are kept within the correct range. When rats were fed the same diet and the same daily caloric intake under the same conditions and allowed to live out their lives, research showed that ungrounded rats had consistently elevated blood glucose levels compared to rats kept in a grounded environment. The ungrounded rats had higher levels of triglycerides and higher blood glucose levels and became 10% heavier when compared to the grounded rats. This was in spite of the same diet and conditions throughout their lives. This is an indicator that metabolic syndrome may well be an inflammation-related disorder rather than only a metabolic disorder and may well be a direct result of electron deficiency syndrome. It must be noted that heart disease and cancer are also probably a result of electron deficiency syndrome.

4. Fourthly, it is believed that high blood pressure can be lowered by Earthing™ through three possible mechanisms. An article in the Earthing Institute's blog (207) discussed three reasons why Earthing™ reduces hypertension. The article states:

   1) Earthing generates a shift in the nervous system away from sympathetic (the vigilant, fight-or-flight mode) mode associated with stress and toward the parasympathetic (calming, relaxing) mode.

2) Earthing exerts a normalizing effect on cortisol, the stress hormone.

3) It promotes better sleep and reduces inflammation and pain. All of these, and likely other factors as well, can benefit blood pressure.'

Earthing™ book coauthor and cardiologist, Dr Stephen Sinatra, is convinced that grounding definitely reduces blood pressure. He advises hypertensive individuals that are taking blood pressure medication must advise their physician if they wish to start Earthing™ as they may have to lower their hypertensive medication dosage. Sinatra advises, 'Track your own blood pressure with a monitoring device at home and bring to your doctor a written record of your readings before and after grounding sessions.' As with any beneficial lifestyle changes you make on your roadway to improved health, a healing crisis often occurs. Regarding the release of accumulated toxins, the article (207) advises that Earthing™ tends to accelerate the detoxification process, and some individuals feel worse for a few days before they feel better and more energized. Individuals should consume more water to help the elimination of released toxins.

46.17 Wild Animals Don't Have Inflammation, and They Don't Develop Cardiovascular Disease

An interesting outcome of Ober's research is regarding house pets. Wild animals don't have inflammation, and they don't develop cardiovascular disease or even have plaque on their teeth. They don't have Multiple Sclerosis or Lupus or fibromyalgia. However, domestic animals kept in the house or in apartments eventually develop the same health disorders as their owners. Animals kept in ungrounded barns or coups need a supply of light or heat to keep them comfortable, while grounded animals in the wild do not need such 'home comforts' to remain well. There is empirical anecdotal confirmation of this difference between wild and domesticated animals. It may well be that the epidemic of overweight is also an artifact of electron deficiency syndrome. Time and ongoing research will tell.

46.18 It Is Critical that We Are Grounded While We Are Sleeping

Bedding and nightclothes often generate invisible static electricity. To generate the visible and audible discharges that result from static electricity, the voltage must be 3,000–5,000 V. These electrical charges affect us adversely. People that are grounded at night generally fall asleep very quickly, and it has been observed that individuals with sleep difficulties usually sleep very effectively when they are grounded at night. This is very replenishing to our energy level and allows healing and rejuvenation to be optimized as Nature intended. Sleep is the most important entity in our life, and that is why one third of our life is dedicated to this nocturnal rejuvenation period. Sound revivifying sleep is absolutely essential for good health. Earthing has been shown to neutralize damaging free radicals generated during the day, calm the nervous system (especially the left brain), and also maintain right–left brain synchronization.

46.19 Earthing™ Stimulates Blood Supply to the Peripheral Circulation

Accessing the Earth's energy stimulates our circulation to the peripheries, and subsequently we would be less prone to cold hands and feet and pins and needles, especially in the winter. This is due to the generally clumped blood cells dissociating from their clumped state, allowing the blood to be thinner and more able to enter peripheral capillaries in single file, carrying oxygen and nutrients to maintain optimal circulation and distal warmth. This is related to the fact that unearthed barnyard animals need heat when they are stabled during winter, while free-range animals in contact with the earth and its donation of free electrons have an optimal circulation and hence do not 'feel the cold' as much and are better able to regulate their body temperature. When individuals are grounded, within 10–15 minutes of allowing Mother Earth to donate free electrons, the drawn, tired grayish complexion is reduced, and the facial skin turns pink due to the increased blood flow. Their cold fingers and toes will start to turn pink and warm up thanks to the increased circulation. While their circulation increases, their pain reduces, and they have a sense of joy and wellness.

46.20 Free Electron Deficiency Syndrome (FEDS)
The body is a highly complex self-healing machine and is designed to stay healthy. Environmental toxins in the air, water, and food and electromagnetic pollution certainly take their toll on your endeavour to remain healthy. If you are failing to heal sufficiently, then free radicals causing stress in your body are not being neutralized, and Free-Electron Deficiency Syndrome (FEDS) is most probably your problem.

46.21 Modern Mankind Is Essentially Ungrounded Most of the Time
Swimming in an uncontaminated, pristine ocean is the best form of grounding. Modern mankind is essentially ungrounded most of the time. Some individuals may spend years being ungrounded, never once touching their bare feet on the ground. Since mankind discovered the light bulb, we have been like moths to the flame and have totally lost appreciation for the simple fact that barefoot contact with the Earth is essential for optimal health and wellness. Touching a tree or plants will also allow the free electrons to bridge the gap and flow into your body through your hands. This could account for the fact that Hippies in the 1960s were 'tree huggers' and had so much joy and love! Marijuana may possibly have also contributed to their general demeanor. The beneficial effects of grounding is probably also why so many people love gardening and feel invigorated even after they have worked hard cutting, pruning, and trimming plants and bushes or digging soil and planting seedlings. The free electrons have entered their body and neutralized the harmful free radicals, while they have been totally oblivious to the whole grounding process.

46.22 The Earth Actually Pushes Beneficial Free Electrons into You When You Touch Ground
'An ounce of prevention is worth a pound of cure' is certainly true when it comes to being grounded on a daily basis. All hospitals ensure that, during surgery, the patient is grounded so that no static charge can interfere with sophisticated diagnostic equipment. Personnel

working in the electronics industry must be grounded so they do not damage sensitive electronic chips and equipment. The fireworks, petroleum tanker distribution, explosives, and gas industries are all highly aware of the necessary practice of grounding to prevent explosions from static sparks. While the average person sleeps, they are exposed to approximately 2,000 millivolts (mV) due to wiring in the wall space behind the bedhead. This is 20 times more than the preferred maximum background voltage 'noise' of 100 mV. Your entire body is thus adversely affected by the electromagnetic fields from such cables and electrical equipment.

When you are standing or lying on a nonconductive insulating material (for example, bitumen, a carpet, timber or tiled floor, or an unearthed bed located even only inches above the Earth), you are ungrounded and completely isolated from the Earth, essentially existing above it in 'free space'. Relative to the Earth, you are infinitely small and as such are completely subject to and affected by any electromagnetic fields in your proximity, which have an adverse, perturbing effect on your health. However, because the Earth is infinitely large compared to you, when you touch an Earthing™, pad you become energized by the Earth, and that electron flow from the Earth neutralizes the effect that local electrical wiring has on you. So when you are barefoot on the Earth or somehow grounded to the Earth, you essentially are an extension of the Earth because the infinitely large Earth allows enough of the infinitely large reservoir of free electrons to flow into your conductive body and synchronize you to its state. The Earth actually pushes the free electrons into you when you allow this transfer by Earthing. The primary benefit of grounding is protection from electrical wattage exposures from equipment connected to the power grid. Because mobile phone microwaves are so high in frequency, they have a short wavelength, and neutralization by free electrons from the Earth just can't work that fast. So it is believed that you can't neutralize high radio frequencies from mobile phone microwaves and protect the body from them by grounding. Other safety practices must be applied.

46.23 Grounding Has a Rapid Calming Influence within the Autonomic Nervous System (ANS)

In the 2012 article titled 'Earthing, Inflammation, and Aging—Something to Think About', the Director of the Earthing Institute, Gaétan Chevalier, PhD (208), pointed out that perhaps one of the most overlooked Earthing™ dividends which is so beneficial in these stressful times is the rapid calming influence that takes place within the autonomic nervous system (ANS), which regulates functions like heart and respiration rates, digestion, perspiration, urination, and even sexual arousal. The effect begins pretty much instantly when you connect with the Earth. The ANS shifts from the typically overactive sympathetic mode associated with stress into a parasympathetic, calming mode. Chevalier reported that more than two-thirds of over 58,000 articles listed in the US National Library of Medicine (PubMed) website on chronic inflammation have been published in the last 12 years. Obviously, there is massive interest in research into the numerous diseases caused by chronic inflammation.

46.24 Contacting the Earth Has a Dynamic Anti-Inflammatory Effect throughout the Body

Regarding the effect that grounding has either by being barefoot outside or by sleeping or sitting on conductive Earthing products indoors, Chevalier (208) noted that whether individuals contact the Earth directly by sitting on green grass or by being barefoot outside or if they sleep or sit on Earthing™ products indoors, there is a strong argument that inflammation is quenched due to the Earth's infinite supply of negatively charged electrons flowing into the body and neutralizing the free radicals, which cause chronic inflammation. Chevalier pointed out that the theory that the Earth acts as an anti-inflammatory is a novel idea that has not been proposed previously by the scientific community and that while it is still only a theory, there is no other known explanation to explain the reduction of inflammation throughout the body when human contact is made with the Earth. He has participated in numerous Earthing studies and believes that daily Earthing™ is a profoundly simple, natural, and effective way to reduce stress and inflammation and improve health and healing. Chevalier (208) proceeded to state:

> There is much support in the research community about chronic stimulation of the immune system leading to exhaustion and for the free radical theory of aging as the leading explanation of the aging process. Earthing has a big impact on both. Earthing counteracts free radical damage by providing the body with an abundance of electrons. Electrons are the power behind antioxidants, and so they consequently prevent or reduce inflammation. Such effects produce a relaxation of the body, and thus a reduction in physiological stress. Stress resistance = ability to survive, and by reducing stress, Earthing improves the ability of the body to survive.

46.25 Local Environmental and Climatic Variability Causes Daily Grounding-Energy Variance

In the book written by Clinton Ober et al. (209) titled *Earthing: The Most Important Health Discovery Ever?*, it is posited that the ground beneath our feet is the 'original anti-Inflammatory.' In other words, the biggest anti-inflammatory on the planet is, well, the surface of the planet itself. If proven true, that's a pretty neat design by Nature.

Ober believes that the Earth is rebounding with energy and constantly changing and that this varying Earth energy is transmitted through the Earthing equipment to anybody that is grounded. Especially when the sun is shining directly on the area of the grounding rod, there is a growing liveliness due to the sunshine generating free electrons and electromagnetic currents with the part of the Earth where the grounding rod connects. When we connect to the earth, this daily grounding energy differential helps to calibrate and harmonize our circadian rhythm, and Ober believes this accounts for Earthing being so powerful in helping individuals overcome jet lag when they touch down at their new destination.

46.26 Grounding Improves Heart Rate Variability (HRV) and Reduces the Stress Response

In a 2011 research article titled 'Emotional Stress, Heart Rate Variability, Grounding, and Improved Autonomic Tone: Clinical Applications', the authors (Gaétan Chevalier, PhD, and Dr Stephen T. Sinatra (210), who is a cardiologist and assistant clinical professor of medicine at the University of Connecticut School of Medicine) commented that when an individual is grounded, the balance of the sympathetic and parasympathetic nervous system is improved. Prior human investigations have confirmed that several biological parameters generally show a marked improvement within 20 to 30 minutes of grounding. For some grounded subjects, positive changes in biological parameters may take several days, while a few individuals show a drastic change in biological parameters within two seconds. The most immediate and profound changes in biological parameters occur for skin conductance tests and during electroencephalographic and electromyographic recordings. The study revealed a positive trend in HRV, with values improving continuously up until the end of the 40-minute grounding trial. This suggests that the HRV value continues to improve with an extended period of grounding. The article noted:

> In patients who experience anxiety, emotional stress, panic, fear, and/or symptoms of autonomic dystonia, including headaches, cardiac palpitations, and dizziness, grounding could be a very realistic therapy. These patients may see positive effects most likely within 20 to 30 minutes and in almost all cases in 40 minutes. Negative emotions such as panic, depression, anxiety, and hostility have all demonstrated reduced HRV (heart rate variability). Grounding has the potential to help support HRV, reduce excessive sympathetic overdrive, balance the ANS (autonomic nervous system), and, thus, attenuate the stress response.

The authors stressed that these research findings have important medical ramifications particularly due to the strong association between depression and the increased risk of cardiovascular events that has been repeatedly confirmed in both patients with established cardiovascular disease and the healthy population as well.

46.27 Minor Variations in the Heart's Beat-to-Beat Interval (HRV) Improved by Grounding

You may not be familiar with the medical parameter called Heart Rate Variability (HRV), so here's a quick synopsis, courtesy of Dr Sinatra (210). In cardiology, a balanced ANS supports good HRV and is an indicator of better heart health. Conversely, the presence of disturbed ANS and HRV means an increased risk of arrhythmias, coronary artery spasm, and sudden cardiac death. HRV has also become a reliable clinical tool for measuring survival potential after a heart attack. HRV is not the same thing as heart rate. Your heart beats faster when you exert yourself or become stressed, and it slows down when you relax—but that's not HRV. HRV refers to the imperceptible variations in the heart's beat-to-beat interval that result from the basic breathing process. We refer to it technically as respiratory sinus arrhythmia. You can't feel the difference, but when you breathe in, your heart rate increases just ever so slightly. When you breathe out, it decreases ever so slightly. We can see these fluctuations on an electrocardiogram, and we've learned even more from sophisticated computer analyses of beat-to-beat intervals. To explain the importance of HRV more fully, Sinatra said:

> Let me give you an example here. Have you ever had an electrocardiogram exercise stress test on a treadmill? If so, you may recall that you remain hooked up for a bit of time after the test. The reason is that your doctor wants to see what your heart rate is doing after exertion on the treadmill. Your heart rate may rise to 150 beats a minute during the test. Afterward, it should drop down into a de-accelerated zone of say 100 beats or less in about six minutes. If not, and it stays higher, you have a classic indicator of disturbed HRV, and increased likelihood of cardiovascular trouble. You may, for instance, have a beat rate of 112. That's not good enough. I remember as a young cardiologist years ago waiting even a half-hour or so for some patients' heart rates to drop down sufficiently.

46.28 The Remarkable Healing Effect that Grounding Has on Stress and Cardiac Conditions

Dr Sinatra (210) says this about stress: 'As a cardiologist, I have repeatedly treated the human wreckage that stress—acute or chronic sympathetic overdrive—can exact. In trying to rebuild and restore the wreckage, I have applied the best tools that both conventional and alternative medicine has to offer.' Showing the remarkable healing effect that grounding has on stress and cardiac conditions, Sinatra continues, 'Reconnecting the body to the Earth offers perhaps the most natural tool available anywhere. I've seen this simple remedy do some amazing things.'

46.29 Free Electrons from Grounding Differentiate the 'Good' and the 'Bad' Free Radicals

A very good question arose regarding how free electrons that enter the body through grounding are able to discriminate between the 'good' free radicals, which participate in protective body functions, and the 'bad' free radicals involved in chronic inflammation. Grounding electron expert and biophysicist James Oschman, PhD (211), answered this question in the article titled 'Earthing vs. "Good" and "Bad" Free Radicals'. He explained that in spite of their bad reputation, free radicals or reactive oxygen species (ROS) are necessary for life processes. Beneficial, good free radicals destroy pathogenic bacteria and other pathogenic organisms and also digest injured cells. The good free radicals are not deactivated by electrons from grounding because the white blood cells deliver very high concentrations of the benign ROS at a precise localized site of inflammation, or their electron-seizing activity is specifically dedicated to processing damaged cells and pathogens. When an individual is ungrounded, an inflammatory barricade can result due to free radicals emanating from a site of injury becoming harmful and this cascade can damage surrounding tissues. Oschman believes that a grounded individual will have a healthy connective-tissue matrix that possesses copious electrons to counteract those leaking bad free radicals and subsequently avert collateral damage to healthy surrounding tissues. He advises that simply eating and breathing generate dangerous ROS and that the prevailing theory of aging, called the free-radical theory of aging, suggests that aging occurs as a result of cumulative damage produced by free radicals derived from highly reactive oxygen molecules in the tissues produced by normal metabolic processes and from environmental toxins. There are three ways that these free radicals are believed to exacerbate aging—namely, (1) damage-causing DNA mutations and cancer, (2) mitochondrial damage leading to less energy production, and (3) protein cross linkage, which damages enzymes and causes skin wrinkles to result. Oschman states, 'We believe that all these mechanisms of age-related free radical damage are slowed significantly by grounding, and that in the grounded body the healthy connective tissue matrix will be populated with enough electrons to neutralize free radicals wherever they form from whatever cause.' He stressed that a grounded body will not allow inflammatory pockets to form around sites of injury and subsequently injuries will heal quickly. Further, a healthy connective-tissue matrix is more conductive to free electrons and consequently will be more effective at quickly neutralizing free radicals wherever they form in the body and from whatever cause.

Note how James Oschman believes that the many benefits associated with grounding are directly dependent upon a healthy connective-tissue matrix being able to neutralize free radicals wherever they form from whatever cause. As the connective-tissue matrix pervades the entire body, every single cell is afforded protection from unwanted damaging free radicals. I propose that this protective role is also afforded by the Triple-Energizer Metasystem, which is intimately associated with the Connective-Tissue Metasystem.

Summary of Chapter 46
Note that biophysicist James Oschman stated, 'We believe the healthy connective tissue matrix of a grounded person will have abundant electrons to neutralize those leaking free radicals and thereby prevent "collateral damage" to healthy tissues.' Here, Oschman believes that the healthy connective-tissue matrix should automatically have abundant electrons and thus prevent damage to healthy tissues. In TCM, maintaining biological and biochemical order throughout the body is a function of the Triple-Energizer Metasystem, which

I believe is intimately connected with the PVS/Connective-Tissue Metasystem, or connective-tissue matrix, as Oschman calls it.

In the final section above, Oschman again emphasizes, 'The connective tissue in the body that is free of inflammatory pockets will be more conductive to electrons (inflammatory barricades are poorly conductive) and therefore will be better able to neutralize free radicals quickly wherever they form from whatever cause.' This epitomizes the protective function of Wei Qi associated with the San Jiao. Remember that the San Jiao is omnipresent, ubiquitous throughout the body, that it is nothing but membranes, and that membranes cover like a bag every cell, structure, and organ within the body. Notice too that Oschman shows that when the connective-tissue matrix is grounded, inflammation is negated. I have grounding pads on my treatment tables so that every patient is grounded during the entire treatment. I trust that doing so undoubtedly benefits and soothes the connective-tissue matrix (as Oschman calls it) or the San Jiao or the Triple-Energizer Metasystem (as I call it). Thus, I believe that being grounded as much as possible during every day is one of the most powerful and inexpensive ways of harmonizing the Triple Burner and subsequently all the other organs because the San Jiao nourishes every cell, structure, and organ with the numerous fluids and especially with Yuan Qi.

That the many organs of the body can benefit from grounding can be seen from the reference above (210), where cardiologist Dr Sinatra reported, 'Patients who experience anxiety, emotional stress, panic, fear, and/or symptoms of autonomic dystonia, including headaches, cardiac palpitations, and dizziness, grounding could be a very realistic therapy.' Remarkably, Sinatra reports, 'These patients may see positive effects most likely within 20 to 30 minutes and in almost all cases in 40 minutes.' Further to that, Sinatra states, 'Negative emotions such as panic, depression, anxiety, and hostility have all demonstrated reduced HRV (heart rate variability).' This is thanks to grounding. My TCM training suggests strongly that in these subjects, the Heart, Liver, and Gallbladder Fire that is Rising is being 'descended' and sedated.

I regularly ground patients with similar conditions through Kidney 1 either with a needle or with a sticky, conductive earthing patch. Note that Kidney 1 is the only acupoint on the bottom of the foot and is the major acupoint that connects us to the earth—*but* only when we are barefoot. And what are the properties of Kidney 1? It is used for 'grounding an individual'. Fancy that! It helps to calm the mind and relax our patients. The acupoint is used to treat dizziness, palpitations, nausea, headaches and migraines, tinnitus, insomnia, poor vision, nosebleeds, hypertension, chronic sore throat, dry mouth, hot flushes, night sweats, etc. In all these conditions, Kidney 1 'descends excess from above'. This is why the Chinese name of Kidney 1 (Yongquan) has the most appropriate name which in English means 'Gushing Spring'. When our bare feet touch the surface of the earth, a gushing spring of multitudinous free electrons are pushed upward from the earth and gush into our body through the 'hole' called Kidney 1. These beneficial free electrons then sedate and neutralize the offending free radicals that caused the pathological disorder in the first place.

CHAPTER 47

The Many Forms of Qi Associated with the Triple Energizer

Introduction
There are numerous forms of Qi and fluids generated and circulated throughout every part of the body by the ubiquitous Triple-Energizer Metasystem. Rather than reinvent wheels regarding the description of the various forms of Qi that empower the body, I searched the Internet for authoritative articles that accurately described the various forms of Qi and fluids and their functions in detail. The following articles in their entirety appear numerous times throughout cyberspace, and I could not determine the original eloquent author(s). Subsequently, I chose to cite the SacredLotus.com website articles (212, 213) below to describe the various constituents produced by the San Jiao.

47.1 Prenatal Jing (Essence) (Preheaven Essence or Congenital Jing)
Jing (translated as Essence) is an essential and precious substance in Chinese Medicine and should be protected and not squandered or misused. The following information is derived from the SacredLotus.com website article titled 'Jing (Essence)—Vital Substances in Chinese Medicine' (213). At conception, the Prenatal Jing is transferred from each of the parents to the embryo. Prenatal Jing, in harmony with energy derived from the Kidneys of the mother, continues to sustain and nourish fetal development during pregnancy.

The transferred Prenatal Jing controls and defines the constitution, strength, and vitality of the newborn. Prenatal Jing is a quantified absolute that is entrenched at birth and is unable to be added to. This Prenatal Jing is stored in the Kidneys. Prenatal Jing can be conserved and protected by conservative and balanced lifestyle measures. For example, moderation in diet, avoiding excesses, avoiding low-quality junk foods, not overworking, and having sufficient rest and sleep all help to conserve Jing (Essence). Excessive sexual activity and 'maladjusted affairs of the bedroom' should be avoided—this one always made classmates smile during lectures as lecturers evaded supplying a definition of 'maladjusted affairs of the bedroom'. Specific breathing exercises, Tai Chi, and Qi Gong all help to conserve and fortify Prenatal Jing integrity.

47.2 Postnatal Jing (Postheaven Essence or Acquired Jing)
After parturition, the neonate must initially take its first independent breath. Subsequently, the child must continue eating, drinking, and breathing independently to survive. The article (213) states 'Its Lungs, Spleen and Stomach then begin functioning to extract and refine Qi from the food and drink it consumes and the air it breaths. Postnatal Jing is the complex of essences thus refined and extracted. Postnatal Jing depends on the functions of Stomach and Spleen.' Subsequently, acquired jing is often called the jing of the organs. Nutrients derived from fluids and foods become nutrient essence or acquired jing in the Spleen. This nutritional substrate is circulated to nourish organs and allow them to perform their functions. Any overflow of Postnatal Jing is stored in the Kidneys and is on standby for organ requirements as needed. So the standby Postnatal Jing is constantly being used and then replenished by fresh Postnatal Jing manufactured after each meal. If the Prenatal Jing is weak or faulty, then organ dysfunction will occur, and poor quality Postnatal Jing will result such that nutrient malabsorption and a weakened constitution will result.

47.3 The Nature of Kidney Jing
The vitality of Kidney Jing is essential to ensure homeostatic biochemical and physiological processes throughout the entire

body. Like Prenatal Jing, Kidney Jing is hereditary and defines and controls constitutional aspects of the individual. Kidney Jing can be reinvigorated and partially replenished by the Postnatal Jing. Kidney Jing resides in the Kidneys but has a 'fluid nature and circulates all over the body, especially in the Eight Ancestral (Extraordinary) Vessels' (213).

47.4 Kidney Jing Governs Growth, Development, Sexual maturation, and Reproduction

Fluid-like Kidney Jing travels in extended, slow cycles and supervises the major stages of development in life. Initially during childhood, Kidney Jing is responsible for controlling the development of bones, teeth, hair; brain enlargement; and sexual maturation at puberty. When Kidney Jing is frail, poor bone growth occurs, and weak chalky teeth develop. There may be underdeveloped growth and mental retardation (213). Kidney Jing controls the seven-year cycles of females and the eight-year cycles of males. The article (213) advises that Kidney Jing supervises fertility and normal reproductive development into adulthood. Kidney Jing controls the processes of conception and the developmental stages of pregnancy. Feeble Kidney Jing results in gynecological weakness—for example, amenorrhea and infertility and recurring miscarriages. With advancing age, Kidney Jing naturally declines, and the many signs of aging result, including loss of hair, weakening of teeth, impairment of memory, and general frailty.

47.5 Kidney Jing Is the Basis for Kidney Qi

Jing is fluid-like and therefore is more Yin in nature. Kidney Jing can be considered as an aspect of Kidney Yin. It is due to the heating and invigorating action of Kidney Yang that Kidney Jing engenders the derivation of Kidney Qi (213). The article notes that using Kidney Jing as a catalyst, Kidney Yang warms Kidney Yin to form Kidney Qi. Subsequently, if catalytic Kidney Jing is deficient, the following medical conditions can result: chronic weak or sore lower back, weak and unstable knees, loose teeth, urinary incontinence, tinnitus and/or deafness, impotence, infertility, etc. All these conditions are potentially signs of weak Kidney Qi and/or weak Kidney Yang.

47.6 Kidney Jing Produces 'Marrow'

Interestingly, the marrow of TCM has no exact counterpart in Western allopathic Medicine. Kidney Jing produces Marrow, and the Marrow produces bone marrow, the brain, and the spinal cord. In TCM, the Brain is called the Sea of Marrow. Subsequently, if the caliber of Kidney Jing is weak, the brain will be undernourished, which will result in poor memory, poor concentration, a fuzzy or cotton wool or empty feeling in the head, dizziness, etc.

47.7 Kidney Jing Determines Our Vitality, Integrity, and Constitution

The strength and integrity of our Defensive (Wei) Qi determines our immunological strength and resistance to attack from external pathogens. Nevertheless, the vitality of Kidney Essence also impacts our strength and resistance. If Kidney Essence is weak or squandered, the person may have poor immunity to exogenous pathogenic influences and may be chronically ill with cold, influenza, allergies, and other debilitating medical conditions.

47.8 The Yuan (Source) Point of an Ailing Organ Is Used to Treat Pathology of That Specific Organ

In the commentaries on the 66th Difficult Issue on pages 568–569 of Unschuld's (1) translation of the *Nan Ching*, regarding the yuan points (origin or source points) on the 12 meridians, scholar Hua Shou states:

> 'The rapids [holes] where the Triple Burner passes [its influences] are the origin [holes]' means the following. The moving influences below the navel and between the kidneys constitute man's life; they are the source of the twelve conduits. Consequently, the Triple Burner is the special envoy [transmitting] the original influences. It is responsible for the passage of the upper, central, and lower influences through the body's five depots and six palaces. '[The Triple Burner is responsible for] the passage of the three influences' means, according to Mr. Chi, [the following]: The lower [section of the Triple] Burner is

endowed with the true primordial influences; these are the original influences. They move upward and reach the central [section of the Triple] Burner. The central [section of the Triple] Burner receives the essential but unrefined influences of water and grains and transforms them into constructive and protective [influences]. The constructive and protective influences proceed upward together with the true primordial influences and reach the upper [section of the Triple] Burner. 'Origin' represents an honorable designation for the Triple Burner, and all the locations where [its influences] stop represent origin [holes] because [the movement of the influences of the Triple Burner] resembles the arrival of the imperial herald, announcing the places where [the Emperor] will pass by and rest. When any of the five depots or six palaces has an illness, it is always appropriate to remove it from these [holes].

The Yuan Qi is circulated through the meridians by the Triple Burner and emerges in a larger amount at each of the Yuan (Source) points. Because of the dynamic nature of Yuan Qi, if any organ is suffering from a pathology, needling the Yuan point of that ailing organ will tonify and ameliorate the pathology of that specific organ.

47.9 Yuan Qi Is Jing in Motion

In the SacredLotus.com website article titled 'Forms of Qi—Vital Substances in Chinese Medicine' (212), regarding the primal nature of Yuan Qi, it is stated, 'Yuan Qi is said to be Essence that has been transformed into Qi, or Jing in motion. Yuan Qi has its root in the Kidneys and is spread throughout the body by the San Jiao (Triple Burner). It is the foundation of all the Yin and Yang energies of the body. Yuan Qi, like Prenatal Jing, is hereditary, fixed in quantity, but nourished by Postnatal Jing.'

47.10 The Functions of Yuan Qi within the Body

Regarding the functions of Yuan Qi, the article titled 'Forms of Qi—Vital Substances in Chinese Medicine' (212) explains that Yuan Qi is the basis of vitality and is responsible for motivating the intrinsic properties of the internal organs. It perfuses the entire body inside the channels via the San Jiao infrastructure. Yuan Qi is the foundation of Kidney Qi and occupies the region between the two Kidneys at the Gate of Vitality, also known as Ming Men. Yuan Qi enables and empowers Qi transformation; for example, Yuan Qi causes the transformation of Zong Qi into Zhen Qi and of Gu Qi into Blood. Yuan Qi surfaces and resides at the 12 Source points.

47.11 The Functions of Gu Qi (Food Qi) within the Body

Regarding the functions of Gu Qi within the body, the article titled 'Forms of Qi—Vital Substances in Chinese Medicine' (212) explains that when food enters the Stomach, it is initially 'rotted and ripened' there before being directed to the Spleen, where the incomplete intermediate Gu Qi is derived. Intermediate Gu Qi is directed upwards from the Middle Burner (housing the Stomach and Spleen) to the Upper Burner, which houses the Lungs and Heart. One portion of the Gu Qi reacts with air to generate Zong Qi, while the other portion of Gu Qi is also directed to the Lungs; from there it is transferred to the Heart, where both Yuan Qi and Kidney Qi will help transmute it into Blood.

47.12 The Functions of Zong Qi (Gathering Qi) within the Body

Regarding the functions of Zong Qi within the body, the article titled 'Forms of Qi—Vital Substances in Chinese Medicine' (212) explains that the Spleen directs Gu Qi upwards to the Lungs, where both Yuan Qi and Kidney Qi help the Gu Qi combine with air and generate Zong Qi. Zong Qi provides nourishment for the Heart and Lungs and bestows the instinctive utilities of heartbeat and respiration. Zong Qi or 'Gathering Qi' supports the respiratory function of the Lungs and allows the Lungs to control Qi. Because Zong Qi supports the Heart's role of governing the Blood and the Blood Vessels, the extremities, particularly the hands, will be feeble or cold whenever the Zong Qi level is low. Because Zong Qi accumulates in the throat and due to its Heart/Lung involvement, weak Zong Qi will result in speech disturbance and poor circulation to the hands due to the Heart involvement and a weak voice due to the Lung involvement. The fact that emotional problems readily disturb Zong Qi explains why grief adversely affects the Lungs and disperses the energy in the chest, causing the emptiness and fatigue associated with grief. Examples of how Yuan Qi and Zong Qi cooperate and mutually support each other include the role of Zong Qi flowing downward to aid the Kidneys, while Yuan Qi flows upward to support the respiratory function and the synthesis of Zong Qi.

47.13 The Functions of Zhen Qi (True Qi) within the Body

Regarding the functions of Zhen Qi within the body, the article titled 'Forms of Qi—Vital Substances in Chinese Medicine' (212) explains that Zhen Qi is also called True Qi or Normal Qi. Zhen Qi is derived from Zong Qi with the assistance of Yuan Qi. Zhen Qi is the culmination of the transmutation and refinement of Qi, and it is the Qi derivative that circulates in the channels throughout the body and nourishes all the organs. Zhen Qi circulates throughout the body in two distinct forms, Ying Qi and Wei Qi.

47.14 The Functions of Ying Qi (Nutritive Qi) within the Body

Regarding the functions of Ying Qi within the body, the article titled 'Forms of Qi—Vital Substances in Chinese Medicine' (212) explains that Ying Qi provides nourishment to the internal organs and to the whole body. Ying Qi is closely correlated to Blood and courses with the Blood inside the blood vessels as well as in the acupuncture channels. It is the Ying Qi that is motivated and mobilized when an acupuncture needle is inserted. Following the Chinese Clock, Ying Qi moves through all twelve channels over a twenty-four-hour period. It spends two hours in each channel, nourishing and maintaining the dedicated organ assigned that specific two-hour time period.

47.15 The Functions of Wei Qi (Protective Qi) within the Body

Regarding the functions of Wei Qi within the body, the article titled 'Forms of Qi—Vital Substances in Chinese Medicine' (212) advises that Wei Qi is more Yang than Nutritive Qi (Ying Qi) and that it is fast-moving, slippery, and is easily motivated. Wei Qi is located primarily on the exterior of the body in the skin and muscles. Wei Qi travels both inside and outside the channels. Protective Wei Qi flows primarily in the superficial layers of the body, especially in the Tendino-Muscular meridians. Wei Qi shields and protects the body from external pathogenic influences and attacks, including the adverse influences and stimuli from Wind, Cold, Heat, and Dampness. Protective Wei Qi nourishes, moisturizes, and warms the skin and muscles and controls the body temperature via sweating by regulating the opening and closing of the pores. While the Lungs regulate the circulation of Wei Qi to the skin, they also distribute fluids to moisten the skin and muscles. The function of perspiration is dependent on the capacity of the Lungs to circulate fluids and Wei Qi to the surface of the body. Subsequently, if there is a deficiency of Wei Qi, then the pores do not open and close correctly, and the fluids leak out, causing spontaneous sweating. When an external pathogen (for example Wind-Cold) invades the Exterior, the pathogen can obstruct the pores and inhibit the regulatory function of the Wei Qi and prevent the cooling function of sweating. In this case, the practitioner would have to restore the dispersing function of the Lungs and reinforce the Wei Qi to produce sweating and expel the pathogen. This protocol of sweating therapy (diaphoresis) is beneficial in the early stages of a Wind-Cold pathogenic invasion to eliminate the pathogen.

47.16 The Complex Circulation Pattern of Wei Qi (Protective Qi) within the Body

Regarding the complex circulation pattern of Wei Qi within the body, the article titled 'Forms of Qi—Vital Substances in Chinese Medicine' (212) explains that Wei Qi circulates through 50 cycles during a 24-hour period—25 times during daytime and 25 during the nighttime. During the daytime, the Wei Qi circulates throughout the Exterior part of the body, but during the nighttime, it relocates deeper throughout the Interior part of the body and perfuses the Yin Organs. The Wei Qi exteriorizes from midnight to noon and is at its maximum vitality on the Exterior at midday. Then from noon through until midnight, the Wei Qi gradually retracts to the Interior to afford protection of the Yin Organs. Because at nighttime the defensive Wei Qi is not on guard at the Exterior, exogenous pathogens can penetrate the body more readily at nighttime and inflict a blow. For example, sleeping under an open window at nighttime is more likely to result in a head cold than doing so during the daytime.

47.17 The Functions of Zhong Qi (Central Qi) within the Body

Regarding the functions of Zhong Qi within the body, the article titled 'Forms of Qi—Vital Substances in Chinese Medicine' (212) explains that Zhong Qi or Postnatal Essence is the Qi form that is derived from consumed food by the Stomach and Spleen and is associated with the potential of epigenetics. The vitality and integrity of the Postnatal Essence or Central Qi is an indicator of the Qi of the Middle Jiao, which embodies the Center. When the Central Qi is weak, the holding function of Spleen is compromised, and organ prolapses (hemorrhoids, prolapsed bowel, etc.) can occur.

47.18 The Functions of Zheng Qi (Upright Qi) within the Body

Regarding the functions of Zheng Qi within the body, the article titled 'Forms of Qi—Vital Substances in Chinese Medicine' (212) states:

- A general term to describe the various forms of Qi that protect the body from exogenous pathogens.
- Usually only used when contrasting the strength of the body's Qi with the strength of the invading pathogen.

47.19 The Nature and Circulation of 'Jin' throughout the Body

The article titled 'Body Fluids (Jin Ye)—Vital Substances in Chinese Medicine' (214) states that Jin 'fluids are clear, light, thin and watery, and circulate in the exterior of the body (skin and muscles) with the Wei Qi'. The Lungs in the Upper Burner control the transformation and movement of the Jin fluids throughout the body to moisturize and partly nourish the skin and muscles. Jin fluid is exuded as sweat but is also used to produce tears and saliva and is similarly used in the manufacture of Blood to keep it thin and flowing so that stasis and coagulation stuckness does not occur.

47.20 The Nature and Circulation of 'Ye' throughout the Body

Ye fluids are body fluids that are more turbid, dense, and heavy than Jin fluids, and they circulate throughout the interior of the body with the Ying (Nutritive) Qi. Ye moves relatively more slowly than Jin fluids. Ye fluids are transformed by and under the control of the Spleen and Kidneys and are circulated and excreted by the Middle and Lower Burners. The function of Ye fluids is to moisten the joints, spine, brain, and the bone marrow. Ye is also used to lubricate the openings of the sensory organs, including the nose, ears, mouth, and eyes (214).

47.21 The Origin of Jin Ye (Body Fluids)

Jin Ye (Body Fluids) are derived from the processing of consumed food and drink, which enter the body via the Stomach, which is also called the origin of fluids. The SacredLotus.com article (214) states, 'The Fluids are transformed and separated into "pure" and "impure" (turbid) parts several times.' This occurs via a complex sequence of purification procedures. The Pure body fluids are directed upward to the Lungs for distribution to the exterior of the body. Impure fluids flow downward and are finally excreted via the bladder.

47.21.1 Involvement of Stomach with Jin Ye

The Stomach is called the origin of fluids. Consumed fluids and foods first enter the Stomach, where they are transformed and separated into pure and impure fluids. From the Stomach, the 'pure' portion is sent to Spleen, while the 'impure' portion is sent to the Small Intestine to undergo further separation. The Stomach is said to be the source of body fluids because it absorbs water and complex fluids from drinks consumed plus the often large volume of inherent fluids present in foods—e.g. many salad vegetables contain in excess of 90% water.

47.21.2 Involvement of Spleen with Jin Ye

The Spleen is vitally important with regards to the physiology and pathology of Body Fluids. The Spleen manages the flow of Body

Fluids, directing pure portions upward and impure portions downward during all stages of the transformation process. Subsequently, the Spleen is treated for all pathologies involving Body Fluids (214).

47.21.3 Involvement of the Lungs with Jin Ye
The Lungs are responsible for the dispersion of the pure portion of Body Fluids coming from the Spleen. The Lungs direct these pure fluids to the space under the skin. In TCM, the Lungs are responsible for regulating the water passages of the body and as such direct a portion of the fluids down to the Kidneys and Bladder. (214)

47.21.4 Involvement of the Kidneys with Jin Ye
The Kidneys are extremely important in the manufacturing and physiology of Blood and Body Fluids. Kidney Fire evaporates some of the fluids they receive and sends them upwards to the Lungs so that the Lungs are moisturized. The SacredLotus.com article (214) reports that Kidney Yang controls many stages of the transformation of fluids within the body, including (1) providing heat so that the Spleen can metamorphose Body Fluids, (2) helping the Small Intestine to separate the pure and impure Body Fluids, (3) providing Qi to the Bladder to allow for appropriate Qi transformation to occur, and (4) assisting the Triple Burner (San Jiao) with its role of transformation and excretion of fluids throughout the entire body.

47.21.5 Involvement of the Bladder with Jin Ye
The Bladder allows fluids to be separated into pure and impure and stores and excretes impure urine with the help of Kidney Yang.

47.21.6 Involvement of the San Jiao (Triple Burner) with Jin Ye
The San Jiao assists with the transformation, transportation, and excretion of various types of fluids throughout the body.

1. Upper Burner is compared to a mist, whereby the Lungs disperse fluids to the space under the skin.

2. Middle Burner is like a muddy pool (or a foam) because the Stomach churns foods and fluids and directs the impure portion to the Small Intestine and the pure portion to the Spleen.
3. Lower Burner is compared to a drainage ditch or swamp because the Small Intestine separates the pure from the impure, while the Bladder and Kidneys transform, separate, and excrete fluids.

47.22 Relationship between Qi and the Jin Ye within the Body

Qi causes the transformation and transportation of fluids throughout the body, or otherwise, the fluids would pool and accumulate. Qi also holds the body fluids in place as it does with the blood. Should the Qi become deficient, then the body fluids can lose their containment and leak out. The SacredLotus.com article (214) suggests the following examples: Spleen Qi Deficiency causing chronic vaginal discharge, Kidney Yang Deficiency producing enuresis, and Lung Qi Deficiency triggering episodes of spontaneous sweating. Conversely, Qi integrity is also dependent on the presence of Fluids, and a Fluid deficiency can precipitate a Qi deficiency. Examples include a loss of Defensive (Wei) Qi concurrent with the Fluid loss in the case of excessive sweating. Similarly, excessive vomiting (e.g. hyperemesis gravidarum during pregnancy) depletes Qi.

47.23 The Relationship between Blood and the Jin Ye within the Body

Regarding the mutual nourishment between Blood and the Jin Ye, the SacredLotus.com article (214) notes that Body Fluids are used to continuously build Blood and keep it thin and fluid so that it does not coagulate. When Body Fluids are harmonious, they are transformed into red Blood. Conversely, Blood also nurtures and promotes the formation of Body Fluids. Subsequently, a loss of Body Fluids (e.g. excessive perspiration) can cause the condition of Deficient Blood. On the other hand, chronic Blood loss can engender a loss of Body Fluids with resultant Dry conditions. Noting well that Blood and Body Fluids have the same origin and mutually enrich and nourish each other, the article (214) concludes that with these principles in mind, never induce sweating when a patient is bleeding or if a Blood Deficiency exists. Likewise, never perform bloodletting while a patient is sweating.

47.24 The Properties of and Relationship between Yuan Qi and Zheng Qi in the Body

Regarding the properties of and the relationship between Yuan Qi and Zheng Qi throughout the body, in the article titled 'The Location and Function of the Sanjiao', the authors, Qu Lifang and Mary Garvey (43), explain that because Yuan Qi originates from the gate of life (ming men), it is the principal Qi responsible for ensuring a healthy and vital existence. The Yuan Qi is circulated throughout the body via the Sanjiao's infrastructure and bathes the Zangfu to ensure they optimally perform their physiological roles, and the Yuan Qi also perfuses the couli spaces and cavities and muscles to ensure that Pathogenic Qi is vanquished wherever it enters. In situations where serious injury occurs from Pathogenic Qi and the Zheng Qi (antipathogenic qi or upright qi) is overwhelmed, Yuan Qi is summoned to assist with the attack. Subsequently, Yuan Qi may be injured and compromised as a result of serious illness. Thus, Yuan Qi constitutes an important component of Zheng Qi.

47.25 The Properties of Zhen Qi within the Body

Regarding the properties of Zhen Qi within the body, in the article titled 'The Location and Function of the Sanjiao', the authors, Qu Lifang and Mary Garvey (43), describe how Zhen Qi is distributed predominantly inside the established channels and how it also flows inside the internal branches. Zhen Qi is responsible for the communication of physiological information between the various zang or fu organs to ensure that they maintain optimal functionality. If the Qi vitality of any one of the zangfu is weakened, the Zhen Qi within that respective zangfu channel will also be compromised and weakened, and consequently, Pathogenic Qi is able to enter into the weakened channel and penetrate more readily and directly attack the respective zang or fu organ.

47.26 The Nature of Ying Qi

Regarding the nature of the Ying Qi of TCM, the author of the book *Gua Sha: A Traditional Technique for Modern Practice*, Arya Nielsen (20), on page 34, states:

> *Like connective tissue, it is said there is nothing the San Jiao does not envelope*, including the vessels that hold the Blood and conduct the Ying Qi. Ying is the nourishing, constructive aspect. Ying flows in the blood vessels and channels, Wei Qi flows outside the channels. Ying suffuses the entire body through the vascular system and the meridian system. According to Ross (1985) Ying and blood are often synonymous. The Blood carries Ying, but Ying is not contained only in Blood. Ying Qi is the Qi activated when a needle is inserted in an acupuncture point (Maciocia 1989). (Emphasis is mine)

Remember that as blood courses through the capillaries, it diffuses into the interstitial spaces, whereupon it becomes lymphatic fluid, so these two fluids have a common origin. Nielsen states above that 'Ying suffuses the entire body through the vascular system and the meridian system'. Recent research has confirmed this possibility, showing that the Primo Vascular System (PVS) ducts flow both inside and outside blood vessels and lymph vessels. It makes perfect sense that these thread-like PVS channels will be determined with future research to actually flow within the acupuncture meridians, which I believe are hollow connective tissues probably predominantly composed of collagen. I love the fact that the citation above by Nielsen states, 'Like connective tissue, it is said there is nothing the San Jiao does not envelope'.

Summary of Chapter 47

The nature of this Chapter is such that I feel that no Summary would suffice as the practical information involves every section of the Chapter.

CHAPTER 48

Practical Application of Acupuncture Points to Treat Problems Associated with the Triple Energizer

Introduction
The San Jiao is known within TCM literature to be the major organ in charge of controlling the correct flow of Qi, Blood, and diverse body fluids throughout the entire body. In this section, I will discuss the practical application of using the acupuncture points of the Triple Burner and how to optimize treatment outcomes by means of acupuncture point combinations as outlined by very authoritative sources.

48.1 The Control of Qi and Blood throughout the Body
Regarding the internal/external relationship of the Pericardium and the San Jiao, in the 1990 article titled 'The Use of the Sanjiao Theory in Clinical Acupuncture' (215), Lao explains that the Sanjiao controls Qi, and the Pericardium controls blood throughout the body. The acupoint *Waiguan* SJ-5 connects the San Jiao channel with the Pericardium channel via the *Yangwei* extraordinary channel, and the acupoint *Neiguan* P-6 connects the Pericardium channel with the San Jiao channel via the Yin Wei extraordinary channel. Communication between the inner and the outer occurs via these two points. This characteristic exemplifies the TCM concept of the interdependence

of Yin and Yang and Qi and Blood within the body. Lao stresses the importance of the theory of the San Jiao as its machinations explain the intricacies regarding the physiological phenomena as per the TCM paradigm. While the San Jiao channel itself is important in disease treatment protocol, the San Jiao also supplies Yuan Qi at the 12 Yuan points of the 12 main channels.

48.2 The Clinical Application of the Yuan Points in Acupuncture

Regarding the clinical application of the Yuan points in acupuncture, Lao (215) explains that it is at the Yuan points present on each of the 12 Zangfu channels, where the Yuan Qi externalizes and lingers. The Yuan acupoints regulate and tonify the Yuan Qi, and fortify the Zheng Qi and disperse exterior pathogenic factors (Xie). They are the main points to remedy diseases present in the Zangfu and invigorate and strengthen the body.

48.3 Practical Application of Some Yuan Points in Acupuncture Treatments

As a practical treatment example, the brilliant article by Lao (215) advises that Professor Yang often combines the three Yuan points (*Taiyuan* LU-9, *Taixi* KID-3, and *Taibai* SP-3) to treat cases of deficiency of Lung, Kidney, and Spleen respectively, where the San Jiao fails to drain fluids and water and damp overflow and where symptoms of unregulated urination and edema of the limbs appear. The article (215) eloquently explains that Yuan acupoints motivate the driving force of the Yuan Qi and subsequently calm the flow of Qi within the organs and ensure the smooth progression of Blood and Qi inside the channels. Yuan points rectify perverse conditions caused by Qi obstruction in the organs, and Yuan points can reduce pain by amending the rebellious and disharmonious flow of Qi and blood. Further examples of using Yuan points in the article include the use of Taiyuan LU-9 to treat rebellious Qi in chest Bi and distended pain in the breasts, and it is advantageous for old age. *Daling* P-7 is significant in its treatment of shortness of breath and pain in the flanks and chest and pronounced hiccups. *Qiuxu* GB-40 is beneficial to treat aches and pains in the flanks and in the chest.

48.4 Important Combinations of Yuan Points with Other Special Points

48.4.1 Combination of Yuan Points and Luo Points

Regarding a dynamic combination to use in the clinical application of the Yuan points in acupuncture, in the 1990 article titled 'The Use of the Sanjiao Theory in Clinical Acupuncture', Lao (215) explains that when there is a disease of one specific channel and simultaneously there are also symptoms of its exteriorly/interiorly related channel, the main treatment point chosen would be the Yuan point of the first channel, and the guest point would be the Luo point of the exteriorly / interiorly related channel. As an example of this combination, let's assume that the predominant problem is a disorder of the Gallbladder channel, with pain in the flanks and chest and a bitter taste and dryness in the mouth. Simultaneously, symptoms of the Liver channel, such as dizziness and blurred vision, occur. In this example, the main point would be the Gallbladder Yuan point Qiuxu GB-40, and the Liver Luo point *Ligou* LIV-5 would be used as the guest point.

The author of the article (215) also explains how Professor Yang treats cases of chronic illnesses causing deficiency and cases of chronic illnesses where the pathogen entered the Luo via the Jing. The author advises to efficiently treat a chronically diseased channel, to choose the Yuan point of that channel to tonify the deficiency, and to restore the Zheng Qi. Professor Yang will fortify this treatment by using the Luo point of the same channel to open and penetrate to the Luo channel. The cited example was for treating chronic cough. The author advises to needle the Yuan point Taiyuan LU-9 in the direction of the Luo connecting point *Lieque* LU-7. To treat chronic palpitation and pain in the chest, one would needle the Yuan point Daling P-7 in the direction of the Luo point Neiguan P-6.

48.4.2 Combination of Yuan Points and Back-Shu Points

In this combination, the Yuan point of a diseased organ is combined with the Back-Shu point of the same organ. The article (215) states this strengthens the organ because the Yuan point reinforces and tonifies the Yuan Qi of the diseased organ, while the Back-Shu point also tonifies deficiency and invigorates the Zheng Qi of the organ. The article continues:

> This combination is mutually promoting, and is especially suitable in cases of cold and Xu in the organs. Professor Yang commonly uses Taiyuan LU-9 in combination with Feishu BL-13, the Lung Back-Shu point to treat insufficiency of the Lung with symptoms such as cough and shortness of breath, or Taibai SP-3 in combination with Pishu BL-20 and Weishu BL-21, the Spleen and Stomach Back-Shu points, to treat deficiency of Spleen and Stomach with poor appetite and listless limbs. When Xu and cold are very accentuated, moxibustion is applied.

48.4.3 Combination of Yuan Points and He-Sea Points

Lao (215) differentiates this principle into three subcategories, where the combination of Yuan and He-Sea, or lower He-Sea points, are varied:

> 48.4.3A—the exteriorly / interiorly related channels,
> 48.4.3B—the same channel, or
> 48.4.3C—of different channels.

48.4.3A Combination of Points on Exteriorly/Interiorly Related Channels

In this combination of points of exteriorly/interiorly related channels, the Yuan point of the interior channel is combined with the He-Sea or Lower He-Sea point of the exterior channel to positively influence a disease of both the exterior and interior channel. An example of this combination is needling the Yuan point Taibai SP-3 combined with the He-Sea point of the Stomach channel, Zusanli ST-36. This combination 'strengthens the Spleen and harmonizes the Stomach, and ascends the clear and descends the turbid'. This treatment could be applied when Stomach and Spleen are in disharmony, with symptoms such as distended abdomen, no appetite, vomiting, and diarrhoea. Another example would be 'Liver / Gallbladder fire rising, causing pain and fullness in the chest and flanks, bitter taste in the mouth and blurred vision'. Here, the Yuan point *Taichong* LIV-3 would be used in combination with the He-Sea point of the Gallbladder channel, *Yanglingquan* GB-34. This combination would dredge the Liver and regulate the Gallbladder.

48.4.3B Combination of Yuan and He-Sea Points of the Same Channel

Lao (215) explained that a frequently used example of the grouping of Yuan point and He-Sea point on the same channel would be the combination of the Yuan point *Hegu* LI-4 with the He-Sea point *Quchi* LI-11 in an effort to clear and regulate the upper Jiao, e.g. for sinus problems.

48.4.3C Combination of Yuan and Lower He-Sea Points of Different Channels

Lao (215) further explains that the grouping of the Yuan point and the He-Sea point of different channels could, for example, include the use of Hegu LI-4 and Shangjuxu ST-37, the lower He-Sea point of the Large Intestine channel, to remove heat and open the intestines. Another example cited involves the use of Taixi KID-3 combined with Weiyang BL-39, the lower He-Sea point of the Sanjiao channel, to tonify the Kidneys and normalize the San Jiao.

48.5 Special Notes Regarding the Clinical Use of the Points of the San Jiao Channel

The San Jiao organ complex is Yang in nature and is one of the six Fu. Subsequently, the main indications for use of points on the San Jiao channel are determined by their being points on a Yang channel. San Jiao points mostly treat disorders on the exterior of the body and conditions of Shi and heat. However, the San Jiao channel also pertains to the Kidney and allows the distribution of Yuan Qi; therefore, its points do have special properties.

48.6 The Use of San Jiao Points for the Treatment of Auricular Disease

Regarding the use of San Jiao points for the treatment of auricular diseases, in the article, Lao (215) explains that to treat auricular diseases irrespective of the aetiology, Professor Yang selects the main points from the San Jiao channel. Lao notes that because Jing-Well points are used to disperse wind and clear heat, bleeding *Guanchong* SJ-1 removes wind-heat conditions of the ear. Similarly, because Ying-Spring points are especially beneficial for clearing heat, *Yemen* SJ-2 is beneficial for treating ear conditions due to flaring heat. As the Yuan point has the property of tonifying and strengthening the Yuan Qi, the Yuan point *Yangchi* SJ-4 is very beneficial for ear conditions due to Kidney deficiency. Further to the preceding acupoints, Lao noted that Waiguan SJ-5 is the major acupoint for remedying any form of auricular disease.

48.7 The Use of San Jiao Points for the Treatment of Qi-Related Disorders

In the article, Lao (215) notes that all the acupoints along the San Jiao channel tonify the Yuan Qi within the San Jiao and can harmonize and smooth the flow of Qi and unblock and dredge the channels. Subsequently, San Jiao acupoints are especially indicated in conditions where there is irregular movement of Qi and Blood or where the correct flow of Qi is in rebellion. The article further notes that San Jiao channel acupoints are excellent for treating medical conditions that occur proximal to the channel. Examples include pain in the little and ring finger, elbow, upper arm, shoulder, neck, behind the ear, the cheek, and at the outer corner of the eye. San Jiao points also regulate and build up the flow of Qi and blood throughout the entire body. The article further states:

> Guanchong SJ-1 treats acute gastroenteritis, Tianjing SJ-10 treats chest Bi and pain in the Heart. According to the *Zhenjiu Da Cheng* Waiguan SJ-5 treats 'pain and numbness in the 5 fingers, when they can't move things', Zhigou SJ-6 treats 'listless limbs, gastroenteritis and vomiting,' and Tianjing SJ-10 treats 'injuries to the lower back and hip pain'. Today's clinical practice uses Tianjing SJ-10's functions of regulating Qi and opening the chest in conditions of oppressed and painful chest, applies Naohui SJ-13 to regulate Qi and dispel phlegm, and Zhigou SJ-6 to regulate Qi and penetrate the Fu in constipation.

48.8 The Use of the Luo Point of the San Jiao to Treat Alternating Cold and Heat

In the article, Lao (215) notes that acupoints along the San Jiao and the Gallbladder channels are used to eliminate pathogenic factors within the Shao Yang and to remedy conditions of alternating cold and heat and, in extreme cases, malaria, which is vexing the Shao Yang. Because Waiguan SJ-5 is the Luo acupoint on the San Jiao channel, the Confluent (Meeting) point of the Yangwei channel, and also the meeting point of the *Shaoyang* and *Jueyin* channels of the arm, Waiguan SJ-5 is the primary point of the San Jiao channel, dredges and clears the channels, regulates Qi and moves blood, and penetrates and regulates the San Jiao.

48.9 Rules for Treating Conditions of the San Jiao

As the San Jiao is the root of the Yuan Qi, the main emphasis when treating the San Jiao must be directed to adjusting the Yuan Qi throughout the body. In the article, Lao (215) notes that Professor Yang employs both distal and local points, differentiating his treatment protocols according to the upper, middle, and lower Jiao. The local points treated are predominantly the acupoints of the *Renmai* (Conception) channel; for example, *Qihai* REN-6 is used for disorders of the lower Jiao, Zhongwan REN-12 is chosen to treat disorders of the middle Jiao, and *Shanzhong* REN-17 is selected for conditions of the upper Jiao. The article continues, 'Distally used points are the Yuan points according to the involved organs, for example for afflictions of the upper Jiao, Daling P-7 and Taiyuan LU-9 are used, for the middle Jiao Taibai SP-3 is used, and for the lower Jiao Taixi KID-3 or Taichong LIV-3 are used.' Lao points out that because 'the Renmai channel basically relates to the Kidney, and penetrates the San Jiao, therefore its points can influence the San Jiao, and the Yuan points regulate the Yuan Qi. When the two of them are combined, they mutually promote each other's actions of dispersing and tonifying'. Lao further noted that, in the case of a Xu condition, adding the relevant Back-Shu points would potentiate their tonifying action. Conversely, where a Shi condition exists, the inclusion of Front-Mu points would potentiate the effect of expelling the pathogenic influence.

48.10 Relaxing the 'Membranes' and Regulating the Cou Li Space According to Maciocia

In his article 'The Triple Burner as a System of Cavities and a Three-Fold Division of the Body', Maciocia (18) states:

> To regulate the Triple Burner in the abdomen and relax the Membranes, one can use Ren-5 Shimen (Front-Mu point of the Triple Burner), Ren-6 Yuan point of the Membranes (Huang) and BL-22 Sanjiaoshu (Back-Shu point of the Triple Burner).
>
> To regulate the Cou Li space one needs to regulate the Triple Burner and the Wei Qi with points such as LU-7 Lieque, LU-9 Taiyuan, L.I.-4 Hegu, ST-36 Zusanli and BL-13 Feishu. For example, to consolidate the Cou Li space one can use LU-9 Taiyuan, L.I.-4 Hegu, BL-13 Feishu and ST-36 Zusanli. To 'relax' the Cou Li space, one can use LU-7 Lieque and L.I.-4 Hegu.

Note from the first paragraph above that the points Ren-5 and Ren-6 are the Mu points of the Triple Burner and Yuan point of Membranes respectively and are side by side. I find it intriguing that these two points are side by side and that, remarkably, Ren-6 is the only point on the Conception Vessel meridian that is only one half a cun away from its neighboring points Ren-5 and Ren-7, while all the other Conception Vessel meridian points are at one cun intervals. That makes Ren-5 and Ren-6 very close. I propose that this amazing closeness of points is because the Triple Burner is intimately related to and connected with the Huang membrane complex of the body and that these two points express the oneness of the Triple-Energizer Metasystem and the connective tissue/membrane metasystem, that I call the Connective-Tissue Metasystem. I subsequently suggest that the point combination of Ren-5 (Front-Mu point of the Triple Burner) and Ren-6 (Yuan point of the Membranes (Huang)) when combined with Triple Energizer 4, which is the Yuan-Source point of the San Jiao Meridian, is very powerful for boosting the San Jiao. Remember that Li Chiung stated, 'The Triple Burner represents nothing but membranes.'

48.11 Specifics to Think about When You See Problems in the TE Channel

In 2012, acupuncturist Kimberly Thompson (46) posted an article entitled 'What the Heck Is a Triple Energizer Anyway?' on the MiridiaTech.com website. In the article, she noted that all the many fluids associated with the Triple Energizer initially stemmed from and eventually returned to the Blood via the capillaries. Thompson suggested that the Triple Energizer can be used to treat medical conditions in four different categories. The Triple Energizer is involved in metabolic disorders due to the fluid interchange at the cellular level because the TE manage the removal and transportation of cellular waste products. The TE's portfolio includes digestion because it manages the fluids surrounding all the organs and it also regulates peristaltic movement throughout the entire alimentary canal. As hormones are circulated throughout the body via body fluids and the TE controls the flow and distribution of body fluids, the TE subsequently regulate the hormonal system of the body. The TE is also instrumental in the development of obesity as it regulates the membranous greater omentum, which forms a fatty apron that hangs over the lower abdomen; voluminous fatty deposits can be stored there, leading to obesity.

48.12 Clinical Application of Triple Burner Acupuncture Points by Giovanni Maciocia

Regarding the clinical application of Triple Burner acupuncture points, in the 2011 article titled 'The Triple Burner (2)', Giovanni Maciocia (10) advised that there are many acupuncture points that can be used to stimulate the Triple Burner's function of transformation and excretion of fluids. He listed them according to each of the three burners. Maciocia stated:

> Upper Burner: Ren-17 Shanzhong, LU-7 Lieque, L.I.-6 Pianli, L.I.-4 Hegu, Du-26 Renzhong (also called Shuigou, i.e. 'Water ditch').

> Middle Burner: Ren-12 Zhongwan, ST-21 Liangmen, Ren-9 Shuifen, Ren-11 Jianli, ST-22 Guanmen, BL-20 Pishu.

Lower Burner: Ren-5 Shimen, BL-22 Sanjiaoshu, ST-28 Shuidao, BL-23 Shenshu, Ren-6 Qihai, SP-9 Yinlingquan, SP-6 Sanyinjiao, KI-7 Fuliu, BL-39 Weiyang.

48.13 Treatment of Dampness, Phlegm, and Edema using Triple Burner Acupoints

Regarding the practical clinical application of Triple Burner acupuncture points, Maciocia (10) continued:

> When fluids stagnate, Dampness, Phlegm or oedema may arise. In order to activate the Triple Burner to move fluids, I activate each Burner using some of the points above, using more points from the Burner where the fluids stagnate. For example, if there is Dampness in the Lower Burner, I would use several points from those of the Lower Burner plus one or two from the Upper and Middle Burner. This usually means that when I stimulate the transformation, transportation and excretion of fluids, I do not hesitate in using more points than I would normally.
>
> For example, if there is Dampness in the Lower Burner causing a urinary problem I would use these points:
> —Lower Burner: Ren-3, Ren-5, BL-22, BL-28, SP-9, BL-39.
> —Middle Burner: Ren-9.
> —Upper Burner: LU-7.
>
> To give an example from the Middle Burner, if there was Phlegm in the Stomach, I would use these points:
> —Middle Burner: Ren-12, Ren-9, ST-21, BL-20, Ren-11.
> —Upper Burner: L.I.-4.
> —Lower Burner: ST-40, SP-9.

48.14 Interesting Relationship between the Shu and Mu Points of Triple Burner and Kidneys

Regarding the relationship between the Triple Burner and the Kidneys, in the 2010 article titled 'Triple Burner and Yuan Qi: Bladder 22 and Ren 5', Giovanni Maciocia (216) states, 'The location of the points BL-22 Sanjiaoshu above BL-23 Shenshu and of Ren-5 Shimen above Ren-4 Guanyuan is interesting and it presents interesting parallels.' He explains, 'This location can be understood only by reference to Chapter 66 of the *Classic of Difficulties* (*Nan Jing*).' Chapter 66 of this text states: 'The Original Qi is the Motive Force [Dong Qi] situated between the two kidneys, it is life-giving and it is the root of the 12 channels.' Maciocia further explains that the Triple Burner has the role of transforming the Original Qi into the numerous forms to be utilized for the many unique and different functions throughout the body. The Triple Burner manages the Original Qi (Yuan Qi) that arises from between the kidneys and governs its transformation and circulation to the five Yin and six Yang organs and their associated channels and the appropriate excretion of waste fluids. All this fluid transformation and transmutation of the Triple Burner throughout the entire body is dependent on cooperation from the heat of Kidney-Yang and the presence of the Yuan Qi.

48.15 BL-22 Powerfully Treats Pathological Dampness in the Lower Burner

Regarding the relationship between the Triple Burner and the Kidneys, Maciocia (216) eloquently explains that the Triple Burner Back-Shu Point, BL-22, is the cardinal point to ensure that the Water passages in the Lower Burner are open and the functions of transformation and transportation of fluids perform correctly and that dirty fluids are excreted. When fluid processing and waste excretion are correctly managed, Dampness does not accumulate in the Lower Burner. He points out that when fluid-handling pathology does occur, BL-22 (Sanjiaoshu) resolves Dampness and opens and clears the Water passages, especially in the Lower Burner. Subsequently, Bladder 22 is used to treat conditions related to fluid imbalance in the Lower Burner, including urinary retention, turbid urine, blood in the urine, difficult or painful urination, edema in the lower abdomen, and edema of the legs.

48.16 BL-22 Is Efficacious for Treating Shao Yang (Lesser Yang) Syndrome Symptoms

Because BL-22 is immediately above the Back-Transporting point of the Kidneys (BL-23), this is the location from where the Triple Burner assists the Original Qi to surface from the Kidneys and transfuse to the Internal Organs. Due to this connection with the Triple Burner, BL-22 is efficacious for treating Shao Yang (Lesser Yang) Syndrome symptoms, including blurred vision, dizziness, bitter taste in the mouth, digestive issues (nausea, vomiting, bloating, poor appetite, and stomachache), hypochondriac pain, headaches, irritability, insomnia, and alternating chills and fever.

48.17 Properties of the Front Mu Point of the Triple Burner

Regarding issues pertaining to fluid transformation of the Mu point of the Triple Burner, in the 2010 article titled 'Triple Burner and Yuan Qi: Bladder 22 and Ren 5', Maciocia (216) describes the properties of Ren-5 (*Shimen*), which means 'Stone Door'. He notes that Ren-5 opens the Water passages and promotes the transformation and excretion of fluids in the Lower Burner and strengthens Original Qi. He points out it is indicated for difficult urination, retention of urine, painful urination, dark urine, edema, diarrhea, genital itching, swelling of scrotum, swelling of the vulva, and swelling of the penis.

48.18 The Mu Point (Ren-5) Is Very Beneficial Treating Lower Burner Fluid Pathology

Regarding the function of Ren 5 in connection with the relationship between the Triple Burner and Original Qi (Yuan Qi), in the 2010 article titled 'Triple Burner and Yuan Qi: Bladder 22 and Ren 5', Maciocia (216) explains that because Original Qi (Yuan Qi) emerges from between the Kidneys and then the Triple Burner manages its dispersal to the five Yin and six Yang organs, the Front Collecting (Mu) point of the Triple Burner, Ren-5, ensures that the Yuan Qi flows to all the organs and circulates through all the channels. Consequently, Ren-5 is used to increase the available Original Qi in weakened individuals with a poor constitution due to

Kidney deficiency. Because the function of the Lower Burner is to transform fluids and excrete wastes, the Front Collecting (Mu) point of the Triple Burner, Ren-5, is very beneficial in treating diarrhoea, abdominal edema, difficult urination, urinary retention, and vaginal discharge.

48.19 Clinical Applications of Triple Burner/Pericardium in the Mental–Emotional Spheres

In the 2011 blog article entitled 'The Triple Burner: Relationship with Pericardium', Maciocia (217) believes there are practical clinical applications for the Triple Burner and Pericardium channels with regard to the mental–emotional sphere. Maciocia states, 'Emotional stress makes the physiological Minister Fire of the Kidneys rise and become pathological: when the Minister Fire is pathological, a person has "Heat". All emotions tend to cause Qi stagnation first and stagnant Qi easily generates Heat: that is why, in mental-emotional stress, the tip of the tongue (reflecting the Heart and the Shen) is red.' Maciocia states, 'The Triple Burner and Pericardium channels affect the mental-emotional state because emotional stress makes the Minister Fire [same as Ming Men] rise towards these two channels; therefore when the Minister Fire is aroused by emotional problems and it rises towards the Pericardium and Triple Burner channels, points of these channels can be used to clear Heat and calm the Mind.' In this situation, Maciocia uses P-6 Neiguan to encourage the movement of the Hun for depression or the P-7 Daling to tranquilize the Shen for anxiety. While the Triple Burner (Shao Yang) acts as the 'hinge' between Tai Yang and Yang Ming, the Pericardium (Jue Yin) acts as the hinge between Tai Yin and Shao Yin. With regards to the hinge property of the Triple Burner and Pericardium channels, he notes another thought-provoking clinical application. Maciocia states, 'Being the "hinge" on a psychological level means that these channels are "mediators" in the sense that they can affect a person's capacity to relate to other people and points of these two channels can therefore be used especially for depression.' He supplies the example of treating depression, using a combination of TB-3 Zhongzhu and P-6 Neiguan.

48.20 San Jiao Organ and San Jiao Channel Pathology

Regarding the pathologies of the San Jiao organ and the San Jiao channel, Yongping Jiang (8), the author of the *Journal of Chinese Medicine* article titled 'The San Jiao: Returning to the Nei Jing (A Modern Explanation of Original Theory)' states, 'San Jiao pathology in the *Nei Jing* is described in terms of two different disorders, organ and channel disorder, which are discussed as follows.'

48.20.1 San Jiao Organ Disorder

Disorders of the San Jiao organ manifest as dysfunction of the movement and metabolism of fluids. Chapter 4 of the *Ling Shu* describes how dysfunction of the San Jiao can lead to an accumulation of water, which manifests as abdominal fullness, difficult or urgent urination, and edema. It suggests using Weiyang BL-39 as the main point for such indications. Not only does Weiyang BL-39 serve as the lower He-sea point of the San Jiao, it is also a Urinary Bladder point. Because such an important point for treating San Jiao pathology is located on the Urinary Bladder channel, it further illustrates the point that the San Jiao 'belongs' to the Urinary Bladder system and emphasizes the idea in the *Nei Jing* that the San Jiao is highly related to the passage of water throughout the body.

48.20.2 San Jiao Channel Disorder

If the San Jiao channel is diseased, the resulting signs and symptoms will follow the trajectory of the primary and luo-connecting channels of the hand shaoyang (San Jiao). According to Chapter 10 of the *Ling Shu*, 'When the San Jiao hand shaoyang channel is disordered, the symptoms are deafness, sore and swollen throat, sweating, pain in the outer canthus and pain along the pathway of the channel, including the face, behind the ear, and down the outside of the shoulder, arm, elbow, forearm and ring finger.' Note that amongst these clinical manifestations of San Jiao channel disorder, no mention is made of problems involving fluid metabolism. Because San Jiao channel disorder does not involve fluids, there is therefore no need to use the lower He-sea point for the San Jiao. Treatment instead involves choosing points along the San Jiao channel. The same chapter of the

Nei Jing Ling Shu also states, 'If the hand shaoyang luo channel has excess, the symptoms are spasm of the arm and rigidity. If the luo channel is deficient the arm is loose and flaccid.' For either of these conditions, Waiguan SJ-5 is the point of choice for treatment, needled with reinforcing technique in cases of deficiency or reducing the point in cases of excess.

San Jiao channel and organ disorders therefore manifest in very different ways. To treat San Jiao organ disorder, we use the lower He-sea point of the San Jiao to directly affect water metabolism, whereas for the channel disorder, we use points along the San Jiao channel that have no function in terms of water metabolism.

48.21 The Location of Body Fluid Pathologies Elucidates the Elusive Morphology and Location of the San Jiao

Substantiation of the location, shape, and structure of the Sanjiao can be elicited from the clinical observation of various body fluid pathologies and their locations. In the *Journal of Chinese Medicine* article titled 'The Location and Function of the Sanjiao', the authors, Qu Lifang and Mary Garvey (43), explain that the Water circulating through the Sanjiao must be in the qihua condition and possess the integrity to ensure that the correct fluid transformations and metabolism occurs throughout the body. When the homeostatic qihua condition becomes dysfunctional, the intrinsic fluidity of Water is lost, and the perverse condition generates 'stuckness', and damp pathogenic conditions (including accumulations, stagnations, and edema) result. The resulting sticky dampness and phlegm coagulate within the Sanjiao's cavities and spaces and hinder or prevent the precise physiological processes of the Sanjiao from occurring. The authors note that Zhang's analysis and differentiation of such disrupted qihua conditions involving accumulation or stagnation of fluids include 'wind water' (which is comparable to acute glomerulonephritis in western medicine), *pi* water (which is related to common edematous conditions or chronic glomerulonephritis in western medicine), and *yiyin* (which is also a form of edema). They further note that *xuan yin* is produced by water accumulation in the pleural cavity and is equivalent to hydrothorax in western medicine, while *zheng shui* and *shi shui* are related to forms of ascites, where water accumulates in the abdominal cavity.

Various body fluid pathologies occur with fluids circulating throughout the San Jiao system when it is out of the qihua condition. Such a qihua condition occurs when transformational harmony is lost. What is important to note is that when the fluids have lost their correct qihua condition, the resultant stagnations and accumulations of perverse dampness, edema, and phlegm occur throughout the body at the locations associated with the fluid disharmony. Thus, the resultant problem could be edema in the feet, glomerulonephritis in the kidneys, ascites (water stagnation) in the abdominal cavity, or puffiness under the eyes. As these San Jiao fluid pathologies can occur from the head to the feet, this confirms that the physical morphology and location of the San Jiao occupies membranous tissue types also present from the head to the feet. The San Jiao is truly an omnipresent organ, and as stated by Li Chiung while discussing the 38th Difficult Issue in the *Nan Ching*, 'the Triple Burner represents nothing but membranes'.

Summary of Chapter 48
The nature of this Chapter is such that I feel that no Summary would suffice as the practical information involves every section of the Chapter.

Glossary of Chinese Medical Terms Used Throughout This Book

This Glossary is based on terminology recommended by the World Health Organization (WHO) (3) presented in 'WHO International Standard Terminologies on Traditional Medicine in the Western Pacific Region'. As there are 263 pages of general terminologies, please forgive me if I have missed any of them in the following Table. I have omitted the most basic terms (e.g. *qi, yin, yang, fire, wood, heart, liver, wind, cold*) for expediency.

Original Term	WHO Term	Chinese	WHO Definition/Description
traditional Chinese medicine	traditional Chinese medicine	中醫學; 中醫	the traditional medicine that originated in China, and is characterized by holism and treatment based on pattern identification/syndrome differentiation
five elements theory	five phase theory	五行學說	one of the philosophical theories of medical practice in ancient China, concerning the composition and evolution of the physical universe, epitomized by the nature and the inhibition-generation relationships of the five phases, wood, fire, earth, metal and water, serving as the guiding ideology and methodology of physiology, pathology, clinical diagnosis and treatment, also known as five elements theory

essence	essence	精	(1) the fundamental substance that builds up the physical structure and maintains body function; (2) reproductive essence stored in the kidney
yuán qi	innate essence	先天之精	the original substance responsible for construction of the body and generation of offspring, often referring to the reproductive essence, also called prenatal essence
postnatal qi or later heaven qi	acquired essence	後天之精	the essential substance acquired from the food after digestion and absorption, and used to maintain the vital activities and metabolism of the body, the same as postnatal essence
kidney essence	kidney essence	腎精	the original essence stored in the kidney
gate of life (ming men)	life gate	命門	(1) the place where qi transformation of the human body originates, serving as the root of life; (2) right kidney; (3) acupuncture point (GV4)
kidney yang	life gate fire	命門之火; 先天之火	innate fire from the life gate, a synonym of kidney yang
heart fire	sovereign fire	君火	another name for heart fire, in contrast to the ministerial fire
ministerial fire	ministerial fire	相火	a kind of physiological fire originating in the kidney and attached to the liver, gallbladder and triple energizer, which, in cooperation with the sovereign fire from the heart, warms the viscera and promotes their activities. If this fire is hyperactive, it is also harmful to the body
prenatal qi	innate qi	先天之氣	the qi that exists from birth and is stored in the kidney, also the same as prenatal qi
post-natal qi	acquired qi	後天之氣	the qi that is acquired after birth and is formed from the food in combination with the fresh air inhaled in the lung, also the same as post-natal qi

normal qi/ genuine qi	healthy qi	正氣	a collective designation for all normal functions of the human body and the abilities to maintain health, including the abilities of self-regulation, adaptation to the environment, resistance against pathogens and self-recovery from illness, the same as normal/genuine qi
true qi	genuine qi	眞氣	the combination of the innate qi and the acquired qi, serving as the physical substrata and dynamic force of all vital functions, also known as true qi
yuan/original/ primordial qi	source qi	原氣; 元氣	the combination of the innate qi and the acquired qi, serving as the most fundamental qi of the human body; the same as original/primordial qi
pectoral qi	ancestral qi	宗氣	the combination of the essential qi derived from food with the air inhaled, stored in the chest, and serving as the dynamic force of blood circulation, respiration, voice, and bodily movements, the same as pectoral qi
wei/defensive qi	defense qi	衛氣	the qi that moves outside the vessels, protecting the body surface and warding off external pathogens, the same as defensive qi
nutritive qi	nutrient qi	營氣	the qi that moves within the vessels and nourishes all the organs and tissues, the same as nutritive qi
kidney qi	kidney qi	腎氣	essential qi of the kidney, the physical substrata and dynamic force of the functional activities of the kidney
middle heater qi	middle qi	中氣	qi of the middle energizer, the physical substrata and dynamic force of the functional activities of the spleen, stomach and small intestine, including digestion, absorption, transportation, upbearing of the clear and downbearing of the turbid
collateral qi	meridian qi	經氣; 經絡之氣	the qi that flows through the meridians, the same as collateral qi

qi transformation	qi transformation	氣化	a general term referring to various changes through the activity of qi, namely the metabolism and mutual transformation between essence, qi, blood and fluids
qi dynamic or qi mechanism	qi movement	氣機	movement of qi, including ascending, descending, exiting and entering as its basic forms, also known as qi dynamic/qi mechanism
blood	blood	血	the red fluid circulating through the blood vessels, and nourishing and moistening the whole body
sweat	sweat	汗	the fluid that exudes from sweat glands; the humor of the heart
triple burners	triple energizers	三焦	a collective term for the three portions of the body cavity, through which the visceral qi is transformed, also widely known as triple burners
upper burner	upper energizer	上焦	the chest cavity, i.e., the portion above the diaphragm housing the heart and lung, also known as upper burner
middle burner	middle energizer	中焦	the upper abdominal cavity, i.e., the portion between the diaphragm and the umbilicus housing the spleen, stomach, liver and gallbladder, also known as middle burner
lower burner	lower energizer	下焦	the lower abdominal cavity, i.e., the portion below the umbilicus housing the kidneys, bladder, small and large intestines, also known as lower burner
extraordinary organs	extraordinary organs	奇恒之腑	a collective term for the brain, marrow, bones, blood vessels, gallbladder and uterus. They are called extraordinary because their morphological and physiological properties are different from the ordinary bowels and viscera
marrow	marrow	髓	an extraordinary organ including bone marrow and spinal marrow, both of which are nourished by the kidney essence
vessel	vessel	脈	the conduit through which qi and blood pass
blood vessel	blood vessel	血脈	the vessels in which blood circulates

separation of the clear and turbid	separation of the clear and turbid	泌別清濁	the small intestine's function, by which the clear (the food essence and water) is absorbed while the turbid (the waste matter) is passed to the large intestine
inborn	innate	先天	possessed from birth, relating to the natural endowment, in contrast to acquired after birth, the same as inborn
meridians or channels	meridian and collateral	經絡	a system of conduits through which qi and blood circulate, connecting the bowels, viscera, extremities, superficial organs and tissues, making the body an organic whole, the same as channels and networks; meridians or channels, in short
channel vessel	meridian vessel	經脈	the main pathways of qi and blood coursing vertically, composed of the twelve regular meridians and the eight extra meridians, the same as channel vessel
triple energizer meridian (TE)	triple energizer meridian (TE)	手少陽三焦經	one of the twelve regular meridians which runs from guanchong (TE1) at the ulnar side of the ring finger, travels along the midline of the posterior side of the arm and through the regions of the shoulder, neck, ear and eye, and terminates at sizhukong (TE23) at the lateral aspect of canthus. A branch is sent from the supraclavicular fossa to the pericardium and down through the thorax and abdomen, linking the upper, middle and lower energizers. There are 23 acupuncture points on either side of the body
governor vessel (GV)	governor vessel (GV)	督脈	one of the eight extra meridians which originates in the lower abdomen and exits at changqiang (GV1), a point at the back of the anus, sending one branch forward to huiyin (CV1). The main portion of the meridian/channel ascends along the midline of the back to the top of the head and then descends along the midline of the face down to yinjiao (GV28), a point between the upper lip and the upper gum in the labia frenum, also called governing vessel

conception vessel (CV)	conception vessel (CV)	任脈	one of the eight extra meridians which originates in the lower abdomen, exists at huiyin (CV1), a point in the center of perineum, and ascends the midline of the abdominal wall and chest to chengjiang (CV24), midpoint of the mentolabial sulcus. The internal portion of this meridian/channel ascends from chengjiang (CV24), encircling the mouth and traveling to the eyes. Another branch travels internally from the pelvic cavity and ascends the spine to the throat, also called controlling vessel
collateral vessel	collateral vessel	絡脈	the small branches of the meridians, serving as a network linking the various aspects of the body
sweat pore	mysterious mansion	玄府	another name for sweat pore. It is so named because it is too minute to be visible
sweat pore	qi gate	氣門	another name for sweat pore
cinnabar field	cinnabar field	丹田	three regions of the body to which one's mind is focused while practicing qigong: the lower cinnabar field – the region located in the upper 2/3 of the line joining the umbilicus and symphysis pubis; the middle cinnabar field – the xiphoid area; and the upper cinnabar field – the region between the eyebrows

References

1. Unschuld, P. U., *Nan Ching: The Classic of Difficult Issues* (e-book edn, Los Angeles: University of California Press, 1986), 771. With commentaries by Chinese and Japanese authors from the third through the twentieth century.
2. Morant, G. S. de, *Chinese Acupuncture* (Brookline, Massachusetts: Paradigm Publications, 1994).
3. World Health Organization, *WHO International Standard Terminologies on Traditional Medicine in the Western Pacific Region* (updated 2007). Available from <http://www.wpro.who.int/publications/who_istrm_file.pdf>.
4. Maciocia, G., *The Foundations of Chinese Medicine: A Comprehensive Text for Acupuncturists and Herbalists* (Edinburgh: Churchill Livingstone, 2002), 502.
5. Grossman, G., 'Cultural Reference for Increased Understanding of the San Jiao'. Available from: http://med-vetacupuncture.org/english/articles/sanjiao.html.
6. Watch Tower Bible and Tract Society of New York, *New World Translation of the Holy Scriptures* (Brooklyn, New York: Watch Tower Bible and Tract Society of New York, 1984).
7. Soh, K. S., K. A. Kang, D. K. Harrison, *The Primo Vascular System: Its Role in Cancer and Regeneration* (Springer, 2012).
8. Jiang, Y., 'The San Jiao: Returning to the Nei Jing (A Modern Explanation of Original Theory)', *Journal of Chinese Medicine*, 91 (2009), 46–50.

9. Ye-Tao, G., A. G. Chen, D. Dadi, 'Gross Conception of Anatomical Structure of the Triple Burner in Huangdi Neijing', *Journal of Accord Integrative Medicine*, 6/1 (2010), 43–58.
10. Maciocia, G., 'The Triple Burner (2)' (updated 30 August 2011). Available from <http://maciociaonline.blogspot.com.au/search?updated-max=2011-12-04T19:05:00-08:00&max-results=7>.
11. Larre, C., and E. Rochat de la Vallée, *Heart Master Triple Heater* (Norfolk: Monkey Press, 1998).
12. Anonymous, 'The Kidney Network and Mingmen: Views from the Past'. Available from <http://www.itmonline.org/5organs/kidney.htm>.
13. Wikimedia Foundation Inc., 'Lymphatic System' (updated 16 September 2014). Available from <http://en.wikipedia.org/wiki/Lymphatic_system>.
14. Anonymous, *Fire Element: San Jiao Theory, Meridians & Points* (on 19 October 2008). Available from <http://www.scribd.com/doc/7336429/SJiaoTheoryMeridsPoints> api-3719759.
15. Dharmananda, S., 'Triple Burner (Sanjiao) with Reference to Treatment of Sjögren's Syndrome' (September 2010). Available from <http://www.itmonline.org/articles/triple_burner/triple_burner.htm>.
16. Maxwell, D., 'An Interview with Edward Neal', *Journal of Chinese Medicine*, 105 (2014), 37–49.
17. Vilain, S., Fact or Fiction—Chinese Medicine's San Jiao Theory?' (updated 12 January 2012). Available from: <http://sam.vilain.net/sci/medicine/occidental/sanjiao_energetics.html>.
18. Maciocia, G., 'The Triple Burner as a System of Cavities and a Three-Fold Division of the Body' (2011). Available from <http://maciociaonline.blogspot.com.au/2011/10/triple-burner-as-system-of-cavities-and.html>.
19. Anonymous, *The Nei Ching (Su Wen and Ling Shu) & Nan Ching Medical Classics* (Miami, Florida: Occidental Institute of Chinese Studies Alumni Association, 1979), 188.
20. Nielsen, A., *Gua Sha: A Traditional Technique for Modern Practice* (New York: Churchill Livingstone, 1995).

21. Anonymous, 'Glossary' (Sheng Zhen Organisation, 2014). Available from <http://www.shengzhen.org/glossary.php>.
22. Anonymous, 'Triple Burner: Fire-Energy Yang Organ'. Available from <http://lieske.com/channels/5e-sanjiao.htm>.
23. Barbour, L., 'San Jiao' (post 2007). Available from <http://kathleenleavy.com/onlineclassroom/wp-content/uploads/2009/07/article-sj-barbour.pdf>.
24. Maciocia, G., 'The Triple Burner' (updated Tuesday, 2 August 2011). Available from <http://maciociaonline.blogspot.com.au/2011/08/triple-burner.html> updated Tuesday, 2 August 2011.
25. Maciocia, G., 'The Triple Burner (3)' (25 September 2011). Available from <http://maciociaonline.blogspot.com.au/search?updated-max=2011-12-04T19:05:00-08:00&max-results=7>.
26. Anonymous, 'Acupuncture' (2014). Available from <http://index-china.com/main/tcm/acupuncture.php>.
27. Pollack, G., *The Fourth Phase of Water: Beyond Solid, Liquid, and Vapor* (Kindle edition, Seattle: Ebner and Sons Publishers, 2014).
28. Myers, T., 'Fascial Fitness: Training in the Neuromyofascial Web', *IDEA Fitness Journal*, 8/4 (2011).
29. Wikimedia Foundation Inc., 'Zang-Fu' (2014). Available from <http://en.wikipedia.org/wiki/Zang-fu>.
30. Wikimedia Foundation Inc., 'Pericardium' (updated 20 May 2014). Available from <http://en.wikipedia.org/wiki/Pericardium>.
31. Clogstoun-Willmott, J., 'Stomach 30: St30. QiChong: Rushing Qi' (2015). Available from <http://www.acupuncture-points.org/stomach-30.html>.
32. Anonymous, 'Blood (Xue)—Vital Substances in Chinese Medicine'. Available from <http://www.sacredlotus.com/theory/substances/blood.cfm>.
33. Wikimedia Foundation Inc., 'Stomach' (updated 24 July 2014). Available from http://en.wikipedia.org/wiki/Stomach.
34. Soh, K. S., Kang KA, Y. H. Ryu, '50 Years of Bong-Han Theory and 10 Years of Primo Vascular System' (cited 2013). Available from <http://www.hindawi.com/journals/ecam/2013/587827/>.

35. Pollack, G., 'The Fourth Phase of Water: Central Role in Biology' (2013). Available from <http://www.youtube.com/watch?v=Y--L6BoH3Ug&noredirect=1>.
36. Sircus, W., 'Human Digestive System' (updated 4 March 2014). Available from <http://www.britannica.com/EBchecked/topic/1081754/human-digestive-system/242914/Gastric-secretion>.
37. Calder, V., 'Newton Ask a Scientist—Stomach Absorption Rates' (2002). Available from <http://www.newton.dep.anl.gov/askasci/zoo00/zoo00366.htm>.
38. Mateljan, G., 'Do You Know Where in the Digestive System Vitamins and Minerals Enter the Bloodstream?' (2014). Available from <http://whfoods.org/genpage.php?tname=dailytip&dbid=36>.
39. Hills, B. A., 'Gastric Surfactant and the Hydrophobic Mucosal Barrier', *Gut*, 39 (1996), 621–4.
40. Wikimedia Foundation Inc., 'Pulmonary Surfactant' (updated 17 June 2014 at 14.27). Available from <http://en.wikipedia.org/wiki/Pulmonary_surfactant>.
41. O'Neil, D., 'Blood Components' (2013). Available from <http://anthro.palomar.edu/blood/blood_components.htm>.
42. Garvey, M., 'Emotions, Desires and Physiological Fire in Chinese Medicine, Part One: The Pericardium and Lifegate', *Australian Journal of Acupuncture and Chinese Medicine*, 7/1 (2012), 16–22.
43. Lifang, Q. and M. Garvey, 'The Location and Function of the Sanjiao', *Journal of Chinese Medicine*, 65 (2001), 26–32.
44. Robertson, J. D., 'Yangsheng and the Channels', *Journal of Chinese Medicine*, 107 (2015), 5–12.
45. Stecco, C., V. Macchi, A. Porzionato A, F. Duparc, R. DeCaro, 'The Fascia: The Forgotten Structure', *Italian Journal of Anatomy and Embryology*, 116/3 (2011), 127–38.
46. Thompson, K., 'What the Heck Is a Triple Energizer Anyway?' (12 July 2012). Available from <http://www.miridiatech.com/news/2012/07/what-the-heck-is-a-triple-energizer-anyway/>.
47. Ji, S., 'Research Confirms Sweating Detoxifies Dangerous Metals, Petrochemicals' (updated 21 January 2015). Available from <http://www.theartofhealing.com.au/research_research_confirms.

html?utm_source=21+January+2015+Weekly+e-Alert&utm_campaign=21+January+2015&utm_medium=email>.
48. Findley, T. W. and M. Shalwala, 'The Fascia Research Congress from the 100 Year Perspective of Andrew Taylor Still' (2012). Available from <https://fasciaresearchsociety.org/sites/default/files/frc/still-100yrs-findley.pdf>.
49. Bastin, S. and K. Henken, 'Water Content of Fruits and Vegetables' (1997). Available from <http://www2.ca.uky.edu/enri/pubs/enri129.pdf>.
50. McCance, R. A. and E. M. Widdowson, 'The Composition of Foods' (London: Her Majesty's Stationery Office, 1960).
51. Ji, S., 'Could Melanin Convert Radiation into Harmless, Even Useful Energy?' (2016). Available from <http://www.theartofhealing.com.au/research_could_melanin.html?utm_source=February+2016+Monthly+e-Newsletter&utm_campaign=February+2016+enewsletter&utm_medium=email>.
52. Buch, V., A. Dubrovskiy, F. Mohamed, M. Parrinello, J. Sadlej, A. D. Hammerich, et al., 'HCl Hydrates as Model Systems for Protonated Water', *The Journal of Physical Chemistry*, 112/11 (2008), 2144–61.
53. Anonymous, 'Oxidanyloxidanium' (2014). Available from <http://www.lookchemical.com/product/oxidanyloxidanium/16740169.html>.
54. Chaplin, M., 'Hydroxide Ions' (10 July 2014). Available from <http://www1.lsbu.ac.uk/water/ionisoh.html>.
55. Saladin, K. S., 'Connective Tissue' (2014). Available from <http://www.biologyreference.com/Ce-Co/Connective-Tissue.html>.
56. Bailey, R., 'Bone Marrow' (2014). Available from <http://biology.about.com/od/anatomy/ss/bone-marrow.htm>.
57. O'Rahilly, R., *Basic Human Anatomy*, ed. R. Swenson (Dartmouth Medical School, 2008).
58. Anonymous, 'The Human Microbiome: Me, Myself, Us', *The Economist* (2012). From the print edition: Science and Technology.
59. Campbell-McBride, N., *Gut and Psychology Syndrome: Natural Treatment for Autism, Dyspraxia, A.D.D., Dyslexia, A.D.H.D.,*

Depression, Schizophrenia (York, Pennsylvania: Medinform Publishing, 2012), 392.
60. Mercola, J., 'The Fourth Phase of Water: What You Don't Know About Water, and Really Should' (18 August 2013). Available from <http://articles.mercola.com/sites/articles/archive/2013/08/18/exclusion-zone-water.aspx>.
61. Ho, M. W., 'Super-Conducting Liquid Crystalline Water Aligned with Collagen Fibres in the Fascia as Acupuncture Meridians of Traditional Chinese Medicine', *Forum on Immunopathological Diseases and Therapeutics*, 3/3–4 (2012), 221–36.
62. Lo, S. Y., 'Dr. Shui Yin Lo' (2013). Available from <http://www.stablewatercluster.net/dr-shui-yin-lo/>.
63. Gann, D. L., 'Double Helix Water Simplified' (2010). Available from <http://doublehelixwater.com/double-helix-water-simplified/>.
64. Jaxen, J., 'The 4th Phase of Water: A Key to Understand All Life' (2014). Available from <http://www.jeffereyjaxen.com/blog/the-4th-phase-of-water-a-key-to-understand-all-life>.
65. Pauling, L., Kamb, B., *Linus Pauling: Selected Scientific Papers*, Volume II (2001). Available from <http://books.google.com.au/books?id=2QduA19d_X8C&dq=hydrate-microcrystal+theory+of+anesthesia&source=gbs_navlinks_s>.
66. Wikimedia Foundation Inc., 'Ignaz Semmelweis' (16 July 2014 at 18.59). Available from <http://en.wikipedia.org/wiki/Ignaz_Semmelweis>.
67. Ponikau, J. U., D. A. Sherris, E. B. Kern, H. A. Homburger, E. Frigas, T. A. Gaffey, et al., 'The Diagnosis and Incidence of Allergic Fungal Sinusitis', *Mayo Clinic Proceedings*, 74/9 (1999), 877–84.
68. Charles, A., 'Vasodilation Out of the Picture as a Cause of Migraine Headache', *The Lancet Neurology*, 12/5 (2013), 419–20.
69. Mercola, J., 'Misdeeds, Not Mistakes, Behind Most Scientific Retractions' (2012). Available from <http://articles.mercola.com/sites/articles/archive/2012/10/17/scientific-research-retractions.aspx>.
70. Mercola, J., 'Why Your Doctor's Advice May Be Fatally Flawed' (2012). Available from <http://articles.mercola.com/sites/articles/archive/2012/07/12/drug-companies-on-scientific-fraud.aspx>.

71. Huff, E. A., '"Sacred Cow" of Industry Science Cult Should Be Slaughtered for the Good of Humanity, BMJ Editor Says' (13 May 2015). Available from <http://www.naturalnews.com/049694_BMJ_peer_review_science_journals.html>.
72. Pollack, G., 'The Fourth Phase of Water: A Role in Fascia?' (2013). Available from http://www.wholenhealthy.com.au/uploads/6/7/3/6/6736920/the_fourth_phase_of_water.pdf.
73. Wikimedia Foundation Inc., 'Parietal Cell' (29 July 2014 at 07.36). Available from <http://en.wikipedia.org/wiki/Parietal_cell>.
74. Bischof, M., *Biophotons: The Light in Our Cells* (Frankfurt: Zweitausendeins, 1998), 522.
75. Sun Y., C. Wang C, J. Dai, 'Biophotons as Neural Communication Signals Demonstrated by In Situ Biophoton Autography', *Photochemical & Photobiological Sciences* (2010), 315–22.
76. Vanwijk, R., 'Bio-Photons and Bio-Communication', *Journal of Scientific Exploration*, 15/2 (2001), 183–97.
77. Anonymous, 'Science & Spirituality: Living Light: Biophotons and the Human Body—Part 1', *Enlightening Entertainment*, episode 3892007, p. 22.
78. Rahnama, M., J. A. Tuszynski, I. Bókkon, M. Cifra, P. Sardar, V. Salari, 'Emission of Mitochondrial Biophotons and Their Effect on Electrical Activity of Membrane via Microtubules', *Journal of Integrative Neuroscience*, 10/1 (2011), 65–88.
79. Bajpai, R. P., 'Quantum Coherence of Biophotons and Living Systems', *Indian Journal of Experimental Biology*, 41/5 (2003), 514–27.
80. Mercola, J., 'McDonald's and Biophoton Deficiency' (updated 21 August 2002). Available from <http://articles.mercola.com/sites/articles/archive/2002/08/21/biophoton.aspx>.
81. Mercola, J., 'Five Principles that Can Heal Virtually Any Illness, Part 2' (3 June 2008). Available from <http://articles.mercola.com/sites/articles/archive/2008/06/03/five-principles-that-can-heal-virtually-any-illness-part-2.aspx>.
82. Ho, M. W., and D. P. Knight, 'The Acupuncture System and the Liquid Crystalline Collagen Fibres of the Connective Tissues: Liquid Crystalline Meridians', *American Journal of Complementary Medicine* (in press) (2014).

83. Mercola, J., 'Global Health Problems Reflect Our Disconnection from the Earth' (2015). Available from <http://articles.mercola.com/sites/articles/archive/2015/02/28/how-modern-lifestyle-caused-global-problems.aspx?e_cid=20150228Z3_DNL_B_art_1&utm_source=dnl&utm_medium=email&utm_content=art1&utm_campaign=20150228Z3_DNL_B&et_cid=DM70305&et_rid=858080303>.
84. Stefanov, M., M. Potroz, J. Kim, J. Lim, R. Cha, M. H. Nam, 'The Primo Vascular System as a New Anatomical System', *Journal of Acupuncture and Meridian Studies*, 6/6 (2013), 331–8.
85. Stutchbury, B., 'DNAmazing!' (updated 22 January 2013). Available from <http://thatsinteresting.scienceblog.com/2013/01/22/dnamazing/>.
86. Mercola, J., 'The Power of Biological Light in Healing' (updated 27 August 2010). Available from <http://articles.mercola.com/sites/articles/archive/2010/08/27/the-power-of-biological-light-in-healing.aspx>.
87. Mercola, J., 'Your Body Literally Glows with Light' (updated 15 August 2009). Available from <http://articles.mercola.com/sites/articles/archive/2009/08/15/Your-Body-Literally-Glows-With-Light.aspx>.
88. Mercola, J., 'Eat Your Food Uncooked? Here's the Really Raw Truth' (updated 21 March 2009). Available from <http://articles.mercola.com/sites/articles/archive/2009/03/21/Eat-Your-Food-Uncooked-Heres-the-Really-Raw-Truth.aspx>.
89. Popp, F. A., 'Properties of Biophotons and Their Theoretical Implications', *Indian Journal of Experimental Biology*, 41/5 (2003), 391–402.
90. Armstrong, K., 'Biophoton Light Resonance Therapy' (2015). Available from <http://in-side-out.com/biophoton-light-resonance-therapy/>.
91. Eslinger, R. A., 'Resurgence of "Old" Therapy Offers New Hope: Ultraviolet Blood Irradiation' (2014). Available from <http://hbmag.com/resurgence-of-%E2%80%9Cold%E2%80%9D-therapy-offers-new-hope-ultraviolet-blood-irradiation/>.
92. Wikimedia Foundation Inc., 'Blood Irradiation Therapy' (2014). Available from <http://en.wikipedia.org/wiki/Blood_irradiation_therapy>.

93. Dünya, U., 'Morphogenetic Field (Body Field)—Rupert Sheldrake, Ph.D, University of Cambridge' (2012). Available from <https://www.youtube.com/watch?v=4BYR32N04sE>.
94. Najemy, R., 'Our Causal Body' (2011). Available from <http://www.armonikizoi.com/2011/a42-our-causal-body-this-article-offers-some-scientific-evidence-declaring-the-existence-of-a-personal-and-species-wide-causal-body>.
95. Becker, K., 'Scientists Never Expected to Discover Antibiotic-Resistant Bacteria Here . . .', (09 June 2015). Available from <http://healthypets.mercola.com/sites/healthypets/archive/2015/06/09/antibiotic-resistance-superbugs.aspx?e_cid=20150609Z3_PetsNL_art_1&utm_source=petsnl&utm_medium=email&utm_content=art1&utm_campaign=20150609Z3&et_cid=DM78658&et_rid=985772731>.
96. Jóhannsdóttir, A. L., 'Kínverskar Lækningar' (year unknown). Available from <http://nalastungur.is/greinar/greinar_sanjiao.htm>.
97. Beck, J. M., V. B. Young, G. B. Huffnagle, 'The Microbiome of the Lung', *Translational Research*, 160/4 (2012), 258–66.
98. Anonymous, 'Boobies and Breastmilk Microbiota' (updated Saturday, 21 September 2013). Available from <http://drbganimalpharm.blogspot.com.au/2013/09/why-we-are-sick-and-fat-calories-in.html>.
99. Anonymous, 'Fascia: The Under Appreciated Tissue' (2005). Available from <http://www.fisiokinesiterapia.biz/download/fascia>.
100. Weber, D., 'New CT Scans Reveal Acupuncture Points' (22 August 2014). Available from <http://www.panaxea.com/wrr/au/wrr2014082106.html>.
101. Margulis, R. K., 'Possible Scientific Basis for Tandem Point Therapy and Acupuncture for Pain Relief',. (updated 17 March 2000). Available from <http://www.tandempoint.com/p17.htm>.
102. Margulis, R. K., 'Acupuncture Theory' (17 March 2000). Available from <http://www.tandempoint.com/p3.htm>.
103. Finando, S., D. Finando, 'An Introduction to Classical Fascia Acupuncture', *Journal of Chinese Medicine*, 106 (2014), 12–20.
104. Maciocia, G., 'Articles—The Heart Channel: Connection with Uterus'. Available from <http://www.giovanni-maciocia.com/articles/heart.html>.

105. Wikimedia Foundation Inc., 'Primo-Vascular System' (2014). Available from <http://au.wow.com/wiki/Primo-vascular_system>.
106. Avijgan, M. and M. Avijgan, 'Does the Primo Vascular System Originate from the Polar Body', *Integrative Medicine International*, 1/2 (214), 108–18.
107. Vlasto, T., 'New Scientific Breakthrough Proves Why Acupuncture Works' (2009). Available from <https://www.actcm.edu/news/new-scientific-breakthrough-proves-why-acupuncture-works/>.
108. Shin, H. S., H. M. Johng, B. C. Lee, S. I. Cho, K. S. Soh, K. Y. Baik, et al., 'Feulgen Reaction Study of Novel Threadlike Structures (Bonghan Ducts) on the Surfaces of Mammalian Organs', *The Anatomical Record Part B: The New Anatomist*, 284B/1 (2005), 35–40.
109. Soh, K. S. and J. S. Yoo, 'A Transformative Approach to Cancer Metastasis: Primo Vascular System as a Novel Microenvironment for Cancer Stem Cells', *Cancer Cell & Microenvironment*, 1/3 (2014).
110. Myers, T., 'Fascia & Tensegrity' (2014). Available from <http://www.anatomytrains.com/fascia/>.
111. O'Connell, J. A., 'Bioelectric Responsiveness of Fascia: A Model for Understanding the Effects of Manipulation', *Techniques in Orthopaedics*, 18/1 (2003), 67–73.
112. Vogelstein, B. and C. Tomasetti, 'Variation in Cancer Risk among Tissues Can Be Explained by the Number of Stem Cell Divisions', *Science*, 347/6217 (2014), 78–81.
113. Drapeau, C., 'The Stemtech Story' (2014). Available from <http://www.stemtech.com/AU/CompanyStory.aspx>.
114. Wikimedia Foundation Inc., 'Adipose Capsule of Kidney' (2014). Available from <http://en.wikipedia.org/wiki/Adipose_capsule_of_kidney>.
115. Anonymous, 'Saturated Fats and the Kidneys' (2000). Available from <http://www.westonaprice.org/health-topics/saturated-fats-and-the-kidneys/>.
116. Jamie, 'Fat, Lard, Suet, Tallow, Leaf Lard, Back Fat, Caul Fat, Kidney Fat' (2014). Available from <http://grassfood.me/2014/01/21/fat-lard-suet-tallow-leaf-lard-back-fat-caul-fat-kidney-fat/>.
117. Anonymous, 'Kidneys'. *Aid to Bible Understanding* (Brooklyn, New York: International Bible Students Association, 1971), 1696.

118. Stendal, R. M., *The Holy Scriptures (Jubilee Bible 2000)* (ANEKO Press, 2000).
119. *The Holy Bible, New International Version* (Essex: New York International Bible Society, 1979), 321.
120. Anonymous, *The New Living Translation* (Tyndale House Publishers, 1996).
121. Anonymous, *The Bible Revised Standard Version* (Glasgow: William Collins Sons & Co. Ltd., 1971).
122. Avijgan, M. and M. Avijgan, 'Can the Primo Vascular System (Bong Han Duct System) Be a Basic Concept for Qi Production?', *International Journal of Integrative Medicine*, 1/20 (2016), 1–10.
123. Thompson, J., 'The Body Cavities' (2001). Available from <http://apbrwww5.apsu.edu/thompsonj/Anatomy%20&%20Physiology/2010/2010%20Exam%20Reviews/Exam%201%20Review/Fig%201.9%20%20The%20Body%20Cavities.htm>.
124. Mader, S. S. and M. Windelspecht, *Human Biology* (10th edn, McGraw-Hill, 2011).
125. Peluffo, E., 'Pi Wei Xiang Biao Li and the Trajectory of Zuyangming', *Chinese Medicine*, 5/1 (2014).
126. Acland, R., 'Greater and Lesser Omentum', Wilkins LW, editor, Acland's Video Atlas of Human Anatomy (2013).
127. Anonymous, 'Gastrosplenic Ligament' (Salt Lake City, Utah 84107 2014). Available from <http://www.anatomyexpert.com/structure_detail/9308/>.
128. Wikimedia Foundation Inc., 'Greater omentum' (updated 23 May 2014). Available from <http://en.wikipedia.org/wiki/Greater_omentum>.
129. Gold R. 'Abdominal Explorations: The Omentum' (2003). Available from <http://www.pacificcollege.edu/pcom_static/alumni/newsletters/summer2003/abdominal.htm>.
130. Wikimedia Foundation Inc., 'Connective Tissue' (2014). Available from <http://en.wikipedia.org/wiki/Connective_tissue>.
131. Liptan, G. L., 'Fascia: A Missing Link in Our Understanding of the Pathology of Fibromyalgia', *Journal of Bodywork and Movement Therapies*, 14/1 (2009), 3–12.
132. Turchaninov, R., 'Research & Massage Therapy, Part 2: Why Does Massage Benefit the Body?', *Massage Bodywork* (2003).

133. Goldman, L., and A. I. Schafer. *Goldman's Cecil Medicine* (24th edn, 2012), 2672.
134. McFarlane, B., 'Notes on Anatomy and Physiology Function of the Thoracolumbar Fascia Part 2' (updated 07/12/2010). Available from <http://ittcs.wordpress.com/2010/07/12/notes-on-anatomy-and-physiology-function-of-the-thoracolumbar-fascia-part-2/>.
135. Poletti, G., 'More about Connective Tissue and Myofascial Release'. Available from <https://sites.google.com/a/seattlemfr.com/www/moreaboutconnectivetissueandmyofascialre>.
136. Traino, J., 'What Is Fascia?' (2012). Available from <http://www.dynamicmassageandbodywork.com/What-Is-Fascia.html>.
137. Caldwell, H., 'Did You Know Your Kidneys Work Like Batteries in your body?' (2012). Available from <https://www.facebook.com/AcupunctureBodyworkPC/posts/222374291222886>.
138. Anonymous, 'The Kidney Network and Mingmen: Views from the Past'. Available from: <http://www.itmonline.org/5organs/kidney.htm>.
139. Anonymous, 'Ming-Men: An Acupressure Point with Power-Full Implications' (2014). Available from <http://taichibasics.com/ming-men-an-acupressure-point-with-power-full-implications/>.
140. Anonymous, 'Jing (Essence)—Vital Substances in TCM' (2000). Available from <http://www.sacredlotus.com/go/foundations-chinese-medicine/get/jing-essence-vital-substance>.
141. Jovinge, S., 'The Heart: Our First Organ' (2012). Available from <http://www.eurostemcell.org/pl/node/24859>.
142. Wikimedia Foundation Inc., 'Kidney Development' (2013). Available from <http://en.wikipedia.org/wiki/Kidney_development>.
143. Lamb, A., and J. Sarfati, 'The Human Umbilical Vesicle ("Yolk Sac") and Pronephros—Are They Vestigial?' (updated 2 May 2009). Available from <http://creation.com/the-human-umbilical-vesicle-yolk-sac-and-pronephros-are-they-vestigial>.
144. Hill, M.A., 'Embryology Uterus Development' (2014). Available from https://embryology.med.unsw.edu.au/embryology/index.php/Uterus_Development.

145. Wikimedia Foundation Inc., 'Paramesonephric Duct' (2014). Available from <http://en.wikipedia.org/wiki/Paramesonephric_duct>.
146. Wikimedia Foundation Inc., 'Human Embryogenesis' (2015). Available from: <http://en.wikipedia.org/wiki/Human_embryogenesis>.
147. Wikimedia Foundation Inc., 'Blastocyst' (2014). Available from <http://en.wikipedia.org/wiki/Blastocyst>.
148. Jacob, M., F. Yusuf, H. J. Jacob, 'The Human Embryo' (2012). Available from <http://cdn.intechopen.com/pdfs-wm/30625.pdf>.
149. McFarlane, B., 'Notes on Anatomy and Physiology Function of the Thoracolumbar Fascia Part 1' (07/08/2010). Available from <http://ittcs.wordpress.com/2010/07/08/notes-on-anatomy-and-physiology-function-of-the-thoracolumbar-fascia-part-1/>.
150. Park, E. S., H. Y. Kim, D. H. Youn, 'The Primo Vascular Structures Alongside Nervous System: Its Discovery and Functional Limitation', *Evidence-Based Complementary and Alternative Medicine* (2013), 1–5.
151. Park, S. H., E. H. Kim, H. J. Chang, S. Z. Yoon, J. W. Yoon, S. J. Cho, et al., 'History of Bioelectrical Study and the Electrophysiology of the Primo Vascular System', *Evidence-Based Complementary and Alternative Medicine* (2013), 14.
152. Oschman, J. L., 'The Development of the Living Matrix Concept and It's Significance for Health and Healing', Science of Healing Conference (2009), 1–12.
153. Gil, H. J., K. H. Bae, L. J. Kim, S. C. Kim, K. S. Soh, 'Number Density of Mast Cells in the Primo Nodes of Rats', *Journal of Acupuncture and Meridian Studies* (2015).
154. Anonymous, 'Your Gut Has Taste Receptors', *ScienceDaily* (2007).
155. Mercola J., 'The Surprising Food Flavor That Can Help You Shed Pounds' (updated 8 September 2014). Available from <http://articles.mercola.com/sites/articles/archive/2014/09/08/umami-flavor.aspx?e_cid=20140908Z3_DNL_art_2&utm_source=dnl&utm_medium=email&utm_content=art2&utm_campaign=20140908Z3&et_cid=DM55382&et_rid=651096725>.
156. Anonymous, 'Can the Stomach Taste?' (2011). Available from <http://doctorsofweightloss.com/can-the-stomach-taste-5122>

157. Trivedi, B. P., 'Neuroscience: Hardwired for Taste', *Nature*, 486/7403 (2012).
158. Mercola, J., 'Studies Show Diet and Lifestyle Choices of Both Parents Have Multigenerational Health Effects' (2014). Available from <http://articles.mercola.com/sites/articles/archive/2014/12/29/parents-lifestyle-children-health.aspx?e_cid=20141229Z3_DNL_art_1&utm_source=dnl&utm_medium=email&utm_content=art1&utm_campaign=20141229Z3&et_cid=DM63362&et_rid=780813295>.
159. Diep, F., 'Sensing Calories without Taste', (updated 22 April 2013). Available from <http://www.the-scientist.com/?articles.view/articleNo/35077/title/Sensing-Calories-Without-Taste/>.
160. Simpson, T., 'Taste Buds: In Your Tongue & Gut—Cats Can't Taste Sweets' (13 February 2013). Available from <http://www.yourdoctorsorders.com/2013/02/taste-buds-trigger-food-cravings/>.
161. Mercola, J., 'How Artificial Sweeteners Confuse Your Body into Storing Fat and Inducing Diabetes' (2014). Available from <http://articles.mercola.com/sites/articles/archive/2014/12/23/artificial-sweeteners-confuse-body.aspx?e_cid=20141223Z3_DNL_art_1&utm_source=dnl&utm_medium=email&utm_content=art1&utm_campaign=20141223Z3&et_cid=DM62968&et_rid=773867334>.
162. Welsh, J., 'Mysterious Taste Sensors Are Found All Over the Body' (2 July 2013). Available from <http://www.businessinsider.com.au/taste-receptors-in-testes-and-fertility-2013-7>.
163. John, S., and C. Hoegerl, 'Nutritional Deficiencies after Gastric Bypass Surgery' (20 July 2009). Available from <http://jaoa.org/article.aspx?articleid=2093757>.
164. Greenwood, V., 'The Startling Sense of Smell Found All Over Your Body,' (updated 10 July 2013). Available from <http://www.bbc.com/future/story/20130710-how-our-organs-sniff-out-smells>.
165. Mercola, J., 'How to Build a Healthy Microbiome, before, during, and after Birth' (2014). Available from <http://articles.mercola.com/sites/articles/archive/2014/12/29/healthy-gut-microbiome.aspx?e_cid=20141229Z3_DNL_art_2&utm_source=dnl&utm_medium=email&utm_content=art2&utm_campaign=20141229Z3&et_cid=DM63362&et_rid=780813295>.

166. Mercola, J., 'Fiber Provides Food to Your Gut Microbes that They Ferment to Shape Your DNA' (30 March 2015). Available from <http://articles.mercola.com/sites/articles/archive/2015/03/30/fiber-fermentation-gut-health.aspx?e_cid=20150330Z3_DNL_B_art_1&utm_source=dnl&utm_medium=email&utm_content=art1&utm_campaign=20150330Z3_DNL_B&et_cid=DM71100&et_rid=895005487>.
167. Neufeld, K. M., N. Kang, J. Bienenstock, J. A. Foster, 'Reduced Anxiety-Like Behavior and Central Neurochemical Change in Germ-Free Mice,' *Neurogastroenterology & Motility* 23/3 (2011), 225.
168. Daniells, S., 'Breakthrough Study Shows Personalised Nutrition Future for Probiotics', (updated 15 September 2010). Available from <http://www.nutraingredients.com/Research/Breakthrough-study-shows-personalised-nutrition-future-for-probiotics>.
169. Diaz-Heijtz, R., S. Wang, F. Anuard, Y. Qiana, B. Björkholmd, A. Samuelssond, et al, 'Normal Gut Microbiota Modulates Brain Development and Behavior,' *PNAS*, 108/7 (2011), 3047–52.
170. Diaz-Heijtz, R., S. Wang, F. Anuar, Y. Qian, B. Björkholm, A. Samuelsson, et al., 'Bacteria in the Gut May Influence Brain Development' (2013). Available from <http://ki.se/en/news/bacteria-in-the-gut-may-influence-brain-development>.
171. Wikimedia Foundation Inc., 'Microbiome', (updated 23 March 2014 at 04:10). Available from <http://en.wikipedia.org/wiki/Microbiome>.
172. Mercola, J., 'How Bugs Become Instantly Resistant to Insecticide by Swallowing Bacteria' (09 May 2012). Available from <http://articles.mercola.com/sites/articles/archive/2012/05/09/how-bug-becomes-instantly-resistant-to-insecticide-by-swallowing-bacteria.aspx?e_cid=20120509_DNL_art_2>.
173. Blaser, M., 'Why Antibiotics Are Making Us All Ill,' *The Observer* (Sunday, 1 June 2014).
174. Wikimedia Foundation Inc., 'Lung Microbiome,' (updated 23 March 2014 at 04:37). Available from <http://en.wikipedia.org/wiki/Lung_microbiome>.
175. Malka, V., 'Oil Pulling the Key to Oral Health' (2015). Available from <http://www.wellbeing.com.au/article/Features/Body-Health/Oil-well_1562>.

176. Mercola, J., 'Gut Bacteria and Fat Cells May Interact to Produce "Perfect Storm" of Inflammation that Promotes Diabetes and Other Chronic Disease', (05 June 2014). Available from <http://articles.mercola.com/sites/articles/archive/2014/06/05/chronic-inflammation.aspx?e_cid=20140605Z3_DNL_art_1&utm_source=dnl&utm_medium=email&utm_content=art1&utm_campaign=20140605Z3&et_cid=DM45962&et_rid=541908700>.
177. Bassler, B., 'How Bacteria "Talk"' (2009). Available from <http://www.ted.com/talks/bonnie_bassler_on_how_bacteria_communicate?language=en>.
178. Paschotta, D. R., 'The Science of Biophotons', *The Photonics Spotlight* (2011).
179. Anonymous, 'Probiotics that SHIME' (cited 2012). Available from <http://www.wellbeing.com.au/newsdetail/Probiotics-that-SHIME_000648>.
180. Baquero, F., and C. Nombela, 'The Microbiome as a Human Organ', *Clinical Microbiology and Infection* (2012).
181. Mercola, J., 'What Secrets Are Revealed in City Sewage about Health' (4 July 2015). Available from <http://articles.mercola.com/sites/articles/archive/2015/07/04/sewage-microbes.aspx?e_cid=20150704Z3_DNL_art_2&utm_source=dnl&utm_medium=email&utm_content=art2&utm_campaign=20150704Z3&et_cid=DM78483&et_rid=1019036386>.
182. Myhill, S., 'Fermentation in the Gut and CFS' (updated 19 May 2014 at 09:34). Available from <http://www.drmyhill.co.uk/wiki/Fermentation_in_the_gut_and_CFS>.
183. Nishihara, K., 'Disclosure of the Major Causes of Mental Illness—Mitochondrial Deterioration in Brain Neurons via Opportunistic Infection', *Journal of Biological Physics and Chemistry*, 12 (2012), 11–8.
184. Myhill, S., 'Faecal Bacteriotherapy' (updated 19 March 2014 at 10:35). Available from <http://www.drmyhill.co.uk/wiki/Faecal_bacteriotherapy>.
185. Bercik, P., A. J. Park, D. Sinclair, A. Khoshdel, J. Lu, X. Huang, et al., 'The Anxiolytic Effect of *Bifidobacterium longum* NCC3001 Involves Vagal Pathways for Gut-Brain Communication', *Neurogastroenterology & Motility*, 12 (2011), 1132–9.

186. Anonymous, 'The Amazing Healing Properties of Fermented Foods: The Art of Healing' (2014). Available from <http://www.theartofhealing.com.au/news_the_amazing_healing.html?utm_source=July+2014+e-Newsletter+2&utm_campaign=July+2014+enewsletter&utm_medium=email>.
187. Mercola, J., 'Antiperspirants Can Make You Smell Worse by Altering Armpit Bacteria' (2015). Available from <http://articles.mercola.com/sites/articles/archive/2015/01/28/antiperspirants-alter-armpit-bacteria.aspx?e_cid=20150128Z3_DNL_RTLC_art_1&utm_source=dnl&utm_medium=email&utm_content=art1&utm_campaign=20150128Z3_RTLC&et_cid=DM68007&et_rid=820658671#_edn5>.
188. Hadhazy, A., 'Think Twice: How the Gut's "Second Brain" Influences Mood and Well-Being', *Scientific American* (2010).
189. Anonymous, 'Lymphoid Tissue' (1 February 2014). Available from <http://www.britannica.com/EBchecked/topic/352830/lymphoid-tissue>.
190. Choi, I., S. Lee, Y. K. Hong, 'The New Era of the Lymphatic System: No Longer Secondary to the Blood Vascular System', *Cold Spring Harbor Perspectives in Medicine*, 2/4 (2012).
191. Mercola, J., 'Tips for Reducing Cellulite with a Power Plate' (2014). Available from <http://fitness.mercola.com/sites/fitness/archive/2014/06/13/cellulite-removal-power-plate.aspx?e_cid=20140613Z3_DNL_art_1&utm_source=dnl&utm_medium=email&utm_content=art1&utm_campaign=20140613Z3&et_cid=DM46417&et_rid=552201837>.
192. Anonymous, 'Lymphatic System' (updated 13 January 2014). Available from <http://www.betterhealth.vic.gov.au/bhcv2/bhcarticles.nsf/pages/Lymphatic_system>.
193. Zimmermann, K. A., 'Lymphatic System: Facts, Functions & Diseases' (updated 8 February 2013). Available from <http://www.livescience.com/26983-lymphatic-system.html>.
194. Ehrlich, A., A. Harrewijn, E. McMahon, 'The Lymphatic System' (updated 28 July 2012). Available from <http://www.lymphnotes.com/article.php/id/151/>.
195. Yarrow, D., 'Heal Your Lymphatic Ocean' (1989). Available from <http://www.dyarrow.org/lymph.htm>.

196. Nordqvist, C., 'What Is Brown Fat? What Is Brown Adipose Tissue?' (updated 15 September 2014). Available from <http://www.medicalnewstoday.com/articles/240989.php>.
197. Doheny, K., 'The Truth About Fat' (updated 13 July 2009). Available from <http://www.webmd.com/diet/features/the-truth-about-fat>.
198. Cannon, B., J. Nedergaard, 'Brown Adipose Tissue: Function and Physiological Significance', *Physiological Reviews*, 84/1 (2004), 277–359.
199. Wikimedia Foundation Inc., 'Brown Adipose Tissue' (updated 10 May 2014). Available from <http://en.wikipedia.org/wiki/Brown_adipose_tissue>.
200. Sacks, H., M. E. Symonds, 'Anatomical Locations of Human Brown Adipose Tissue: Functional Relevance and Implications in Obesity and Type 2 Diabetes', *Diabetes*, 62/6 (2013), 1783–1790.
201. Mercola, J., 'Researchers Find Healthy Brown Fat Regulates Your Blood Sugar' (updated 08 August 2014). Available from <http://fitness.mercola.com/sites/fitness/archive/2014/08/08/brown-fat-blood-sugar.aspx?e_cid=20140808Z3_DNL_art_2&utm_source=dnl&utm_medium=email&utm_content=art2&utm_campaign=20140808Z3&et_cid=DM53213&et_rid=613282231>.
202. Anonymous, 'Adipose Tissue' (2015). Available from <http://www.yourhormones.info/glands/adipose_tissue.aspx>.
203. Mercola, J., 'Fat Beneath Skin May Ward Off Infections' (2015). Available from <http://articles.mercola.com/sites/articles/archive/2015/01/17/skin-fat-may-ward-off-infections.aspx?e_cid=20150117Z3_DNL_BuyerA_art_2&utm_source=dnl&utm_medium=email&utm_content=art2&utm_campaign=20150117Z3-BuyerA&et_cid=DM66966&et_rid=807904609>.
204. Mercola, J., 'Grounding Is a Key Mechanism by Which Your Body Maintains Health', (updated 2 August 2014). Available from <http://articles.mercola.com/sites/articles/archive/2014/08/02/grounding-earthing.aspx?e_cid=20140802Z3_DNL_art_1&utm_source=dnl&utm_medium=email&utm_content=art1&utm_campaign=20140802Z3&et_cid=DM52281&et_rid=606117103>.
205. Stevenson, R., 'Barefoot Earthing', *WellBeing* (2012).

206. Mercola, J., 'Clint Ober Discusses Earthing Research' (22 March 2012). Available from <http://www.youtube.com/watch?v=p7gWnRG8gfk>.
207. Anonymous, 'The Blood Pressure Connection' (2011). Available from <http://www.earthinginstitute.net/?p=597>.
208. Chevalier, G., 'Earthing, Inflammation, and Aging—Something to Think About' (2012). Available from <http://www.earthinginstitute.net/?p=496>.
209. Ober, C., S. T. Sinatra, M. Zucker, *Earthing: The Most Important Health Discovery Ever?* (Laguna Beach, California 92651: Basic Health Publications Inc., 2010).
210. Chevalier, G., S. T. Sinatra, 'Emotional Stress, Heart Rate Variability, Grounding, and Improved Autonomic Tone: Clinical Applications', *Integrative Medicine*, 10/3 (2011), 16–21.
211. Anonymous, 'Earthing vs. "Good" and "Bad" Free Radicals' (2010). Available from <http://www.earthinginstitute.net/?p=603>.
212. Anonymous, 'Forms of Qi—Vital Substances in Chinese Medicine', Available from <http://www.sacredlotus.com/theory/substances/qi_forms.cfm>.
213. Anonymous, 'Jing (Essence)—Vital Substances in Chinese Medicine'. Available from <http://www.sacredlotus.com/theory/substances/jing.cfm>.
214. Anonymous, 'Body Fluids (Jin Ye)—Vital Substances in Chinese Medicine'. Available from <http://www.sacredlotus.com/theory/substances/jinye.cfm>.
215. Lao, T. L., 'The Use of the Sanjiao Theory in Clinical Acupuncture', *Journal of Chinese Medicine*, 34 (1990), 20–22.
216. Maciocia, G., 'Triple Burner and Yuan Qi: Bladder 22 and Ren 5' (16 March 2010). Available from <http://maciociaonline.blogspot.com.au/2010/03/triple-burner-and-yuan-qi-bl-22-and-ren.html>.
217. Maciocia, G., 'The Triple Burner—Relationship with Pericardium' (updated 30 December 2011). Available from <http://maciociaonline.blogspot.com.au/2011_12_01_archive.html>.

Index

A

abdominal artery 347
abdominal viscera 192, 422, 425
abdominopelvic cavity 47, 411-12
accumulation illnesses 423
acetaldehyde 589, 603
Achilles tendon 444
acidic 88, 101, 104, 145, 202, 244, 592
acidophilus 566
Acinetobacter 572
Acland, Robert 418
actine 441, 455
Actinobacteria 593, 597
AcuGraphTM 154
acupuncture xii-xiii, 152-4, 219-23, 320-4, 326-34, 346-7, 350-2, 357-8, 376-7, 397-406, 508-11, 518-20, 525-34, 684-7, 713-15
acupuncture Meridian System 223, 270, 327-8, 346, 357, 377, 383, 397-9, 403-6, 448-9, 503, 510-11, 525, 527-8, 532
adipocytes 190, 340, 634, 637, 639-41, 643
adiponectin 639, 644
adipose tissues xi, 7, 348, 390-2, 404, 502, 505, 507, 627-37, 639, 642-4, 724
adrenal fatigue 575
agriculture 268
Agrobacterium spp. 579
Aid to Bible Understanding 393
air 42, 59, 100, 105-6, 139, 141, 265, 269, 271, 286, 309, 464, 661, 676, 702-3
Alcian blue 140, 344, 348-9, 369, 507, 511
algae 221, 389, 570, 601
alkaline 88, 212, 218, 244
allergens 321
allopathetic 562
allopathetic medicine 562
allopathic 27, 38, 148, 211, 293, 296, 438, 562, 674
allopathic medicine 27, 148, 211, 293, 438, 562, 674
Alzheimer's 549
amenorrhea 84, 558, 673
American Association of Anatomists 358

American Heart Journal 295
amino acids 91, 98, 101-2, 255, 373, 385, 441, 455, 545-6, 550, 553, 569, 592, 594, 600
ammonium 178, 208
amniotic fluid 54, 126, 473
amniotic sac 493, 502
amperometric 329
analogy 67, 129, 311, 317, 333, 427
Anatomical Record 358, 716
anatomy trains 332
anemia 556
anesthesia 230-1, 327
anesthesiology 230
anesthetics 229-31, 251, 327
annihilation 202, 234, 236, 447
anorexia 74
anti-inflammatory interleukin 572-3, 601
anti-innovatory 240
Anti-Müllerian Hormone (AMH) 488, 490
antiatherosclerotic hormone 639
antibiotic resistance 307-8
antibiotics 235, 292-3, 296, 307-8, 563, 565, 571, 573, 582, 584, 591, 593, 601, 721
antimicrobial peptide 641
antimicrobial peptides 572-3, 601, 640-1
antioxidants 232, 645, 655-6, 664
antiseptics 571
anus 26, 80, 346, 552, 605, 705
anxiety 83, 566, 596, 665, 669, 697, 721
AOBiome 597, 604
aorta 415, 491, 501, 632-4, 643
apoptosis 374, 490
appendices testis 488

aromatherapy 561
aromatic 252, 255, 555
arrhythmias 635, 666
arsenic 146-7, 151
arteries 7, 216-17, 265, 330, 391, 419-20, 454, 608, 614, 616-17, 622, 633-4, 643
arterioles 341, 617
arthritis 294, 413, 575-6, 588, 623, 626
artificial sweeteners 539-40, 542, 546, 549-50, 720
ascend 5, 8-9, 63, 65-7, 70, 87-8, 91, 110, 114-15, 346, 353, 359, 519, 533, 705-6
aspirin 101-2, 232
asthma 294, 574-6, 602, 623, 626
atoms 125, 185, 194-5, 223, 271, 399, 461, 495, 516-17, 655
ATP synthase enzyme 630
Australia 233, 279, 295-6, 552, 613, 648, 653
Austria 295-6
autoimmune diseases 292, 294, 296, 567
autoimmune disorders 294, 646
autonomic dystonia 665, 669
autonomic nervous system (ANS) 47, 341, 663, 665-6
avenue 50
Avijgan 377, 381, 405, 407-8, 504-5, 508, 512-13, 534, 716-17
axillary region 635
axon 425

B

B-group vitamins 582, 584
babushka doll 368

baby 8, 126, 343, 468, 473, 493, 498, 502
bacteria 251, 293-4, 307-8, 313-15, 342-3, 563-70, 572-4, 577-80, 582-4, 586-7, 589-604, 606-7, 613-14, 641, 721-3
bacterial infection 235
bacteriophages 570, 572, 593, 601
bacteriotherapy 593, 722
Bacteroides 311, 586, 592-5, 602, 604
Bacteroidetes 593
bags 31-4, 73, 75, 80, 120-1, 132, 148, 226, 417, 428, 435-6, 473, 498, 522-3, 653-4
Bankei 113, 122, 365, 420-1, 432, 457-8, 493-4
Bankei, Katō 113, 122, 365, 420-1, 432, 457-8, 493-4
Bao Luo 343-4, 462
Bao Mai 194, 343-4
Barber's Pole worm 104
Barbour, Linda 49, 190, 193, 247, 318, 431
bariatric surgery 543
basophilic 350
Bassler, Bonnie 578-80, 602, 722
Batteries 77, 80, 161, 179, 208, 463
beer 125, 310-11
Beige Adipose Tissue 627
Benveniste, Jacques 206
beta-carotene 656
bias 37, 311
Bible 3, 6, 393-5, 707, 716-17
bicarbonate 244, 592
Bifidobacteria spp. 311, 593, 595
Bifidobacterium longum 596, 722
binds 244, 422
Bio-Photons 252, 257, 261-3, 713

bioaccumulated 146-7
biochemically 186, 209
bioelectrical 270, 330, 373, 383, 399, 514, 526, 719
bioinformation 251
bioluminescence 578
biomedical 237, 279, 358, 376, 511, 551, 649
biomolecular sources 370
biophoton 108, 249-51, 256, 258-60, 262, 264, 272-3, 277-80, 285-6, 288, 290, 293, 299, 400, 713-14
Biophoton Deficiency 264, 272-3, 285, 713
biophoton emissions 249-50, 262, 264, 277-8, 280, 371, 400
biophotons 107-8, 110-11, 168-70, 174-6, 249, 251-9, 261-4, 271-2, 274-81, 283-6, 288-92, 296-300, 399-401, 523-5, 713-14
biostructures 264
Bisphenol A (BPA) 147, 151
bitter 540-4, 546-8, 550, 552, 555, 559-60, 687-8, 696
bitter (T2R) 541
bitter receptors 546, 559-60
bitter substance 542
bitter taste 541-2, 546, 548, 559, 687-8, 696
BL-13 688, 692
BL-20 688, 693-4
BL-21 688
BL-22 144, 692, 694-6
BL-22 Sanjiaoshu 692, 694-5
BL-39 135, 144, 689, 694, 698
bladder 2, 7-9, 28-9, 43-7, 49-50, 66-7, 70-1, 111-16, 126, 131, 134-5, 143-5, 149-50, 680-1, 695-6

blastocyst 79, 384, 473-4, 483-8, 498, 500, 522, 719
blatantly 239, 342, 516
blockages 57, 161, 292, 317
blood 5-7, 82-6, 89-99, 144-7, 190, 292-5, 347-50, 385-7, 608-18, 620-5, 632-7, 646-7, 657-60, 684-6, 703-5
blood cells 27, 85, 181, 190, 216, 292, 294, 352, 386, 446, 451, 608-11, 613-14, 624, 657-8
blood clots 646
bloodstream 94, 98, 102, 109, 190, 244, 318, 425, 545-6, 548, 575, 589, 591, 604, 638-9
body consciousness 324-8, 530
bone 82-3, 85, 190, 219-20, 324, 334, 338, 389-90, 435-6, 441, 450-2, 472, 513-14, 608-9, 673-4
bone marrow 82-3, 85, 190, 386, 389-90, 472, 608-9, 624, 674, 680, 704, 711
bones 26, 43, 63, 86, 93, 138-9, 159-60, 219-20, 229, 427-9, 435-6, 441, 450-2, 458-9, 513-14
Bong-Han 139, 156, 303, 346-7, 349, 351-2, 387-9, 405, 506
Bong-Han system 156, 303, 346-7, 387-9, 506
Bonghan 139, 344-9, 351-5, 358, 373, 385, 400, 403, 505-6, 510-11, 520, 532, 716
Booster Amplifiers 330
boosting 315, 342, 692
born xiii, 184, 219, 314, 344, 458, 473, 487, 498, 500, 571, 587
bowels 3, 47, 571, 593, 606-7, 704-5

brain 47, 230-1, 254-5, 325-8, 346, 410, 425-6, 547-50, 565-7, 582-3, 588-9, 591-2, 603, 605-7, 673-4
bread 169, 207
breastmilk 314-15
British Medical Journal 238-9
bronchitis 576, 623, 626
brook 523
brown adipose tissue 627-32, 636-7, 642-3, 724
brown adipose tissue (BAT) 627-38, 642-3
brown fat 627-9, 634, 636, 638
bulk 65, 69, 107-8, 110-11, 174-6, 183, 185-9, 194-5, 200-1, 209-10, 218-19, 222, 243-4, 288-90, 328
bundles 420
bupivacaine 230
butyl alcohol 589, 603
butyraldehyde 589, 603
butyrate 582
butyric acid 582, 591, 604

C

cadmium 146-7, 151
calcium 370, 452, 556
calming 306, 401, 653, 658, 663
Cambridge University 305
Campbell-McBride, Natasha 203, 566, 711
cancer 141, 240, 250-1, 280, 344, 352, 355-6, 388-90, 404, 434, 506, 515, 587-8, 595, 614
cancer metastasis 344, 352, 389, 716
Candida albicans 588

carbohydrate metabolic derivatives 355, 363-4, 520
carbohydrates 65-6, 97, 101-2, 107-8, 546, 550-1, 554, 569, 573, 584-5, 589-92, 603
carbon dioxide 140, 244, 622
cardia 45
cartilage 190, 229, 324, 381, 427, 436, 445, 452, 459
cascade 458, 478, 486, 560, 667
catalyst 67, 156, 197, 301-2, 673
catecholamines 345, 637
cathelicidin antimicrobial peptide 641
caudal parts 488, 490
cauldron 289
cavitation 379, 485
cavities 2, 15, 46-7, 53-4, 58, 127-8, 135-6, 148, 190-1, 396-7, 410-13, 421, 423, 432, 708
cavity xvii, 42-3, 47, 49-51, 136, 191, 295, 410-14, 421, 423-4, 485, 574, 576, 699-700, 704
cell 187-91, 250-3, 257-9, 261-4, 274-5, 277-9, 283-5, 351-2, 373-5, 385-6, 405, 449-51, 455-6, 492, 512-14
cell death 261, 374
cellular 85, 186, 195, 257, 261-2, 274, 335-6, 399-400, 446, 449-50, 453, 485-8, 617-18, 656-7, 693
cellulite 610-11, 624, 723
cerebrospinal fluids 54, 346, 348-50, 403-4, 414, 556
cervical 476, 535-6, 635
Chaga 174
channel 56, 60, 80, 112, 126, 159, 267, 325, 343-4, 372, 400, 683, 685-91, 698-9, 705-6

charge 17-18, 89, 108-10, 159-61, 178-81, 183-4, 186-8, 193-7, 207-9, 214, 224-6, 228-32, 245-7, 646-7, 649-50
chemical 27, 65, 132, 173-4, 186, 195-7, 243-4, 250-3, 269, 278-9, 309, 560, 578-81, 588-9, 602
chemiluminescence 263
chemiosmosis 630
Chernobyl 173
Chevalier, Gaétan 652, 663-5, 725
Ch'i Po (Baron of Qi) 45-6, 166
child 458, 468, 544, 672
chin 9, 29, 88, 111, 115, 118, 144-5, 170, 201-2, 204, 354
Chinese xiii, 3, 13-16, 37-40, 45-6, 126-7, 135-8, 226-7, 333-7, 390, 474-5, 670-1, 675-9, 707-10, 725
cholecystokinin 542, 550
cholecystokinin (CCK) 542-3, 550
cholesterol 235, 290, 392, 553, 563, 620
chondrocytes 190
Chondroitin sulfate 448
Chong Mai 56, 78, 84, 289-90, 298, 359, 363-4, 533
choroid plexus 414
Chrome-hematoxylin 140, 149, 344
chronic diseases 336, 567, 573-4, 577, 601-2, 646, 722
chronic fatigue 294, 574, 577, 602
chronic rhinosinusitis 235-6
chronopharmacology 203
chyle 43, 609, 612, 624
chylomicrons 612, 620
cilia 559-60
cinnabar field 171, 175, 287, 297-8, 706

circadian 59-60, 63, 168, 171-2, 203, 278, 282, 288, 331, 652, 664
circuits 124, 326, 387, 566
circulate 18, 53-4, 63, 65, 83, 156-7, 166, 183, 302-3, 317, 331-3, 418-19, 519, 523-4, 677-80
Circulation 129, 333, 660, 678-80
circulatory system 25-6, 128, 204, 295, 344-5, 355, 358, 387, 403, 437, 445, 454-5, 475, 610-13, 616-17
Clegg, James 180, 207, 225
Clostridia 592, 594
Clostridium difficile 591
clouds 9, 25, 88, 115, 144, 170, 178, 184, 200-1, 205, 209
clowns 211
clusters 198, 223
CO2 89, 232
coding 274, 569, 600
coffee 199-200
coherence 250, 252-3, 256-8, 264, 271-2, 299, 321, 400
coherent 219, 222-3, 250, 264, 271-3, 275, 324, 371, 399-400
collagen 188-9, 219-22, 226, 265-6, 271, 320-8, 331-3, 370-1, 398-9, 430-1, 441-2, 446-7, 452-3, 455, 526-8
collagenous-based 436
collarbones 615-16
collaterals 347
colloid osmotic pressures 341
commentator 23, 167, 170
communicate 125, 129, 254, 259, 263-4, 274, 285, 326, 348, 403, 450, 456, 511, 529, 607

communication 123-4, 127-8, 250-1, 262, 267-71, 273, 290-1, 321-2, 338, 355-6, 430-1, 453-6, 517-18, 525-6, 577-81
communication system 260, 271, 277, 321, 335, 510, 514
complexion 144, 647, 660
computerised tomography (CT) scans 328, 715
concept 1-3, 13, 21-2, 25, 40-1, 48, 52, 67, 69, 74, 80, 278-9, 400-1, 507-8, 647
concrete 15, 121, 178-9, 208, 505, 645, 649, 653
conduction 251, 257, 321, 324-5, 327-8, 332, 383
conduits 5, 92, 95, 99, 149-50, 167-8, 289-90, 343-4, 361-4, 486-7, 507-8, 519-20, 524-6, 533, 555
confusing 16, 52
confusion 23, 25, 49, 72, 112-13, 116, 280
conjunctivitis 159, 161
connective tissue 132-3, 189-90, 220-2, 334-5, 370-2, 380-1, 405-6, 408-9, 420-3, 432-6, 438-40, 449-53, 459-60, 512-18, 528-9
connective tissue proper 381, 459
connective tissues 189-90, 265-7, 319-24, 370-2, 380-1, 397-8, 405-6, 408-9, 420-3, 432-5, 438-40, 448-53, 459-60, 512-17, 522-9
constructive influences 5-6, 92-4, 97, 387, 538-9, 548, 550-1
container 27-8, 198, 218, 443
contemporary 39-40, 113
CoQ10 581-2

cord 346, 350, 410, 414, 421, 436,
    450, 461, 516, 565, 599, 605,
    607, 633, 674
Cordyceps Sinensis 390
corn syrup 445
coronary artery disease 575
cortisol 659
cough 63, 687-8
couli 2, 10, 53, 127-8, 132, 137-44,
    149-51, 160, 524, 628, 683
cranial cavity 410
cravings 342, 549, 569, 589, 600
creator 393, 657
crystalline mesophases 266-7, 323
CV 4 362, 520
CV 12 7, 155-7, 391, 510, 525
cyanobacteria 221, 389
cyberspace 400, 671
cycle 59-60, 88, 111, 115, 172, 251,
    288, 290, 331, 363, 385-6, 405,
    468, 524, 617
cyclical stages 450
cytoskeletons 222, 255, 335, 449,
    492, 528

D

Daling 686-7, 691, 697
damp 318, 570, 601, 686, 699
Dampness 678, 694-5, 699-700
Dampness in the Lower Burner
    694-5
Dan Tien 488
DC 220, 320-1, 324, 326-8, 330-1
defence 140-1, 435, 560
defense 70, 139, 524, 567, 640
Defensive Qi 1, 57, 61, 64, 68, 100,
    313

demystify xiv, 178, 208
dense connective tissue 438
Department of Primary Industries
    213
depot 21, 23, 29, 31, 33-4, 75, 113,
    118, 120-1, 316, 467, 568, 632
depots 21, 23, 28-31, 33-5, 71-3, 75,
    113, 118-23, 166-7, 420-3, 432,
    467, 551, 555-6, 631-3
depression 61, 583, 588, 665, 669,
    697, 712
derivative 64, 94, 270, 381, 409, 459,
    494, 512, 518, 677
dermatan sulfate 448
dermatology 640
detoxing 564
diabetes 394, 539-40, 544, 574-5,
    588, 602, 631, 639, 644, 655,
    658, 720, 724
diaphragm xvii, 2, 5, 9-10, 17-18,
    45-6, 50, 53, 63, 92, 106, 163-5,
    182, 411-12, 704
dielectric 266, 322
diffusion 35, 59, 61, 63, 106, 472
digestion 2, 15, 19, 42-3, 47, 59, 61,
    101, 107, 158, 231, 244, 584-5,
    605, 702-3
digestive system 26, 300, 380, 454,
    545-6, 554, 581, 583, 609, 616,
    620
dinoflagellates 570, 601
dipalmitoylphosphatidylcholine 106
discharge 10, 105, 461, 516, 613, 682,
    697
dispersed 136, 148, 197, 375, 465,
    504
dispersing 172, 287-8, 678, 691
dissemination 271

distention 161
ditch 29-31, 49, 118-20, 316, 342, 423, 515-16, 682, 693
ditches 6, 8, 29-31, 54, 78, 87, 114, 118, 120, 178, 207, 316, 423, 451
divergent channels 333-4, 529
dividing zygote 378, 408
dizziness 159, 161, 665, 669-70, 674, 687, 696
DNA 250, 258-9, 263-4, 271-5, 279-80, 285, 305, 367, 369-74, 399-401, 482, 527-8, 530-1, 563-4, 599-600
doctor of philosophy (PhD) 223, 229, 242, 304, 320, 539, 629, 640, 642, 647, 663, 665, 667
dogma 202-3, 206-7, 212, 233
domicile 360, 375, 504
Dong Qi 695
dopamine 549, 566, 590, 594
dorsal 140, 149, 205, 344, 369, 410, 412, 414, 421, 511
Double-Helix Water 223, 226-7
DPPC 103, 106, 110
drinks 4, 6, 8-9, 68-9, 77, 80-2, 86-7, 97-9, 107, 110-11, 114, 289, 565-7, 599-600, 680
drive 9, 186, 188, 203-4, 209, 277, 399, 444, 641
Drost-Hansen, Walter 180, 207, 225
drugs 238, 553, 563, 591, 598, 637
dry 70, 83, 161, 169, 179, 275, 593, 647, 670
DU-16 346
Du-26 693
ducts 86, 99, 216, 345, 350, 355-6, 369-70, 403-4, 476, 478, 481, 488-90, 499, 507-8, 612

dynamic 128, 188, 211, 250, 253-4, 279, 285-6, 318, 321, 328, 334, 338, 425-6, 513-14, 703-4

E

E. coli 593-4
Earth Deficiency Syndrome 649
Earth Deficient 650
Earthing 645, 648, 652-4, 657-60, 662-4, 667, 724-5
Earthing Institute 658, 663
Eastern medicine 352, 418, 523
eating 86, 169, 174, 235, 280-1, 286, 306-7, 543, 546-7, 550, 557, 623, 667, 672
ebullition 213
ectoderm 378, 380, 382, 384, 407, 480, 504, 512, 522
ectodermal derivatives 381, 409, 459, 512
edema 341-2, 516, 614, 686, 694-700
edematous 318, 699
edifice 180, 207
editor 39, 212, 238-40, 637
efficacious 245, 292, 294, 696
eggs 35, 169, 235, 300, 357, 468, 480, 497, 540
Eigen ion 183
eight extra meridians 364, 705-6
eight-year cycles of males 673
eight-year life cycles in males 59, 288
eject 461, 516
elasticized tensional milieu 444
elastics 436, 439
electrical system 260
electrodynamical field 321

electromagnetic field   253, 266, 279-81, 286, 290, 350, 372, 527, 662
electron deficiency syndrome   645, 654-5, 657-9
electron-enriched earth   645
electronegative   232, 244, 645-6
electronegativity   89, 214, 232
electrophysiological   433, 441-2
Electrophysiologists   649-50
electrophysiology   535, 719
element   18, 56, 75, 109, 111, 134, 147, 217, 267, 269, 328, 531, 656
elusive morphology   32, 619-20, 625
embryogenesis   160, 375, 378-80, 383, 395, 402, 407, 409, 459, 473-5, 481, 483, 502-5, 520-1, 621
embryologic plasticity   381, 459, 461, 494, 516
embryological vestige   375, 504
embryology   79, 318, 467, 471, 474-6, 481, 537, 710, 718
'Embryology Uterus Development'   481, 718
Embryonic   377-80, 382, 402-3, 407-8, 459, 467, 475-8, 481, 483, 486, 494, 499-500, 504-6, 533-6
embryonic heart   477, 499
emissary   10-11, 543
emissions   10, 29-30, 75, 118, 120, 249, 262, 264, 277-8, 280, 282, 292, 370-1, 400
emitted   10-11, 140, 143-4, 245, 259, 271, 273, 279, 294-5, 400, 559
emotion   61, 267
emotional strain   394
employer vessel   359, 533
empowering   160, 162

empty space   530-1
encapsulated   77, 80, 431, 438, 523
encloses   30, 32-3, 37, 62, 73, 76, 80, 90, 93, 119-20, 245, 353, 406, 416-17, 422
enclosing   20, 22-3, 25, 29-31, 33-5, 75, 113-14, 118-20, 126, 415-16, 444, 462, 467
encompassing fascia   444
encyclopedia   393-4
endocrine   xi, 26, 28, 33, 147-8, 151, 191, 345, 357, 377, 388, 406, 425, 542-3, 643
endoderm   378, 380, 382, 384, 407, 479, 504, 512, 522
endodermal   381, 409, 459, 479, 512
endometrium   379, 384, 468, 485, 497, 522
endomysium   436, 450
endoplasmic reticulum   191
energy   171-5, 186-9, 193-6, 203-5, 209-10, 214-26, 229-32, 243-7, 281-90, 331-3, 462-4, 527-8, 594-5, 629-32, 647-9
engender vibrant   360
Engenders   96, 128, 139, 229, 267, 281, 360, 373-4, 376, 380, 459, 467, 474, 480, 483
enhanced lubricity   397
enigmas   178, 195, 208
enmeshes   381, 405, 409, 422-3, 433, 512-13
enteric nervous system   596, 605-7
enterocytes   582, 620
entheses   429
entry   10, 58, 79, 105, 132, 575, 629, 649
entwined   375, 504

environment   101, 103-4, 128, 205, 232, 243, 267, 269-70, 298, 335-6, 338, 450, 517-18, 559, 599
envoy   362, 368-9, 520-1, 674
enzyme   195, 197, 222, 244, 568, 578-80, 590, 594, 602, 630, 656
epiblast cells   378, 407, 504, 512
epidemic   237, 658-9
epigenetics   530, 563, 598, 679
epigenome   563, 598
epimysium   422, 436, 447, 450
epithelium   413, 435, 479, 481, 585
Epstein-Barr virus   614
equipment   249, 327, 404, 654, 661-2, 664
erythrocytes   85-6, 190
essence   59, 70, 86-7, 90, 96, 107, 171-2, 175, 287-90, 312, 365-6, 481-4, 671-2, 702, 704-5
essential influences   8, 87-8, 90, 110, 114, 145
Eureka   275, 529
Europe   284, 593
evaporates   9, 88, 111, 115, 144-5, 170, 201-2, 354, 681
evidence   161, 179, 208, 215, 228, 234-6, 239-40, 243, 251, 254, 261, 270, 290, 323, 327
evil   29, 31, 93-4, 119-20, 142-3, 145, 291, 360-1, 365, 374, 387, 424, 570, 601
excimers   263-4
exciplexes   264, 399
excrement   66-7, 70
excreted   107, 144, 147, 150-1, 680, 695
excretion   2, 15, 18, 42, 45, 89, 132, 146, 681, 693-6

exhalation   164, 359, 458, 494, 533
exogenous   104, 131-2, 674, 679
experience   39-40, 179, 199, 213, 263, 322, 326, 332, 360, 558, 641, 654, 657, 665, 669
experts   53, 164, 356, 464
extends   27, 45, 50, 80, 220, 226, 266-7, 343, 415, 419, 470, 492, 631, 642
external palace   72-3, 122, 316, 420
external skin   403, 511
external wall   73, 122, 383, 409, 420-2, 432-3, 619, 625
externally   2, 27, 30-1, 33, 37-8, 76, 119-20, 142-3, 268-71, 273, 338, 416-17, 420-2, 432, 437
extracellular fluid   610, 617, 624
extracellular matrices (ECM)   349, 427, 436, 438-9, 446, 449-50, 456
extracellular matrix   220, 226, 340-1, 427, 436, 438-9, 446, 449
extraembryonic PVS   379, 382-3, 408-9
extraembryonic tissue   379, 382
extraordinary channels   333, 529
extremities   126, 133, 136, 156, 297, 458, 635, 676, 705
EZ water   80-1, 87-8, 106-11, 176-8, 183-9, 209-10, 213-18, 221-2, 224-6, 245-9, 289, 298-300, 302-3, 523-5, 527-9

F

fascia   132-3, 317-19, 332-8, 371-2, 427-31, 433-9, 441-4, 449-56,

459-62, 491-2, 494-6, 501, 512-17, 526-8, 715-19
fascial membranes   434, 474, 498, 654
fascial metasystem   335, 337, 437, 443
fascial pockets   438, 448
fascial sac   428
fat tissue   371-2, 406, 639, 643
father   ix, 7, 203, 221, 359, 375, 395, 403, 504, 533, 544-5, 562, 586, 602
fatty membrane   8, 23, 37, 62, 186, 209, 353, 391-2, 406, 416-17, 422, 432-3, 443-4, 519-20, 534
feet   4, 28, 31, 44, 72, 95, 106, 118, 120, 167, 218, 268-9, 491-2, 660-1, 700
Feishu   688, 692
female   59, 74, 288, 306, 481-2, 484, 488, 490, 500
fermentation   310-11, 313, 582, 595
fertilization   375-6, 378-9, 384, 395-6, 404, 408, 460, 468-9, 480, 482-5, 497, 499-500, 504-5, 517, 521-2
fertilized ovum   378, 468, 497
fetal   379, 457-8, 474-5, 493, 565, 599, 671
fetus   392, 468, 480-1, 499, 517
fiber optic   399, 430, 508, 526
fibers   189-90, 219-22, 226, 265-6, 320-1, 324, 326-8, 331-4, 341, 430-1, 435-6, 438-9, 446-8, 491-2, 606-7
fibrinolysis   294
fibroblast   439
fibroblastic cells   262

fibroblasts   189, 340, 420, 427, 438-9, 446, 451-2, 469, 472-3, 482, 497-8, 517-18
fibromyalgia   294, 438, 577, 588, 659, 717
fibrous   190, 334, 338, 375, 413, 422, 431, 459-60, 486, 494-5, 507, 512-13, 515
filtered plasma   611
finger   690, 698, 705
Fire   8-9, 35, 53-4, 66-7, 87-8, 91, 111-16, 246-7, 300-2, 353-4, 462, 470-3, 497-8, 531-2, 701-2
Firmicutes   593
flat-earth society   233
flavored   51
flesh   6, 9-10, 29-32, 63-4, 118-20, 126, 136-40, 142-3, 148-50, 169, 461, 515-17, 551, 556, 628
floating   87, 94, 345-8, 354, 386, 403-4, 505-7, 511, 532, 622
flow   54-6, 82-4, 89-94, 139-41, 193-4, 216-17, 340-2, 350-1, 385-7, 516, 523-5, 615-17, 622-4, 660-2, 683-6
fluctuations   282, 666
fluids   54, 88-9, 136-40, 142-5, 158-62, 203-4, 298-303, 317-19, 340-2, 418-22, 515-17, 609-17, 620-6, 678-82, 693-700
fluorescence   196, 222
fluorescent nanoparticles   7, 140, 149, 344, 348, 391, 525
fluoride   102, 292
fluoroquinolones   308
fog   10, 29-31, 63, 105, 118-20, 310, 312
foggy brain   589, 603

folic acid 584, 594
food 1, 17-18, 42-3, 64-9, 97-100, 107-9, 174-5, 268-9, 281-6, 540-1, 546-8, 557-8, 569-70, 581, 596-8
forces 77, 80, 132, 155-7, 172, 207-8, 216-17, 301-2, 305, 334-6, 384, 494-6, 522, 562, 703
formative causation 304, 309
formless 34, 37, 192
forms 18-19, 53-4, 57-9, 125, 150, 209-10, 221-2, 244, 288, 381-2, 387, 424, 443, 528-9, 671
Formula 183-4, 229, 509
foul 555, 597
fourteen conduits 289
FOXP3 572-3, 601
framework 256, 376, 381, 383, 417-18, 433, 435, 522-3, 525, 609
fraud 237, 240-1
fraudulent 237, 240-1
free electrons 130, 222, 269, 524, 645, 647-50, 654-6, 660-2, 664, 667-8, 670
free radicals 173, 194-5, 645, 650, 655-7, 660-1, 663, 667-70, 725
Fresh 228
frowzy 555
frozen shoulder 430
fructose 445
fruits 125, 169, 229, 284, 286, 547, 711
fu 2, 15-19, 28, 32, 34, 36-8, 54-5, 71-5, 122, 134-5, 137-8, 152-3, 462, 465, 470-1
fungi 26, 125, 173-4, 235, 308, 313, 389-90, 564, 570, 573, 583, 601, 641

furniture 168, 175, 229
fuse 481, 484, 765
Fusobacterium 572

G

G protein 539, 541
gallbladder xvii, 2, 15, 22, 28, 33, 38, 43, 46, 112-13, 116, 157-8, 161, 688-9, 704
gametes 480, 484
Gascoyne, Peter 527
Gastric 101-4, 202, 244, 419-20, 510, 543, 546, 549-51, 555-6, 587, 720
gastric bypass 549, 551, 555-6
gastric motility 510
gastrointestinal chemosensation 539, 551
gastrosplenic 419
gastrosplenic ligament 419
gastrulation 376, 378, 382, 384, 407, 460, 485-6, 494, 507, 512, 522
gate 359-62, 365-6, 394-5, 457-8, 466-9, 472-7, 479-80, 482-4, 486-7, 493-4, 496-500, 517-18, 522, 537, 554-5
Gate of Destiny 466, 496
gate of exhalation and inhalation 359, 458, 494, 533
Gate of Life 71, 360-2, 365-6, 394-5, 457-8, 466-9, 472-7, 479-80, 482-4, 486-7, 493-4, 496-500, 517-18, 522, 537
Gate of Power 466, 496
gelatine desserts 228, 243
gels 109, 181, 185-6, 193, 214, 222, 229, 436, 439, 454, 476, 486

gender   133, 480, 573
genetic inheritance   359
genetics   219, 268, 320, 460, 606
Genkan, Tamba   361-3, 423-4, 520
genki   463
Genuine Qi   463, 703
geriatric   295
germ   234, 274, 314, 351, 381, 384, 409, 459, 479-80, 494, 499, 512, 522, 566, 621
Germany   251, 257-8, 284, 286, 290, 293, 295-6, 595
Gestaltbildung   291
ghrelin   546, 550
GI motor abnormalities   606
glands   26-7, 128, 151, 194, 216, 314-15, 342, 622, 704, 724
glomerulonephritis   699-700
glucagon-like peptide 1 (GLP-1)   542-3, 545, 550
glucose   101-2, 173, 539-40, 542, 545-6, 548-50, 554, 630, 636, 638, 658
glutamate   540, 554
glutamic acid   540
glycosaminoglycans   448-9
goal   40
gonads   479-80, 621
Gracovetsky, Serge   202, 234, 236, 447
grains   5-6, 10-11, 28-30, 77-8, 80-2, 86, 89-91, 93, 97-8, 107-10, 118-19, 168-70, 182, 204, 538-9
grants   207, 241
granules   139, 345, 348, 350-1, 372, 399-401, 407, 508, 531
great sea   297, 299-300

greater omentum   418-19, 424, 633-4, 643, 693
gross   41, 47, 535, 628
ground   6, 218, 221, 317, 325, 335, 338, 341, 435, 438, 451-2, 469, 497, 517-18, 645-7
grounding   218, 232, 645-51, 654-5, 657-9, 661-70, 724-5
growth   128, 174, 185, 195, 250-1, 257, 261-2, 291, 324-5, 355-6, 363-4, 425, 448, 583, 673
Gu Qi   465, 496, 676
Guanchong   690
gullet   46
gushing   5, 91, 155, 670
gushing spring   155, 670
gustducin   539, 541
gustotopic map   543
gut microbiome   xi, xiv, 37-8, 313, 342, 563-4, 577, 580-1, 583, 585-6, 590-1, 596, 598-600, 602-3, 607
GV4   366

H

$H(H_2O)_3^+$ 104-5
H_2O 65, 77, 81, 98, 104-7, 180, 183-5, 194, 207-8, 213-14, 218, 225, 229, 247, 300
Haemophilus influenzae   572, 588
hair   26, 32, 63, 106, 121, 125, 139-40, 144, 150, 172, 353-4, 390, 505, 515, 673
hallucinogenic LSD   590
hand   4, 18, 23, 48, 51, 60, 72-4, 83, 106, 113, 167, 529, 555, 579, 698-9

Hardy, William 225, 228, 243, 247
harmony 45, 58, 83, 124, 142, 273-4, 303, 328, 338, 372, 380, 408-9, 508-9, 555-6, 657
has a name but no form 20, 23-4, 36, 40, 49, 55, 72-3, 177, 192, 402, 443, 456
Hawking, Stephen 373
He-Sea points 80, 135, 144, 688-9, 698-9
healing crisis 576, 659
heart 22-6, 28-31, 44-7, 58-69, 82-3, 85-8, 117-20, 163-5, 411-13, 415-16, 467-70, 475-8, 535-7, 610-13, 665-6
Heart Master Triple Heater (Larre and Rochat de la Vallée) 15, 35-6, 44, 58-9, 62-4, 66-8, 96, 106, 123-4, 134, 152, 160, 317-18, 469-70, 472-3
heart rate variability (HRV) 647, 665-6, 669, 725
heart valves 87, 353-4, 505
heat 18, 21, 67, 129, 159, 161, 216-17, 401, 628-32, 634-6, 638-9, 642-3, 659-60, 689-91, 697
heaters xiv, xvii, 13, 15, 19, 36, 43-4, 51-3, 61, 75, 123-4, 134, 153, 161, 311
heavenly yang 9, 88, 111, 115, 170, 201, 354
heavy 146-7, 151, 200, 223, 591, 680
heavy metals 146-7, 151
Hegu 689, 692-3
Helicobacter pylori 104, 211, 233, 567, 587
helix 219, 264, 266, 272, 274-5, 325, 446, 712

hematopoietic organs 386
hemopoietic activity 479
hence 44, 96, 104, 122, 221-2, 309, 322, 332, 420, 464, 551, 556, 597, 660
heparan sulfate 448
heretic 203
hertz 258
high blood pressure 159, 161, 565, 637, 658
Hippocrates (Greek physician) 46, 282, 562-3, 609, 618, 624
Hippocratic Oath 282
Histological 7, 343-4, 387, 391, 438, 525
history xiii, 14, 20, 22, 107, 205, 292, 295, 564, 599, 646
hitchhikers 342, 569-70, 578, 600
Ho, M. W. 265-6, 320-8, 338, 514, 530
Ho, Mae-Wan 219-21, 226, 320, 528
hollow 17, 28, 34, 71, 86, 99, 135-7, 149-50, 312-13, 396, 403-5, 421, 427-8, 432, 632-3
holy scriptures 3, 6, 281, 394-5, 463, 707, 717
homeodomain transcription 621
homeostasis 128, 220, 250, 288, 380-1, 408, 459, 509, 648
homeostatic 337, 445, 454, 672, 699
horizontal gene transfer 564, 599
horizontal movement of Qi 56
hormonal systems 269, 376-7, 402, 407, 454, 511, 693
hormones 26, 262, 294, 356, 385, 388, 425, 435, 451, 510, 540, 543, 545, 547, 616-17
hospital 146, 556, 591, 606

hostility 665, 669
(H3O)⁺ 194
[H3O2]⁻ 77, 81, 98, 130, 178, 183-4, 207-9, 246, 300, 302, 478, 532, 537
Hua Tuo 268-9, 273-4, 338-9, 437
Huang Wei-san 37, 62, 69, 186, 209, 353, 405-6, 416-17, 422, 432, 443, 519
Human Genome Project 563, 598
human immunodeficiency virus (HIV) 293
humoral immunity 294
hunger hormone ghrelin 546, 550
hyaline 452
hyaluronic acid 345, 353-4, 373, 385, 397, 420-2, 433, 441, 446, 455, 509
hydrate-microcrystal theory of anesthesia 230
hydrated 184, 208, 246, 327, 332, 532
hydration 77, 80, 127, 184, 208, 322, 523, 573, 575
hydrochloric acid 88-9, 183, 202, 211, 233, 244
hydrogen 104, 185, 194, 221, 230, 244, 324, 327, 338, 514, 595
hydrogen sulfide 595
hydronium 188, 193-4, 203-4, 243-4
hydronium ions 188, 193-4, 203-4, 243-4
hydrophilic 65-7, 103, 105, 109-10, 184-9, 193-4, 196-7, 209-10, 216-17, 242-3, 245-7, 274-5, 289-90, 298-300, 447-9
hydrophobic 103-5, 110, 274-5, 289
Hydroxide 178, 184, 208, 246, 532
hydroxide hydrate anion 246, 532

hydroxyoxidanium 184
hydroxyoxonium 184
hydroxyperoxonium 184
hygienic 234
hypersensitive 325
hypertonic muscles 652
hypodermis 133, 356, 369, 511
hypothalamus 47, 554, 565, 630
hypothesis 14, 48, 177, 190, 211, 216, 222-3, 233, 236, 304, 313, 372, 378, 407-8, 531
hypothesized 185, 403, 506, 531, 533

I

IDEA Fitness Journal 72, 442, 444-6, 709
Iliotibial band 444, 452
illnesses 138, 423, 575, 687
image 21, 155, 310, 318, 323, 351-2, 628
immaterial 171-2, 174-5, 287, 297
immersed 126, 322, 403, 435, 440, 511, 532
immune 26-8, 191-2, 294, 313-15, 336-7, 376-7, 424-5, 465, 509-11, 531, 565, 583-4, 602-4, 608-10, 655-7
immune system 22, 26-7, 37, 139, 191-2, 294, 313-14, 348, 360, 465, 567, 574-5, 583-4, 602-3, 655-7
immunofluorescent 344
immunoglobulins, IgM 314
immunologically defensive 387
imperial herald 369, 675
impure 43, 176, 202, 204, 680-2
Inborn Qi 463

inconceivable 180, 208
incretins 545
index 60, 183, 200, 213, 718
indirectly 84, 140, 150, 168, 175, 244, 300, 391, 426, 508, 515
industrial 103, 213, 268, 550, 654
infection 27, 235, 413-14, 422, 424, 432, 575-6, 588, 594-5, 614, 623, 626, 641, 643
infectious colitis 596
inferior vena cava 416, 491, 501, 633, 636
infertility 544, 553, 558, 673
inflammation 439, 531, 575-7, 645-7, 649-50, 653, 655-9, 663-4, 667, 669, 722, 725
inflammatory bowel disease 575, 589
influences 8-12, 29-33, 57-62, 77-80, 84-97, 110-11, 113-15, 118-20, 163-8, 170-2, 297-300, 359-69, 387, 555-8, 674-5
infrared 168, 175, 183, 195-6, 199, 201, 213-17, 224-5, 228, 243, 245, 247, 251, 254, 329
infrastructure 99, 127, 137, 139, 153, 265, 326, 356, 376, 381, 384, 404, 453-4, 471-2, 512
infuses 286, 450-1
inhalation 9, 88, 111, 115, 164, 170, 354, 359, 458, 463, 493-4, 502, 533
inhaling 468
inherited genes 388
injecting 7, 347, 391, 404, 525, 545
injured 29, 119, 161, 218, 369, 453, 521, 556, 667, 683
innate qi 463, 702-3
Inner Classic 397-8, 526

insufficient lactation 313
insulating material 436, 439, 662
insulation 336-7
insulin 539, 542, 544-7, 549-50, 636, 639, 644
insulin-sensitizing 639, 644
integral 27, 132, 148, 337, 403, 438, 453, 465, 467, 496, 506, 520, 532
Integrative Medicine International 345, 377-9, 382, 716
integrins 335, 449-50, 455-6, 528
integrity 64, 83, 97, 154, 185, 234, 237, 253, 255, 273, 336, 445, 452, 455, 674
interact 175, 179, 247, 380, 419, 571, 577, 649, 656, 722
intercommunication xii, 220, 226, 265-7, 271, 321-2, 324-5, 328, 338, 514, 530
interconnected xii, 265, 267, 270, 282, 288, 322, 327, 333-4, 338, 373, 408, 417, 427, 444
intermediary 36, 152, 381, 409, 459, 471-2
internal organs 50, 138-40, 148-9, 158-60, 348-9, 355, 363, 369, 383, 403-4, 411, 413, 423, 511-12, 676-7
International Institute of Biophysics 251, 257-8
interpenetrates 334, 338, 513-14
interpretation xi, xv, 1, 16-17, 20, 23-4, 51, 203, 265, 305, 372, 466, 472, 483, 496
interscapular 634
intersection 78, 365

interstitial fluid   26, 186, 209, 231, 340, 342, 405, 428, 515-16, 556, 582, 610-12, 616, 624
interstitial hydrostatic   341
intestines   xvii, 2, 15, 28, 33, 38, 44, 46, 49-50, 311, 313-15, 411-12, 424-5, 554, 582-6
intraembryonic   381-3
intramembrane vessels   357, 406
intramuscular   334, 398, 407, 439, 513
intrinsic   59, 85, 137, 168, 170, 174-5, 221, 246, 280, 300, 325, 437, 452, 587, 612
Intrinsic Factor   85
ions   104, 183-4, 187-8, 193-4, 203-4, 208, 214, 230, 232-3, 243-6, 253, 324-5, 338, 340, 485
irrigates   57
irrigation   1, 3, 15, 54, 68, 108, 127-8, 149, 160, 226, 231, 236, 317, 319, 462
Isoflurane   230
isolated   69, 173, 178, 307-8, 351, 434, 439, 531, 559, 662
Italy   38, 295-6

J

Japan   282
Jehovah   6, 393-4
Jesus   281
jet lag   652-3, 664
jin ye   64, 131, 298-9, 418-19, 679-82, 725
joints   56, 84, 93, 128, 141, 159, 161, 183, 229, 352-4, 413, 421-2, 429-30, 523-4, 651-2

journal   13, 39-40, 48-9, 126-7, 135-6, 236, 334-7, 357-8, 370-2, 377-9, 444-6, 513, 698-9, 707-11, 713-15
Journal of Acupuncture and Meridian Studies   269, 350, 357-8, 370-2, 401, 506, 511, 513, 527, 714
Journal of Chinese Medicine   13, 16, 39-40, 48-9, 80, 126-7, 135-6, 153, 334-7, 513, 523, 698-9, 707-8, 710, 715
juice   202, 427, 609, 624

K

karyolysis   374
ke-shu   90, 92
keratin   441, 455
key   25, 117, 213, 220, 224, 226, 238, 278, 288, 331, 382, 429, 573-5, 601-2, 646
kidney qi   673, 676, 703
kidneys   82-5, 171-2, 287-90, 297-302, 359-63, 365-8, 391-5, 462-71, 475-9, 487-93, 498-503, 531-7, 670-7, 695-7, 702-4
kilogram   192, 587, 603
Kimberly Thompson   138, 158, 191, 231, 693
kinesthesia   447
Klinghardt, Dietrich   258-9, 280, 288
knife   396
Knight, D. P.   265-6, 320-8, 338, 514, 530
Kyoto University   282
Kyungrak System   370

L

L. casei 566
L. rhamnosus 566
laboratory 213-14, 232, 242, 245, 309, 314, 547
lactase 342, 569, 582, 594, 600
lacteals 313, 609, 620
Lactobacilli spp. 595
Lactobacillus 311, 314, 565, 572
Lactococcus 572
lactose 582, 584, 594
Lancet, The 240, 712
large intestine 5, 36, 43-5, 60, 66-7, 70-1, 75, 80, 101, 126, 204, 313, 331, 412, 581-2
laser 215, 217, 223, 258, 264, 271-4, 294-5, 401, 508
lattice 185, 187, 194, 266
lecithin 103, 235
legs 56, 79, 126, 231, 491, 615, 618, 695
lesser omentum 418, 425, 717
leukocytes 190, 608, 624, 640
L.I.-4 *Hegu* 692-4
libido 557-8
Lidocaine 230
Lieque 687, 692-3
ligament 7, 391, 418-20, 437, 439, 444, 489, 491, 501, 525, 717
ligamentum flavum 491, 501
light 166-8, 170-2, 183, 195-6, 212-18, 224-5, 248-55, 258-60, 262-5, 271-82, 284-8, 294-5, 351, 578, 713-14
light conductive 259
lightning storms 649
Ligou 687
like a bag 32, 75, 226, 417, 423, 433, 435, 440, 522, 619, 625, 654, 669
Lingering Pathogens 570, 601
link 112, 134, 198, 336, 438, 470, 530, 554, 564, 573, 601, 717
lipids 105-6, 191, 230, 255, 323, 385, 612, 620, 628, 630
liquid crystalline 80-1, 176-8, 184, 209-10, 219-23, 226, 228, 245, 265-7, 270-1, 320-3, 325-8, 497-8, 526-30, 712-13
liquids 5-9, 64-7, 69-70, 88, 109-11, 144-5, 176-8, 198-204, 219-29, 245, 265-7, 320-8, 353-5, 397-9, 526-30
little toe 444, 449, 455
LIV-3 688, 691
liver 2, 22, 43, 46, 58, 60-1, 75, 84-5, 92, 112, 157, 161, 369, 521, 687-9
living water 98, 108, 130, 219, 247, 478, 537
Lo, Shui Yin 223, 226, 712
Loose Connective Tissue 340, 342, 371-2, 406, 413, 420-2, 514-16
loose teeth 673
lovastatin 563
low electrical obstructions 398
lower 5-9, 29-30, 43-7, 49-51, 56-8, 66-8, 70, 87-90, 118-20, 133-7, 144-5, 580-1, 634-6, 688-91, 693-9
LU-7 687, 692-4
LU-9 686-8, 691-2
lubricate 75, 353-4, 405, 413, 421, 425
lumbar region 144, 491

luminescent 272
lunar 59, 172, 288
lung immunity 313
lungs 2, 5-6, 10-12, 42-6, 60, 62, 85-8, 96-7, 103, 105-6, 139-41, 163-5, 313, 572, 676-82
lupus 641, 659
lymph 7, 26, 93-4, 96, 190, 194, 314, 334, 338, 341, 404-5, 420, 513-14, 608-17, 620-4
lymphatic capillaries 341-2, 516, 615, 617-18
lymphatic endothelial cells 618
lymphatic system 26, 43, 49, 231, 312, 355, 390, 404, 417, 419, 426, 445, 455, 608-26, 723
lymphatic vessels 26, 94, 96, 344-5, 349, 364, 386, 403, 609, 611-13, 620
lymphocytes 341, 609-10, 614, 624
lymphoid tissues 608-9, 624, 723
lysergic acid diethylamide (LSD) 590

M

maceration 19, 49, 65, 67-8, 134, 175, 198
machinations 189, 210, 257, 528, 686
Maciocia, Giovanni 1, 15, 40, 45-6, 50, 52, 55-8, 89, 343, 417, 423, 465, 692-7, 707-9, 725
macro 336, 418, 436, 439, 523
macromolecules 219, 221-2, 448
macrophage collections 424
macrophages 190, 425, 438, 609, 624, 640
magnetic resonance imaging (MRI) 329

maintained 127, 157, 268-9, 273-4, 321, 338, 437, 455, 459, 577, 630, 642, 656
maladjusted affairs of the bedroom 672
male reproductive organs 481, 500
mammalian organs 349, 354, 716
man 6, 71-2, 86, 90, 92-5, 97, 167-8, 171, 175, 287, 359, 361-2, 394-5, 482-4, 533-4
management 67, 125, 134, 198, 352, 592
mansion 457-8, 493, 706
manufacture xiv, 21, 43, 53, 58, 82-3, 85, 89-90, 96, 98, 107-8, 110-11, 195, 238, 386
Margolskee, Robert F. 539-41, 543, 545, 550, 553
Margulis, Rena K. 330-4, 338, 514, 529, 715
marrow 43, 82-3, 85, 190, 386, 389-90, 425, 472, 608-9, 624, 674, 680, 704, 711
Marshall, Barry James 202, 211, 233
masculine uterus 490
massage 132, 245, 401, 441, 508, 717
mast cells 427, 438, 446, 531, 640, 719
mast cells (MC) 531
masters 10-11, 78, 97, 144-5, 366, 436, 468, 560
mastitis 313
matrices 220, 224, 226-7, 325, 334-5, 338, 340-1, 438-9, 444-6, 448-9, 486-7, 513-15, 528, 530-1, 667-9
matryoshka 368

matter 18-19, 32, 36, 51, 74, 101, 145, 178, 205, 222, 229, 325, 377-8, 393-4, 534

maze 153, 226, 246, 315, 328, 417, 433, 522, 530, 532, 577, 619, 625

McDonald's 168, 264, 272-3, 285

McFarlane, Bruce 449-50, 491

measuring 154, 666

mechanical stresses 323, 336

mechanics 89, 255, 366, 431, 444, 468

mechanosensitive 337, 434

mediastinum 411-12, 436, 632

mediated 217, 223, 230, 253, 268, 270, 274, 518, 539, 579, 596

medications 27, 102, 562-3, 565, 657-8

megabytes 259, 279

Megasphaera 572

melanin 172-5, 205, 711

membrane 17-18, 37, 122, 163-5, 182, 186-7, 214, 229, 390-2, 410-13, 415-24, 432-3, 519-20, 534-5, 630-1

membrane conduit system 418

membranes 67-8, 79-80, 186-7, 189-94, 209-10, 221-2, 226, 371-2, 413-21, 423-4, 431-4, 474-5, 630-1, 642, 692

membranous bags 473, 498

memory ix, 206, 250, 254, 259, 280, 304, 325-6, 453, 528, 530, 544-5, 548, 566-7, 673-4

meninges 191, 334, 413-14, 421, 513, 633

Mercola, Joseph 211-13, 237-8, 258-60, 272-5, 277-82, 284-6, 540, 564-5, 599-600, 610-11, 636-8, 640-1, 646-7, 712-14, 719-25

mercury 146-7, 151, 260, 292

merge 136, 148

meridians 152-5, 219-24, 320-2, 326-33, 346-7, 357-9, 368-73, 375-7, 383-6, 395-408, 448-9, 502-6, 510-11, 525-34, 705-6

mesenteric 432, 474

mesoderm cells 378, 380-4, 407, 409, 476, 479-80, 489, 504, 512, 517, 522, 533, 536, 621

mesodermal 381, 409, 459, 494, 512

mesonephric kidney 478

mesonephros 476, 478, 481, 489, 499, 536

metabolic disorder 658

metabolic enzymes 259-60, 285

metanephros 476, 478, 536

metaphors 397-8, 526

metasystems 28, 156-7, 269-71, 298-303, 337-9, 383-4, 402-5, 421-3, 427-30, 432-3, 442-5, 502-6, 511-16, 518-22, 532-3

micro 66, 329, 345, 401, 407, 436, 439, 508, 571

microbes 313, 564-5, 569, 571, 582, 588, 590, 592-3, 598-600, 614, 721-2

microbial infections 291

microbiome xiv, 27, 37-8, 192, 268, 313, 342, 562-4, 567-9, 571-5, 580-91, 596, 598-600, 602-3, 720-2

microbiomology 587

microbiota 313-14, 565-7, 587-9, 599-600, 715, 721

microflora 27, 125, 311, 313-15, 342-3, 348, 567-70, 573, 581, 583, 587, 589-91, 594, 600-1, 607
micrometers 246, 532
microorganisms 145, 173, 194-5, 268, 313-15, 342, 565-7, 570-3, 575-7, 580-7, 592-3, 598-600, 603-4, 610-11, 613-14
microsensing apparatus 328-9
microspheres 181, 232
microtubules 137, 149, 254-5, 257, 280, 298, 303, 367-8, 403, 462, 495, 517, 533, 713
middle kidney 478, 481, 499
migrate 345, 380-1, 409, 479-80, 535
minerals 65-6, 91, 102, 107-8, 268, 289, 555, 710
Ming dynasty 246
ming men 58, 76-7, 123, 141, 155-6, 289, 301-2, 457-8, 465-7, 469-74, 486-8, 491-2, 495-6, 498, 501-2
mingmen 16-17, 466-7, 483, 496, 534, 708, 718
Mingmen abound 467, 496
miniscule 110, 376, 507, 512-13, 526
minister 8, 21, 50, 53, 87-8, 91, 112-16, 246, 326, 462, 472, 697
Minister Fire 53, 88, 112-13, 115-16, 246, 697
mist 10-11, 19, 29, 31, 49, 63-4, 67, 105-6, 118, 120, 134, 141, 198-9, 310, 312
mistletoe 277
mites 570, 601
mitochondria 255, 264, 581, 590, 594-5, 630-1, 639, 642
mitosis 388, 399

mixing 470, 497
model iv, 96, 148, 183, 185, 219, 222, 245, 259, 267, 310, 376, 380, 399, 459
moistens 10-11, 83-4
molecular biologist 274
Molecular Weight 184, 247, 596
molecules 101-2, 105, 107-9, 179-81, 184-5, 196-7, 199, 208-9, 212-14, 229-31, 242-5, 247, 278-80, 578-81, 602-3
monkeys 306-8
mononucleosis 614
mononucleotides 373, 385
moon 59, 166-7, 170-2, 174-5, 287-8, 290, 297, 300
Moraxella catarrhalis 572
Morgagni sinus 490, 501
morphic field 304-5
morphic resonance 304-5, 309
morphogenetic fields 304-9, 715
morphological 22, 32, 74, 376, 402, 407, 458, 486, 504, 511, 521, 704
morphology xi, 19, 25, 32, 34-5, 48, 68, 73, 127, 136, 328, 344, 620, 625, 699-700
mortality 233, 314
morula 483-5, 487-8, 500
mother ix, 79, 82, 84, 307, 313-15, 343, 359, 375, 395-6, 449, 468, 493, 502, 533
Motive Force 156-7, 301-2, 695
moulds 125, 570, 601-2
moxibustion 20, 46, 220, 329, 688
Mu point 362, 692, 696
mucopolysaccharides 447
mucous 413, 421

mucus 103-4, 183, 235, 299, 413, 421, 447, 572, 622-3, 626
muddy pool 49, 682
Müllerian duct 488
multicirculatory system 373
multitude 201, 203, 269, 573, 601, 605
muscular 25-6, 80, 189, 415, 447, 588, 628, 633-4, 643, 678
mushrooms 235, 540-1
Mutaflor 595
mutations 283, 388-9, 668
myeloid cells 355
Myers stresses 512
myofascial 333, 336, 430, 453, 718
myosin 441, 455
myriad 48, 474, 498
myristic acid 392
myths 235, 241

N

Nafion 181, 193, 197, 222
Najemy, Robert 305-6
Naohui 690
nasal 235-6, 295, 570-1, 601
National 212, 320, 349, 539, 547, 553, 558, 663
National Genetics Foundation 320
nature xiv, 12-14, 35-7, 63-6, 113-17, 124, 180-1, 195-6, 264-6, 320-3, 444-5, 483-4, 528, 672-3, 684
nausea 548, 576, 670, 696
navel 7, 44, 171, 287, 297-8, 300, 359, 362, 365, 367-8, 391, 622, 674
necrotizing 314, 589
Needham, Joseph 45, 323
needles 154, 350, 388, 660

neighboring 220, 262, 442, 449, 692
nematodes 570, 601
Neonatal Intensive Care Unit (NICU) 314
nephridia 478, 499
nephrotomes 476, 536
nerve 187, 229, 251, 334-5, 374, 425, 431, 438-9, 442, 461, 495, 513-14, 516-17, 596, 606-7
nerve impulse 229
nervous system 26, 250-1, 254, 321, 324-6, 337-8, 356-7, 374-5, 406, 431, 445, 454-5, 605-7, 663, 665
network 29-31, 33-4, 113-14, 118-20, 127-8, 135-7, 139-41, 148-50, 167-8, 324-7, 346-8, 427-8, 461-2, 513-17, 523-4
neural canal 491, 501
neural pathways 466, 496, 535
neuroendocrine 565, 600
neuromyofascial 429, 442, 709
neurons 253-5, 325, 559, 590, 596, 605-7, 722
neurophysiology 327
neurotoxins 570, 601
neurotransmitters 262, 425, 566, 568, 578, 580-1, 586, 588, 590, 602, 606-7
New Anatomical System 269, 350, 356-8, 370-2, 376, 383, 401, 506, 511, 513, 527, 714
new viewpoint 370
Nielsen, Arya 45, 131, 155, 183, 298, 301, 512, 684
night 63, 95, 280, 543, 651, 660, 670
Nitrogen 125, 373
nitrosamines 582-3
Nitrosomonas eutropha 597-8, 604

no form 19-20, 22-4, 31-6, 40-1, 52-3, 55, 72-4, 117, 120-1, 133-4, 191-2, 416, 443-4, 456, 465
noble gases 230
noninvasive 250, 638
nonprotein nitrogen 373
norepinephrine 629
North Korean 344, 346
nothing but membranes 37, 186, 209, 226, 248, 256, 260, 276, 337, 406, 417-18, 432, 522, 619, 625
nourished 127-8, 268-9, 273-4, 338, 437, 595, 675, 704
nourishing 43, 50, 54, 83, 85, 92, 101, 161-2, 396, 557, 583, 616, 677, 684, 704
nourishment 10, 128, 183, 286, 299, 396-7, 434, 457-8, 583, 676-7, 682
nucleic acids 252, 278, 369-70
nutraceuticals 389
nutrients 11-12, 90-1, 99-102, 107-9, 172, 174, 286-9, 425-6, 475, 535, 546-7, 550-1, 555-6, 581-3, 615-17
nutrition 234, 335, 547, 554, 584, 590, 648, 721
nutritive qi 1, 53, 57, 61, 68, 108, 677

O

Ober, Clint 645, 651, 654, 659, 664, 725
official 1, 3, 6, 8, 18, 78, 87, 108, 114, 178, 207, 226, 231, 316, 319
official in charge of irrigation 1, 108, 226, 231, 247, 319, 326, 527
Oil Pulling 573-7, 601-2, 721
oleic acid 392
oligosaccharides 448
omentum 191, 418-19, 424-6, 633-4, 643, 693, 717
omnipresence 123, 143, 195, 264, 339, 371, 427, 434, 450, 461, 513, 516
omnipresent 25-8, 127, 184, 209-10, 226-7, 246-7, 269-70, 275-6, 298-300, 337-9, 427-30, 436-7, 504-8, 512-13, 522-5
100,000 chemical reactions per second 252
1 percent theory 306
opening 5, 37, 45, 50, 74, 142, 160, 167, 186, 209, 226, 317, 406, 417-18, 457-8
operate 275, 285, 458, 493, 502
optimal 6, 57, 88, 128, 156-7, 172, 214, 265, 274, 281, 285-8, 290, 570-1, 648, 660-1
orchestrates 32, 257, 259, 270, 319, 458, 474, 498
organ xiii-xv, 1-3, 12-19, 32-4, 36-8, 46-51, 334-5, 402-9, 443, 464-5, 474-8, 498-500, 533-7, 585-9, 698-700
organ complexes 477, 498, 537
organelles 250, 255, 367, 418, 523
organism 8, 10-11, 21, 30-1, 33, 77-8, 94-6, 119-20, 219, 249-50, 266-7, 322, 367, 428, 581
organized 137, 141, 150, 192, 219, 259, 272, 384, 427, 446, 452, 524, 587, 603, 608
organogenesis 378, 407, 618
organs 32-8, 43-4, 49-51, 71-6, 137-40, 148-50, 156-60, 311-12,

348-9, 354-7, 380-4, 403-4, 411-13, 416-27, 522-6
origin   297-9, 361-2, 368-9, 377-8, 402, 407-8, 456-8, 473-4, 481-4, 493-6, 505, 519-21, 533-5, 620-1, 674-5
origin of the Triple Burner   8, 353, 359, 391, 422, 452, 456-8, 493-4, 519-20, 534-5
original   16-17, 72, 122-3, 135-8, 152-3, 155-7, 297-303, 359-63, 365-9, 373-5, 486-8, 520-1, 533-5, 695-6, 701-3
original influences   72, 77, 113, 122-3, 129, 156, 297-300, 359-63, 365-9, 373-5, 482-3, 486-7, 520-1, 533-4, 674-5
original qi   17, 69, 82, 85, 123, 152-3, 319, 463, 493, 502, 568, 695-6
Oschman, James L.   334-5, 338, 514, 527-31, 667-9, 719
osmosis   45, 66, 485
osteoporosis   556, 575, 594-5, 606
Otto Madsen Dairy Research Laboratory   213
overflow   95, 178, 208, 672, 686
oxidanyloxidanium   184, 711
oxygen   85, 106, 125, 140, 185, 194-5, 217-18, 221, 269, 274-5, 294, 329, 458, 594, 616-17
oxygenated   6, 10-11, 85, 468, 613, 617, 623-4

P

P-6   685, 687, 697
P-6 *Neiguan*   697
P-7   686-7, 691, 697
P-microcells   94, 156, 299, 303, 351, 360-1, 364-5, 367-9, 374, 403-4, 506, 521, 533
packaging wrap   417
pain   79, 161, 230-1, 280, 313, 330-2, 336, 448-9, 453-4, 647, 651-3, 659-60, 686-8, 690, 698
paired pronephri   476, 533, 536
palace of uniqueness   28-30, 32, 118-19, 342, 430, 515
palaces   14, 28-31, 34-5, 62, 71-3, 75, 86, 113, 118-23, 166, 353, 416-17, 420-3, 432, 551
palatable   307
palindromic   466, 483, 534
palmitic acid   392
palpitations   657, 665, 669-70
Panaxea   328
pancreas   42, 107, 380, 412, 427, 541, 544, 552, 559, 633
panic   665, 669
paper   104, 172, 239, 330, 332-3, 434
paradigms   21, 39-40, 178, 194, 202-3, 211, 234-6, 241, 296, 352, 388, 400, 404, 444, 562
paramecium   125, 570, 601
paramesonephric ducts   481, 490
parasites   27, 104, 573, 589, 592, 603
parasympathetic   645, 658, 663, 665
parents   155-6, 301, 396, 463, 468, 480, 499-500, 517, 586, 671, 720
parturition   8, 473, 483, 491, 498, 501, 672
passages   1, 6, 21, 23-4, 54-5, 57-8, 78, 122-4, 134-5, 160, 316-17, 368, 570-1, 674, 695-6

passageway 59, 127-8, 135-6, 142-3, 149, 312-13
passageways of water 9, 29-31, 77, 87, 114, 118, 120, 127, 178, 207, 299, 312, 316, 342, 515-16
path 11, 140, 149, 344, 363, 369, 397, 511, 590, 623
pathogenic 113, 127-8, 132, 141-3, 148, 293, 567, 572-5, 579, 582-3, 593-5, 667, 678, 683, 691
pathogens 109, 139, 336, 350, 570, 572-3, 588, 593, 601, 616, 640-1, 655, 667, 679, 703
paths 15, 97, 353-4, 520
Pauling, Linus 230, 712
peanuts 169
pecks 29, 118
pectoral qi 100, 703
peculiar 192, 370, 587, 603
pelvic 43, 51, 136, 411-12, 612, 706
pelvic cavity 43, 51, 136, 411-12, 706
penetrating 14, 84, 100, 168, 175, 289, 363, 434
peptide YY (PYY) 543
perceive 24, 279, 282, 326, 539, 552
pericardium 2, 18-19, 33-4, 38, 46, 49, 53, 60-1, 74-5, 112-13, 115-16, 415-16, 685, 697, 709-10
Pericardium channel 60, 685
perinephric fat 392, 466, 496, 535
periodontal disease 343, 577
periosteum 438, 450
peripheral circulation 660
peripheral nerves 436
peripheral neuropathy 556
periphery 128, 179, 634
peritoneum 164-5, 191, 413, 418-19, 422, 436

permeates 6, 11, 68, 135, 157, 226, 245-6, 285, 334, 417, 433, 435, 438, 451, 469
permeation 149
perspiration 146, 663, 678, 682
persulfate 178, 208
Pethig, Ron 527
petrochemical 147
Peyer's patches 609
pH 88-9, 101, 103, 105, 176, 186-7, 202, 232, 266, 322, 715
phagocytic 190
phase 18, 65, 67, 77, 177-8, 180-1, 199-201, 212-13, 222-9, 242-3, 245, 249, 322, 709-10, 712-13
Phenylthiocarbamide 542
philosophical 350, 701
phone 129, 285, 662
phospholipids 103-6, 620
photographs 200, 224, 227, 651
photon 196, 249, 251-4, 256-7, 259, 261-3, 277, 279-80, 287, 370-2, 399, 527
photon emissions 249, 277, 370
photonic 171, 205, 287
photosynthesis 195-6, 221, 243, 285, 388
photosynthesize 229
phthalates 147
physical properties of EZ water 183
physiological 12, 18, 21, 43, 69, 101, 113, 173, 245, 251-2, 256, 335, 337, 683, 724
physiology xi, 19, 50, 64, 84, 136, 233, 335, 337, 347, 357, 563, 587-8, 680-1, 718-19

piezoelectric 300, 332, 399, 430, 441-2, 449, 453, 455, 466, 496, 514, 526, 528, 535
piezoelectricity 130, 441, 453, 455, 508, 528, 533
pioneering 218, 240
pioneers 378, 408
Pishu 688, 693
placenta 79, 485, 565, 599
plants 174, 196, 204-5, 221, 229, 232, 243, 245, 249, 251, 254, 268-9, 285, 300, 661
plaque 576, 659
plays 58, 60, 217, 341, 375, 396, 429, 504, 531, 616
plethora xi, 296, 568
pleura 413, 436
pleural cavity 411, 699
pluripotent 351, 360-1, 369, 378, 407, 521
Pluripotent Adult Stem Cells 351
pluripotent stem cell (PSC) 378, 407
pneumonia 292, 294, 296, 623, 626
polar bodies 377-9, 382, 408
Polar Body 345, 377-80, 382
polarization 324, 329
poles 471
Poletti, Glenda 450-3, 469, 497, 517
Pollack 65-7, 77, 80-1, 88-9, 108-10, 177-89, 193-206, 208-10, 212-18, 221-2, 224-34, 241-7, 298-9, 301-2, 528-9
pollution 292, 661
Polygonum multiflorum 390
polymers 109-10, 181, 654
polywater 185, 205
Popp 252-3, 258, 261, 277-80, 286, 288, 291-2, 296, 400, 580-1, 714

Popp, Fritz-Albert 258, 277-8, 288, 292, 296
pores 29-30, 118-19, 131, 140-1, 143-5, 148, 150, 166, 524, 678
portions xvii, 5-8, 43, 86, 91, 182, 195, 204, 316, 430, 681, 704
posterior perirenal fascia 491, 501
Postnatal Jing 82, 672-3, 675
potential 132, 139, 142, 154, 173, 186-8, 204-5, 209, 214, 229-30, 232-3, 363-4, 424-5, 657-8, 665-6
power 66, 77, 80, 129, 142, 182, 277-81, 290, 295-6, 466-7, 470-1, 492, 496, 610-11, 723
practical xi, xv, 204, 290-2, 332, 418, 429, 523, 585, 652, 684-6, 694, 697, 700
Practice 45, 131, 155, 298, 301-2, 512, 574, 684, 708
pre-natal qi 463
predominant 135, 188, 202, 217, 260, 334, 399, 452, 514, 526, 687
pregnancy 8, 468, 473, 497, 517, 575, 585, 671, 673, 682
pregnant 194, 315, 343, 565
prenatal essence 464, 702
Prenatal Jing 83, 468, 479-80, 499-500, 517, 671-3, 675
pressure 37, 47, 159, 161, 198, 204, 207, 216-18, 245, 298, 322, 329, 341, 647, 658-9
Prevotella 572, 595
primo micro cells (PMC) 345, 378, 407
Primo Nodes 7, 139, 346, 348, 356, 384, 386, 391, 525, 531, 719

Primo Vascular System 139-41, 149-50, 268-70, 344-7, 349-52, 356-8, 367-72, 374-86, 389-91, 398-404, 497-8, 502-3, 505-9, 511-15, 517-18
Primo Vascular System (PVS) 98-9, 139-40, 149-50, 268-70, 298-9, 343-50, 352, 354-8, 367-86, 389-91, 395-409, 502-15, 517-22, 524-7, 533-4
primo-vessels 344
primordial 155-7, 368, 375-6, 401-2, 404, 407, 469, 472-3, 475-6, 479-81, 493, 497-9, 502, 504-5, 517-19
primordial influences 368, 675
primordial qi 404, 463, 493, 502, 703
'Primum non nocerum' 562-3
prions 125, 570, 601
probiotic 565, 567, 594-6, 599
probiotics 314, 565-6, 571, 594-5, 721-2
Proclamation Gate 466, 496
pronephric kidney 478
pronephros 475-8, 498-9, 536-7, 718
proprioception 337, 447
propyl alcohol 589
propylaldehyde 589, 603
prostatic utricles 488, 490
protection 128, 183, 299, 337, 380, 393, 408, 413, 421, 432, 435, 505, 509, 567, 662
protective 29-30, 90-1, 93-7, 104, 118-19, 142, 165-8, 368, 387, 414, 421, 447-8, 538, 667-9, 677-8
Protective Qi 677-8
protein 156, 181, 185, 197, 219, 230, 232, 251, 265, 325, 448, 528, 539-41, 611, 630

protein core 448
proteins 27, 65-6, 97, 101-2, 105, 107-9, 186, 222, 323, 327, 345, 447-8, 459-60, 541-2, 553-4
Proteobacteria 593
proteomic 355, 362, 520
protonated 183, 193, 202, 244, 711
protons 193, 197-8, 202, 216, 229, 321, 630
prove xiv, 51, 194, 263, 289, 475
proximal 313, 466, 492, 496, 535, 616, 690
Pseudomonas 572, 579
psoriasis 641
psychological health 565, 567, 599
pulmonary trunk 416
pumps 77, 80, 176, 187, 189, 193, 210, 215-17, 224, 233, 245, 298-9, 350, 478, 536-7
purified 64, 66-7, 264, 613, 624
puzzles 207

Q

qihua 144, 299, 699-700
Qiuxu 686-7
Quackwatch 211-12, 233, 295
quantum 219-20, 222-3, 252, 255-6, 261, 399-400, 531, 713
quinine 548

R

radiant 195-6, 205, 216, 225, 243, 245, 249, 253, 273, 286
radiant energy 196, 205, 216, 225, 243, 245, 249, 253, 273, 286

radical 194, 211, 233, 355, 463, 598, 646, 649, 656, 664, 667-8
radio 129, 662
rain 9, 88, 115, 125, 170, 200-1, 297, 312, 588
rainbow worm 219-20
raw 169, 185, 259, 281, 284, 286, 289-90, 363, 451, 714
Rayleigh-Bénard 199
react 144, 179, 539, 546, 570, 656
receptacle 67, 312
rectum 412
red blood cells 85, 294, 386, 614, 646, 657
reflected 72, 172, 287, 358
reflects 72, 568
regenerate 156, 303, 387
regeneration 69, 250, 345, 351, 356, 360, 369, 385-6, 404-5, 425, 510, 515, 521, 707
regenerative medicine 352, 356
regions 17, 88, 105, 109, 126, 181, 204, 255, 330, 425, 430-1, 490, 543, 612, 705-6
regulate 9, 36, 87, 114, 141, 158, 233, 261, 284, 288-9, 311, 566, 585, 636, 689-93
Reichmanis, Maria 331
rejuvenation 390, 404-5, 519, 525, 660
Ren-5 692, 694-7
Ren-6 692, 694
Renzhong 693
reproductive organs 390, 411-12, 481-2, 490, 500
repulsive 188, 198
resistance 139, 154, 216, 293, 307-8, 326, 330, 397, 421, 510, 664, 674, 703

resonance 132, 148, 267, 272, 277, 279, 286-7, 290, 304-5, 309, 329, 400, 714
respiration rates 663
respiratory 216, 263, 380, 413, 421, 428, 454, 572, 575, 584, 588, 608, 624, 666, 676-7
restored 161-2, 336, 455
reticular cells 609
reticulin 446
retinaculae 334, 513
retrorenal fascia 491, 501
revolution 122, 278, 420
rheumatoid arthritis 294, 413, 588
rhinitis 623, 626
rhythm 59-60, 62, 69, 278, 280, 288, 331, 652, 664
rhythmic 282, 584
rib cage 411, 614
ribonucleic acid 373
ribonucleic acid (RNA) 156, 272-4, 303, 345, 370
rice 169, 300, 310, 563
ring 96, 100, 166, 690, 698, 705
ripened 284, 676
Rolfing 332
root 8, 17-18, 90, 171, 175, 287, 362, 466, 468-9, 474, 482-3, 486, 496-7, 518-20, 533-5
root of the Triple Burner 366, 466, 468-9, 482-3, 486, 496-7, 518, 534
rosacea 641
round 72, 163, 179
Royal Society 239
rule 172, 281, 283

S

Saccharomyces 311
sacs 436, 439
sagacious 246, 282, 532
saliva 45, 183, 299, 556, 573, 576, 582, 679
Salmonella 588, 594
Salmonella enterica 588
San Jiao xiv-xv, 1-3, 13, 16, 19, 36-8, 48-52, 131-2, 134-5, 153, 155-7, 318-19, 684-6, 689-92, 698-700
Sanal-Cell Cycle 385-6
sanal cells 377-8, 407
sanals 345, 351, 360, 369, 374, 377-8, 385-6, 401, 404-5, 407, 508, 519, 521, 525, 533
sanitisers 571
sanjiao xvii, 2, 13-14, 37, 112-13, 126-8, 135-7, 139, 141-4, 150, 523-4, 683, 685, 699, 708-10
Sanjiao channel 689
Sanjiaoshu 144, 692, 694-5
Saturated Fats and the Kidneys 392, 716
scaffolding 417, 421, 432-3, 492, 522
scalpel 445
scent receptors 558-60
schizophrenia 588, 712
science ix, xi, xiv, 180, 205-7, 212-13, 223, 233-4, 237-41, 292-3, 342, 358, 445, 539, 713
scientific xiii-xv, 51-3, 202-3, 206-9, 211-13, 233-5, 237-41, 247-8, 290-1, 328-30, 358, 618-20, 636-7, 649-51, 712-13
scientists 178-80, 205-7, 211-12, 225, 233, 240-1, 243, 257-8, 277, 306-7, 320-1, 441, 597-9, 604-7, 627-8
sclerosis 280, 588, 659
scorched 52, 464
screen-gate 554
scripture 3, 57, 71, 77, 87, 93, 97, 110, 114, 298, 360-1
scrotum 412, 489, 696
sea 11, 62, 64-5, 69, 179, 289, 297-300, 306-7, 363-4, 366-7, 426, 482-3, 486, 688-9, 698-9
Sea of Blood 90, 92, 289, 363
Sea of Marrow 426, 674
Sea of Water 78, 289, 354
second brain 605-7, 723
seconds 271, 665
secretory immunoglobulin A 572-3, 601
seed 309
selenium 656
semiconductor 173, 183, 187, 528
Semmelweis, Ignaz 202, 211-12, 233-4, 712
sensory 251, 327, 337, 442, 447, 455, 541, 550-1, 559, 680
separates muscles 422
serotonin 425, 566, 569-70, 583, 590, 595, 600, 606-7
Serratia spp. 579
serum 146-7
seven-year cycles of females 673
seven-year life cycles in females 59, 288
sexual arousal 663
Shangjuxu 80, 329, 689
Shao Yang 53, 55-6, 138, 158-62, 317, 465, 691, 696-7
shape xiv, 25-6, 34, 36-7, 62, 73, 77, 138, 160, 191-2, 256, 265, 274, 278, 392

sheaves 142
sheets 184, 418, 436, 439, 452, 653
Sheldrake, Alfred Rupert 304-5, 308, 715
Shigella 594
Shimen 692, 694-6
shock 415, 425, 639
shrink 461, 495, 516
shu 10-11, 29-30, 56, 70-2, 90-2, 97-8, 117-19, 134-5, 316, 346, 472-4, 550-1, 555-7, 687-8, 698-9
significance 14, 393, 466, 492, 496, 573, 601, 629, 719, 724
Sinatra, Stephen 659, 665-7, 669, 725
sinewy stuff 446
sinus pocularis 490, 501
sinuses 346, 559-60, 570-1, 601
SJ-1 690
SJ-5 685, 690-1, 699
SJ-10 690
SJ-13 690
skin 9-10, 29-33, 118-21, 125-6, 131-2, 136-41, 143-51, 168, 172-3, 175, 219-20, 352-5, 597-8, 639-41, 678-9
slab 245
sleep 231, 639, 647, 649, 651-3, 659-60, 663, 672
slide 184, 422, 430, 447
Small 16, 36, 43, 48-9, 54, 60, 66, 80, 108-9, 113, 135-6, 149, 313-14, 581, 680-2
small intestine 16, 36, 43, 48-9, 54, 60, 66, 80, 108, 135, 149, 157, 313-14, 581, 680-2
smells 558, 561, 720
sodium lauryl sulfate 598

Soh, Kwang-Sup 7, 354, 386, 390-1, 505, 507
soil 9, 88, 111, 115, 125, 170, 173, 201, 268-9, 308, 314, 597, 645, 661
solid 5, 28, 43, 65, 74-5, 109, 180-1, 266, 322, 397-8, 421, 432, 471, 555-6, 632-3
solitary palace 316, 420
sore throats 623, 626
source 48, 72, 152-7, 171, 174-5, 198, 231, 287, 297-8, 366-9, 399, 407-8, 474, 521, 674-6
source points 2, 36, 53, 152-3, 155-6, 368-9, 377, 402, 408, 505, 521, 534, 674, 676
SP-3 686, 688, 691
spatial 253
sperm 35, 378, 384, 404, 468, 477, 480, 482, 484, 497, 499, 522, 543-4, 553, 558-9
spermatozoa 480, 489, 499, 553, 558
Sphingomonas 572
spicy foods 211, 233
spinal 251, 255, 346, 350, 410, 414, 421, 436, 450, 545, 605, 607, 633, 674, 704
spirit-turtle 297-8
spirits 3-4, 29, 75, 86, 119, 124, 182, 285, 297-8, 360-1, 365, 396
spirochetes 570, 601
spleen 42-4, 81-5, 87-8, 90-1, 109-11, 114-15, 125-6, 139-40, 418-20, 550-1, 556-8, 613-14, 623-4, 679-82, 688
spoiling 91
spring 267, 285, 670, 690, 723
ST-36 329, 510, 688, 692

ST-37 80, 329, 689
Staphylococcus 572, 598, 641, 643
Staphylococcus aureus 598, 641, 643
steam 8-9, 63, 65, 88, 106, 111, 115, 144-5, 170, 199, 201-2, 353-4, 427, 505, 519-20
steams inside of the membrane 122, 420-1, 432
stearic acid 392
Stecco, Carla 133, 420, 422, 429-30
stereomicroscopic 351-2
sterile 313-14, 410-12, 553, 558, 565, 599
stiffen 453-4
Still, Andrew Taylor 164, 434, 442, 461, 495, 516, 711
stimulated 30, 76, 90, 93, 119, 132, 256, 261-3, 329, 374, 538, 543, 550, 629
stockpiled 285, 639
stomach 15-18, 44-50, 64-6, 68-71, 75-83, 85-91, 97-111, 244, 288-90, 417-20, 522-4, 538-9, 543-8, 554-7, 679-80
storehouse 259, 372, 464
stores 23, 75, 84-5, 92, 170-1, 279, 281, 286-8, 290, 317, 365-6, 463-4, 467-8, 482-4, 639-40
storing 35, 59, 171, 175, 274, 287, 290, 297, 300, 372, 401, 407, 481, 500, 508
straw 26
street 5, 8, 76, 78, 280
strengthening 265, 309, 690
strengthens ligaments 422
Streptococcus 572, 595
Streptococcus pneumoniae 572
Streptomyces 572

stress 211, 233, 280, 340, 429, 454, 528, 565, 573, 646, 652-3, 656-9, 661, 663-7, 697
stretching 492, 547
strings 436, 439
structural 17-18, 149, 185, 189, 194, 227, 255, 266, 324, 329, 385, 435-6, 446, 452, 609
structural elements 149, 385, 405, 446
structures 47-8, 132-3, 183-5, 304-5, 328-31, 346-50, 352-7, 379-84, 396-8, 419-22, 428-36, 472-4, 487-93, 498-510, 618-19
struts 436, 439
stuck blood 646-7
students 112, 203, 282, 716
Su Wen 1, 54, 131, 148, 247, 418-19, 708
subarachnoid 346, 356
subatomic 223
subclavian 612, 615-16, 624, 633-5, 643
subclavian veins 612, 615-16, 624
substances 65-6, 85-6, 101-3, 146-7, 181-2, 308-9, 340-1, 352-4, 451-4, 469, 516-18, 542-3, 581-3, 675-9, 725
substrate 77, 197, 222, 242, 383, 398, 401-2, 405-6, 425, 446, 453, 486, 510-11, 514, 594-5
subtle 5-6, 64, 70, 91, 222, 321, 325, 457, 534, 631, 643, 645, 650
succinctly 192, 202, 245, 302, 312, 417, 429, 451, 586, 603, 615
sucralose 542
suet 393, 716

sugar   274-5, 373, 394, 539-40, 546-7, 549-50, 569, 573, 576, 589, 603, 636-7, 643, 658
sulfonic   181
sun   9, 17-18, 59, 88, 111, 166-8, 170-2, 174-5, 195-6, 201, 215-17, 283-8, 300, 524, 648
sunlight   168, 170, 173, 175, 195, 221, 285, 300, 464
Sunrays   285, 290
superficial   58, 126-8, 132-3, 136, 138-42, 148-50, 158, 161-2, 333, 336, 344, 350, 356, 392, 404
supervisor   78, 346, 364
supplied   4, 53, 77-8, 90, 128-9, 167, 253, 312, 315, 461, 473, 498, 516, 647-8, 656
surface-active phospholipid (SAPL)   103-4, 106
surging   78, 129
swallowed   65, 211, 464, 542, 576
swamp   49, 682
sweat   9, 27, 88-9, 106, 111, 115, 128, 141, 143-7, 151, 170, 201-2, 353-4, 597-8, 706
sweating   127, 131, 142-7, 150-1, 678, 682-3, 698, 710
sweet potatoes   306-7
swell   461, 495, 516, 610, 614, 624
Switzerland   295-6
symbiotic   570, 586
symmetry   382
sympathetic nerves   420, 629
symphony   267, 405
symptoms   83, 142, 161, 231, 344, 570, 576, 601, 614, 653, 665, 669, 686-8, 696, 698-9
synchronizer   299

synchronizing   299, 362, 371, 654
synergize   526
synovial   54, 138-9, 141, 148, 159-60, 229, 413, 421, 524
synthesized   68, 98, 145, 244, 308-9, 452, 586, 622
synthetic   103, 308, 405, 540, 563, 650, 654-5
Szent-Györgyi, Albert   40, 180, 203, 207, 221, 225, 234, 527

T

T-cells   294
T1R3   539, 541, 545-6, 550
Taibai   686, 688, 691
Taichong   688, 691
Taixi KID-3   686, 689, 691
Taiyuan   686-8, 691-2
tangible   21, 23, 33-4, 37, 72, 74, 177, 318, 619, 625
taste receptors   538-45, 550-4, 556, 558-60, 719
TB-3 *Zhongzhu*   697
teachers   19
Teflon   103, 181, 198, 289
television   306, 654
tendons   84, 190, 265, 334, 427, 436, 438-9, 446, 453, 513
tenfold   187, 237
tensegrity   265-6, 322, 325, 375, 427, 429, 437, 459-60, 486, 492, 494, 507, 512, 716
tension   103, 106, 265, 341, 431, 492, 649, 652-3
textbooks   22, 357, 406

texts xi, xiv, 1, 15-16, 19, 31, 36, 38-40, 44, 46, 52, 120, 312, 442, 470-1
textures 127, 137-8, 142, 144, 150
theoretical 39, 179, 223, 291, 323, 373, 418, 523, 714
theories 2-3, 48-51, 153-4, 156-7, 177-8, 211-12, 227-8, 232-4, 244-5, 303-6, 349-52, 387-8, 474-5, 663-4, 707-9
therapeutic 132, 294-5, 354-5, 389, 403, 510, 565, 594, 631-2
thermal imaging 651
thermogenesis 367, 629-31, 634, 642
thick 29-31, 118-20, 143, 164, 222, 341, 355, 370, 379, 403, 420, 446-7, 452, 507, 511
thixotropic 454-5
thoracic 47, 51, 136, 356, 411-13, 491, 612, 633
thoracolumbar aponeurosis 444
thoracolumbar fascia 449, 466, 491-2, 496, 501, 535, 718-19
thoroughfare 141, 524
threadlike structures 346, 349, 354-5, 716
Three Amigos 243, 513
three-dimensional 139, 325, 334, 338, 513-14
Three Heater 36
three qi 127-8, 141-2, 523-4
three ruling centers 62, 69, 416
throat 2, 5, 46-7, 161, 289, 359, 363, 419, 533, 560, 614, 670, 676, 698, 706
through-way vessel 5, 289, 359, 361-2, 364, 533
thrusting vessel 363

thymus 190, 411, 608, 610, 613-14, 624
Tianjing 690
tie between the kidneys 8, 353, 391-2, 422, 487-8, 519-20, 534
tissue injury 439
tongue 5, 45, 353, 520, 539, 541-4, 546-8, 550, 552-4, 558-9, 574, 647, 697, 720
tonify 85, 157, 675, 686-7, 689-90
tonsils 608, 614
tool 291, 666-7
totipotent 355, 403, 506, 533
toxic 145-7, 151, 173, 194-5, 260, 293, 426, 451, 475, 535, 542-3, 546, 550, 592, 611
toxins 27, 144-5, 260, 292, 294, 548, 574-6, 579, 584-5, 610-11, 616, 623-4, 626, 659, 661
traditional Chinese medicine (TCM) xiii-xv, 1-3, 13-16, 48-50, 52, 82-5, 107-9, 160-1, 202-3, 267, 350, 426, 480-3, 498-500, 684-6
Traino, John 453-4
transduce 205, 450
transducer 225
transformation 42-3, 57, 61, 87-9, 111, 114-15, 122, 201, 301-2, 361-2, 364-5, 465, 496, 681-2, 693-6
transformed 5, 9, 86-8, 90-1, 97-8, 111, 115, 144-5, 188-9, 201-2, 298-9, 353-4, 518-20, 554-8, 680
transformed into influences 9, 88, 111, 115, 144, 170, 201-2, 244, 354, 555
transmembrane 335

transmission 15, 58, 220-1, 330-1, 334, 336, 338, 349, 398, 421, 430, 513, 515
transmitted 47, 86, 93, 97, 182, 244, 255, 285, 316, 330, 336, 366, 387, 543-4, 555-6
transmute 225, 676
trauma 424, 430, 453-4
trees 204, 343, 649
trichrome 7, 391, 525
trigger 253, 331-2, 487, 500, 543, 545-6, 548, 636
triglycerides 620, 658
Triple Burner 12-16, 18-25, 28-37, 44-8, 53, 57-62, 72-80, 113-15, 117-23, 361-9, 420-3, 464-70, 518-23, 534-5, 692-7
Triple Energizer (TE) xvii, 8-10, 12-13, 19-20, 40-3, 158-61, 167-70, 177-9, 188-92, 207-10, 231, 415-18, 486-8, 692-3, 705
Triple-Energizer Metasystem 28, 52-4, 156-7, 269-71, 298-303, 337-9, 402-5, 421-3, 427-30, 432-3, 435-7, 442-5, 502-6, 518-22, 532-3
trophoblast 379-80, 408, 485, 508
True Qi 677
truth 203, 206, 237, 241-2, 284, 628, 642, 657, 714, 724
Trypan blue 348, 382
tryptophan 252, 569, 595, 600
Tsang 71, 75, 157
tubes 80, 193, 216-17, 351, 413, 461, 477-8, 481, 484-5, 488-9, 495, 499, 516-17, 537, 612-13
tubulin 259, 280
tumor cells 262, 340-1, 355-6

turbid 53, 70, 86, 163-5, 181-3, 299, 311, 316, 680, 688, 695, 703, 705
Turchaninov, Ross 441

U

UBI Therapy 293-4
ubiquitous xvii, 25-7, 125, 130, 141-2, 147, 150-1, 156, 177, 226-7, 242, 371-2, 375-6, 444, 524
ubiquity 123, 195, 245, 264
ulcer 413, 549
ultrasound 196, 329
ultraviolet 173, 175, 195, 214, 216, 225, 292-3, 295, 351, 714
ultraviolet (UV) 172, 183, 214-15, 251-2, 277, 351
Ultraviolet Blood Irradiation (UBI) 292-6, 714
ultraweak 249, 251
umami 540-1, 544, 553-4, 558, 561
umbilical 473, 477, 479-80, 565, 599, 718
umbilical vesicle 473, 477, 479-80, 718
umbilicus xvii, 2, 44, 46, 50, 487, 491, 502, 704, 706
Undaria pinnatifida 389
undigested 101
unfertilized ova 468, 497
unhealthy 285, 388, 454-5, 647
unified 153, 257, 335, 338, 444-5, 515, 579
unique 13, 28, 32, 38, 73, 96, 123, 135-6, 329, 396, 430, 491, 509, 573, 581
uniqueness 28-30, 32, 118-19, 342, 430, 515

units 187, 194, 200, 256, 261, 284, 286, 304, 314, 335, 402, 448, 587, 591, 603
unity 35-6, 44, 267, 427, 470-1
University of California–Los Angeles (UCLA) 223, 606
unprejudiced 238
unravelled 444
unstructured 247
urination 127, 143, 663, 686, 695-8
utensil 366, 468, 482-4, 500
Uterus Channel 343-4
Uterus Vessel 343-4
utriculus prostaticus 490

V

vacuum 233, 250, 417
vagina 27, 481, 488, 490, 501, 569-71, 589, 601
vagina masculina 490, 501
vagus nerve 47, 554, 582, 586, 596, 606-7
valves 87, 352-4, 477, 499, 505, 613
vegetables 125, 169, 286, 540, 542, 594, 680, 711
Veillonella 572
venous blood 244, 612, 615-17, 624, 635-6
ventral 191, 410-12, 489
versatile organic system 418, 523
vertebral cavity 410
vertebrates 384, 481, 488
vesica prostatica 490, 501
vessels 25-6, 83-6, 93-6, 101, 166-7, 344-5, 349-50, 357, 373-5, 386-7, 403-6, 420, 560, 611-13, 684
Vestigial 477, 488-90, 718

vibrates 258, 279
Vibrio fischeri 578
Vibrio spp. 579
vigilant 466, 496, 535, 658
Vioxx 238
virions 570, 601
viruses 27, 125, 293-4, 413, 564, 567, 570, 572-3, 583, 587, 590, 593, 601, 614, 641
viscera 3, 192, 336, 388, 419, 422, 425, 702, 704-5
viscerosomatic 336
visible 106, 183, 195, 200, 212, 214-15, 221, 251, 278, 351-2, 382, 397, 526, 660, 706
vital 6, 54, 64, 70, 203-4, 259-60, 283-6, 362-3, 456-8, 461, 494, 527, 675-9, 702-3, 725
vitality 96-7, 127, 268, 281, 284-6, 360, 464-5, 468, 492, 495, 517, 672, 674, 676, 678-9
vitally 174, 286, 429, 479-80, 680
vitamin B 85, 342, 544, 556
vitamin C 180, 207, 656
vitamin D 168, 175, 290, 556
vitamin E 656
vitamin K 584, 594
in vivo 202, 256, 349
voltage 650, 660, 662
vortexing 178, 208, 218

W

Waiguan SJ-5 685, 690-1, 699
wall 65-6, 79-80, 102-3, 107-10, 175, 383-4, 390-1, 412-13, 420-2, 432-3, 449, 479, 490-1, 522-3, 582

walls 99, 168, 173, 175, 179, 228-9, 265, 295, 345, 412-13, 419, 427, 439, 605, 612
Wáng Jūyì 397-8, 526
wastes 1, 43, 66, 70, 88, 101, 129, 134-5, 143, 417, 426, 611, 616-17, 622-3, 695
watchdog 211
water 65-7, 77-8, 86-9, 101-11, 114-15, 125-30, 169-70, 174-89, 193-232, 241-50, 297-303, 315-19, 324-8, 523-9, 698-701
water dipoles 324, 338, 514
waterways 6, 53-4, 78, 95, 108, 127, 160, 310, 316-17, 333, 451, 509
wavelengths 213-15, 225
Weber, Daniel 328-9
Weber organ 490, 501
Weishu 688
Weiyang BL-39 135, 144, 689, 698
WellBeing 648, 721-2, 724
wellness 6, 53, 157, 292, 360, 429, 562, 583, 595, 652, 660-1
Wén Huì (prince) 396
Wernicke encephalopathy 556
WHO International Standard Terminologies on Traditional Medicine in the Western Pacific Region xvii, 463-4, 707
wine 125, 310-11, 546, 657
wires 436, 439
Wolffian ducts 481, 488-9
womb 79, 365-6, 395-6, 467-8, 482-4, 497, 500, 564-5, 599
wood 77, 84, 229, 267, 701
wooden dolls 368

work ix, 3, 40, 51, 96, 188, 193, 205, 229-31, 245-6, 254-5, 309-10, 330, 352, 606-7
wound isolation 424
wrapping 7, 33, 80, 226, 384, 391, 417, 423, 428, 433, 447, 455, 522-3, 525
wreak havoc 563

Y

yang 8-11, 55-6, 72-5, 88, 94-5, 111, 115, 143-4, 158-62, 170-2, 174-5, 287-8, 297-301, 686-91, 695-7
Yanglingquan 689
'Yangsheng and the Channels' (Robertson) 128, 267, 396-7, 710
Yangwei channel 691
yeasts 125, 569-70, 572, 587, 589, 592, 601, 603
yeh 9, 29, 88, 111, 115, 118, 144-5, 170, 201-2, 204, 354
Ying Qi 43, 50, 57-8, 61, 63, 83, 183, 299, 512-13, 677, 684
yolk sac 379, 382, 473, 477, 479-80, 718
Yuan 1-2, 53-5, 152-7, 298-9, 301-3, 359-63, 365-9, 383-4, 463-5, 495-6, 519-22, 674-7, 683, 686-92, 695-6
yuan points 152-4, 157, 368, 377, 402, 408, 505, 521, 534, 674, 686-8, 691
Yuan Qi 2, 53-5, 152-7, 301-3, 359-63, 365-9, 383-4, 463-5, 495-6, 519-22, 675-7, 683, 686-7, 689-91, 695-6

Z

Zang 2, 19, 36-8, 57, 74, 100, 141, 152, 167, 201, 465, 619, 625, 709
zeta potential 657-8
Zhang Jiebin 112, 116, 124-5, 462, 472
Zhen Qi 127, 136, 141-2, 149-50, 465, 496, 523-4, 676-7, 683
Zhongwan 50, 691, 693
Zhuangzi 396
zona hatching 378-80, 408, 508
zona pellucida 378-9, 408
Zong Qi 57, 99, 465, 496, 676-7
Zuckerkandl's fascia 491, 501
Zusanli 329, 688, 692
zygote 378-9, 408, 483-4, 487-8, 500, 568

About the Author

From the age of 8, Louis Gordon was entertaining local kids with his chemical experiments, making luminous paints, rockets, smoke bombs, and Hydrogen balloons fitted with a fuse that exploded 300 m up, causing what resembled a sonic boom that shook windows and brought mothers racing outside to see what the explosion was. Louis graduated from USQ as a Biological Laboratory Technician, and later, while working in government chemistry and bacteriological laboratories in his early twenties, Louis expanded his education to include classical philosophical concepts associated with Traditional Chinese Medicine (TCM) into his scientific repertoire.

Western medical paradigm does not accept that traditional Chinese philosophy should be so deeply entrenched and instrumental in determining the optimal protocols associated with the truly holistic outcomes accompanying the practice of traditional Chinese medicine. The author finds it incongruous that Western medicine practitioners segregate medical conditions to specific locations of the body as if one portion of the body operates independently of all other body systems and organs.

Louis demonstrates that ancient Chinese philosophers were truly knowledgeable when it came to understanding exactly what makes the human body tick. Ancient Chinese philosophers discussed in eloquent detail what modern scientific researchers are only now uncovering thousands of years later. When Louis graduated as a Chinese Medicine Practitioner more than three decades ago, he was ridiculed for believing in the supposedly non-existent San Jiao organ. The World Health Organisation (WHO) now defines it as the Triple Energizer. After over three decades of scientific research and thanks to the ancient philosophical literary classic the Nan Ching, Louis marries ancient Chinese philosophy and modern western medical science and provides concrete proof of the physical existence of the Triple-Energizer Metasystem and defines its intricate location, morphology, and how it works.

CPSIA information can be obtained
at www.ICGtesting.com
Printed in the USA
BVHW03s0158190318
510899BV00021B/31/P